The Latest in Crystal Growth

The Latest in Crystal Growth

Edited by **Sharon Levine**

NY RESEARCH
PRESS

New York

Published by NY Research Press,
23 West, 55th Street, Suite 816,
New York, NY 10019, USA
www.nyresearchpress.com

The Latest in Crystal Growth
Edited by Sharon Levine

International Standard Book Number: 978-1-63238-438-6 (Hardback)

Printed in China.

Contents

Preface

The purpose of the book is to provide a glimpse into the dynamics and to present opinions and studies of some of the scientists engaged in the development of new ideas in the field from very different standpoints. This book will prove useful to students and researchers owing to its high content quality.

This book presents latest advances in the field of crystal growth technology. Crystal growth is the basic step for a wide number of crucial applications. The advancement of novel devices and products in various fields varying from the traditional microelectronic industry to pharmaceutical industry and many others, relies heavily on crystallization processes. The purpose of this book is to present crucial selected topics, applied to organic and inorganic systems. Authors who have contributed in this book are important researchers in their area of specialization, working in reputed universities and research labs across the world. The book covers biological systems like proteins, bone and ice crystallization, applications in inorganic systems and explains common growth techniques like chemical vapor crystallization and electrodeposition. This book has been compiled to meet the requirements of students studying in the field of crystal growth and for scientists and engineers in the fields of crystalline materials, crystal engineering and the industrial applications of crystallization processes.

At the end, I would like to appreciate all the efforts made by the authors in completing their chapters professionally. I express my deepest gratitude to all of them for contributing to this book by sharing their valuable works. A special thanks to my family and friends for their constant support in this journey.

Editor

Section 1

Biological and Other Organic Systems

Crystal Growth of Inorganic and Biomediated Carbonates and Phosphates

Antonio Sánchez-Navas, Agustín Martín-Algarra,
Mónica Sánchez-Román,
Concepción Jiménez-López, Fernando Nieto and
Antonio Ruiz-Bustos

Additional information is available at the end of the chapter

1. Introduction

Precipitation of carbonate minerals is tightly linked to water chemistry. After hydration of dissolved carbon dioxide, two pH-dependent partitioning-reactions govern the abundance of chemical species (H_2CO_3, HCO_3^- and CO_3^{2-}) formed in aqueous solution:[1,2]

$$H_2CO_3 \leftrightarrow HCO_3^- + H^+ \leftrightarrow CO_3^{2-} + 2H^+$$

where the O-H covalent bond in the oxyacid makes carbonate salts moderately soluble. The most common metal cations forming carbonate minerals are Ca^{2+}, Mg^{2+}, Mn^{2+}, Fe^{2+}, Pb^{2+}, Sr^{2+}, Co^{2+}, Ni^{2+}, Zn^{2+}, Cd^{2+} and Cu^{2+}. Continental and marine waters are enriched in Ca and Mg and are known to be saturated with respect diverse Ca-Mg carbonates such as calcite ($CaCO_3$), aragonite ($CaCO_3$) and dolomite ($MgCa(CO_3)_2$).[3] The concentration of the phosphate species (H_3PO_4, $H_2PO_4^-$, HPO_4^{2-}, and PO_4^{3-}) is also a function of pH, and their respective oxyacids are stronger than those of carbonic acids.[2] Because of this, phosphates are more stable than carbonates at low pH (<5). Chemical composition of phosphate minerals is more variable than that of carbonate minerals, and crystalchemical substitution of the PO_4^{3-} group by CO_3^{2-}, OH^-, F^-, Cl^-, etc, is rather common. In addition, numerous metals as Ca^{2+}, Mg^{2+}, Fe^{2+}, Na^+, Sr^{2+}, Ce^{3+}, La^{3+}, Ba^{2+}, and Pb^{2+} can be incorporated into the structure of the phosphate minerals.

Limestones and dolostones constitute the most important carbonate reservoirs on the Earth, but phosphates are much more diluted within the Earth crust, as phosphate rocks are much less common that carbonate rocks. Nevertheless, many organisms form $CaCO_3$ shells, and Ca-carbonates and phosphates are the main constituents of normal bone and of pathologically calcified organic tissues.[4] In many cases, the precipitation of carbonate and phosphate minerals at Earth-surface conditions is a organic matrix- biomediated process related to shell formation. Indeed, carbonate and phosphate crystals form sophisticated composite materials (*e.g.*, shells, spicules, bones, teeth), where they appear embedded within an organic framework, which apparently controls the final texture of the mineralized material. When these organisms die, Ca-carbonate shells accumulate to form limestone. Precipitation of Ca-carbonate often occurs in domestic and industrial installations where untreated natural waters are used.[5]

The study of the origin of life on Earth, and of the possible existence of extraterrestrial life is also tightly associated with carbonate and phosphate precipitation, as it is recorded in microbial accretions, frequently forming stromatolites and oncoids.[6-9] Ca- and Mg- carbonate and phosphate minerals similar to those found in ancient and modern stromatolites are also usually obtained in laboratory bacterial culture experiments.[10-13]. The occurrence of amazing complex textural features in relation to these minerals in the geologic record has prompted many authors to consider them as biomarkers.[14]

Different formation mechanisms have been proposed for bacterially mediated carbonate and phosphate minerals.[15] The metabolic activity of the bacteria alters the physico-chemical parameters of their surrounding habitats, allowing mineral precipitation. It has been suggested that microbes influence nucleation and that, in general, they control the kinetics of precipitation of carbonate and phosphate minerals, and therefore their morphology. Spheroidal precipitates (spherulitic bioliths) are commonly obtained in bacterial culture experiments.[13]

Some crystal-growth studies have shown that the microstructure of carbonates and phosphate minerals precipitated by bacteria at high supersaturation conditions has particular signatures.[7,13,16] They nucleate as nanocrystal units after the precipitation of amorphous precursors, and then, start to grow, developing spherulites and dumbbells. Singular crystal-growth features of aragonite and calcite occur in diverse species of molluscs as result of the inhibition of growth of specific crystallographic faces by organic molecules.[17] However it is well known that the habit of the crystals formed at high supersaturation is completely different from the crystal growth features observed in precipitates obtained at lower supersaturation.[18] Hence, spheroidal morphology may results from different origin. In addition, some crystal aggregates that normally occur in shells can be easily explained by competitive crystal growth processes.[17]

In this chapter, we describe similar crystal growth features in inorganic and biogenic carbonates and phosphates formed in natural environments and in laboratory experiments. We focus on the importance of the kinetics on the crystal habit of carbonates and phosphates precipitated in biological and abiotic systems. We also discuss the influence of the nature and composition of the precipitation medium, as well as the structural control on crystal habit of Ca-Mg carbonates.

2. Material and methods

2.1. Natural Ca-phosphate samples

Crystal growth studies of natural Ca-phosphate have been performed on ancient phosphate stromatolites and on modern calcified tissues. Stromatolites come from Upper Jurassic and Lower Cretaceous limestones of the Betic Cordilleras, Southern Spain (Almola Sierra and Peñón del Berrueco, respectively). Calcified tissues correspond to enamel, dentine and bones from laboratory rats.

2.2. Inorganic experiments

The inorganic synthesis of amorphous calcium phosphate was obtained mixing 300 ml of 0.04 M calcium salt ($CaCl_2.6H_2O$) with 0.036 M ammonium phosphate dibasic salt ($(NH_4)_2HPO_4$) in 400 ml of constantly stirred 0.15 M buffer. All solutions were thermally equilibrated at 25ºC before mixing. The amorphous precursor phase was removed immediately after mixing.

Inorganic growth of carbonates in laboratory experiments was performed by using aqueous solutions of diverse metal cations (mainly Ca^{2+}, Mg^{2+}, Fe^{2+} and Pb^{2+}) and carbon dioxide or highly soluble bicarbonate/carbonate salts as source for carbonate ions. Inorganic solution growth of carbonates was carried out by bulk crystallization (free drift experiments), where two solutions of relatively easily dissolved salts of the diverse metals and sodium bicarbonate/carbonate were mixed: $Pb(CH_3COO)_2.3H_2O$, $CaCl_2.6H_2O$, $Ca(NO_3)_2.4H_2O$, $MgCl_2.6H_2O$, $FeCl_3.6H_2O$, $Fe(ClO_4)_2.nH_2O$, Na_2CO_3 and $NaHCO_3$ (see results for the specific concentrations used for the precipitation of the different obtained carbonate minerals). The resulting solution becomes supersaturated with respect to the less soluble metal carbonate. Temperature was usually 25ºC. Gel growth of Ca-Mg carbonates was performed by the counter-diffusion of carbonate solutions versus solutions containing diverse metal salts through a column of silica gel (stock solutions, 1 M of Na_2CO_3, 1 M $CaCl_2.6H_2O$, and variable concentrations of $MgCl_2.6H_2O$; see below). The silica gel with a pH 5.5 was prepared by acidification of sodium silicate solution with 1 M HCl. It was poured into a U-tube and allowed to polymerize to a solid gel.

2.3. Bacterially mediated precipitation

A liquid culture medium with a natural bacterial consortium mainly belonging to the genus *Acetobacter* sp., was used for biomediated precipitation of Ca, Fe and Pb phosphates and carbonates. This consortium was obtained from vinegar dregs after natural degradation of untreated wine. Aerobic *Acetobacter* grown in wine makes volatile acetic acid in the form of vinegar, and oxidizes acetic acid to carbon dioxide and water. Bacteria metabolize nitrogenated organic matter (*e.g.* proteins) with the subsequent production of CO_2 and NH_3. Metal cations were added to the liquid culture as soluble salts (50 mM $CaCl_2.6H_2O$, 50 mM $FeCl_3.6H_2O$ and 5 mM $Pb(NO_3)_2$) to obtain phosphate and carbonate minerals. Yeast extract (1%) was the carbon source for bacteria growth.

Concerning phosphate, bacterial phosphate precipitates of Fe and Pb were obtained, but with only minor amounts of lead phosphate, since bacterial tolerance to Pb is very low. Concerning carbonates, NH_3 production by *Acetobacter* increases the basicity of the precipitating medium (thus favouring the formation of carbonate minerals) but vinager acidity obviously hinders carbonate precipitation. So the culture medium was supplemented with bicarbonate (50 mM $NaHCO_3$), and the pH adjusted to 5.5-6 with 1 M NaOH to favour the precipitation of Ca-carbonates.

2.4. Analytical techniques

Natural mineral samples and laboratory precipitates (both biotic and inorganic) were analyzed by X-ray-diffraction (XRD) using a PANalytical X'Pert Pro diffractometer (CuKα radiation, 45kV, 40mA) equipped with an X'Celerator solid-state lineal detector and on line connection with a microcomputer. The diffraction patterns were obtained by using a continuous scan between 3-50 º2θ, 0.01 º2θ of step size and 20 s of time step. Data processing was performed using Xpowder®.[19]

Carbonates and phosphates were studied with a high-resolution Field-Emission Scanning Electron Microscope (FESEM) LEO GEMINI 1525, equipped with an energy dispersion X-ray (EDX) spectroscopy microanalysis Inca 350 version 17 Oxford Natural samples and inorganic (abiotic) and bacterial (biotic) precipitates were studied under Transmission Electron Microscopy (TEM). Representative samples of the natural material (stromatolites and calcified tissues) were ion-milled after extraction from selected areas of thin sections. The same powder samples of bacterial and inorganic precipitates used for XRD analyses were embedded in an epoxy resin, then sectioned by ultramicrotome following the methodology of Vali and Koster for clays,[20] and finally carbon coated. The samples were examined using a LIBRA 120 PLUS EFTEM (Energy Filtered TEM) instrument equipped with in-column corrected OMEGA Filter, operating at an acceleration voltage of 120 kV. Optimum amplitude contrast was achieved using lens aperture of 30 mm, and afterward removing inelastic electrons by zero-loss filtering in the elastic scattering image (ESI) mode. Electron energy loss spectroscopy (EELS) was also performed with this instrument. HRTEM and analytical electron microscopy study was performed using a Philips CM20 instrument (Philips) operated at 200 kV and equipped with an EDX system EDAX for microanalysis.

3. Results

3.1. Crystal growth features in phosphate stromatolites and calcified tissues

Ca-phosphate mineral occurring in stromatolites is francolite ($Ca_5(PO_4)_{2.5}(CO_3)_{0.5}F$). Francolite is a low crystalline variety of apatite.[7,21] XRD analyses show that francolite from phosphate stromatolites is more crystalline than other phosphate minerals precipitated by us in the laboratory (Fig 1). Secondary electron images of the phosphate stromatolites show the occurrence of accretion structures with columnar and arborescent morphologies defined mainly by francolite (Fig. 2A). Phosphate laminae alternate with particulate sediment layers, and appear surrounded by a sediment matrix. These sediments are made of cryptocrystalline carbonate (micrite) formed by bioclastic particles (Figs. 2B-C). Sediment particles were trapped by microbes (heterotrophic bacteria) within the biosedimentary structure. Gelly-like appearance of arborescent morphologies and other mineralized structures, such as tubular bridges with smooth surfaces connecting phosphatic microcupolae (Figs. 2C and 2D), constitute the fossil microbial mat.

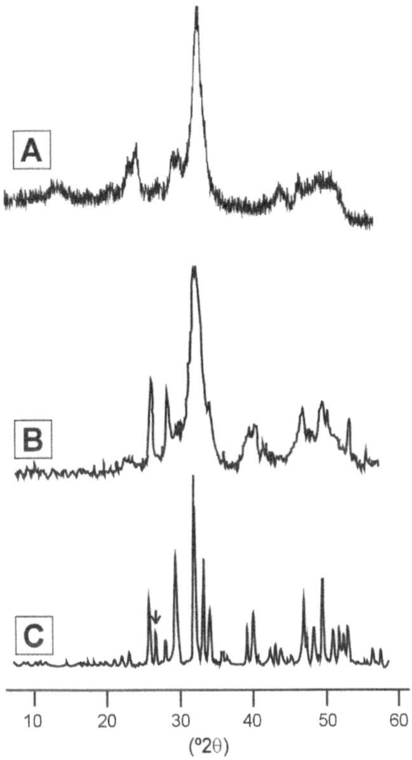

Figure 1. XRD patterns corresponding respectively to Pb-phosphate from biomediated laboratory experiments (A), Ca-phosphate from inorganic precipitates (B) and Ca-phosphate from phosphate stromatolites (C). Arrow in figure C corresponds to quartz.

The authigenic laminae within stromatolites are defined by the occurrence of neoformed minerals containing P, Fe, Mn and Al, as deduced from X-ray maps of figure 3. Amorphous precursors of francolite, calcite, iron oxyhydroxides and clay minerals, mainly smectites are still preserved in small areas within these authigenic laminae as evidenced in TEM images (Figs. 4 and 5). The authigenic minerals within microbial laminae normally surround sediment particles such as terrigenous grains (detrital clay and quartz), micron-sized fossils (benthic and

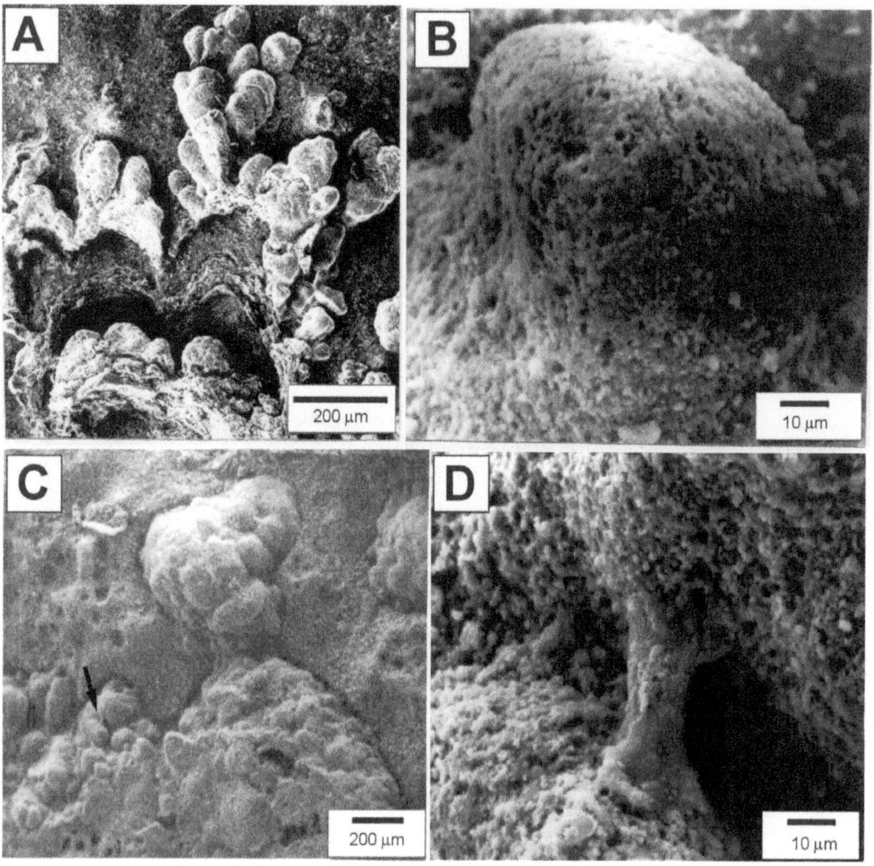

Figure 2. Secondary electron images from phosphate stromatolites. (A) General view of arborescent microstromatolites (etched sample). (B) Cupola formed by francolite. (C) Carbonate-rich sediment located among the phosphate cupolas (arrow points to close-up shown in D). D) Gelly-like tubular bridge between two phosphate domes; it is made of poorly crystalline P-rich and Si-Al-Fe-rich substances.

plancktonic foraminifera and coccoliths) and bioclasts that constitute the fine-grained marine, mainly carbonate sediment (Fig. 5). Francolite crystals from these laminae are nanometre-sized prisms with hexagonal basal sections and elongated along the *c* axis. Lattice fringes at 0.8 nm are observed (Fig. 4A).

Crystal growth features described for francolite in stromatolites are very similar to those observed for Ca-phosphates from calcified tissues (enamel, dentine and bone), which mainly correspond to hydroxyapatite. Dentine from adult rats shows accretion patterns defined by alternating layers composed by hydroxyapatite crystals and collagen fibrils (Fig. 6A). Honey-

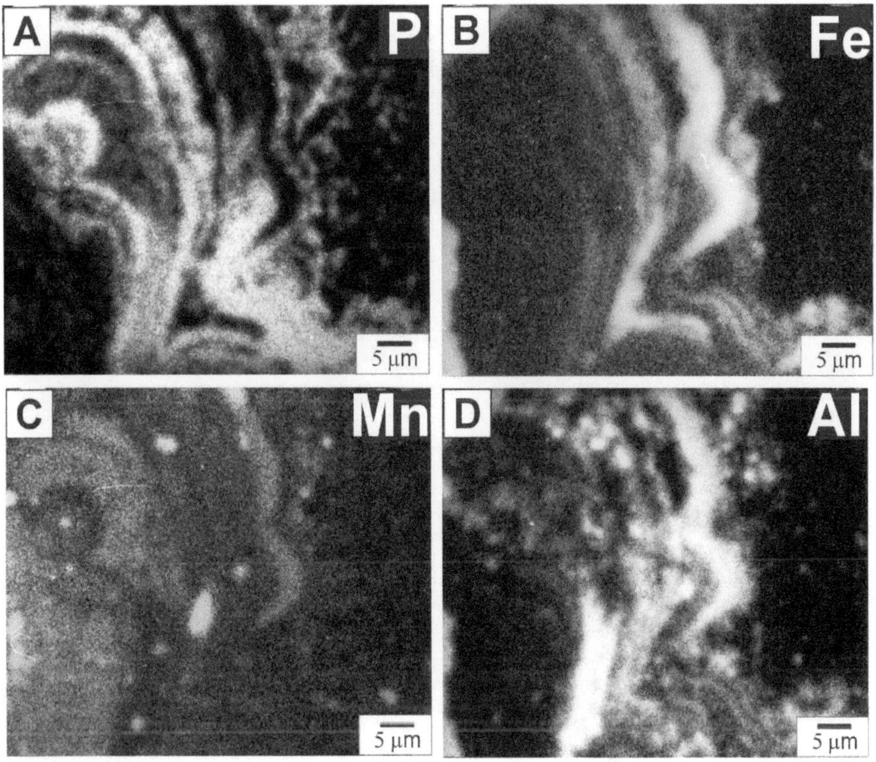

Figure 3. X-ray images of P (A), Fe (B), Mn (C) and Al (D) defining lamination in stromatolites. P-rich laminae are made of francolite. Fe and Mn rich laminae correspond to Fe and Mn oxyhydroxides, respectively. Authigenic and detrital clays are represented by the Al-rich areas. Most black areas in these images correspond to carbonate sediment, which is mainly calcite.

comb-like aggregates of small hexagonal apatite crystals of dental enamel (Fig. 6B) are roughly equivalent to parallel aggregates of prismatic phosphate crystals observed in phosphate stromatolites (Fig. 7 in Sánchez-Navas and Martín-Algarra).[7] Hydroxyapatite crystals in dentine have lower sizes than in enamel, and appear embedded in an organic framework (Fig. 7). They are usually closely packed in subparallel aligments, with the long dimension *c* axis being attached to collagen fibrils.[23,24]

Precursory phases of the Ca-phosphate have been observed in the studied calcified tissues, as evidenced by TEM images corresponding to the incipient stages of mineralization in dentine, enamel and bone of young rats and embryos (Fig. 8). In the case of enamel from embryos, amorphous calcium phosphate forms rounded masses of more electron dense material within an organic matrix (Fig. 8A). Nanometre-sized acicular crystals can be differentiated from amorphous Ca-phosphate forming electron dense regions with massive texture in TEM

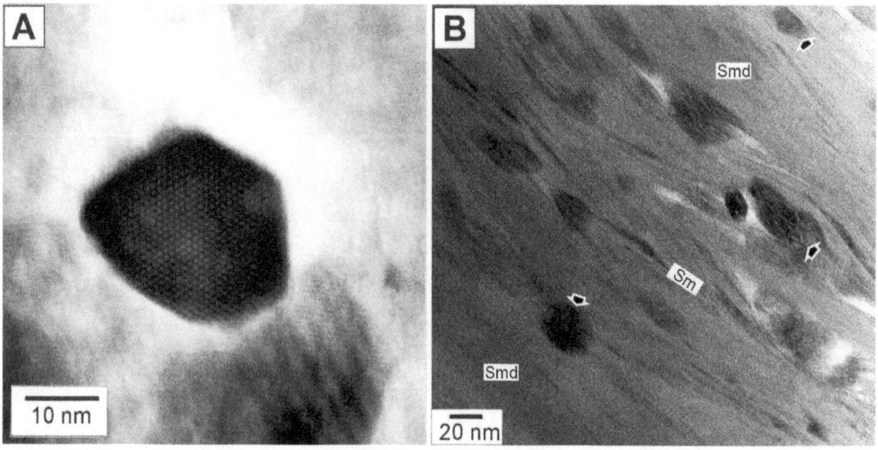

Figure 4. (A) HRTEM image of a francolite nanocrystal from an authigenic phosphate stromatolite lamina. It corresponds to a (001) hexagonal basal section. Three equivalent sets of crystallographic planes can be observed: (100), (010) and (110), intersecting at 120º, with an interplanar spacing of 0.82 nm. (B) Spindle-like packets 20-50 nm thick of smectite (Sm) with layer terminations and wavy layers from the same stromatolite lamina as A. Packets without fringe contrast (Smd) result from slight misorientation of the layers, mostly with 1 nm *d*-spacing. Fe-rich amorphous substances (arrows) are intergrown between smectite packets.

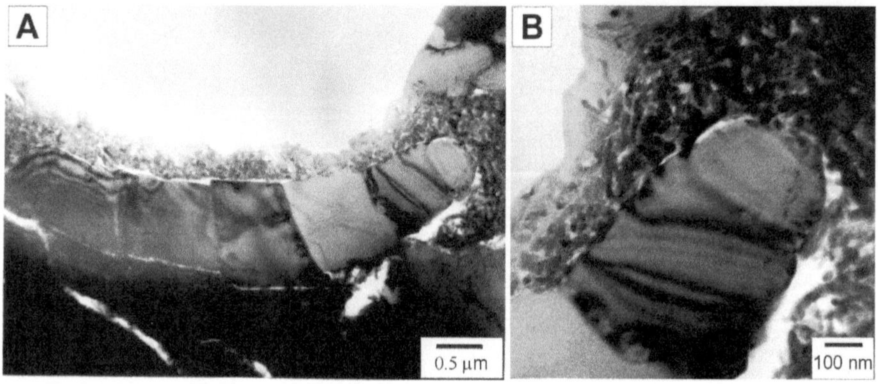

Figure 5. (A) TEM image of a coccolith bioclast surrounded by iron oxyhydroxides within an authigenic phosphate stromatolite lamina (compare this image with the transmission electron micrographs of cocoliths of Figure 17 in de Vrind-de-Jong and de Vrind, 1997).[22]. This bioclast is made of five calcite crystals, each around 1μm in size. (B) Detail of iron oxyhydroxide nanocrystals surrounding the bioclast. The average diameter of these crystals is 30 nm in pseudo-hexagonal sections, and around 70 nm along directions perpendicular to basal section.

images; this is particularly evident in dentine (Figs. 8B and C), and in bone from one-day-old rats (Fig. 8D).

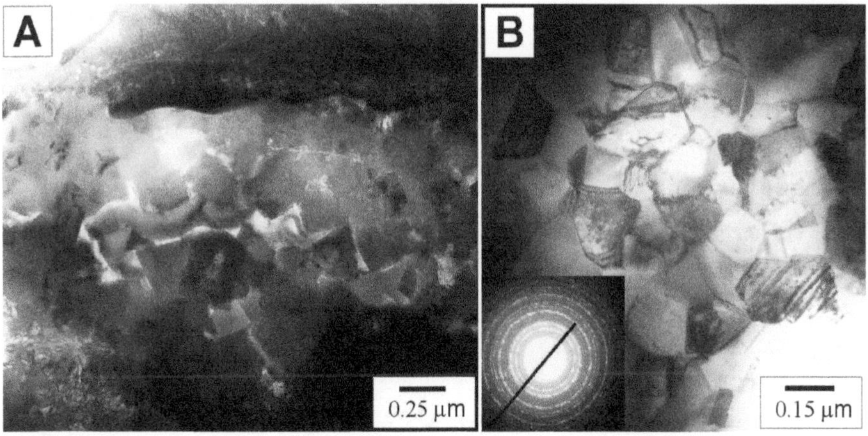

Figure 6. (A) TEM image of hydroxyapatite crystals surrounded by collagen fibrils in dentine from adult rat. (B) Honeycomb-like aggregate of small hexagonal crystals (50–200 nm in average size) in dental enamel of adult rat. Note the concave-convex boundaries among crystals. Inset corresponds to selected area diffraction pattern of the parallel aggregate of hydroxyapatite crystals with their *c* axis parallel aligned to the normal to outer surface.

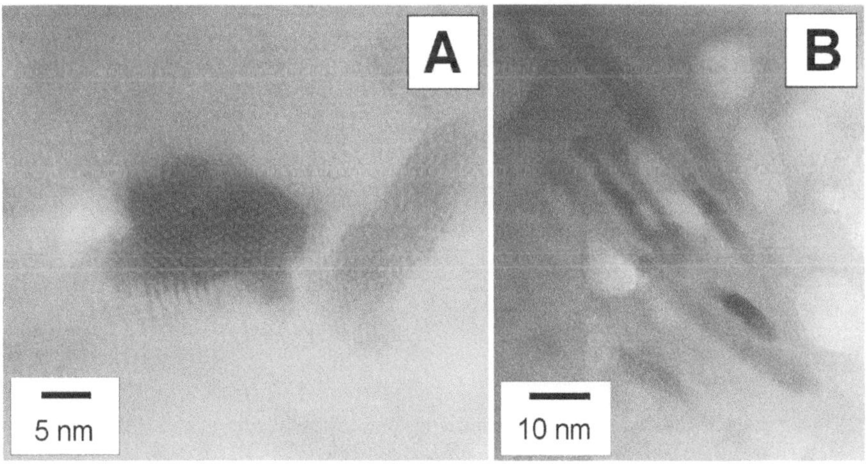

Figure 7. HRTEM images of nanocrystallites of hydroxyapatite in dentine. Crystals appear oriented with their *c* axis normal (A) and parallel (B) to the plane of observation. Lattice-image of figure A corresponds to a hexagonal section of one nanocrystal, for which the three sets of equivalent lattice planes parallel to *c* axis, with interplanar spacings at 0.82 nm, are resolved simultaneously. Crystals have sizes around 10-15 nm in basal section (A) and up to 80 nm in length following the *c* axis (B).

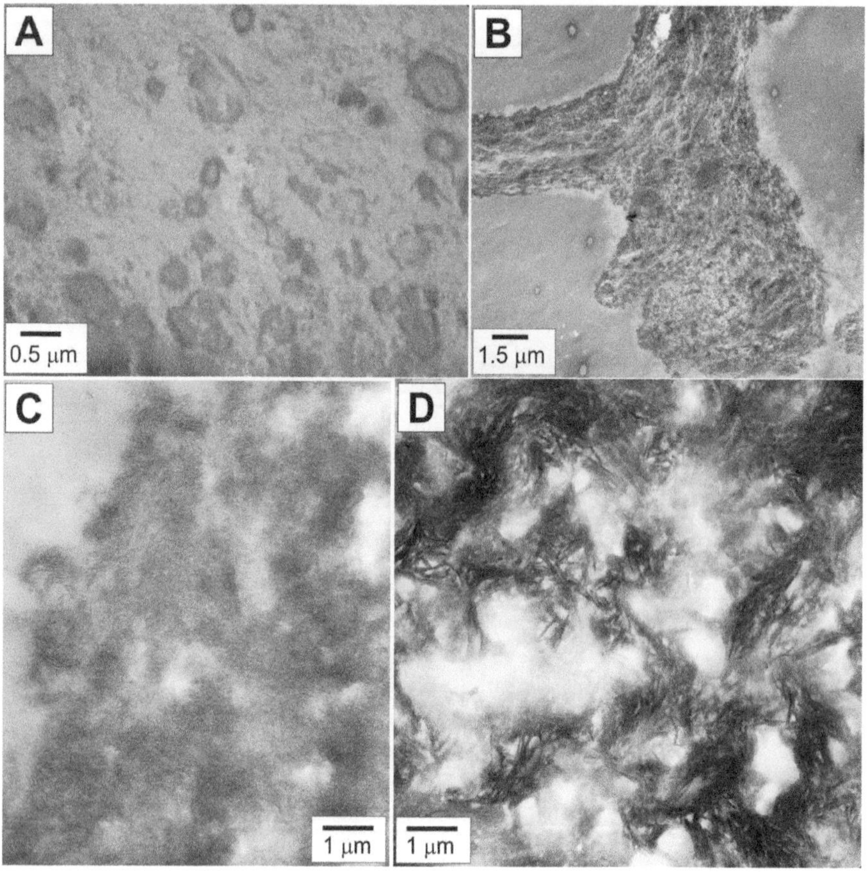

Figure 8. TEM images showing incipient mineralization of hard tissues. (A) Amorphous calcium phosphate forming electron dense masses within a lower electron dense organic matrix in enamel from rat embryo. (B) Electron dense regions formed by amorphous calcium phosphate in dentine from one-day-old rat. (C) Magnified TEM image showing thin nano-sized crystals within the mineralized regions shown in (B). (D) TEM image (ESI mode) needle-shaped crystals in bone from one-day-old rat.

3.2. Inorganic phosphate and carbonate minerals obtained by solution and gel growth

Poorly crystalline Ca-phosphates have been determined after XRD study of inorganic precipitates (Fig. 1B). In the secondary electron images they appear as clumps with a foamy porous structure, where nanometre-sized acicules are visible in the most crystalline areas (Fig. 9A). TEM images confirm the existence of needle-shaped crystallites, which normally appear surrounded by structureless films and "clouds" of amorphous calcium phosphate (Fig. 9B). The acicular crystals are up to 30 nm in length and up to 4 nm in thickness. In some cases, lattice fringes at 0.8 nm corresponding to *d*-spacing of (100) crystallographic planes of apatite are observed (Fig. 9C). There exists a marked resemblance between these TEM images and those previously shown for calcified tissues in one-day-old rats (Figs. 8B-D). Thus, needle-shaped Ca-phosphate nanocrystals surrounded by amorphous calcium phosphate occur both in the non-mature bone (Fig. 8D) and in the inorganic precipitates (Fig. 9C).

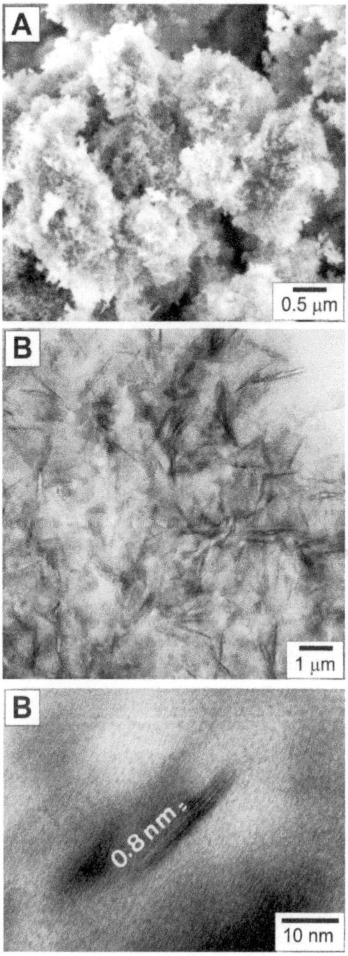

Figure 9. (A) FESEM image of inorganically precipitated phosphate sample showing structureless "clouds" of amorphous calcium phosphate including acicular crystals. (B) ESI-mode TEM image of the inorganically precipitated phosphate sample, where nano-sized needle crystals of apatite are included within amorphous calcium phosphate. (C) Lattice-fringe image of needle apatite crystallite with lattice fringes at 0.82 nm corresponding to the *d*-spacing of (100) or equivalent crystallographic planes.

Rhombohedral micron-sized siderite ($FeCO_3$) crystals and siderite spherulites have been precipitated in oxygen-free water inside an anaerobic chamber from solutions with salt concentrations 50 mM $NaHCO_3$–50 mM $Fe(ClO_4)_2.nH_2O$ and 1 M $NaHCO_3$–1 M $Fe(ClO_4)_2.nH_2O$, respectively (Fig. 10). Siderite spherulites normally show rough surfaces formed by uniaxial aggregates of skeletal crystals with nanometre size (Figs. 10C and D). Both siderite crystals and spherulites appear surrounded by a poorly-crystalline precursory material formed by iron oxyhydroxides (Figs. 10E and 10F).

Crystallization of Ca-Mg rhombohedral carbonates were conducted through gel-growth experiments to favour the incorporation Mg to the calcite structure. The effect of Mg content on the structural parameters and on the crystal habit of calcite was analyzed. Thus, calcite rhombohedrons with dendritic morphologies were formed in Mg-free crystallization experiments (Fig. 11A). Calcite crystals with rough surfaces and crystal morphologies corresponding probably with {021} forms (Fig. 11B) and calcite spherulites (Fig. 11C) grew from 0.1 M and 1 M $MgCl_2.6H_2O$ stock solutions, respectively. The increase of the Mg-content also produced a lattice contraction and a decrease of the crystallinity of the Ca-Mg carbonate, as well as the appearance of metastable phases (monohydrocalcite) as deduced from XRD analyses (Fig. 11D).

Spherulites composed by low-crystalline Mg-rich Ca-Mg carbonates were also studied by SEM and TEM. Secondary images of these spherulites indicate that they grew by adhesion of nanometre-sized particles (Fig. 12A). In addition, TEM study of these samples reveals the occurrence of carbonate nanocrystal building units, with sizes from 40 to 80 nanometres in diameter, which aggregate to form small clusters with approximately the same crystallographic nanocrystals orientation (Fig. 12B).

3.3. Crystal growth features of bacterially-mediated phosphates and carbonates

Bacterially-mediated phosphate and carbonate precipitates show analogous crystal growth features to those of some inorganic (abiotic) precipitates (compare Figs. 10B, 11A and 12A with Figs. 13 and 14). Poorly crystalline Fe- and Pb-phosphates occur in relation to bacterial cells, organelles and biofilms during the initial stages of biomineralization (Figs. 13A and 13B).

Fe-phosphate spherulites and capsules with diverse sizes and textural features were formed after several weeks (Figs. 13C-13E). Voids and channels with widths of 200 nanometres and 1 micron in length are normally observed in SEM images of Fe-phosphate spherulites (Figs. 13C and 13D). Fe-phosphate capsules develop a central hollow (Fig. 13E).

Other bacterially-mediated precipitates are constituted by poorly crystalline Pb-phosphates (Fig. 13F). Their low crystallinity is evidenced by the broad peaks observed in the XRD patterns (Fig. 1A). Secondary electron images obtained from of *Acetobacter* Pb cultures show two different types of precipitates (Fig. 13F). Those of largest dimension correspond to Pb-phosphate spherulites. The smallest precipitates are constituted by irregular clumps of spheroidal Pb carbonates (hydrocerussite) particles around 0.1-0.3 μm in diameter, formed by aggregation of nanoparticles.

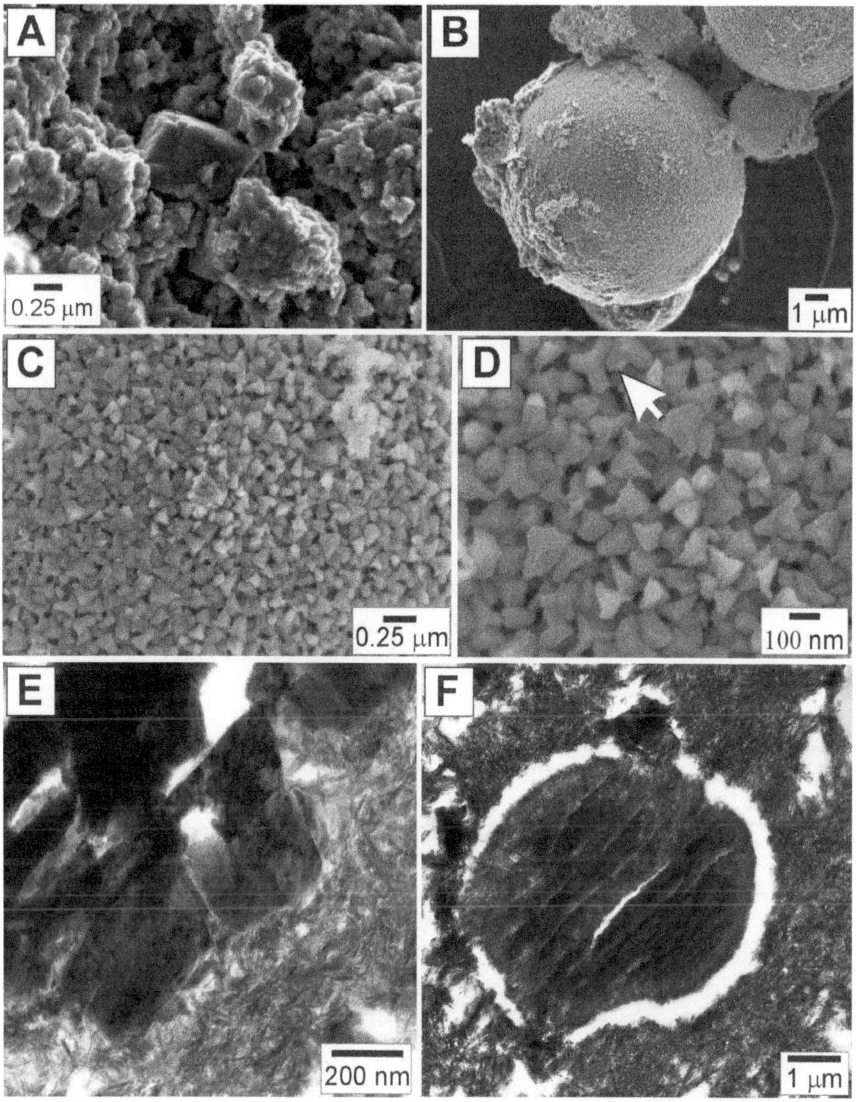

Figure 10. (A) Secondary electron image of siderite rhombohedron surrounded by less crystalline precipitates obtained from solution growth at the following salt concentration: 50 mM $NaHCO_3$, and 50 mM $Fe(ClO_4)_2.nH_2O$. (B) Siderite spherulites partially covered with a thin poorly crystalline Fe-rich layer obtained from solution growth at the following salt concentration: 1 M $NaHCO_3$, and 1 M $Fe(ClO_4)_2.nH_2O$). (C) Uniaxial aggregate of crystals at the rough surface of a siderite spherulite. (D) Close-up of (C) showing the triangular tips of the tripod-like dendritic crystals of siderite. (E) TEM image of rhombohedral siderite crystals surrounded by small needle-like iron oxyhydroxydes. (F) TEM view of a siderite spherulite within a poorly crystalline Fe-rich matrix.

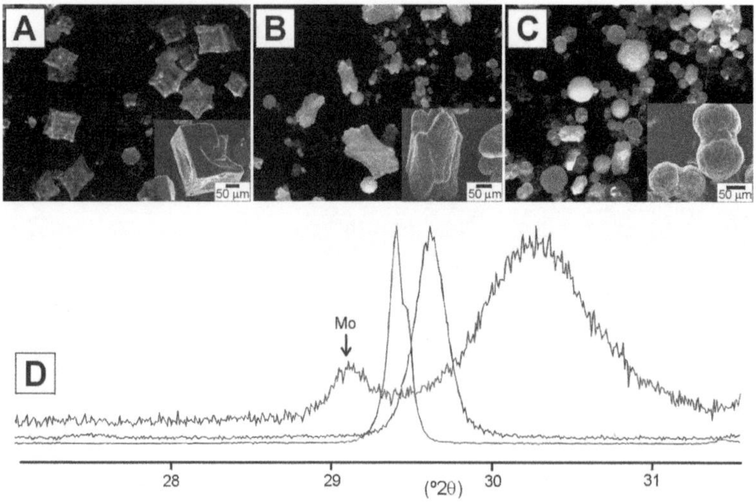

Figure 11. (A) to (C): Optical images and SEM views (insets) of the Ca-Mg carbonates obtained from gel growth inorganic precipitation experiments. (D) XRD features of their (104) peaks. (A) Dendritic calcite crystals precipitated in Mg-free medium. (B) Crystal morphologies of magnesian calcites precipitated from 0.1 M $MgCl_2.6H_2O$ stock solution. (C) Spherulites formed in Mg-richest medium (1 M $MgCl_2.6H_2O$ stock solution). (D) XRD patterns showing the displacement and broadening of the (104) peak with increasing Mg content of the Ca-Mg carbonate from calcite (A) to Mg-calcite (B) and to Mg-kutnahorite (C). Monohydrocalcite (Mo) is also formed together with the Mg-kutnahorite.

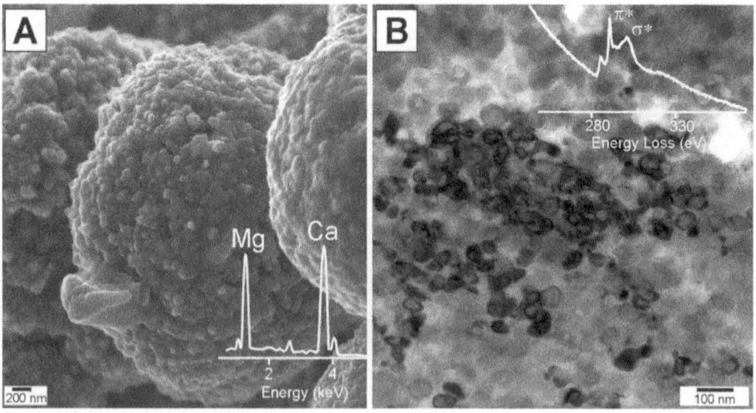

Figure 12. (A) SEM image and EDX spectra of Mg-kutnahorite spherulites obtained from gel growth inorganic precipitation experiments. They are formed by aggregation of rounded growth units, less than 100 nm in size. (B) TEM image (ESI mode) of Ca-Mg carbonate nanoparticles forming Mg-kutnahorites. The two main C K edge peaks in the EELS spectrum of the nanoparticles correspond to π* and σ* antibonding molecular orbitals of CO_3^{2-} cluster (peaks at 290.2 and 301.3 eV respectively [25]). Less intense C K edge peaks from amorphous carbon coating appears also superposed.

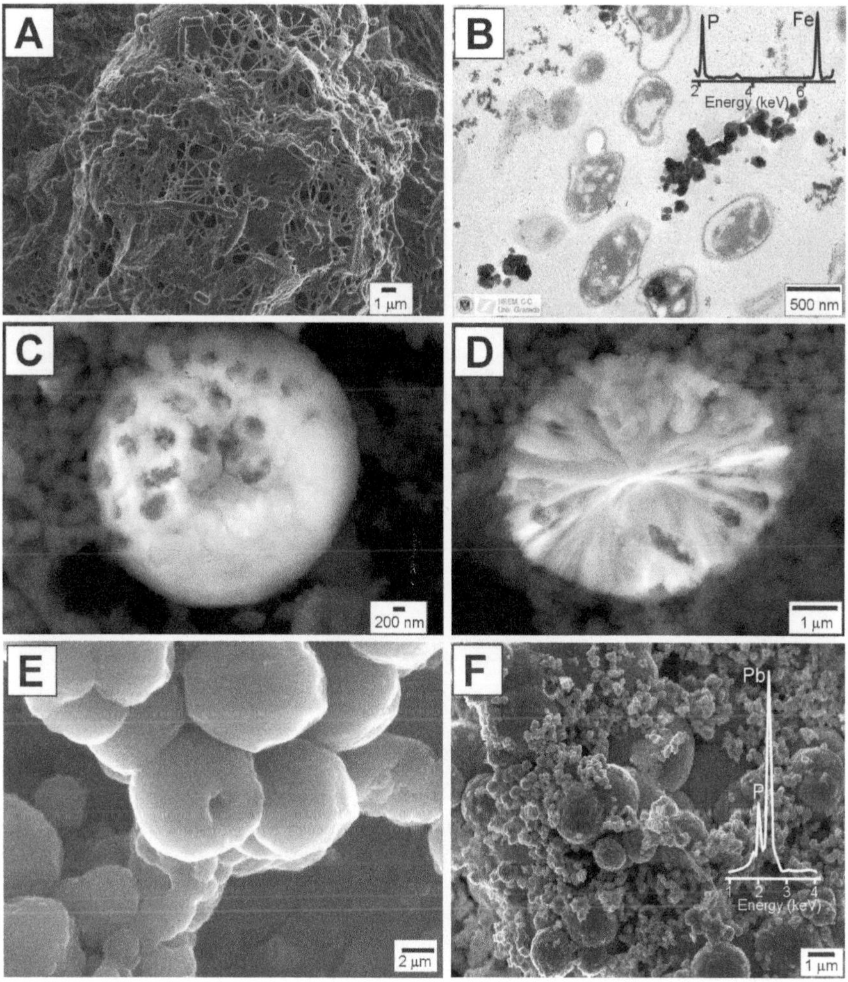

Figure 13. SEM and TEM images of bacterially-mediated phosphate precipitates obtained from precipitation experiments with vinegar dregs. (A) Secondary electron image corresponding to early stages of mineralization of the vinegar dreg. (B) TEM image of bacteria surrounded by iron phosphate nanoparticles, formed at the initial stages of mineralization. (C) SEM image of a Fe-phosphate spherulite with voids of 200 nm of size. (D) View of a sectioned spherulite showing a radial growth pattern and a porous texture. (E) Secondary electron image of Fe-phosphate capsules with smooth surfaces. (F) Pb-phosphate spherulites of very different sizes surrounded by irregular clumps of smaller Pb-carbonate particles.

Bacterially-mediated precipitation experiments with $CaCl_2.6H_2O$ (50 mM), and supplemented with bicarbonate (50 mM $NaHCO_3$), resulted in the formation of calcite spherulites and sheaf-like crystals (Fig. 14A). The latter are elongated following the *c* axis of calcite crystals. They are formed by aggregation of rounded nanoparticles, with sizes smaller than 50 nanometres, at their tips (Fig. 14B).

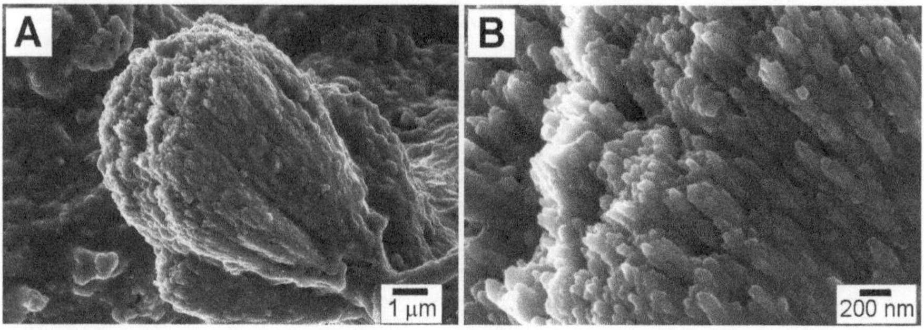

Figure 14. (A) Bacterially-mediated, shrub-like Ca-carbonate precipitate. (B) Close-up of (A), showing the parallel aggregate of calcite crystals elongated along the *c* axis. They grow by aggregation of rounded nanoparticles usually less than 50 nm in size.

4. Discussion

4.1. Kinetics constraints on crystal growth of carbonate and phosphate minerals

Metal phosphates and carbonates occurring in the studied natural samples and in the precipitates obtained from biotic and abiotic experiments develop complex crystal morphologies: from single crystals to crystalline aggregates forming dumbbells and spherulites. In some inorganic experiments, several tens of microns-sized polyhedral forms have been obtained directly (Fig. 11A). However, crystal growth features observed in most samples indicate that phosphates and carbonates here studied can be considered "mesocrystals".[26] These are superstructures formed by the aggregation of nanocrystal building units (Figs. 10E, 12 and 14) that are frequently associated with amorphous precursors (Figs. 1, 4, 5, 8, 9, 10E and 10F). A comparison between the morphology of abiotic and biotic precipitates may help to understand nucleation and crystal growth of new organic-inorganic hybrid materials that are crystallized by oriented aggregation of nanoparticles instead of by ion-by-ion or single molecule attachment. Actually, the occurrence of nanocrystalline particles and aggregation based morphologies within an organic matrix seems be key features of biomineralization.[13, 27]

Precipitation of calcium phosphate and carbonate nanoparticles on diverse substrates in relation to gelly-like substances may be the dominant mechanism of phosphate and carbonate formation in some geologic, microbially-mediated biosedimentary structures (Figs. 2-5)[7, 31] and in some calcified tissues (Figs. 6-8).[23, 24] In phosphate stromatolites, the authigenic clay and phosphate laminae within stromatolites are related to the formation of an extracellular, mucilaginous, bacterial gel, rich in polysaccharides at the living microbial mat.[7, 31] It has been shown that gelly-like media favour the nucleation and the formation of clusters of nanocrystals within hydrogel networks.[27] High supersaturation occurring in gel media leads to the formation of assemblies of nanometer-sized crystalline particles that actually constitute the building blocks of larger crystalline aggregates. These textural features are less frequently formed in crystal-growth from diluted solutions.[28]

Spherulitic morphologies typically form at extremely high growth rates, and result from highly nonequilibrium processes operating both in solution and gel growth.[29] Siderite spherulites have been produced under high supersaturation in solution growth for some inorganic precipitation experiments (Figs. 10B, 10C, 10D and 10 F). In bacterially-mediated precipitation, spherulites of many carbonate and phosphate mineral can be precipitated (Figs. 13 and 14). In this case, the fast attachment of building blocks to the surface of the spherulites frequently preserves bacterial moulds and capsules, and voids are also very common (Figs. 13C, 13D, and 13E). Either obtained in biotic or abiotic experiments, all these spherulites and related morphologies are characterized by a radial texture of nanocrystal aggregates resulting from a geometrical selection process. In this way, fastest crystal-growth directions are perpendicular to the growing surface (Figs. 6, 10C, 10D, and 14). Honeycomb-like aggregates of small hexagonal apatite crystals of dental enamel (Fig. 6B) and equivalent parallel aggregates of prismatic phosphate crystals observed in phosphate stromatolites also result from competitive crystal growth processes.

The obtained carbonate and phosphate nanocrystals constitute a finely divided crystalline matter composed by particles that attach each others (Figs 4, 7, 8C, 8D, 9B, 9C and 12B). However, the aggregation of these nanoparticles is not related to atom-by-atom attachment, as they are still very large when compared to the atomic size scale. In nanometre-size crystals, the density of electronic energy levels, related to the energy of valence electrons responsible of the chemical bonds, varies smoothly between the atomic and bulk limits.[30] In very small nanocrystals the energy absorption spectra develop discrete features, similar to van der Waals or molecular crystals, where the bands in the solid are very narrow. Therefore, the interactions that allow the attachment of nanocrystals during the aggregation-based growth-process producing some of the crystal growth features here described must be very weak.

As shown in this work, and also described elsewhere, [7,31] apatite crystals of the authigenic laminae in phosphate stromatolites are usually surrounded by poorly crystalline smectites and Fe-Si-Al amorphous oxyhydroxides (Figs. 3, 4 and 5). These substances represent the fossil record of a mineralized extracellular mucilaginous bacterial gel that played the same role as bacterial organic matter in biotic laboratory experiments, and an equivalent role to that played by organic framework rich in collagen fibrils of dentine and bone. However, crystal growth features observed in authigenic phosphates (francolite) and smectitic clays of stromatolite laminae are not exclusively explained by the occurrence of gelly-like organic substances in a precursory microbial mat. In addition, the clay-rich material and related amorphous substances in phosphate stromatolites play an analogous role to that of gels in inorganic and biomediated experiments. In such media, bulk flow of a grain boundary fluid containing dissolved solutes does not occur. Within clay-rich media, fluid film is restricted to a monolayer of adsorbed molecules onto the clay particles. This suppresses convective and advective mass transport and favours diffusion and the development of spherulitic instead of polyhedral morphologies (see next section). Therefore clayey material acts as a gel, and its occurrence favours the formation of amorphous precursors and nanocrystals of later precipitates. This is the case of francolite crystallization in phosphate stromatolites, which is preceded by the precipitation amorphous calcium phosphate, as suggested by Sánchez-Navas and Martín-Algarra.[7]

4.2. Considerations related to gel growth and structural control of habit in Mg-Ca carbonates

Gels and related substances are responsible for high supersaturation producing the crystal growth features observed in the solid porous media. High supersaturation gradients arise from the slow transport of the chemical species present in gel growth experiments, and the degree of supersaturation driving the crystallization depends on the diffusion gradient in solid porous media.[28] Hence, high growth rates are expected in porous media where the critical supersaturation for phosphate and carbonate minerals is reached more quickly than in solution growth.

Gel growth also explains why some elements like Mg are incorporated to the structure of calcite. A correlation between Mg-content in calcite and crystallinity may be deduced from the figure 11. Slow diffusive transport media, such as a clay-rich and water deficient environments or porous media, make the transport of reacting material from solution to those of growth be the rate-determining process. Transport properties in porous media controls on the incorporation of elements during the crystal growth.[32] To understand it, we may suppose that the partition coefficient between the concentration of magnesium in carbonate crystal and its concentration in bulk medium ($C^{Mg}_{crystal} / C^{Mg}_{medium}$) is less than 1. When carbonate crystal growth is diffusion controlled, the concentration of Mg at the crystal-medium interface rises above its concentration in the medium. If the partition coefficient remains constant, the Mg concentration in the carbonate increases due to the slower diffusion rate and, in the limit, it is equal to the concentration of Mg in the medium. It makes that the effective partition coefficient $C^{Mg}_{crystal} / C^{Mg}_{medium}$ equals 1 and, therefore, most of the Mg of the medium is incorporated into the structure of the rhombohedral carbonate.

Crystal chemical substitution of Mg for Ca within the structure of rhombohedral carbonates decreases crystallinity, stabilizes nanocrystals and favours an aggregation-based crystal growth. The different size of the cations involved (Ca^{2+} and Mg^{2+}) distorts the calcite crystal lattice. Of these two cations the Mg^{2+} ion has the smallest ionic radius. The lattice contraction due to the substitution of Mg for Ca is evidenced by the displacement of the (104) peak in XRD patterns (Fig. 11). The $d_{(104)}$-spacing and lattice constants of Mg-calcite are shorter than those of calcite, being the *c* axis the most affected by the incorporation of Mg.[33] This lattice contraction causes the CO_3^{2-} anions layers to be pulled closer together along the *c*-axis and produces a lattice deformation around Mg atoms. The incorporation of Mg into the crystal structure increases drastically the volume strain energy. This prevents reduction in free energy for nucleation with size increase and, therefore, further growth of nanocrystals. In addition, lattice contraction also stabilizes attractive interactions among nanocrystals. Lattice contraction carries out nonbonded repulsion between CO_3^{2-} anions within the structure of the rhomboedral carbonates. The more the distance between CO_3^{2-} anions decreases because of the incorporation of the small Mg cation into the structure of calcite, the more repulsion increases. The increase of nonbonded repulsion makes valence electrons feel less potential wells.[34,35] In the case of carbonates, electrons feel less the potential of the oxygen nuclei within CO_3^{2-} anions. The decrease of electron binding energy in O atoms increases oxygen polarizability. The improvement of the so-called polarization, correlation, Van der Waals or long-range forces between CO_3^{2-} enhances nanocrystals attachment during the aggregation-based growth-process.

5. Conclusions

Crystal growth features in biogenic and abiotic carbonates and phosphates occurring in natural environments (phosphate stromatolites and calcified tissues -bone and teeth) and formed in laboratory experiments are described in this work.

Phosphate and carbonate minerals precipitated from abiotic and biotic experiments form commonly spherulites and related morphologies. In most cases they can be considered mesocrystals because, independent of their biotic or abiotic origin, they are superstructures formed by the aggregation of nanocrystal building units and appear associated with amorphous precursors. Parallel aggregates constituted by Ca-phosphate crystals are formed by competitive crystal growth processes and, together with amorphous Ca-phosphates, are observed in dentine and enamel as well as in phosphate stromatolites

Gels in laboratory experiments, organic matrix in calcified tissues and equivalent clay-rich media in phosphate stromatolites favour diffusive transport and supersaturation of carbonate and phosphate minerals. Supersaturation is responsible for the development of spherulitic morphologies and the occurrence of nanometre-sized crystals and related amorphous substances. Gels also explain the incorporation of some elements as Mg to the structure of calcite.

Acknowledgements

We acknowledge support from grants CGL-2009-09249 (DGICYT, Spain) and P11-RNM-7067 of the Junta de Andalucía (Spain), and by the research group RNM-179 (Junta deAndalucía). We would like to thank María José Martínez Guerrero, Alicia González Segura, María del Mar Abad Ortega, Concepción Hernández, Isabel Guerra Tschuske and Juan de Dios Bueno (Centro de Instrumentación Científica-CIC, University of Granada) for guidance with SEM-EDX and TEM-AEM-EELS studies. Agustín Martín Rodríguez, José Suarez Valera, María Navas Moreno and Elena Beatriz Sánchez Martín significantly helped with laboratory precipitation experiments.

Author details

Antonio Sánchez-Navas[1], Agustín Martín-Algarra[2], Mónica Sánchez-Román[3], Concepción Jiménez-López[4], Fernando Nieto[1] and Antonio Ruiz-Bustos[1]

1 Departamento de Mineralogía y Petrología, Facultad de Ciencias, Universidad de Granada, Granada, Spain

2 Departamento de Estratigrafía y Paleontología, Facultad de Ciencias, Universidad de Granada, Granada, Spain

3 Centro de Astrobiología, INTA, Torrejón de Ardoz, Madrid, Spain

4 Departamento de Microbiologia, Facultad de Ciencias, Universidad de Granada, Granada, Spain

Instituto Andaluz de Ciencias de la Tierra. C.S.I.C., Armilla, Granada, Spain

References

[1] Capewell SG, Hefter G, May PM. Potentiometric investigation of the weak association of sodium and carbonate ions at 25 ºC. Journal of Solution Chemistry 1998; 27: 865-877.

[2] Benjamin MM. Water Chemistry. New York: McGraw-Hill; 2001.

[3] Broecker WS, Peng TS. Tracers in the Sea. New York: Lamont-Doherty Geological Observatory; 1982.

[4] Bonucci E. Biological calcification. Berlin: Springer-Verlag; 2007.

[5] Cowan J, Weintritt D. Water formed scale deposits. Houston: Gulf Publ. Co.; 1976.

[6] Allwood AC, Walter MR, Kamber BS, Marshall CP, Burch IW. Stromatolite reef from the Early Archaean era of Australia. Nature 2006; 441: 714-718.

[7] Sánchez Navas A, Martín Algarra A. Genesis of apatite in phosphate stromatolites. European Journal of Mineralogy 2001; 13: 361-376.

[8] Allen CC, Albert FG, Chafetz HS, Combie J, Graham CR, Kieft TL et al. Microscopic physical biomarkers in carbonate hot springs: implications in the search for life on Mars. Icarus 2000; 147: 49-67.

[9] McKay DS, Thomas-Keptra KL, Romanek CS, Gibson Jr EK, Vali H. Evaluating the evidence for past life on Mars. Science 1996; 274: 2123-2124.

[10] Allwood AC, Walter MR, Kamber BS, Marshall CP, Burch IW. Stromatolite reef from the Early Archaean era of Australia. Nature 2006; 441: 714-718.

[11] Chivas AR, Torgersen T, Polach HA. Growth rates and Holocene development of stromatolites from Shark Bay, Western Australia. Australian Journal of Earth Science 1990; 37, 113-121.

[12] Walter MR, Golubic S, Preiss WV. Recent stromatolites from hydromagnesite and aragonite depositing lakes near the Coorong Lagoon, South Australia. Journal of Sedimentary Research 1973; 43: 1021-1030.

[13] Rivadeneyra MA, Martín-Algarra A, Sánchez-Román M, Sánchez-Navas A, Martín-Ramos D. Amorphous Ca-phosphate precursors for Ca-carbonate biominerals mediated by Chromohalobacter marismortui. The ISME Journal 2010; 4: 922-932.

[14] Buczybnski C, Chafetz HS. Habit of bacterially induced precipitates of calcium carbonate and the influence of medium viscosity on mineralogy. Journal of Sedimentary Petrolology 1991; 61, 226-233.

[15] Ehrlich HL. Geomicrobiology (4th edn.). New York: Marcel Dekker; 2002.

[16] Sánchez-Navas A, Martín-Algarra A, Rivadeneyra MA, Melchor S, Martín Ramos JD. Crystal-Growth Behavior in Ca–Mg Carbonate Bacterial Spherulites. Crystal Growth and Design 2009; 9: 2690-2699.

[17] Sunagawa I. Morpholgy of Crystals. Tokyo: Terra Scientific Publishing Company; 1987.

[18] Checa González A, Sánchez-Navas A, Rodríguez-Navarro A. Crystal growth in the foliated aragonite of monoplacophorans (Mollusca). Crystal Growth and Design 2009; 9: 4574-4580.

[19] Martín-Ramos J.D., Díaz-Hernández J.L., Cambeses A., Scarrow J.H., López-Galindo A. Pathways for Quantitative Analysis by X-Ray Diffraction. In: Aydinalp C. (ed.) An Introduction to the Study of Mineralogy. Rijeka: InTech; 2012. http://www.intechopen.com/articles/show/title/pathways-for-quantitative-analysis-by-x-ray-diffraction (accessed 20 July 2012).

[20] Vali H, Koster HM. Expanding behaviour, structural disorder, regular and random irregular interstratification of 2–1 layer-silicates studied by high-resolution images of transmission electron-microscopy. Clay Minerals 1986; 21: 827-859.

[21] Jarvis I. Sedimentology, geochemistry and origin of phosphatic chalks: the Upper Cretaceous deposits of NW Europe. Sedimentology 1992; 39: 55-97.

[22] De Vrind-de-Jong E.W., de Vrind J.P.M. Algal deposition of carbonates and silicates. In. Banfield J.F., Nealson K.H. (eds.) Geomicrobiology: Interactions between Microbes and Minerals. Reviews in Mineralogy: Mineralogical Society of America; 1997. p267-307.

[23] Landis WJ, Hodgens KJ, Arena J, Song MJ, McEwen BF. Structural Relations Between Collagen and Mineral in Bone as Determined by High Voltage Electron Microscopic Tomography. Microscopy Research and Technology 1996; 33: 192-202.

[24] Ziv V, Sabanay I, Arad T, Traub W, Weiner S. Transitional structures in lamellar bone. Microscopy Research and Technology 1996; 33: 203-213.

[25] Garvie LAG, Graven AG, Brydson R. Use of electron-energy loss near-edge fine structure in the study of minerals. American Mineralogist 1994; 79: 411-425.

[26] Cölfen H, Antonietti M. Mesocrystals: Inorganic Superstructures Made by Highly Parallel Crystallization and Controlled Alignment. Angewandte Chemie International Edition 2005; 44: 5576-5591.

[27] Grassmann O, Neder RB, Putnis A, Löbmann P. Biomimetic control of crystal assembly by growth in an organic hydrogel network. American Mineralogist 2003; 88: 647-652.

[28] Putnis A, Prieto M, Fernández-Díaz L. Fluid supersaturation and crystallization in porous media. Geological Magazine 1995; 132: 1-13.

[29] Gránásy L, Pusztai T, Tegze G, Warren JA, Douglas JF. Growth and form of spherulites. Physical Review E 2005; 72 (011605): 1-15.

[30] Alivisatos AP. Nanocrystals: building blocks for modern materials design. Endeavour 1997; 21(2): 56-60.

[31] Sánchez-Navas A, Martín-Algarra A, Nieto F. Bacterially-mediated authigenesis of clays in phosphate stromatolites. Sedimentology 1998; 45: 519-533.

[32] Henderson P. Inorganic geochemistry. Oxford: Pergamon Press; 1986.

[33] Goldsmith JR, Graf DL. Lattice constants and composition of the calcium magnesium carbonates. American Mineralogist 1961; 46: 453-457.

[34] Bader RFW, Preston HJT. A critique of Pauli repulsions and molecular geometry. Canadian Journal of Chemistry 1966; 44: 1131-1145.

[35] Bader RFW. Pauli repulsions exist only in the eye of the beholder. Chemistry European Journal 2006; 12: 2896-2901.

Chapter 2

Direction Controlled Growth of Organic Single Crystals by Novel Growth Methods

M. Arivanandhan, V. Natarajan,
K. Sankaranarayanan and Y. Hayakawa

Additional information is available at the end of the chapter

1. Introduction

The green lasers are attractive and highly useful for lot of practical applications. It is more than fifty times brighter when compared to a red laser and thus it can be seen from miles away. Due to these features, the green lasers can be used in high-tech weapons for aiming purposes. Moreover, the green lasers are highly useful for laser televisions and medical applications. Most of the commercially available green lasers are based on the diode pumped solid state frequency-doubled (DPSSFD) laser technology. For the past several decades, researchers and several industries were trying to develop the laser diodes based on the compound semiconductors such as Gallium nitride (GaN) and Indium Gallium nitride (InGaN) especially in the blue and green region of wavelengths [1]. However, it is very difficult to grow bulk crystal of these materials with reasonable quality. Moreover, for the preparation of epitaxial thin films of these materials on substrates, highly sophisticated and expensive techniques like molecular beam epitaxy (MBE) and metal organic chemical vapour deposition (MOCVD) are needed. Apart from these growth aspects, the relatively low power and limited wavelength range restricts their use in important applications [2]. Therefore, laser sources based on Second harmonic generation (SHG) is a better choice for the applications which requires higher powers or longer wavelengths (>400nm). As a consequence, the green laser technology still depends on the nonlinear optical phenomena such as frequency doubling.

In the DPSSFD lasers, a nonlinear optical (NLO) crystal must be placed to halves the wavelength of a solid state laser. In the today's market, inorganic NLO crystals of Potassium Titanyl Phosphate (KTP), Lithium triborate (LBO) are used as frequency doublers. For example, the KTP crystal is used to generate green laser at 532 nm by halving the wavelength of Nd:YAG laser of 1064nm. Organic NLO materials are superior to the inorganic materials both in the speed of response and high NLO susceptibilities. Moreover, they have high laser damage threshold compared to inorganic materials. For frequency doubling applications, the growth direction of NLO crystal has to be controlled towards a phase-matched direction. Therefore, direction-controlled growth is an indispensable technology, especially for the bulk growth of NLO crystals due to its anisotropic nature of NLO properties.

In the literature, different kind of direction-controlled growth technique such as indirect laser heated pedestal growth (ILHPG) method [3], Microtube-Czochralski (μT-CZ) method [4], seed-oriented undercooled melt growth [5], Vertical Bridgman (VB) method [6], and uniaxially solution crystallization (USC) method of Sankaranarayanan-Ramasamy [7] have been reported with the aim of growing unidirectional NLO crystals. Despite the unidirectional organic NLO crystal can be grown by ILHPG [3] and seed-oriented undercooled melt growth [5], the complicated experimental set-up and multistep growth process leads to difficulty in growing large size unidirectional crystal. Whereas the μT-CZ [4] and VB [6] methods are more versatile and bulk directional crystals can be grown by optimizing the growth parameters. On the other hand, the recently reported USC method [7] attracted the researchers by its unique advantage than the other methods such as unidirectional growth at ambient temperature which causes minimum thermal induced grown-in defects using simple experimental set-up with high solute-crystal conversion efficiency and high growth rate.

Benzophenone is a promising organic NLO material and it has nearly six times higher NLO efficiency than that of Potassium dihydrogen phosphate (KDP), a well-known inorganic NLO material [8, 9]. It crystallizes in the non-centrosymmetric orthorhombic space group $P2_12_12_1$ and the lattice parameters are reported as a = 10.26Å, b =12.09 Å and c =7.88 Å [10]. Benzophenone is also known as aromatic ketone which is a particularly interesting material for studying the impact of crystallization conditions on crystal defects and quality [11]. In the present investigation, benzophenone single crystals have been grown by VB, μT-CZ and USC method. Since all the three growth methods are quite different, the crystals grown by these methods may have different crystalline perfection, which may lead to some differences in their physical properties. In order to understand the impact of crystallization conditions on crystal quality, a comparative study has been made on the benzophenone crystals grown by these three different techniques by employing X-ray diffraction (XRD), and high resolution X-ray diffraction (HRXRD).

Further, the USC growth method was extended to grow benzophenone single crystals in three different orientations. The growth rate of the crystal in different orientaitons were measured. Laser damage threshold and hardness of the directional samples were studied. In harmonic generation, the range of conversion is recently extended up to extreme ultraviolet and soft X-ray regions [12]. During the practical operation, the NLO materials are exposed to high intensity laser radiations for harmonic generations. Thus, the NLO crystals must have the ability to withstand high power laser radiations [13]. Laser induced damage studies on NLO crystals are obviously important, since the surface and bulk damage of the crystal by high power lasers limits its performance in NLO applications. The damage threshold of the NLO material must be investigated by multiple shot mode as well as single shot mode since the NLO crystals are generally used for long durations in repetitive mode at various applications. On the other hand, mechanical hardness of a material is also one of the decisive properties especially for post-growth processes and device fabrications. Hence, the laser damage threshold and hardness properties of the unidirectional samples were investigated. The observed laser damage profile and hardness variations in three different growth directions are explained based on the crystal structure of benzophenone. The mechanism for the laser induced damage in benzophenone is discussed.

2. Experiment

2.1. VB growth of benzophenone crystal

Prior to filling the source material of benzophenone, the ampoules were chemically cleaned in HCl : HNO_3 mixture (prepared in the ratio of 1: 1) and kept in electronic grade acetone in order to remove the surface contamination to avoid any possible multi-heterogeneous nucleation. The benzophenone powder was purified by the zone refining method using a movable furnace assembly. Bulk crystals of benzophenone were grown using the indigenously constructed VB system (Figure 1). The VB system consists of three major parts such as transparent furnace, temperature controller and ampoule translation assemby. The transparent furnace consists a central quartz tube which is centrally placed in a glass beaker filled with two immiscible liquids. Sufficient volumes of deionized water and sunflower oil (normally used for cooking) were used for low temperature and high temperature zones respectively since the melting point of benzophenone is ~48°C. Spiral shaped tubular resistive heaters which are fabricated in our laboratory, were encircled the growth tube at hot and cold zones. Commercially bought Eurotherm 818 PID temperature controller with an accuracy of ±0.1°C was employed to control the zone temperatures. Lowering rate of 1-4 mm.h-1 was achieved using an in-house built ampoule lowering system

Direct observation of solid-liquid interface, which is more feasible in transparent furnaces than conventional furnaces, is important for the directional solidification to determine the desired interface shape by controlling the growth parameters. In the case of VB growth of organic material, due to its low thermal conductivity one has to adopt the recommended translation rate of 1-2 $mm.h^{-1}$ [14]. With the aid of thermocouple attached ampoule holder, the in-situ thermal profile analysis was made during the growth and measured a vertical temperature gradient near the solid-liquid interface. In the present work, the growth run was initiated with the translation rate of 1 $mm.h^{-1}$ when hot and cold zone temperatures are kept at 55 and 32°C. The growth experiments were performed with different growth parameters and an experiment was successful with reasonably good quality crystals, when the hot and cold zone temperatures were 51°C and 35°C, respectively while the translation rate was 1 $mm.h^{-1}$. Hence, the temperature gradient of the furnace and lowering rate of the ampoule influence the quality and directionality of the growing crystal. The more details of the growth processes can be found elsewhere [6].

Figure 1. Transparent vertical Bridgman growth system.

2.2. μT-CZ growth of benzophenone crystals

A technical brief of the experimental setup employed in this investigation can be found elsewhere [4]. Highly purified benzophenone material was filled in a circular shape static glass crucible having the size of 60 mm ID and 110 mm height. The source material filled crucible was placed inside a resistive heated furnace. Commercially bought Eurotherm 818 PID temperature controller with an accuracy of ± 0.1°C was employed to control the temperature of the furnace. In the present work, instead of seeding by pre-grown defect free seed crystal, stainless steel microtube of size 8 μm ID has been used for seeding the melt. Since melt wets the inner walls of the fine capillary tube, it rose to a height, which depended by the tube radius, the surface tension of the melt, the melt density and the contact angle of the melt with micro tube. A fine column of melt raised inside the microtube was crystallized first due to heat desipation through seed rod and the grown crystal inside the microtube was acted as a seed for further growth. The growth temperature and the pulling rate of the crystal were optimized for the growth of benzophenone single crystals. The optimized growth parameters for the present investigation are, pulling rate: 1.0 - 1.5 mm.hr^{-1}, seed rotation rate: 5-10 rpm, the cooling rate: 1°Chr^{-1}, length of the microtube underneath the melt surface (l_{ums}): 1.5 mm and the axial thermal gradient: 8°C/cm. The temperature at which the microtube seeding is made (t_{ms}), and l_{ums} play a vital role in deciding the orientation of the growing crystal inside the microtube (will be discussed in the next section) [4]. Once the growth run was completed, the system temperature was reduced to room temperature (31°C) at a predefined cooling rate to avoid the thermal stress in the grown crystal.

2.3. Growth of benzophenone crystal by USC method

Unidirectional benzophenone single crystals have been grown along <110> direction by mounting a dislocation free seed crystal at the bottom of a glass ampoule in such a way that the (110) plane of seed crystal facing towards the saturated solution of benzophenone. Then, the ampoule was filled with saturated solution of benzophenone having optimized solute concentration and porously sealed. The schematic view of the experimental set-up used for the USC growth is shown in Figure 2. The transparent nature of the experimental set-up and the visibility of the solid-liquid interface, facilitate the measurement of growth rate in the particular crystallographic direction. Growth rate of a uniaxial crystal of particular size along a particular growth axis largely depends on the packing density of that plane, purity of the raw materials, degree of supersaturation and the rate of diffusion of the solute in the solvent medium. A comprehensive experimental report can be found in the literature [15]. In USC method, since the crystal is growing in selective growth orientation, the commonly observed growth features in the case of conventional solution grown crystal such as growth sectors, grain boundaries, twins, stacking faults and dislocations are not observed in the X-ray topography [15], indicating that the USC grown sample is relatively free from these defects.

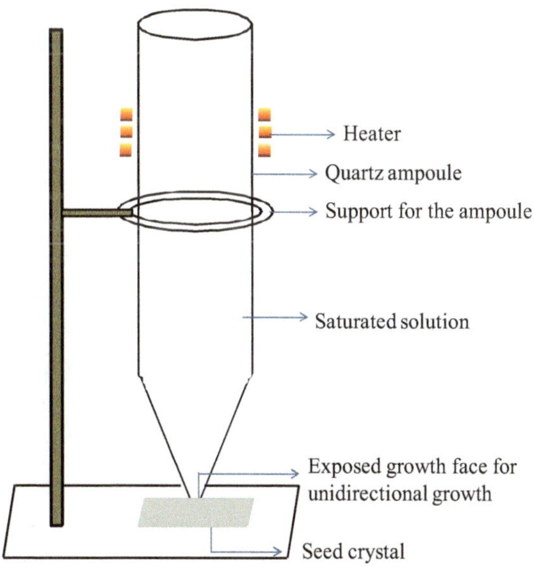

Figure 2. Schematic view of USC esperimental set-up.

3. Results and discussion

3.1. VB growth

The VB growth was conducted for various translation rate of ampoule and temperature gradient of the two zone furnace to control the quality of the crystal. The growth experiment was initiated with the translation rate of 1 $mm.h^{-1}$ and the temperature gradient of 0.75°C/mm. The grown crystal was opaque with larger grains and cracks (Figure 3 a). The major cracks may be attributed due to low thermal conductivity of benzophenone. This result basically confirms the suitability of the thermal profile present in the constructed furnace for the Bridgman growth of benzophenone crystal. In order to study the effect of temperature gradient on the quality of grown crystal, the temperature gradient of the furnace was lowered to 0.5°C/mm by increasing the cold zone temperature. A single crystal with relatively high transparency, small numbers of cracks and few numbers of bubbles were obtained Figure 3 b). Due to the transparency of the furnace and the melt, the solid-liquid interface was visible and found to be concave.

In an attempt to remove the grown-in defects such as cracks, bubbles and to study the influence of translation rate on these grown-in defects, the growth runs were made with the translation rate of 2 and 2.5 $mm.h^{-1}$.The observations made on the resultant material obtained from the growth run with the translation rate of 2 $mm.h^{-1}$ revealed that the transparency of the ingot was increased when compared to the previous growth run and the number of large size cracks and bubbles was reduced. However, the concavity of the solid-liquid interface was maintained. Further increase in the translation rate to 2.5 $mm.h^{-1}$, resulted a bubble-free ingot with good transparency (Figure 3 c). However, very fine cracks were observed in the crystal possibly due to the thermally induced strain and faster growth rate at high translation rate. Also, the high translation rate increases the concavity of the solid-liquid interface.

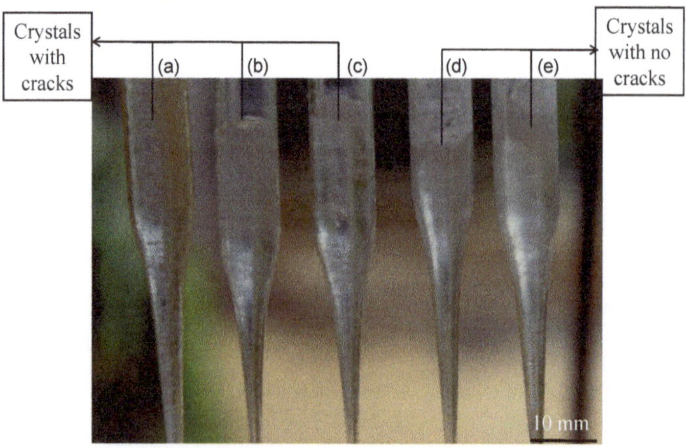

Figure 3. Photographs of the grown crystal in various growth runs.

Even though the obtained single crystal from the growth run conducted with the vertical temperature gradient of 0.5°C/mm and translation rate of 2.5 $mm.h^{-1}$ was free of bubbles and larger cracks, the fine cracks due to thermally induced strains are present. To eliminate the fine cracks, the hot and cold zone temperatures were reduced from 55 to 51°C and 39 to 35°C, respectively. At this temperature, a growth run with the translation rate of 1 $mm.h^{-1}$ has resulted a near flat solid-liquid interface but the grown single crystal was not free of fine cracks. Also, further reduction in the translation rate leads to supercooling of melt. Due to different thermal conductivity and viscosity of the liquids of two zones, a sharp variation of temperature was observed at the interface between the liquids. This variation probably causes the fine cracks in the grown crystal.

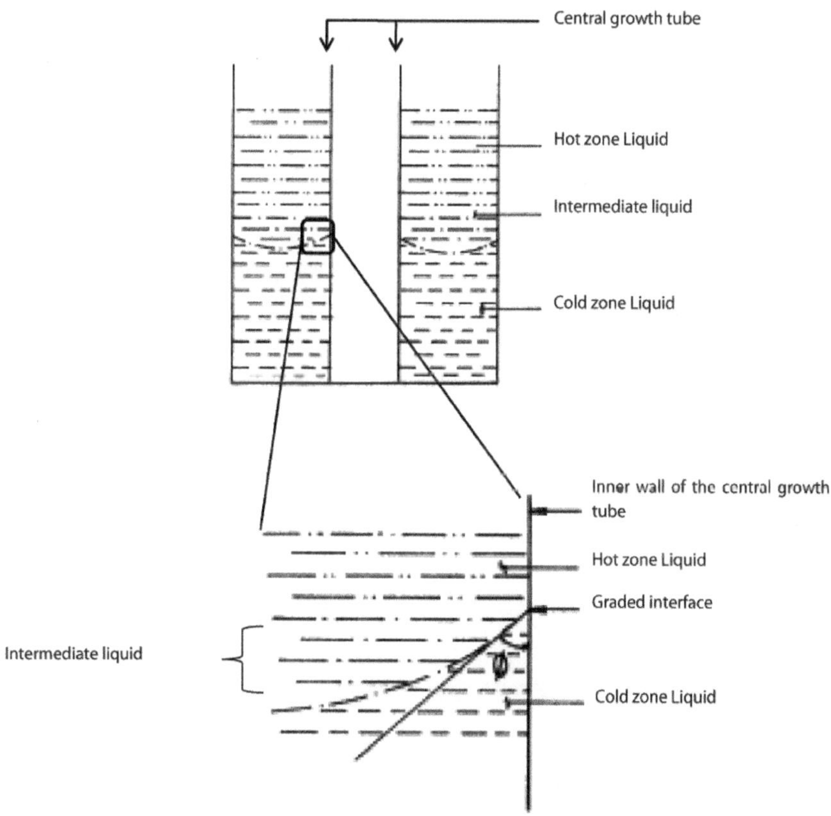

Figure 4. Schematic view of graded interface in VB system.

In order to control the fine cracks, a graded interface was established by introducing intermediate liquid in between the two liquids which has different wetting angle (φ) with the inner growth tube (Figure 4). The density of the intermediate liquid is smaller than the topper liquid and larger than the bottom one. It is expected that by establishing the graded interface, one can control the sharp variation of temperature at the interface between hot and cold zone thereby the fine cracks in the grown crystal. Keeping the zone temperatures and translation rate as constant, growth runs were conducted with different volumes (50, 100 and 150 mL) of intermediate liquid. The result shows that the grown in defects such as fine cracks were suppressed and a transparent, optical quality single crystalline ingot was grown when 150 mL of intermediate liquid was used. Figure 3 d, e illustrates the grown benzophenone ingots free of grown in defects such as thermal induced strains, cracks and bubbles.

3.2. Microtube CZ growth

The selection of microtube for the material to be grown as a bulk single crystal mainly depends on the melting point of the source material, surface tension of the melt, and the chemical reactivity of the melt with the microtube. In the present case, due to the low melting point of the benzophenone (~48°C), stainless steel microtube (ID: 0.8 mm) was used for seeding the melt. The important experimental parameters used for controlling nucleation inside the tube are, radius and rotation rate of the microtube, t_{ms}, pulling rate, axial temperature gradient and l_{ums}.

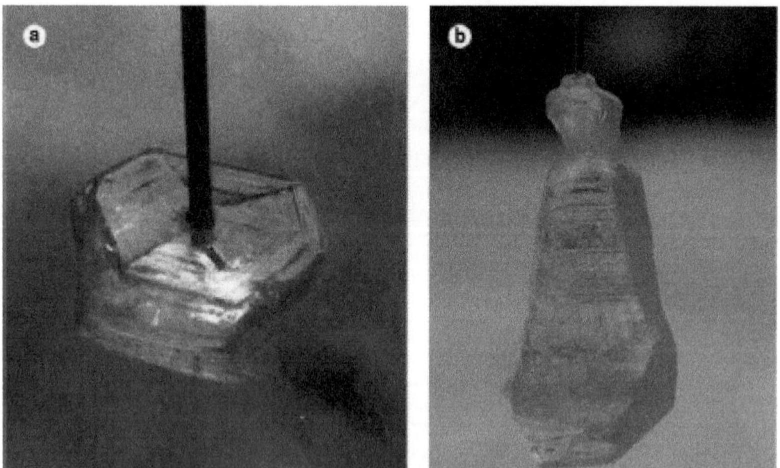

Figure 5. Photographs of the grown benzophenone crystals with (a) hexagonal morphology and (b) cubic morphology.

Further, the seed rotation rate plays a vital role in the initial stage of growth inside the microtube rather than during the growth. Sankaranarayanan et al (1998) [4] have reported that for a fixed radius of microtube there exists a critical rotation rate. Below this, the effect of change of crystal rotation is influential in deciding the morphology of the resultant crystal. If the rotation rate is greater than the critical rotation rate, the effect of changes in rpm leads to very minimal effect since the radius of the microtube is very small. This has also been confirmed in the case of growth of benzophenone. Obvious change in morphology from hexagonal (Figure 5 a) to cubic (Figure 5 b) was observed by varying the seed rotation rate from 5 to 10 rpm at the initial stage of the growth [16].

Jackson et al has analyzed theoretically the structure of solid-liquid interfaces and has demonstrated that the morphology is determined by a factor $'\alpha' = (L_0/kT_e)\ (\eta/z)$ Where L_0 = the enthalpy of fusion, k= Boltzmann's constant, T_e= equilibrium growth temperature, η and z are the number of atoms within the plane and the bulk crystal, respectively [17]. This analysis showed that materials for which α>2 are generally faceted on one or more planes. Moreover, the η/z value depends on the crystal structure of a material. Benzophenone has large α factor (α=6.3) and facetted growth is to be expected [18]. Therefore the observed two different morphologies are probably due to large value of Jakson's α factor. In addition, in order to obtain crystal with morphology as in the conventional Czochralski technique, efforts were made to study the influence of shoulder angle by varying the growth rate just after seeding was conducted. The experimental growth parameters such as t_{ms} = 44°C, l_{ums} = 1.5 mm and seed rotation rate = 10 rpm were kept constant. The pulling rate and the axial temperature gradient were varied in order to vary the growth rate. At the initial stage of seeding and growth, no marked changes were observed until the growth rate was kept at low. A sudden increase in the growth rate leads to higher shoulder angle as evidenced in the figure 6. The growth was continued until the crystal reaches a required size. Thereafter it was pulled at high pulling rate in order to detach the crystal from the melt surface and to analyze the shape of the solid-liquid interface. The solid-liquid interface was found to be concave towards melt which indicates the higher growth rate [19].

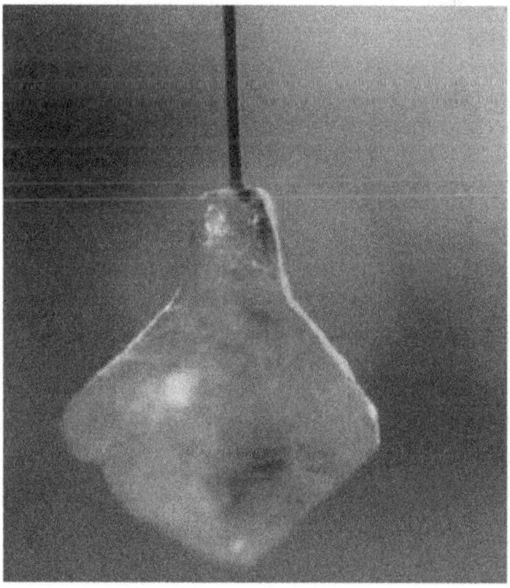

Figure 6. Grown crystal with high shoulder angle.

The morphology of the grown crystal was cubic. It suggested that in μT-CZ technique, the resultant orientation of the growing crystal will be decided by the orientation of the crystsl which emerges out of the microtube, it largely depends on the t_{ms} and l_{ums}. Subsequently the sample (Figure 7) was fabricated from the grown crystal by cut and polishing for structural and optical characterization studies.

Figure 7. Prepared specimen from the μT-CZ grown crystal.

3.2.1. Orientation of the crystal

The nucleation phenomena that happened inside the microtube were explained with the aid of a schematic diagram in figure 8. In capillary rise, most of the organic melts form a convex meniscus. In Figures 8, 'a' and 'b' represent the melt level inside the microtube along both sides and this derivation mainly depends on the planarity of 'a' and 'b'. If the microtube exactly coincides with the center of axis of the system, then 'a' and 'b' lie at the same level. In that case, depending on the pulling rate of the microtube, a fine layer of melt will be retrieved along the wall due to wetting and nucleation will be initiated at 'a' and 'b', simultaneously. The axial temperature gradient and the cooling rate will influence the growth rate but the orientation at 'a' and 'b' need not be the same. If the orientation at 'a' and 'b' is the same (figure 8i), then the value of l_{ums} is not vital.

Suppose, the orientation of crystal is different at 'a' and 'b', then l_{ums} plays a crucial role. The value of l_{ums} should be sufficient enough to allow any one of the grains to proceed at the end of the microtube as the deciding crystal orientation. Since the radius of the microtube is very small, the value of l_{ums} will serve like the "necking" phenomena in melt crystal growth (Figure 8 ii). Owing to practical difficulties, if the microtube is deviated, i.e. as in figure 8 iii, it leads to differences in the thickness of the melt film retrieved. In this case, the resultant crystal orientation depends mainly on the orientation of the film which crystallized first (Figure 8 iv).

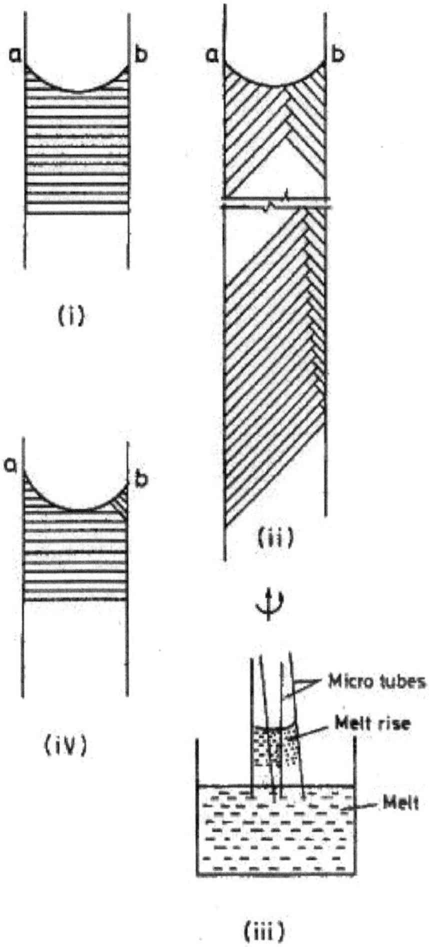

Figure 8. Schematic representation of initial growth of benzophenone inside the microtube.

Even though a crystal with multiple grains was formed inside the microtube initially, at the final stage of crystal growth inside the tube, a perfect single crystal was emerged out from the tube with single orientation facing the melt. The growth direction of the seed crystal was selected inside the microtube. Subsequently, the crystal at the middle part is grown like conventional Czochralski method with the seed crystal grown inside the microtube. Hence it would be possible to grow a single crystal with a particular orientation using a microtube.

3.3. USC growth

In this method, the effective zone width of the solution and the maximum temperature of the ring heater determine the effective evaporation rate of the solvent for a given diameter of the ampoule. Also, overheating the growth solution seems to be the key point to prevent spontaneous nucleation. Due to the transparent nature of the solution and the experimental set-up, real-time close-up observation on the solid-liquid interface was possible. The significant growth parameters of the USC technique are effective heating zone of solution column and temperature of the solution. Depending on the values of these two growth parameters, the solvent evaporation rate can be controlled more effectively. Moreover, the rate of supersaturation was controlled by means of controlling the solvent evaporation rate and thus the growth rate of the crystal can also be controlled. At an optimized growth conditions, maximum growth rate of 10 mm/ day was achieved for the (110) orientation at 44°C of heating zone temperature. Cut and polished samples of USC grown crystal are shown in Figure 9.

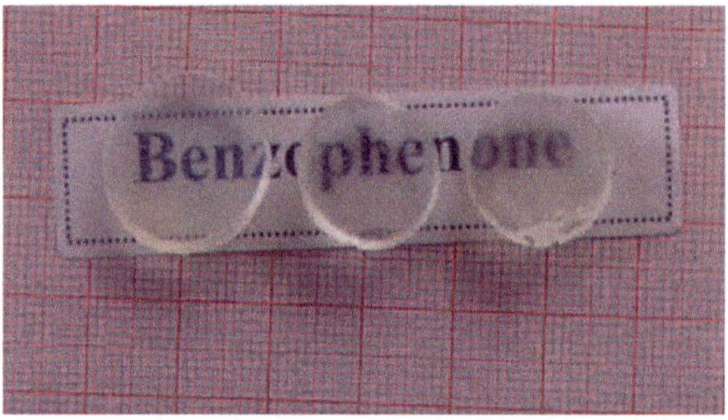

Figure 9. Cut and polished samples from USC grown crystal.

The grown ingots of benzophenone crystals by VB, μT-CZ and USC growth methods were characterized by XRD, HRXRD and LDT measurements and their results have been compared in the following sections.

3.4. X-ray diffraction (XRD) studies

The ingots which were grown under optimized growth conditions using VB, μT-CZ and USC methods were subjected to X-ray diffraction studies. The cut and polished sample prepared from the grown ingots were used for XRD analysis to identify the growth orientation of the crystal. The XRD spectrums were recorded for the respective samples at room temperature in a 2θ range of 10 to 50° using CuKα radiation of wavelength 1.5418 Å and the spectrums were shown in Figures 10 a, b, and c. From the diffraction pattern the d-spacing and hkl values for each diffraction peak in the corresponding spectrum of sample were identified. Using the orthorhombic crystallographic equation, the lattice parameter values of benzophenone were calculated and compared with the reported values [10]. The calculated values are in-line with the literature values.

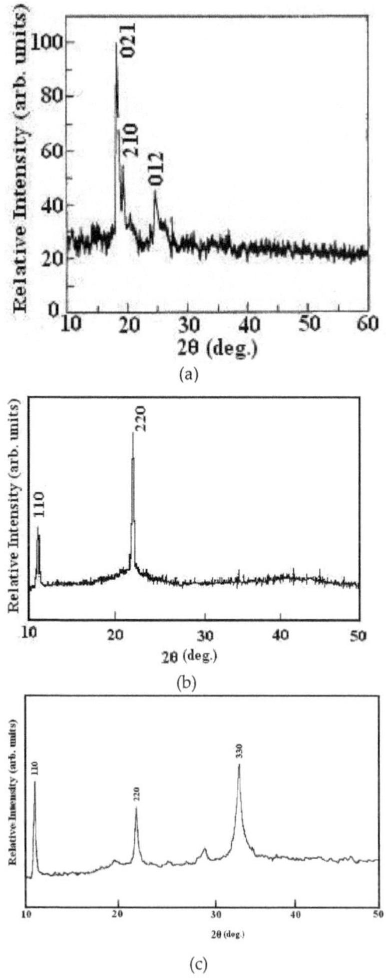

Figure 10. X-ray diffraction spectrum recorded for benzophenone grown by (a) VB (b) μT-CZ and (c) USC method.

Moreover, in Figure 10 a, the presence of narrow and strongest peak along the <021> direction confirms the single crystalline nature of the ingot grown by with VB <021> as a most preferred orientation. In the case of μT-CZ grown crystal (Figure 10 b), the preferred orientation was along <110> direction as evidence from the diffraction peaks at 11.31° and 22.27°. The obtained two different preferred orientations for a material grown from two different growth techniques demonstrate that the shape of the container material (tip of the ampoule in the case of VB growth, microtube in the case of microtube CZ growth) plays a vital role in the initial stage of nucleation. In unseeded crystal growth methods, the melt confinement in a small volume will largely help to control over the formation and number of nucleation. The resultant orientation which emerges out of the tapered end (in the case of VB) or the microtube (in the case of μT-CZ) henceforth decides the growth orientation of the rest of the crystal. Even though spontaneous, multi nucleations were formed initially at both methods, it undergoes a geomentric selection during the restricted growth inside the microtube/cone reagion of the ampoule. Therefore, it is quite feasible to grow a single crystal along a particular direction emerges out at the tapered end or at the end of the microtube if their physical length is sufficient to allow any one of the grain's orientation as a resultant growth direction. This happens when the progression of grains having low growth rate is suppressed or dominated by the progression of grain having the highest growth rate. Hence in the above two methods, it is possible to grow crystal along specific orientation by properly optimizing the growth conditions. In the case of USC method, the seed crystal was selected with the plane (110) as the imposing growth orientation. The recorded spectrum (Figure 10 c) on the grown sample justifies the unidirectional growth along the orientation of seed crystal.

3.5. High-Resolution X-Ray Diffraction (HRXRD) studies

High resolution X-ray diffraction (HRXRD) studies have been carried out using multicrystal X-ray diffractometer (MCD) [20] on the grown samples to evaluate the crystalline perfection. A well-collimated and monochromated MoKα_1 beam obtained from a set of three plane (111) Si monochromator crystals set in dispersive (+,-,-) configuration has been used as the exploring X-ray beam. The specimen crystal was aligned in the (+,-,-,+) configuration. Due to dispersive configuration, though the lattice constant of the monochromator crystal(s) and the specimen are different, the unwanted dispersion broadening in the diffraction curve of the specimen crystal is considerably less.

The HRXRD curves recorded by multicrystal X-ray diffractometer (MCD) revealed that the crystals grown by all the three methods contain internal structural grain boundaries. Figures 11 a and b show the diffraction curves (DC) recorded for the (211) diffracting planes of VB and μT-CZ grown sample using MCD with MoKα_1 radiation in symmetrical Bragg geometry. As can be seen from the Figure 11 a, the DC of the VB grown sample contains multiple peaks in an angular separation of few min., whereas the DC (Figure 11 b) of the μT-CZ grown sample contains only one additional peak at the angular separation of 50 arc sec. The DC of VB grown sample exhibits four main peaks with FWHM of 39, 21, 42 and 30 arc sec having angular separations 61, 48 and 72 arc sec. These multiple peaks illustrates that the samples contain many structural internal low angle ($\alpha \geq 1$ arc min) grain boundaries, whose tilt angles range from 72 to 61 arc sec.

The DC (Figure 11 b) of μT-CZ grown sample shows that it contains two peaks, one main peak and the second peak at lower angle side, which indicates that the crystal contains a very low angle grain boundary ($\alpha < 1$ arc min). The solid line in the figures is the convoluted curve obtained by the Gaussian fit of the observed peaks. The additional peak in the DC is 50 arc sec away from the main peak which corresponds to a very low angle grain boundary. The FWHM of the main peak and the second peak are respectively 42 and 105 arc sec. From this, one can infer that μT-CZ grown benzophenone crystal exhibited better crystalline quality than the VB grown crystal. Moreover, the presence of additional grains in VB grown crystal was not found in μT-CZ grown crystal. The observation of the additional grain in VB grown crystal may be attributed to the differences in thermal expansion of the growing crystal and the ampoule which may lead to occurrence of plastic deformation in the grown crystal during post growth annealing process. Whereas such plastic deformations were absent in μT-CZ grown crystal due to the freestanding nature of the growing crystal. The HRXRD studies demonstrate that the microtube seeding is more reliable than VB growth for the growth of directional crystals with low imperfection.

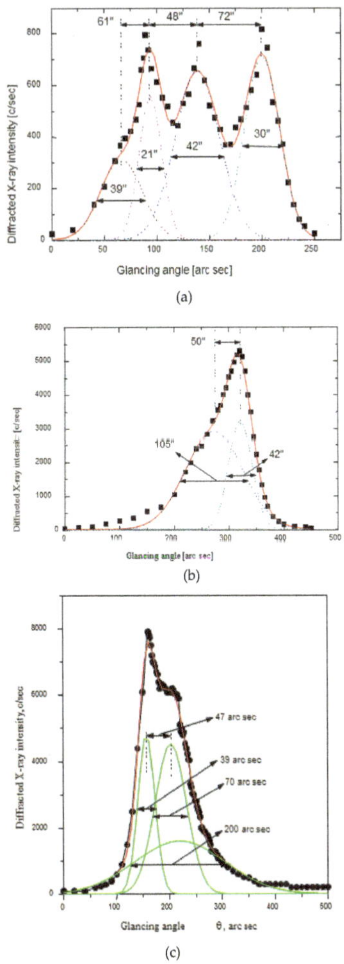

Figure 11. HRXRD curve recorded for benzophenone grown by (a) VB (b) μT-CZ and (c) USC method.

The recorded DC (Figure 11 c) for the USC grown sample is considerably sharp but it contains two peaks: one main peak and second, a hump at higher angle side, which indicates that crystal contains a very low angle ($\alpha < 1$ arc min) grain boundary. The solid line (convoluted curve) is well fitted with the experimental points represented by the filled circles. On deconvolution of the diffraction curve, it is found that there are three peaks. The two main peaks are separated by ~47 arc sec and their half widths are 39 and 70 arc sec, respectively. The third peak is quite broad and having FWHM of 200 arc sec. All these three peaks were merged with the main peak and not possible to identify before deconvolution, which shows that the crystalline quality of the specimen is reasonably good. From the results, one can infer that the unidirectional benzophenone crystal grown by the USC method has relatively good crystalline perfection. The presence of the additional peaks may be attributed to the possible plastic deformation occurred during the restricted growth of the crystal inside the ampoule.

3.6. Laser damage threshold measurements

Optical damage in dielectric materials (NLO materials) may severely affect the performance of high power laser systems as well as the efficiency of the optical devices based on nonlinear processes. Hence, high damage threshold is a significant parameter for nonlinear optical crystal. Two main mechanisms that cause laser induced damage in the wide band gap dielectric materials are dielectric break down and thermal absorptions. The laser induced damage studies have been carried out for the VB, μT-CZ and USC grown samples under identical experimental conditions. The samples were carefully selected from the grown ingots with better quality and low dislocation densities. Then the samples were chemically polished using ethanol just prior to the LDT measurements. A Q-switched Nd:YAG laser operating at 1064 nm radiation was used. The laser was operated at the repetition rate of 10 kHz with the pulse width of 65 ns. For the LDT measurement 1.64 mm diameter beam was focused on the sample with a 10 cm focal length lens. The beam spot size on the sample was 0.51 mm. The multiple shots LDT measurements were made on the VB, μT-CZ and USC grown samples. The samples were irradiated at different spots on the same plane at similar experimental condition (wavelength – 1064 nm; repetition rate – 10 kHz; pulse width- 65 ns; beam diameter- 1.64 mm) and the damage pattern was observed using an optical microscope.

The measured LDT values of benzophenone samples grown by VB, μT-CZ and USC method are 2.3, 3.0 and 7.9 MW/cm^2, respectively. As evidenced from HRXRD studies, the high value of LDT for SR grown sample can be attributed due to the high crystalline perfection when compared with that of VB and μT-CZ grown samples. The threshold for bulk laser damage is in principle a material dependent property and imperfections in a material are depends on the growth parameters which are subjected to control during the growth process. Furthermore, the LDT is a function of pulse duration, maximum pulse power, pulse wavelength, focal point radius, and in the case of multi shot experiments, repetition rate [21]. It is known that in the long-pulse regime ($\tau > 100$ ps), the damage depends on the rate of thermal conduction by the crystal lattice and in the short-pulse regime ($\tau < 100$ ps), the dielectric breakdown and various nonlinear ionization mechanisms (multi-photon, avalanche multiplication) become dominant [22]. Hence, thermal effects become important for the nanoseconds and longer pulse widths. In the present multi shot experiments, relatively longer pulse width (65 ns) and high repetition rate of laser (10 kHz) were used and it may causes the low LDT for all the benzophenone samples. Wang et al [9] reported that the single-shot LDT of benzophenone crystal is 17.6 GW/cm^2 at the wavelength of 1064 nm while 39.5 GW/cm^2 at the wavelength of 532 nm. In the same laser parameters, the reported single-shot LDT of KDP, a well-known NLO crystal is 14 GW/cm^2 at the wavelength of 1064 nm and 17 GW/cm^2 at the wavelength of 532 nm [9]. Hence, it is obvious that benzophenone crystal has higher LDT than that of KDP crystal. The mechanical hardness of the materials also plays a vital role in LDT of the crystals grown in different crystallographic orientations [23]. In the next section, the hardness property of the benzophenone was correlated with the observed damage profile during laser damage threshold measurements.

4. Anisotropy of hardness and laser damage threshold

4.1. Unidirectional benzophenone in different orientation

USC method was extended to grow unidirectional benzophenone single crystals in three different crystallographic directions such as <110>, <010> and <001>. For this experiment, transparent single crystals obtained by slow evaporation method were used as a seed. To grow unidirectional benzophenone crystal along different crystallographic directions, three identical glass ampoules having inner diameter of 20 mm were carefully mounted with respective plane of the seeds by facing normal to the saturated solution of benzophenone. Saturated solution of benzophenone with pre-determined solute concentration was prepared using xylene as a solvent and carefully transferred into the growth ampoule. In this experiment, the solutions were not heated by ring heater to allow the natural evaporation of solvents. Growth was initiated at the seed crystal-solution interface when the supersaturation increased by evaporation of solvent from the solution. The transparent nature of the experimental set-up and the visible elevation of the solid-liquid interface measured at specific intervals facilitated the measurement of growth rate for the three different directions. The measured data are tabulated (table 1) in comparison with the relative growth rates of conventional solution grown and melt grown benzophenone crystals [24]. The grown ingots were sliced perpendicular to growth direction using in-house built wet-thread cutting machine. Figures 12 a and b show the grown crystals and sliced specimens of benzophenone in three different orientations. The samples were chemically polished on a polishing sheet using a mixture of acetone and xylene in the volume ratio 1:2.

Table 1 shows the relative growth rates for benzophenone crystal grown from different crystallographic directions in comparison with the relative growth rates of conventional solution grown and melt grown benzophenone crystals [24]. High growth rate was observed along <001> direction and low growth rate was observed along <110> direction. The observed relative growth rates vary drastically with growth directions and it follows the same tendency with the relative growth rates for melt grown benzophenone [24]. On the other hand, the observed growth rates do not follow the reported growth rates for conventional solution grown benzophenone (toluene solvent) [24]. In conventional solution growth, the rate of diffusion of solute molecules or growth slice can be influenced by solvents. Due to the different vapor pressure of the solvents and the chemical environment (interacting between the solute and solvent) around the growing surface, growth rate can be changed by solvents. Moreover, the solvent in the present experiment and growth mechanism are entirely different from the conventional solution growth and under cooled melt growth.

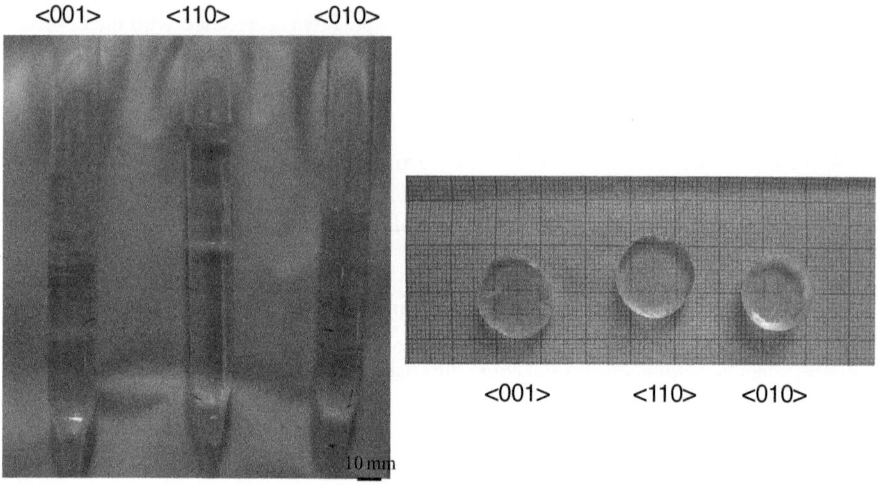

Figure 12. (a) Unidirectional benzophenone crystal in various growth directions and (b) prepared specimens from the respective ingots.

(hkl)	Uniaxial solution growth	Conventional solution growth (toluene solvent) [22]	Melt growth [22]
(110)	1.00	1.00	1.00
(010)	1.75	0.64	1.43
(001)	2.65	0.83	1.66

Table 1. Comparison between the relative growth rates of different growth directions for benzophenone crystals grown from conventional solution growth, melt growth [22], and uniaxial solution growth (this work). [The data are normalized with respected to the (110)]

In conventional solution growth, the growing crystal is fully exposed to the saturated mother solution. As a result, the crystal grows with different faces separated by growth sector boundaries. The shape of a crystal can be described by the distance from the centre of the crystal to the respective crystal faces and these distances are proportional to the relative growth rates of the crystal face [24]. The formation of inclusion in the growing crystal may cause the growth rate despersions (GRD) [25, 26] i.e., a crystal (or crystal faces) of the same size growing under the same environmental conditions may grow at different rates. But, in USC method, unlike to conventional solution growth, the face (hkl) which has to grow is alone exposed to mother solution instead of the whole seed crystal. Hence the formation of inclusion due to solvent and other impurities incorporation is strongly suppressed in the present method. Moreover, in the USC grown crystals, the commonly observed growth features in the case of conventional solution grown crystal such as growth sector boundaries, twins, stacking faults and dislocations are not observed since the crystal was grown along selective growth axis [27].

The uniaxial crystallization process essentially involves gathering of a vast number of molecules together and forms a unique ordered arrangement which is driven by the level of supersaturation of the solution. The degree of supersaturation was controlled for all the experiments at predetermined solute concentration and constant temperature. The purity of the source materials and the solvent were almost the same for all the growth experiments. Hence, the observed variation in growth rate for different growth axis is likely to be the function of molecular packing energy or attachment energy, E_{att} of the respective growth face (eq (5)) [28].

$$R_{hkl} \propto E_{att} \tag{1}$$

E_{att} can be defined as the energy released when one slice of thickness d_{hkl} crystallizes onto a crystal face (hkl). According to the BFDH model [29], the relative growth rate is inversely proportional to d_{hkl} (eq. (6)).

$$R_{hkl} \propto \frac{1}{d_{hkl}} \tag{2}$$

The reported attachment energy variation [29, 30] for the respective growth planes were compared with the growth rates and interplanar spacings (d_{hkl}) (table 2). As shown in table 2, high growth rate was observed for the plane which has small d spacing and high attachment energy. However, the reported attachment energies were calculated by considering the conventional solution growth conditions [29] and undercooled melt growth conditions [30], that means it was considered that crystal grows in all faces simultaneously. In contrast, in the present experiment, the crystal has grown only on the selected growth face. Moreover, the different between the d spacing of (110) and (010) is very small (table 2). But the d-spacing for (001) plane is obviously smaller than other planes and has high growth rates compared to other planes.

(hkl)	Relative growth rates	Interplanar spacing d (hkl) (Å)	Attachment energy E_{att} (kcal/mol)
(110)	1.00	7.82 (110)	10.3[a](9.86[b])
(010)	1.75	8.15 (010)	14.6[a] (18.92[b])
(001)	2.65	6.01 (001)	17.1[a]

ote: [a]Ref. [27] [b]Ref. [28]

Table 2. Comparison of relative growth rates of unidirectional crystals, d spacing (JCPDS data) and reported attachment energy for various growth directions.

4.2. Optical absorption studies

The absorption coefficient of optical radiation in transparent materials is more vital in studies of the process of interaction of radiation with matter. The information on the energy absorbed by a material in the course of damage is an important characteristic to understand laser breakdown mechanism. In general, increase in the absorption is mainly due to the defects in the crystal and the presence of inclusions as well as impurities. For optical applications, the material considered must be highly transparent (low absorption) in the wavelength region of interest. The absorption coefficients of benzophenone are calculated using the following equations.

$$\alpha = -\frac{1}{d} ln \frac{\left\{ \left[B^2 - R^2 \right]^{\frac{1}{2}} - B \right\}}{R^2} \tag{3}$$

$$B = (1 - R^2)/2T \tag{4}$$

$$R = \left(\frac{n-1}{n+1} \right)^2 \tag{5}$$

where d is the thickness of the crystal used for the optical studies and n is the refractive index of the crystal. T is the percentage of transmission and it was measured for 5 mm thick <110> oriented sample as a function of wavelength using UV-VIS-NIR spectrophotometer. The refractive index of the biaxial crystal (benzophenone) can be derived by the following equation:

$$n_{biaxial} = \frac{n_x + n_y + n_z}{3} \tag{6}$$

The reported refractive index data [8] for three different polarization of light parallel to a-axis (n_x), b-axis (n_y) and c-axis (n_z) were used for the present calculations. The absorption coefficient curve of benzophenone is shown in Figure 13.

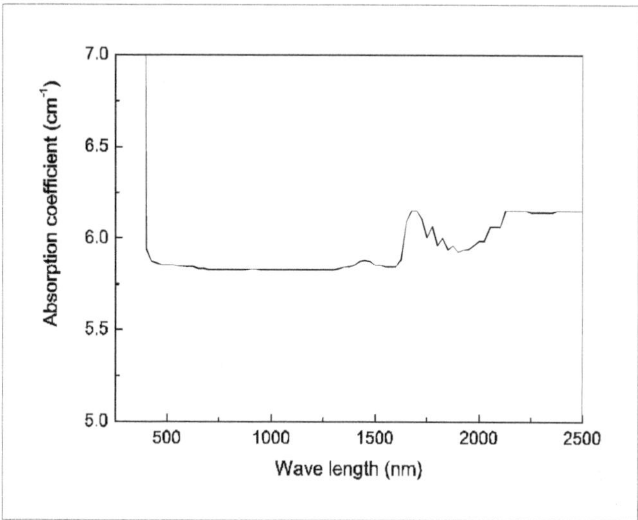

Figure 13. Measured absorption coefficients curve for benzophenone as a function of wavelength.

The presence of inclusion and grown-in defects (which can be controlled by growth process) causes the relative increase of absorption coefficient and thus increasing the over-all heating of the material during laser irradiation. The local heating in the region of an inclusion can lead to suppress the laser damage threshold value of a particular material. The present optical absorption study shows that the benzophenone material has very low absorption in the wavelength range from 400 to 1300 nm.

4.3. Micro hardness of benzophenone

Mechanical hardness of a material is also one of the decisive properties especially for post-growth processes and device fabrications. Load dependence of Vicker's micro hardness was measured on polished (110), (010) and (001) surfaces of 5 mm thick samples which were prepared from the respective unidirectional ingots. The indentation time was maintained as constant at 10 s. The diagonal lengths of the indented impression were measured for different loads varying from 10 to 80 g. The successive indentations were made at different sites of the sample surface. The mean diagonal length was used for the calculation of Vicker's hardness number (VHN). Figure 14 shows the VHN variation as a function of applied load for the (110), (010) and (001) planes of the prepared samples. Moreover, the mechanical hardness is correlated with the laser damage threshold of the materials. As can be seen from the Figure 14, the hardness number increased steeply in the beginning and got saturated after 50 g of load. Larger hardness value was measured for (010) plane whereas small hardness value was measured for (001) plane. The VHN values observed on (010), (110), (001) planes were 16.1, 14.6 and 12.8 at 60 g of load.

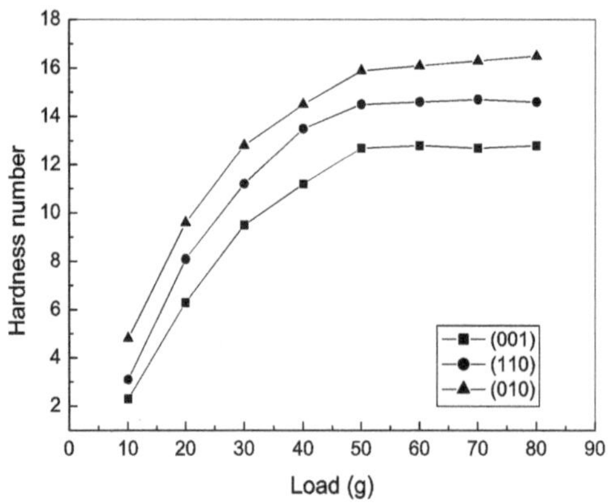

Figure 14. Hardness variation along three different crystallographic planes as a function of applied load.

4.4. Laser damage threshold studies

The LDT was measured on the (110), (010) and (001) surfaces of the unidirectional crystals grown along the respective orientations. The samples used for LDT measurement were carefully selected from the grown ingots with best quality and low defect densities. Then the sample was chemically polished using xylene prior to the LDT measurements. A Q-switched Nd:YAG laser operating at 1064 nm radiation was used for single as well as multiple shot experiment. For the single shot experiment, the laser was operated at the repetition rate of 10 Hz with the pulse width of 10 ns. The single shot LDT measurements were made on the sample and the experiment was repeated at different places of each plane to measure the damage threshold precisely. During the single shot experiment, care was taken to select a fresh region after each shot to avoid the cumulative effects resulting from multiple exposures. For surface damage, the testing plane of the sample is placed at the focus point of the lens and the 50 mm diameter laser beam was focused on the sample using a lens of 20 cm focal length.

The surface damage can be determined by various methods, viz., by observing it with optical microscope, by visual incandescence, or by observing the scattering of helium-neon (He–Ne) laser beam passing through the damaged volume. In the present investigation, the resulting damage pattern is observed by an optical microscope. In order to observe the damage profile more clearly, the LDT experiment was performed on the (110) plane of the respective oriented ingot by multiple shots mode using the same laser with high repetition rate (10 kHz). Since the <110> oriented ingot showed high micro hardness, it was selected for the multiple shot experiment. The sample was irradiated at different spots on the same plane at similar experimental condition and the damage profile was observed by optical microscope.

The respective single shot damage threshold measured for the <110> <010> <001> oriented benzophenone samples was 16.99, 17.23 and 12.95 MW/cm^2 for 1064 nm Nd:YAG laser radiation. In the multiple shots experiment, the LDT value for (110) plane was measured as 7.69 MW/cm^2 using the 1064 nm Nd:YAG laser. For the single shot experiment, 1064 nm Nd:YAG laser with 10 ns of pulse width and 10 Hz of repetition rate was used. Whereas for the multiple shot experiment, the Nd:YAG laser with 65 ns of pulse width and 10 kHz of repetition rate was used. The threshold for bulk laser damage of a material mainly depends on laser parameters such as pulse width, focal spot geometry, and in the case of multi shot experiments, repetition rate of laser [21, 31]. From the single shot experiment, the anisotropic properties of the laser damage threshold were observed in the unidirectional benzophenone crystals. The <010> oriented sample shows high laser damage threshold whereas the <001> oriented sample shows low value of damage threshold. In both the case of single and multi shot experiments, due to large pulse width (τ > 100 ps) and low melting point of the material, thermal effects become the main causes for damage.

Figures 15 a, b and c show the optical micrograph of the single shot damage patterns for 1064 nm laser radiation on (110) (010) (100) planes of respective unidirectional benzophenone crystal. As can be seen from the Figure 15, irrespective of orientations, circular damage profile was obtained for all the samples at small repetition rate (10 Hz) and pulse width (10 ns). However, the diameter of the circular pattern was larger (~1220 μm) for <001> oriented ingots whereas it was very small (~200 μm) in size for <010> oriented ingot. Figure 16 shows the optical micrograph of the multiple shot damage pattern of 1064 nm laser radiation on (110) surface of unidirectional benzophenone crystal. The figure clearly depicts that the core of laser induced damage was at the centre of the pattern with strong cracks in different directions. As can be seen from Figure 16, the cracks were strongly propagated along the crystallographic <001> and <010> directions. It should be noted that the crack was stronger along <001> direction compared to <010> direction. Probably the high repetition rate and large pulse width causes the profound damage with cracks unlike to circular pattern observed in the single shot experiment.

Generally the hardness of the material is directly related to its bonding and crystallographic orientation. One can expect to obtain some information concerning the hardness anisotropy and damage profile observed in benzophenone based on its crystal structure. Benzophenone crystallizes in the non-centrosymmetric orthorhombic crystal structure with space group $P2_12_12_1$ with four atoms per unit cell [9]. A projection of crystal structure along the c axis is shown in Figure17 [30]. The shadowed elliptical chains disposed to each other at an angle of 90° show the rigid layers of the structures parallel to <110> and <101> directions and these directions correspond to predominant {110} prisms of the growth morphology of conventional solution grown crystal [32]. The strong bonding structure which forms rigid layers (shadows in Figure 17) causes higher mechanical hardness in (110) compared to (001) plane. Hence the size of the laser damage pattern was smaller for the ingot grown along <110> direction (Figure 15 a) compared to the ingot grown along <001> (Figure 15 c). On the other hand, when we look at the damage profile of the multi shot experiment (Figure 16) the cracks were strongly propagated along <001> direction compared to <010> direction. Moreover, the intermolecular bonding energy of the respective growth plane plays a major role in hardness properties and laser induced damage. In addition, the magnitude of the cumulative bonding energy of a plane depends on the number of bonding of nearest neighbor of the atoms or molecules of the respective plane. From the hardness and damage profiles, it is obvious that the bonding energy of the (001) plane is small compared to (110) and (010) planes of benzophenone crystal. The intermolecular bonding energy can be related with the attachment energy as follows [28],

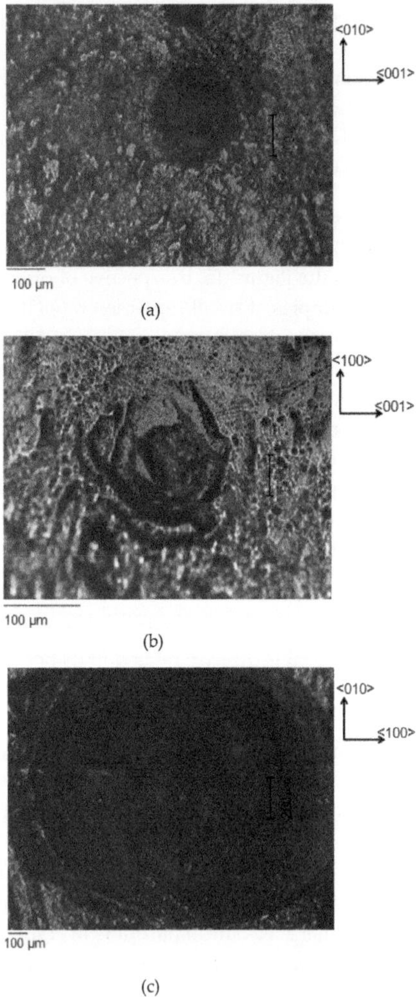

Figure 15. Optical micrograph of the damage pattern. (a) (110) surface, (b) (010) surface, (c) (001) surface.

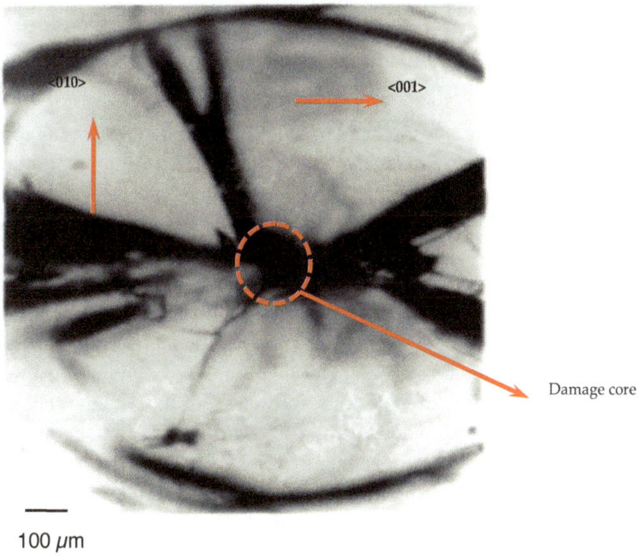

Figure 16. Optical micrograph of a damage pattern observed on the (110) surface for the laser with high repetition rate (10 kHz).

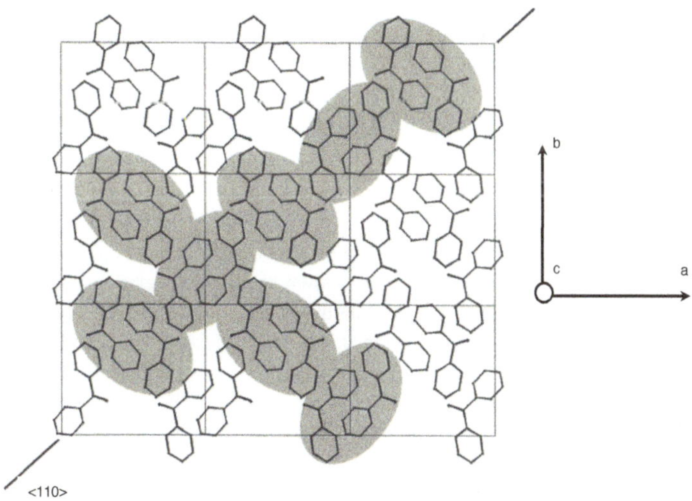

Figure 17. Molecular packing of benzophenone viewed on the orthorhombic (001) plane. The ridged layers parallel to <110> and <101> are shadowed.

$$E_{cry} = E_{sl} + E_{att} \tag{7}$$

Where E_{cry} is the crystallization energy or lattice energy and E_{sl} is the intermolecular bonding energy or slice energy of a growth plane.

According to the theory of Hartman and Bennema [28], the attachment energy is inversely proportional to the intermolecular bonding energy of the respective growth planes. In other words, a growth plane (hkl) which has low intermolecular bonding energy could grow with high growth rate and high attachment energy. As shown in table 2, the ingot grown from (001) plane had high growth rate and high attachment energy [29]. Moreover, small micro hardness value was measured on (001) plane and the size of the laser damage pattern was larger for the <001> oriented sample (Figure 15 c). This is probably because of weak intermolecular bonding existed in the (001) growth plane as high attachment energy was reported for the same plane of benzophenone crystal [29]. On the other hand, the ingot grown from (010) plane has low growth rate compared to ingot grown from (001) plane. High micro hardness value was measured on (010) plane and the size of the laser damage pattern was relatively small for the <010> oriented sample (Figure 15 b). In contrast, the ingot grown from (110) plane which has the lowest growth rate shows small micro hardness than that of ingots grown from (010) plane. It is feasible that the crystal is easily distorted in a line normal to the mechanical fragility, which results in a weak shearing stress of the crystal structure [21]. However, the relationship between the bonding energy of molecules and the shearing stress has not yet cleared well. In addition, the damage profile obtained for the <110> oriented sample under multi shot mode reveals some relation with the crack directions, hardness variations and crystallographic directions. As can be seen from Figure 16, a strong crack was observed along the mechanically weak <001> direction. Whereas the crack along <010> direction is quite weaker. From the results, it is obvious that the obtained damage profile is closed related with hardness anisotropy of the material.

5. Conclusion

Benzophenone single crystals were grown by various growth methods such as VB, μT-CZ and USC method and the structural perfection of the crystals were comparatively investigated. HRXRD studies revealed that the USC grown sample had relatively high crystalline perfection than the samples grown by other methods. On the other hand, VB grown crystal was found to have low crystalline perfection due to the difference in thermal expansion of the growing crystal and the ampoule which may lead to occurrence of plastic deformation in the grown crystal during post growth annealing process. The USC grown sample had high LDT than the crystals grown by other methods, probably due to low dislocation density in the USC grown ingots. Unidirectional benzophenone single crystals were grown along three different crystallographic directions. It was observed that the growth rate of the grown crystals varied with orientation. Moreover, the laser damage threshold was larger for the <110> and <010> oriented crystals compared to <001> oriented crystal at the wavelength of 1064 nm. The result was consistent with the hardness variation observed for the three different crystallographic directions of benzophenone crystal. The optical micrographs of the damage profile and hardness anisotropy were correlated with the internal crystal structure of benzophenone.

Author details

M. Arivanandhan[1], V. Natarajan[2], K. Sankaranarayanan[3] and Y. Hayakawa[1]

1 Research Institute of Electronics, Shizuoka University, Hamamatsu, Japan

2 Aditanar College of Arts and Science, Tirucherdur, Tamil Nadu, India

3 School of Physics, Alagappa University, Karaikudi, Tamil Nadu, India

References

[1] Nakamura S, Senoh M, Nagahama S, Iwasa N, Yamada T, Matsushita T, Kiyoku H, and Sugimoto Y. InGaN Based Multi-Quantum-Well-Structure Laser Diodes. Japanese Journal of Applied Physics 1996; 35: 74–76.

[2] Hollemann G, Peik E, and Walther H. Frequency-stabilized diode-pumped Nd:YAG laser at 946 nm with harmonics at 473 and 237 nm. Optical Letters 1994; 19: 192-194.

[3] Yokoo A, Tomaru S, Yokohama I, Kobayashi H, Itoh H, Kaino T. A new growth method for long rod-like organic nonlinear optical crystals with phase-matched directions. Journal of Crystal Growth 1995; 156: 279-284.

[4] Sankaranarayanan K, Ramasamy P. Microtube-Czochralski technique: a novel way of seeding the melt to grow bulk single crystal. Journal of Crystal Growth 1998; 193: 252-256.

[5] Wang W, Huang W, Ma Y, Zhao J. Oriented growth of benzophenone crystals form undercooled melts. Journal of Crystal Growth 2004; 270: 469-474.

[6] Arivanandhan M, Sankaranarayanan K, Ramamoorthy K, Sanjeeviraja C, Ramasamy P. A novel may of modifying the thermal gradient in Vertical Bridgman-Stockbarger Technique and studies on its effects on the growth of benzophenone single crystals. Crystal Research and Technology 2004; 39: 692-698.

[7] Sankaranarayanan K, Ramasamy P. Unidirectional seeded single crystal growth from solution of benzophenone. Journal of Crystal Growth 2005; 280: 467-473.

[8] Lammers D, Betzler K, Xue D, Zhao J. Optical Second Harmonic Generation in Benzophenone. Physica status solidi (a) 2000; 180: R5-R7.

[9] Wang W, Lin X, Huang W. Optical properties of benzophenone single crystal grown from undercooled melt with oriented growth method. Optical Materials 2007; 29: 1063- 1065.

[10] Masaru Tachibana, Shigeki Motomura, Akira Uedono, Qi Tang and Kenichi Kojima. Characterization of grown–in dislocations in Benzophenone single crystals by X-ray Topography. Japanese Journal of Applied Physics 1992; 31: 2202-2205.

[11] Hammond R B, Pencheva K, Roberts K J. An examination of polymorphic stability and molecular conformational flexibility as a function of crystal size associated with the nucleation and growth of benzophenone. Faraday Discuss 2007; 136: 91-106.

[12] Sekikawa T, Kosuge A, Kanai T, Watanabe, S. Nonlinear optics in the extreme untraviolet. Nature 2004; 432: 605-608.

[13] Carr C W, Radousky H B, Rubenchik A M, Feit M D, Demos S G. Localized Dynamics during Laser-Induced Damage in Optical Materials. Physical Review Letters 2004; 92: 087 401.

[14] McArdle B J, Sherwood J N, Damask A C. The growth and perfection of phenanthrene single crystals: 1. Purification and single crystal growth. Journal of Crystal Growth 1994; 22: 193-200.

[15] Arivanandhan M, Sankaranarayanan K, Ramasamy P. Studies on large uniaxially grown benzophenone single crystals. Crystal Research and Technology 2007; 42: 578-582.

[16] Arivanandhan M, Sankaranarayanan K, Ramamoorthy K, Sanjeeviraja C, Ramasamy P. Microtube-Czochralski (μT-CZ) growth of bulk benzophenone single crystal for nonlinear optical applications. Optical Materials 2005; 27: 1864 -1868.

[17] Jackson K A, Uhlmann D R, Hunt J D. On the nature of crystal growth from the melt. Journal of Crystal Growth 1967; 1: 1-36.

[18] Bleay J, Hooper R M, Narang R S, Sherwood J N. The growth of single crystals of some organic compounds by Czochralski technique and the assessment of their perfection. Journal of Crystal Growth 1978; 43: 589-596.

[19] Arivanandhan M, Sankaranarayanan K, Sanjeeviraja C, Arulchakkaravarthi A, Ramasamy P. Optical frequency doubling in microtube-Czochralski (μT-CZ) grown benzophenone single crystals. Journal of Crystal Growth 2005; 281:596-603.

[20] Lal K, Bhagavannarayana G. A high resolution diffuse X-ray scattering study of defects in dislocation free silicon crystal grown by the floating zone method and comparison with Czochralski-grown crystal. Journal of Applied Crystallography 1989; 22: 209-215.

[21] Yoshida H, Fujita H, Nakatsuka M, Yoshimura M, Sasaki T, Kamimura T, Yoshida K. Dependences of Laser-Induced Bulk Damage Threshold and Crack Patterns in Sever-

al Nonlinear Crystals on Irradiation Direction. Japanese Journal of Applied Physics 2006; 45: 766-769.

[22] Stuart B C, Feit M D, Rubenchik A M, Shore B W, Perry M D. Laser-Induced Damage in Dielectrics with Nanosecond to subpicosecond Pulses. Physical Review Letters 1995; 74: 2248-2251.

[23] Vanishri S, Babu Reddy J N, Bhat H L, Ghosh S. Laser damage studies in nonlinear optical crystal sodium p-nitrophenolate dihydrate. Applied Physics B: Lasers and Optics 2007; 88: 457-461.

[24] Roberts K J, Sherwood J N, Yoon C S. Understanding the Solvent-Induced Habit Modification of Benzophenone in Terms of Molecular Recognition at the Crystal/ Solution Interface. Chemistry of Materials 1994; 6: 1099-1102.

[25] Tanneerger U, Lacmann R, Herden A, Klapper H, Schmiemann D, Becker R A, Mersmann A, Zacher U. The dispersion of growth rate as a result of different crystal perfection. Journal of Crystal Growth 1996; 166: 1074-1077.

[26] Ulrich J. Growth rate dispersion –a review. Crystal Research and Technology 1989; 24: 249-257.

[27] Arivanandhan M Sankaranarayanan, K.; Ramasamy, P. Growth of longest <100> oriented benzophenone single crystal from solution at ambient temperature. Journal of Crystal Growth 2008; 310: 1493-1496.

[28] Hartman P, Bennema P. The attachment energy as a habit controlling factor: I. Theoretical considerations. Journal of Crystal Growth 1980; 49: 145-156.

[29] Roberts K J, Docherty R, Bennema P, Jetten L A M J. The Importance of considering growth-induced conformational change in predicting the morphology of benzophenone. Journal of Physics D: Applied Physics 1993; 26: B7-B21.

[30] Wang W, Wang M, Huang W. Theoretical prediction and experimental study on the growth morphology of benzophenone crystals. Materials Letters 2005; 59: 1976-1979.

[31] Yoshida H, Jitsuno T, Fujita H, Nakatsuka M, Yoshimura M, Sasaki T, Yoshida K. Investigation of bulk laser damage in KDP crystal as a function of laser irradiation direction, polarization, and wavelength. Applied Physics B: Lasers and Optics 2000; B70: 195-201.

[32] Kutzke H, Klapper H, Hammond R B, Roberts K J. Metastable β-phase of benzophenone: independent structure determinations via X-ray powder diffraction and single crystal studies. Acta Crystallographica B 2000; B56: 486-496.

Chapter 3

Proteins and Their Ligands: Their Importance and How to Crystallize Them

Astrid Hoeppner, Lutz Schmitt and Sander H.J. Smits

Additional information is available at the end of the chapter

1. Introduction

The importance of structural biology has been highlighted in the past few years not only as part of drug discovery programs in the pharmaceutical industry but also by structural genomics programs. Although the function of a protein can be studied by several biochemical and or biophysical techniques a molecular understanding of a protein can only be obtained by combining functional data with the three-dimensional structure. In principle three techniques exist to determine a protein structure, namely X-ray crystallography, nuclear magnetic resonance (NMR) and electron microscopy (EM). X-ray crystallography contributes over 90 % of all structures in the protein data bank (PDB) and emphasis the importance of this technique. Crystallization of a protein is a tedious route and although a lot of knowledge about crystallization has been gained in the last decades, one still cannot predict the outcome. The sometimes unexpected bottlenecks in protein purification and crystallization have recently been summarized and possible strategies to obtain a protein crystal were postulated [1]. This book chapter will tackle the next step: How to crystallize protein-ligand complexes or intermediate steps of the reaction cycle?

A single crystal structure of a protein however, is not enough to completely understand the molecular function. Conformational changes induced by for example ligand binding cannot be anticipated *a priori*. The determination of particular structures of one protein, for example with bound ligand(s) is required to visualize the different states within a reaction cycle. Ideally, one would trap an open conformation without any ligand, an open ligand-bound and a closed form with the bound molecule as well as the closed ligand-free protein to visualize the conformational changes occurring during catalysis in detail.

Within this chapter, the structural conformational changes induced by ligand binding with respect to the methods chosen for the crystallization are described. Here three distinct protein families are exemplarily described: first, where one substrate or ligand is bound, second, a protein with two or more bound substrates and finally, the structures of proteins, in which the product of the reaction cycle is present in the active site.

Specific methods or expressions written in **bold italics** are explained in the glossary box at the end of the chapter.

1.1. General approaches to obtain crystals with bound ligands and how to prepare the ground

Often the knowledge of the structure of a protein or enzyme without bound **ligand(s)** is not sufficiently significant since there is no or only little information provided about the catalytic mechanism. To gain further insights, it is important or at least helpful to obtain a **binary** or **ternary structure** of the protein of interest.

In theory there are different approaches to reach this goal even though it can be a difficult task in reality. All of them have in common that the naturally catalysed reaction must not occur. Apart from reporting all possible attempts we would like to give a general overview about several **co-crystallization/soaking** strategies first, followed by selected examples described within this bookchapter.

Possible **co-crystallization** or **soaking** trials:

(In order to keep it simple and coherent the expression „**ligand**" in the following paragraph is used in terms of „**substrate**", „**cofactor**" or „binding partner".)

- first ligand without second ligand
- second ligand without first ligand
- first ligand with product of the second ligand
- product of the first ligand with second ligand
- substrate analogue/inhibitor or non-hydrolysable cofactor
- application of substances that mimic transition state products (e. g. AlF_3 which imitates a phosphate group)
- usage of catalytically inactive mutants with bound ligand(s)
- creation of an environment (i. g. buffer condition) which shifts the equilibrium constant so that the reaction cannot occur

The most important point concerning preparing **co-crystallization** trials is the knowledge of the corresponding kinetical parameters. Proteins bind their natural ligand(s) with high **affinity**, which means in the nM- up to low mM range. To successfully crystallize a protein with the ligand(s) bound, the affinity needs to be determined. There are numerous biophysical techniques to achieve this, for example **Intrinsic Tryptophan Fluorescence, Isothermal Calorimetry, Surface Plasmon Resonance** and many others. In principle the affinity is determined by the size of a ligand as well as the property of the binding site of the protein. As first approximation, one can state that affinity increases with a decrease in ligand size.

The application of a too low concentration of the ligand can lead to an inhomogeneous protein solution, which means that not all of the protein molecules are loaded with ligand (and this can prevent crystallization). It is also possible, that a low level of **occupancy** causes an undefined electron density so that the ligand cannot be placed or which even makes a structure solution impossible. As a rule of thumb the concentration of the ligand(s) should be applied to the crystallization trial about 5-fold of the corresponding K_M value (the Michaelis constant K_M is the substrate concentration at which the reaction rate is half of V_{max}, which represents the maximum rate achieved by the system, at maximum (saturating) substrate concentrations).

Beyond that all requirements for the protein solution itself remain valid as described in [1] in more detail.

2. Binding protein with one ligand – How to crystallize and what can be deduced from the structure

A typical class of a protein binding one **ligand** are **substrate**-binding proteins (SBPs), and substrate-binding domains (SBDs) [2]. They form a class of proteins (or protein domains) that are often associated with membrane protein complexes for transport or signal transduction. SBPs were originally found to be associated with prokaryotic ATP binding cassette (ABC)-transporters, but have more recently been shown to be part of other membrane protein complexes as well such as prokaryotic tripartite ATP-independent periplasmic (TRAP)-transporters, prokaryotic two-component regulatory systems, eukaryotic guanylate cyclase-atrial natriuretic peptide receptors, G-protein coupled receptors (GPCRs) and ligand-gated ion channels [2].

Structural studies of a substantial number of SBPs revealed a common fold with a bilobal organization connected via a linker region [2]. In the ligand-free, open conformation, the two lobes or domains are separated from each other, thereby forming a deep, solvent exposed cleft, which harbors the substrate-binding site. Upon ligand binding, both domains of the SBP move towards each other through a hinge-bending motion or rotation, which results in the so-called liganded-closed conformation. As a consequence of this movement, residues originating from both domains generate the ligand-binding site and trap the ligand deeply within the SBP [3]. In the absence of a ligand, unliganded-open and unliganded-closed states of the SBP are in equilibrium, and the ligand solely shifts this equilibrium towards the liganded-closed state. This sequence of events has been coined the "Venus-fly trap mechanism" [4-6]; it is supported by a number of crystal structures in the absence and presence of a ligand [7, 8] and other biophysical techniques [3].

For the maltose binding protein (MBP) from *Escherichia coli* [9], it has been shown that both domains are dynamically fluctuating around an average orientation in the absence of the ligand [10]. NMR spectroscopy of MBP in solution revealed that the ligand-free form of MBP consists of a predominantly open species (95 %) and a minor species (5 %) that corresponds to a partially closed state; both forms co-exist in rapid equilibrium [11]. The open form of MBP observed by NMR is similar to the crystal structure of the unliganded-open conformation [12]. However, the partially closed species detected by NMR [11] does not correspond to the ligand-bound, fully closed form found in crystallographic studies. Instead, it represents an intermediate, partially closed conformation [13], suggesting that the substrate is required to reach the final, liganded-closed conformation.

Upon **substrate** binding, the closed conformation is stabilized, and the **ligand** is trapped within a cleft in between the two domains [14-16]. In principle one can divide the conformational changes in four (I-IV) states (highlighted in Figure 1). State I is the „open-unliganded" where the protein adopts an open conformation and no substrate is bound to the protein. State II is the „closed-liganded" conformation where the substrate is bound and induced a conformational change of both domains towards each other. This is likely the state within the cell before delivery of the substrate to its cognate transporter. Two other states are known to be present in solution although less frequent and the equilibrium is shifted towards the open-unliganded conformation. These forms are state III, the „closed-unliganded" state and state IV, the „semi-closed-unliganded" state. These are unfavorable conformations of the SBP, which occur due to the flexibility of the linker region in between both domains.

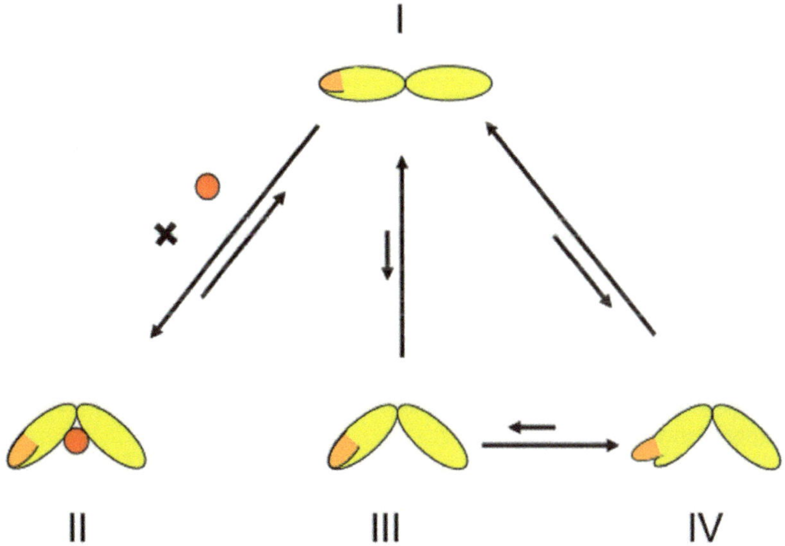

Figure 1. Substrate binding proteins exist in four major conformations: I) unliganded-open II) liganded-closed III) unliganded-closed and IV) unliganded-semi-open. All states are in equilibrium with each other. In solution states I and II occur most frequently. To fully understand the opening and closing mechanism of the protein however snapshots of every state are needed to gain full knowledge.

To fully understand the function as well as the structural changes happening upon **ligand/ substrate** binding it would require structural information of at least states I and II, preferably also of states III and IV.

2.1. Crystallization of the open-unliganded conformation (state I)

The crystallization of an open conformation of a rather flexible protein is not straight forward and most of the success came from „trial-and-error" approaches. After purification of the protein, a reasonable concentration of the protein is taken to set up crystallization trials. Most commonly the **vapor diffusion** method with the *hanging or sitting drop* is used. SBPs mainly exist in the open-unliganded conformation in the absence of the **substrate** whereas only a small fraction is in a closed-unliganded conformation [5, 11, 17]. Thus, basically a standard crystallization approach is used to obtain crystals suitable for structure determination. This is reflected by the large number of structures solved in the unliganded-open conformation (see [2] for a recent summary of the available SBP structures). The open conformation basically gives an overall picture of the protein structures and in the case of SBPs the bilobal fold of the protein can be observed. In this conformation the binding site of the substrate is laid open and a detailed picture on how the substrate is bound cannot be deduced.

Most of the times the open conformation crystallizes differently from the ligand bound state. This is reflected in the different crystallization conditions as well as in changes of the crystal parameters (unit cell and/or spacegroup). One example is given below for the glycine betaine binding protein ProX.

2.2. Crystallization of the substrate bound closed conformation (state II)

The vast majority of **substrate** binding proteins have been crystallized in the closed-**ligand** bound conformation (for a detailed list see [2]). This is mainly due to the fact that the substrate bound protein adopts a stable conformation and possesses a drastically reduced intrinsic flexibility. In principle there are four methods to include the substrate into the crystallization trials: 1) **co-crystallization** 2) ligand **soaking** 3) **micro or macro seeding** 4) endogenously bound ligands.

The first method is **co-crystallization**. Here, normally the **substrate** is added prior to crystallization to the protein solution. As listed in Table 1 this is the method used the most in SBP crystallization trials. Knowledge about the **affinity** of the **ligand** is important, since the bilobal SPBs exist in equilibrium between the open and closed state in solution and the addition of substrate directs this equilibrium towards the latter. Exemplary, 11 SBPs are listed in Table 1 where the affinity of the corresponding ligand(s) as well as the concentration used in the crystallization trials is highlighted. In principle the concentrations used are 10-1000 times above the K_d.

Protein	Organism	Ligand(s)	Open- un-liganded	Closed-liganded	Reso-lution (Å)	Max. affinity	Used Conc.	Method	Ref.
BtuF	*E. coli*	vitamin B12	Y	Y	2	15 nM	5 mM	1	[18]
Lbp	*S. pyogenes*	zinc	-	Y	2.45	~10 µM	-	4	[19]
GGBP	*S. typhimurium*	D-glucose	Y	Y	1.9	0.5 µM	3 mM	1	[20]
MBP	*E. coli*	Oligo-sacharide	Y	Y	1.67	0.16 µM	2 mM	1	[21]
RBP	*E. coli*	D-ribose	Y	Y	1.6	0.13 µM	1 mM	1	
OppA	*L. lactis*	Oligo-peptide	Y	Y	1.3	0.1 µM	0.5-5 mM	1 and 4	[22]
ProX	*A. fulgidus*	glycine betaine, proline betaine	Y	Y	1.8	50 nM	1 mM	1	[23]
PotD	*T. pallidum*	spermidine	-	Y	1.8	10 nM	n.n	2	[24]
SiaP	*H. influenzae*	sialic acid	Y	Y	1.7	58 nM	5 mM	1	[25]
UehA	*S. pomeroyi*	ectoine	-	Y	2.9	1.1 µM	10 mM	1	[26]
ChoX	*S. meliloti*	choline	Y	Y	1.8	2.7 µM	2 mM	1 and 3	[14, 15]

Table 1. Solved structures of selected SBPs. Listed are the proteins, the host organism, the substrate, whether the structure was solved in the unliganded-open and/or liganded-closed state, the highest resolution, the biochemically determined affinity, the used substrate concentration during crystallization and the method used: 1) co-crystallization 2) soaking 3) seeding 4) endogenously bound substrates.

2.2.1. *Co-crystallization to obtain the ligand bound structure*

The method of **co-crystallization** ensures the presence of only the substrate bound conformation of the SBP in solution. One major advantage of co-crystallization is the possibility to add different ligands into the crystallization trial. A prominent SBP member where several crystal structures were solved is the maltose binding protein (MBP). This protein binds a maltose molecule and delivers it to its cognate ABC transporter, which imports maltose into the cell for nutrient purposes. Substrate ranges from maltose, maltotriose, beta-cyclodextrin and many other sugar derivatives. All these structures were solved by using the addition of the substrates to the protein. Another example is the ectoine binding EhuB protein of *S. meliloti* [27]. Here, the structure was solved with both ligands, ectoine and hydroxyectoine, which yielded two high-resolution structures. The different binding modes of the substrates could be detected and the difference in affinity explained. The latter example was only crystallized in the closed-liganded state and no crystals could be obtained when the crystallization solution was depleted of substrate. This highlights the flexibility of the SBPs and the presence of multiple conformations of the SBPs in solution and in presence of the ligand. In many cases the ligand-closed conformation was crystallized under conditions, which differ greatly from the unliganded-open conformation also indicating the flexibility in the protein.

2.2.2. Ligand soaking to obtain the ligand bound state

The second method, which can be used to obtain a **ligand** bound protein structure, is **ligand soaking.** Soaking crystals with ligands is often the method of choice to obtain crystals of protein-ligand complexes owing to the ease of the method. However, there are several factors to consider. The crystals may be fragile and soaking in a stabilization buffer or cross-linking may be required. The soaking time and inhibitor concentration need to be optimized, as many protein crystals are sensitive to the solvents used to dissolve the ligands. Although for other proteins ligand soaking is successfully applied, for SBPs this method is not very commonly used as reflected by the low number of structures solved using this method. This is likely due to the fact that upon **substrate** binding the two domains undergo a relative large conformational change. Since crystal contacts are fragile and are disrupted easily, large conformational changes induced by soaking can damage crystal contacts resulting either in a massive drop in the resolution of the diffraction or the crystals crack/dissolve completely.

2.2.3. Seeding – A method to obtain the ligand bound state with unusual substrates

In some cases the **ligand** used for crystallization cannot be crystallized in a closed conformation. This occurs for example when the ligand is not stable during the time of crystallization. One such example is acetylcholine. During crystallization of the choline binding protein ChoX from *S. meliloti,* it became evident that besides the natural ligand choline also actylcholine is bound by this SBP [8]. To understand the binding properties of ChoX, a structure determination of ChoX in complex with acetylcholine was undertaken. For this purpose the protein was subjected to **co-crystallization** experiments. Acetylcholine presents a chemical compound, which is easily susceptible to hydrolysis especially at non-neutral pH values. Although the crystallization of ChoX was done at low pH values, a co-crystallization with intact acetylcholine was achieved. However, subsequent structural determination showed that the **substrate** was hydrolyzed to choline in the setup during the time of crystal growth. To overcome this limitation, a **micro seeding** strategy was devised. The application of micro seeding helped to crystallize ChoX complexed with acetylcholine within 24 hours. Structural analysis revealed that acetylcholine was not hydrolyzed in the drop during this short period of time required for crystal growth. Thereby, it was possible to solve the structure of ChoX in complex with acetylcholine. The quality of the crystals was good, resulting in diffraction up to 1.8 Å [28]. However, one drawback encountered, when crystals of ChoX were obtained by seeding, was that they all showed a high twinning fraction (up to 50 %). This effect is possibly due to the rapid growth process where crystals reach their final size within a day allowing the formation of merohedral twins, a phenomenon one has to take into account when using the streak seeding method.

2.2.4. *Endogenously bound ligands*

During purification of some proteins with high **affinity** for their ***substrate*** often the ligand is co-purified. Here, OppA from *L. lactis* is an excellent example. OppA belongs to peptide binding subgroup of the family of SBPs and is involved in nutrient uptake in prokaryotes and binds peptides of lengths from 4 to at least 35 residues and with no obvious specificity for a certain peptide sequence. These peptides bind so tightly that they remain associated with the protein throughout purification. The crystallization of the closed-ligand state therefor is relatively easy since the protein will stay only in the closed-liganded conformation. This results in a liganded bound structure. To obtain more different states of the protein one has to remove the ligand first, and afterwards add the wanted substrate. In the case of OppA the peptide was removed prior to crystallization and incubated either with a different ligand or no ligand to obtain a ligand free structure. In the case of OppA, the endogenous peptides can be removed from the protein only by partly unfolding using guanidium chloride, which generates ligand-free OppA. This removal of endogenous peptides was required to allow the binding of defined peptides which was used for crystallization. By this tour de force Bertnsson et al. were able to solve several structures with different ligands bound as well as a ligand free structure, explaining the substrate binding specificity of this protein in great detail [22].

2.3. Crystallization of the closed-unliganded state (state III)

The intermediate states of SBPs have been crystallized as well, although only a couple of structures have been reported. This energetically unfavorable state has been crystallized not on purpose in most cases. The choline binding protein ChoX from *S. melioti* has been crystallized in the absence of a **ligand** via micro seeding to gain structural insights into the open, ligand-free form of this binding protein. These attempts were not successful. Instead, the obtained crystals revealed a closed but ligand-free form of the ChoX protein. Nevertheless many structures are known of **substrate** binding proteins in either their unliganded-open or liganded-closed states [15].

2.4. Crystallization of a semi-open or semi-closed state (state IV)

During our efforts to solve the crystal structure of the choline-binding protein ChoX from *S. meliloti* we used the technique of **micro seeding** [15] to obtain ChoX crystals in the ligand-free form. To our surprise, a ligand-free structure, which was different from those that were expected for the ligand-free closed and/or open forms of SBPs described so far, was obtained. Here, ChoX was present in a ligand-free form whose overall fold was identical to the closed-unliganded structure. This structure however, represented a more open state of the **substrate** binding protein, which had not been observed before. From the crystal parameters such as the dimensions of the unit cell is was already obvious that the conformation of the protein had changed, since one axis of the unit cell appeared to be significantly larger (35 Å) when compared to the unliganded-closed crystal form of ChoX.

The structure revealed that the domain closure upon **substrate** binding does not occur in one step. Rather, a small subdomain in one of the two lobes is laid open and closes only after the substrate is bound. This observation was in line with data observed for the maltose importer system MalFEGK. Here it was observed that the ATPase activity of the ABC transporter was not stimulated by the maltose substrate binding protein when it was added in the unliganded-closed conformation. This is likely due to the fact that the subdomain is not fully closed and rotated outward, which does not activate the transporter. Thus, this biochemical phenomenon could only be explained by the semi-open/semi-closed structure of ChoX [14].

2.5. State I-IV: What do they tell about conformational changes

Substrate binding proteins are flexible proteins, which consist of two domains, which constantly fluctuate between several states of which the open and fully closed state are the most populated ones. Both domains together build up a deep cleft, which harbors the substrate-binding site. As described above the structural work on these proteins has been successful and in the next part a general outcome will be given of what these different states actually tell us about function and mode of action of this protein family.

The unliganded substrate binding proteins are thought to fluctuate between the open and closed state. The angle of opening varies between 26° up to 70° as observed in several open-unliganded structures, suggesting that the extent of opening is likely influenced by crystal packing. This has been observed very nicely for the ribose binding protein of which three different crystal structures have been described. Here the opening of the two domains varies between 43° and 63°. This suggests that the opening can be described as a pure hinge motion. The variation of the degree of opening has been elucidated by NMR in solution for the maltose binding protein MalF. Here 95 % of the protein adopts an open conformation fluctuating around one state with different degrees of opening.

2.5.1. Open and closed - An overall structure view

As an example for the closing movement observed when comparing the open-unliganded and closed-liganded structure the glycine betaine (GB) binding protein ProX from *A. fulgidus* is highlighted in more detail. ProX has been crystallized in different conformations: a liganded-closed conformation in complex either with GB or PB (proline betaine) as well as in an unliganded-open conformation [23]. From the crystallographic parameters it was already anticipated that crystals differ in the conformation of the protein. ProX crystals were grown using the **vapor diffusion** method. The authors attained four different crystal forms depending on the presence or absence of the ligand (hint 1). Liganded ProX crystallized in **hanging drops** using a reservoir solution containing 0.2 M $ZnAc_2$, 0.1 M sodium cacodylate, pH 6.0-6.5, 10-12 % (w/v) PEG 4000 and they belonged to the space group $P2_1$ (crystal form I). In a different setup, liganded ProX crystallized in **sitting drops** equilibrated against a reservoir containing 30 % (w/v) PEG 1500 and belong to the space group $P4_32_12$ (crystal form II). Unliganded ProX crystallized in hanging drops against a reservoir solution containing 0.3 M $MgCl_2$, 0.1 M Tris, pH 7.0-9.0, 35 % (w/v) PEG 4000. The first crystals appeared after

2-3 months, and belong to space group C2 (crystal form III). Again using a different setup, unliganded ProX crystallized in hanging drops equilibrated against a reservoir containing 0.1 M $ZnAc_2$, 0.1 M MES, pH 6.5, 25-30 % (v/v) ethylene glycol. These crystals grew within 4 weeks, reached a final size of 200 × 150 × 20 μm^3, and belong to space group $P2_12_12_1$ (crystal form IV) [23]. Thus, the different crystallization conditions as well as space group already suggested that several different conformations had been crystallized. Initial phases were obtained by two-wavelength anomalous dispersion of ProX-PB crystals of form IV. All other structures were determined by molecular replacement.

In Figure 2 the opening and closing of the glycine betaine binding protein ProX from *A. fulgidus* is highlighted. Here domain II was taken as an anchor point.

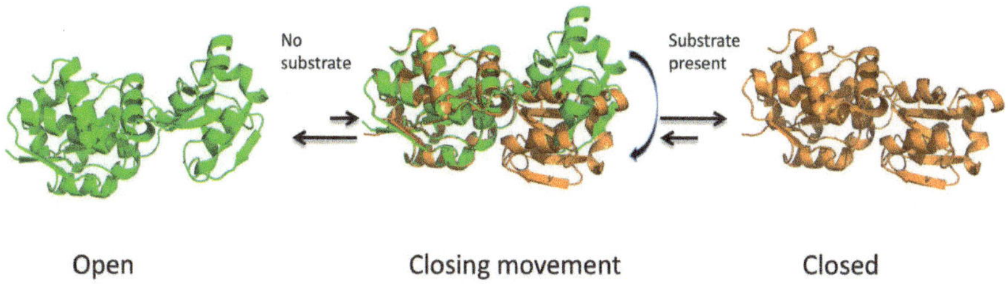

Figure 2. Equilibrium between the open and closed states of substrate binding proteins (ProX from *A. fulgidus).* The unliganded structure (highlighted in green) of an SBP is fluctuating between the open and closed state (highlighted in orange). In the absence of substrate this equilibrium is pointing towards the open conformation. In the presence of the substrate this equilibrium is changed towards the closed conformation. Here the two domains are close together and side chains of both domains bury the substrate in a deep cleft in between them. (PDB entries: 1SW2, 1SW5). All Figures containing structures were prepared with pymol ("www.pymol.org").

Figure 2 highlights the open conformation (green), which is in equilibrium with the closed state although only a small percentage will be present in the closed unliganded state. Upon the addition of glycine betaine a stable closed conformation is reached and the equilibrium is shifted towards this state. Besides the crystal structure of the **substrate** bound state with glycine betaine, proline betaine and betaine as a substrate also the open conformation was crystallized. This allowed a detailed analysis of the closing and opening motion mediated by the hinge region between both domains. The comparison of the ligand-free and liganded conformation of other binding proteins showed an approximate rigid body motion of the two domains highlighting a total rotation of domain II by ~ 58° with respect to domain I (Figure 2). The total rotation has two components: 1) the hinge angle between the two domains of ~ 40° with its axis going through the above-mentioned hinges in the polypeptide and 2) a rotation perpendicular to the hinge axis of ~ 42°. Although the domains behave more or less as rigid bodies, there are a few changes of the **binary complex** in two regions of ProX. If one succeeds in crystallizing several conformation of a protein one can search for and visualize small distinct changes in the overall structure. This has been also observed in ProX, the α-helical conformation (in the open form) of residues 144–148 (domain I) change either to an isolated-β-bridge or to a turn conformation (in the closed form). This conformational change may be caused by the proximity to Arg149, which plays an important role in ligand binding as discussed below. Furthermore, residues 222–225 (domain II), which are in turn and 3_{10}-helix conformation (in the open form), become rearranged to a short α-helix in the closed form. These structural changes highlight an important point in the function of such a protein (more detail below).

2.5.2. *Open and closed - An active site view*

A closer look at the binding site or the amino acids involved in **substrate** binding shows that small but distinct conformational changes of the amino acids involved in ligand binding occur upon substrate binding. Again as an example the glycine-betaine binding protein ProX from *A. fulgidus* is used.

The binding site is located in the cleft between domains I and II and can be subdivided into two parts, one binding the quaternary ammonium head group and the other binding the carboxylic tail of these compounds. The quaternary ammonium head group is captured in a box formed by Asp109 and the four tyrosine residues Tyr63, Tyr111, Tyr190, and Tyr214 being oriented almost perpendicular to each other. The tyrosine side chains provide a negative surface potential that is complementary to the cationic quaternary ammonium head group of GB. The carboxylic tail of GB is pointing outward of this partially negatively charged environment forming interactions with Lys13 (domain I), Arg149 (domain II), and Thr66 (domain I), respectively. Furthermore the structure was solved at a resolution sufficient to locate water molecules. An important water molecule was observed, which was held in place by residues Tyr111 and Glu145, and stabilizes domain closure. Here it is important to mention that this water molecule was not observed in the open unliganded structure and its importance would therefore be easily overlooked when no comparison between the two states were possible.

The superposition of the open-unliganded form and the closed-liganded form of ProX allowed an unambiguous identification of residues of domain II that are involved in ligand binding. They show virtually the same orientation in the open and closed forms (see Figure 3). Residues Tyr63, Tyr214, Lys13, and Thr66 superimpose very well. Only the main chain carbonyl of Asp109 from domain I is slightly out of place compared to the closed form because of the enormous main chain rearrangement between Asp110 and Tyr111 upon domain closure. The residues contributed by domain II behave quite differently. Tyr111 and Tyr190 are not only moved as parts of domain B but they undergo a major conformational change to adopt the conformation of the closed-liganded binding site. The side chain conformation of Arg149 shows only small changes between the open and closed conformations although it undergoes a large movement as part of domain II.

Recently, another structure of ProX was solved in the liganded but open conformation [29]. This conformation represents a state of which only very few structures are known. In other words, the protein has a ligand bound and is on its way to close up the binding site. This structure provided an even more detailed picture on the function of ProX and finally highlighted the crucial role of Arg149. In addition to the direct interaction with GB and residues

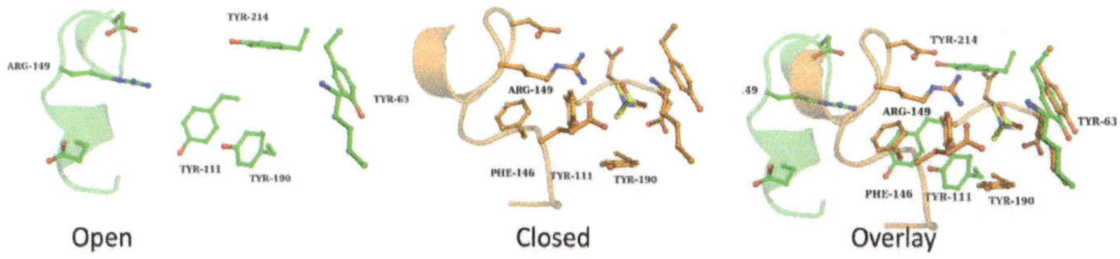

Figure 3. The binding site of ProX is highlighted in the open (depicted in ball and stick in green-left picture) and the substrate bound closed conformation (depicted in ball and stick in orange middle picture). As observed some of the ligand binding amino acid change their conformation. The right picture shows an overlay of both structures to visualize these conformation changes (PDB entries: 1SW2, 1SW5 and 3MAM).

that are part of the substrate-binding pocket (Tyr111, Thr66), Arg149 is a major determinant in domain-domain interactions in the closed structure. As such, Arg149 interacts with Val70 (domain I) and Asp151 (domain II), thereby acting as a linking element between the two domains enforcing stable domain closure. These interactions complement those mediated by Pro172 of domain II, where Pro172 interacts via its Cα-atom and a water molecule with Glu155 of domain II. Together, this provides a further explanation for the crucial role of Arg149 for the stability of the liganded-closed state, which has been observed in mutagenis studies. Here, the binding **affinity** of GB was dramatically lowered when Arg149 was mutated to alanine, a phenomenon that could not be explained since the aromatic cage which dominates the binding affinity was still present to bind glycine betaine. This suggested that Arg149 is the final amino acid to interact with the **substrate** and, thereby, terminate the molecular motions that result in the high affinity closed state of ProX. Besides this crucial role of switching from a low affinity to a high affinity state via the interaction of Arg149 the open liganded structure also shed light on the movement that the amino acids undergo during closure of the protein. In the open-liganded structure the presence of glycine betaine is communicated to Arg149 through interactions of the side chains of Tyr190, Tyr111, and Phe146 via a side-chain network [29]. Interestingly when comparing the open and closed structures of other SBPs, the maltose binding protein (MBP) [9] and the ribose binding protein (RBP) from *E. coli* and the N-Acetyl-5-neuraminic acid binding protein (SIAP) from *H. influenza* [25] a similar network can be identified in these proteins, something which had not been identified before due to the lack of an open-liganded structure.

In summary, the "Venus fly trap" model describes the opening and closing of SBPs. Here the equilibrium between these two conformations is shifted towards the closed state upon **substrate** binding. Many crystal structures of SBPs have been solved in the unliganded-open, liganded-closed, and, more rarely, in the liganded-open or unliganded-closed state [3, 14, 15, 23]. The crystal structure of one of these states will give information on the overall structure of the protein as well as the **ligand** binding site. Several SBPs have been crystallized in two or more states and quite clearly the increasing amount of states will shed a more detailed look on how domain closure is occurring. Thus, although crystallization is trial and error and sometimes tedious, it is worth to search for crystals in the liganded-closed conformation as well as crystals of another state since every stage will visualize the cleverness of nature to use conformational changes for the formation of ligand binding sites.

3. Protein with multiple ligands – How to crystallize the different ligand bound intermediate states

Besides proteins that bind one substrate, a large number of enzymes are binding two or more substrates and convert these into a product. Here, the crystallization of the **apo-enzyme** (protein without any ligand bound) often reveals the binding site of these ligands. However, the exact influence of the binding of these ligands can only be deduced from several structures, where different ligands are bound or one structure with all ligands bound. The different states are called apo-enzyme, when the enzyme is depleted of all ligands, the **binary complex** when the first substrate is bound, the **ternary complex** when the second ligand is bound as well. A quaternary complex would describe the protein with three ligands bound.

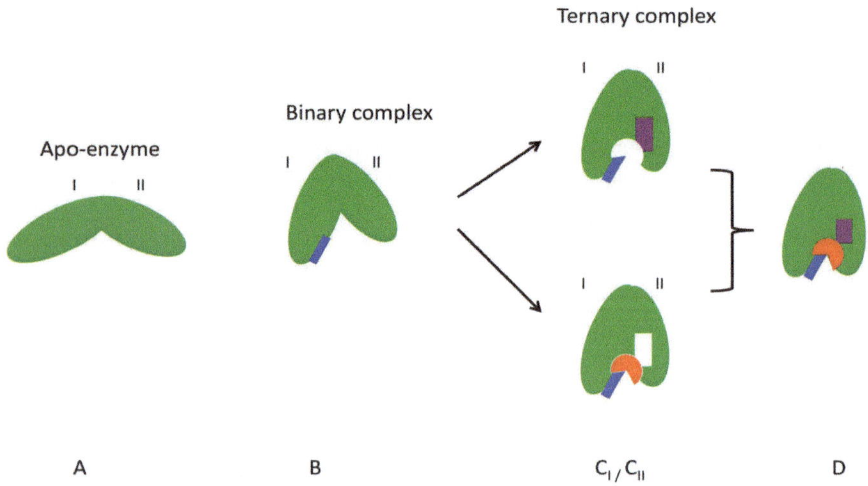

Figure 4. Overview of the conformations a protein can adopt with multiple ligands. A) The apo-enzyme B) binary complex where the first ligand is bound. This ligand with the highest affinity induces a stable conformation of the enzyme which allows the binding of the second ligand (ternary complex C_I or C_{II}). D) Enzyme complex where all ligands are bound.

Most of these proteins are enzymes. In reactions mediated by enzymes, the molecules at the beginning of the process, called **substrates**, are converted into different molecules, called products. Almost all chemical reactions in a biological cell need enzymes in order to occur at rates sufficient for life. Since enzymes are selective for their substrates and speed up only a few reactions from among many possibilities, the set of enzymes synthesized in a cell determines which metabolic pathways are utilized. Obtaining a snapshot of the substrate bound enzyme is difficult, because the enzymatic reaction will proceed immediately after substrate binding. One "trick" mostly used to solve this problem is to inhibit the reaction by either the reaction condition, meaning by varying pH of the buffer to a value where the reaction is not occurring. Another approach often appied in crystallography is to use a mutant, which cannot catalyze the reaction anymore; however it is still capable of binding the substrate. This has been proven to be successful in many cases. For example the catalytic cycle of nucleotide binding domains has been unraveled by such a mutation. In the latter case the ATP hydrolysis, in the wild type the measure for activity, has been abolished by mutation of a crucial amino acid, which still allowed binding of ATP but prevented hydrolysis. Thereby the dimeric state of the protein was stabilized and the active form of the NBD (nucleotide binding domain) could be crystallized in the presence of ATP [30-32].

Below the structural studies of the octopine dehydrogenases (OcDH) from *P. maximus* will be described in more detail. This enzyme catalyses the reductive condensation of L-arginine with pyruvate forming octopine under the simultaneous oxidation of NADH (reduced form of nicotinamide adenine dinucleotide). This oxidation of NADH is the terminal step in the anaerobiosis, meaning the generation of ATP when organisms are suffering from low oxygen levels. A prominent member of these terminal pyruvate oxidoreductases is the lactate dehydrogenase, which catalyzes the transfer of a hydride ion from NADH to pyruvate, with produces NAD^+ (nicotinamide adenine dinucleotide) and lactate. Thereby the redox state in vertebrates is maintained during functional anaerobiosis. OcDH fulfills the same function in the invertebrate *P. maximus*.

This enzyme has been chosen due to the fact that three **substrates** need to be bound simultaneously for the reaction, in contrast to the lactate dehydrogenase, which has only two subtrates, NADH and pyruvate. Furthermore this enzyme was crystallized as wildtype protein and in all substrate bound states (**binary** and **ternary complex** C_I and C_{II}) and the corresponding structures were elucidated. The state where all substrates were present did not yield a structure due to the immediate conversion to the product. However, the other structure allowed a detailed view on how the latter state might look like.

In 2007 Mueller and co-workers achieved cloning and heterologously expression of this enzyme using *E. coli* as expression system [33]. After the purification the enzyme was characterized and the authors proposed a sequential binding mode of the **substrates**. Here, NADH was bound first followed by either L-arginine or pyruvate. The order of the last two was not revealed by the enzymatic analysis. Furthermore, a catalytic triad was proposed consisting of three highly conserved amino acid, building up a protein rely-system for the reduction of NADH. This triad has been observed in the sequence and structure of the lactate dehydrogenase as well. Sequence analysis of different proteins from this family revealed that the protein contained two distinct domains where domain I contained the characteristic Rossmann-fold, a domain responsible for the binding of NADH. Domain II was assigned as octopine dehydrogenase domain, which is specific for this protein family and was suggested to contain the binding site for both L-arginine and pyruvate. Both domains are connected via a linker region of 5-8 amino acids suggesting that these domains might undergo large conformational changes.

3.1. The crystallization of apo-enzyme and the binary complex

Parallel to the biochemical characterization, the crystallization of the enzyme was started. Due to the two-domain structure OcDH can adopt multiple conformations in solution, which prevents crystal formation. However, purified OcDH-His_5 yielded small crystals that appeared to be multiple on optical examinations (Figure 5 A). They diffracted to a resolution of 2.6 Å. However the diffraction showed multiple lattices in one diffraction image and could not be used for structure determination (Figure 5 A) [34]. All attempts to improve these crystals using for example **seeding**, temperature ramping or various crystallization conditions failed. Finally, the primary **ligand**, NADH, was added prior to crystallization. This produced crystals under conditions similar to those in the absence of NADH. Here, the incubation temperature appeared to be critical and needs to be kept at 285 K. The crystals obtained were single and diffracted to 2.1 Å resolution, which allowed processing of the data and subsequent structure determination (Figure 5 B). The structure of OcDH was solved as binary complex with NADH [34, 35].

Cofactors like NADH are often observed to be co-purified. This was assumed to be the case for OcDH as well, however, no activity was ever observed without NADH, but in the presence of the other two substrates. This implies that OcDH is not homogenous and multiple conformations exist as observed in the multiple crystal lattices of the diffraction image. This is in line with the only other available three-dimensional structure of an enzyme of the OcDH superfamily, the apo-form of N-(1-D-carboxylethyl)-L-norvaline dehydrogenase (CENDH) from *Arthrobacter sp.* strain 1C [36]. CENDH catalyzes the NADH-dependent reductive condensation of hydrophobic L-amino acids such as L-methionine, L-isoleucine, L-valine, L-phenylalanine or L-leucine with α-keto acids such as pyruvate, glyoxylate, α-ketobutyrate or oxaloacetate with (D, L) specificity [37]. The structure of the **binary complex** of CENDH with NAD^+ was determined to a resolution of 2.6 Å. Although NAD^+ was added in the crystallization trials the cofactor could not be observed unambiguously in the electron density. This was likely due to the concentration of NAD^+, which was below the K_d. As a result not all proteins had the **substrate** bound, which led to a not very well defnied electron density. Only the nicotinamide ribose moiety was of moderate quality and the density of the nicotinamide ring was very weak. This has been attributed to low NAD^+ **occupancy** in this crystal, hence the **co-factor** has been omitted from the high resolution refinement [36].

This highlights the importance to verify the **affinity** of substrate prior to crystallization. Since NAD^+ is the product of the reaction and to ensure the release of the product, the affinity of NAD^+ must be lower than the affinity of NADH. In a recent study on the OcDH the affinities have been determined to be 18 μM for NADH and 200 μM for NAD^+ [38]. As described above the addition of substrate in crystallization trials need to be at least a 10-fold above the K_d. For OcDH 0.8 mM NADH was used for the crystallization of the binary complex, which represents a 40-fold excess.

The structure of the OcDH-NADH **binary complex** revealed why the initial crystallization attempt of the **apo-enzyme** failed. NADH is bound by the Rossmann-fold located in domain I as well as by an arginine residue in domain II. Thereby the OcDH captured in a state which enables the binding of the other **substrates**, pyruvate and L-arginine (see below) [34, 35].

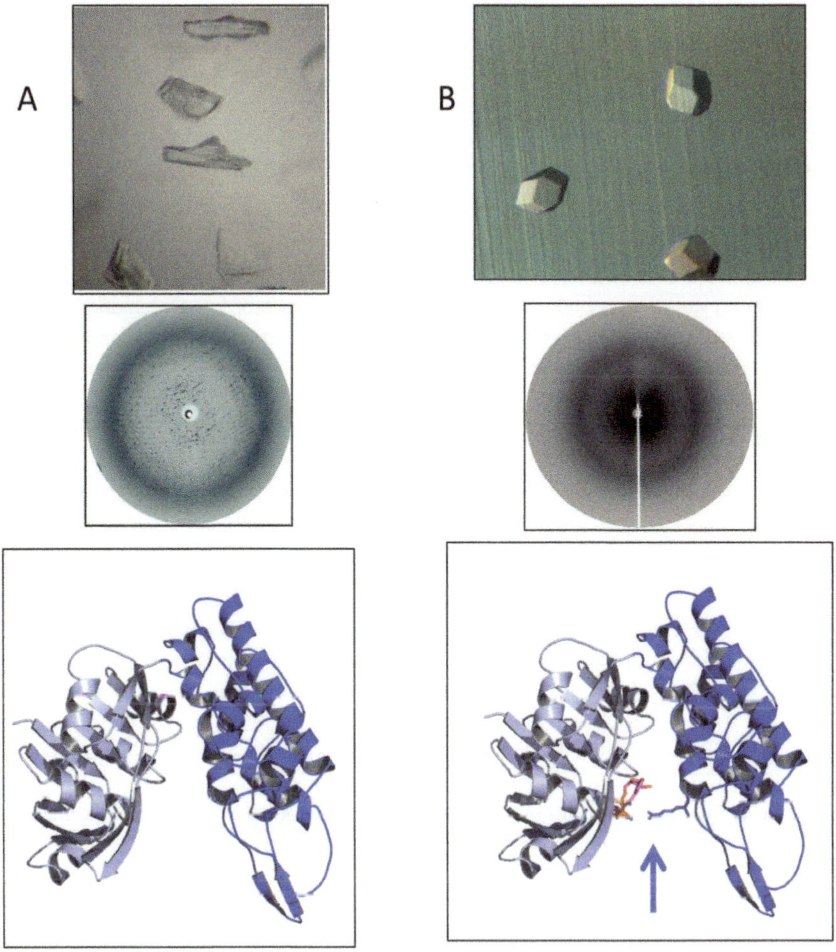

Figure 5. Crystallization of OcDH in the absence and presence of NADH. A) Absence of NADH. The crystals obtained are multiple (upper panel) and the diffraction pattern yielded showed several lattices (middle panel). The structure of OcDH shows two distinct domains connected by a flexible linker, which can rotate freely in the absence of NADH (lower panel). B) Crystals obtained in the presence of NADH (upper panel). The diffraction showed a single lattice diffracting up to 2.1 Å (middle panel). The structure revealed the binding site of NADH and an interaction of an arginine residue from domain II with NADH, which locks OcDH in one stable conformation (lower panel) (PDB entries: 3C7A and 3C7D).

In summary, the apo-state of multiple **ligand** binding enzymes is difficult to crystallize when the enzyme undergoes large conformational changes. In the case of the OcDH only the binary complex in the presence of NADH could be crystallized. Here crystals were of sufficient quality to determine the structure.

3.2. The crystallization of the ternary complexes C_I and C_{II}

OcDH catalyzes the condensation of L-arginine with pyruvate to form octopine under the oxidation of NADH. Biochemical analysis as well as the crystal structure revealed that NADH is the first **substrate** to bind to OcDH. The structure of this **binary complex** exhibited a stable conformation of the protein in solution with an Arg-sensor, which binds NADH, and thereby stabilizes the protein in one conformation (see above).

So the next step was to determine the structure of the OcDH in the presence of the second and third **substrate,** L-arginine and pyruvate, respectively. Initially, the protein and the substrate were mixed and an extensive search for suitable crystallization conditions was started. However, no crystals were obtained for OcDH in the presence of L-arginine and/or pyruvate. Instead only needles were grown which were multiple and very fragile similar to the crystals obtained for the **apo-enzyme**. This is in line with the biochemical data, which highlights the order of substate binding which show that NADH has to be bound prior to binding of L-arginine as well as pyruvate [38, 39]. Here the authors used two other techniques, NMR and **ITC** (isothermal titration calorimetry) repectively, to show that L-arginine only binds after saturation of the **apo-enzyme** with NADH. Pyruvate was shown to be bound only after L-arginine binding to the enzyme. This suggests that OcDH undergoes a conformational change when NADH is bound and thereby the binding site of L-arginine is formed. Furthermore the binding site for pyruvate is only created when L-arginine is bound.

Since crystallization was not successful the next step was to use **co-crystallization** with the OcDH protein and L-arginine and/or pyruvate to obtain structural information of the **ternary complex** (C_I and C_{II}). This yielded crystals of OcDH only in the presence of NADH and no additional density was observed for neither L-arginine nor pyruvate. So, **soaking** the **ligand** into preformed OcDH-NADH crystals was the last method chosen. Crystallization trials were carried out using the **hanging-drop vapor diffusion method** and crystals of OcDH were grown in the presence of 0.8 mM NADH. L-arginine-bound crystals were obtained by soaking NADH-bound OcDH crystals in 100 mM MES pH 7.0, 1.15 M Na-citrate, 0.8 mM NADH containing 10 mM L-arginine for at least 24 hours. Pyruvate-bound crystals were obtained also by soaking the crystals in 100 mM MES pH 7.0, 1.15 M Na-citrate, 0.8 mM NADH and 10 mM pyruvate for at least 8 hours. Both concentrations were chosen relatively high but they resemble the *in vivo* concentration as well as were backed up by the **affinity** observed for both **substrates** in biochemical and biophysical studies, being 5.5 mM L-arginine and 3.5 mM pyruvate, respectively. During soaking a cracking of the crystals was observed after the first minutes. However, the crystals recovered completely from this cracking within the following hours and showed no fissures or other damages after that soaking procedure. Desprite this, the diffraction analysis revealed a loss in diffraction. Initally the crystals diffracted to 2.1 Å. After soaking in L-arginine or pyruvate the diffraction potential was reduced to 3.0 Å and 2.6 Å, respectively. The phenomenon of crystal cracking and decline of the diffraction already was a good indication that the substrates diffused into the crystal. A dataset was collected from crystals where either one of the ligands was soaked in and besides the decrease in diffraction potential also the unit cell parameters changed (see Table 2).

Crystal Complex	Unit cell parameters (a,b,c in Å)
OcDH-NADH	99.8, 99.8, 126.5
OcDH-NADH/L-arginine	95.9, 95.9, 117.9
OcDH-NADH/pyruvate	95.0, 95.0, 120.2

Table 2. Crystallographic parameters of the unit cell of the binary OcDH-NADH complex and after soaking of the ternary complex C_I: OcDH-NADH/L-arginine and C_{II}: OcDH- NADH/pyruvate

The change in unit cell parameters suggested that a conformational change occurred during the **soaking** with the **ligand**. This was further observed after the structure was resolved and electron density was clearly defined for L-arginine in the one and for pyruvate in the other dataset. The structure of OcDH-NADH/L-arginine showed a rotational movement of domain II towards the NADH binding domain I, and a stronger interaction of the Arginine residue with NADH. A domain closure was also observed in the pyruvate bound structure. So stable binding of NADH to the Rossman fold of domain I, the first step in the reaction sequence of OcDH, occurs without participation of domain II. A comparison of the OcDH-NADH (colored light-purple in Figure 6) and the OcDH-NADH/L-arginine complexes revealed a 42° rotation of domain II towards the NADH binding domain (domain I) in the latter complex. This domain closure is triggered by the interaction of Arg324 (domain II) with the pyrophosphate moiety of NADH bound to the Rossman fold in domain I.

A comparison of the two **ternary complexes** suggests that both, pyruvate and L-arginine, are capable to trigger domain closure to a similar extent. However, in the OcDH-NADH/pyruvate complex, pyruvate partially blocks the entrance for L-arginine, while in the OcDH-NADH/L-arginine complex, the accessibility of the pyruvate binding site is not restricted by L-arginine [34, 35]. From these structures it could be deduced that L-arginine binds to the OcDH-NADH complex in a consecutive step and induces a rotational movement of domain II towards domain I. This semi-closed active center, which is further stabilized using the pyrophosphate moiety of the bound NADH and by interactions of L-arginine with residues from both domains is then poised to accept pyruvate and consequently the product octopine can be formed. With regard to the structures it was proposed that instead of a random binding process, an ordered sequence of substrate binding in the line of NADH, L-arginine and pyruvate will occur.

This ordered sequence of **substrate** binding was then biochemically proven by **ITC** studies where the binding affinities of the substrates were measured. Here, the binding of L-arginine was only observed when NADH was bound primarily and the binding of pyruvate only when the complex was preloaded with L-arginine [38, 39]. Furthermore this ordered binding mechanism explains why no lactate is found in side *P. maximus* which is normally formed when NADH and pyruvate is bound by lactate dehydrogenases. Here, it is worth mentioning that the conformational changes induced by **ligand soaking** into the crystal were also observed in NMR studies that were perfomed in solution. So the apparent conformational changes in the crystal resemble the changes the protein undergoes in solution.

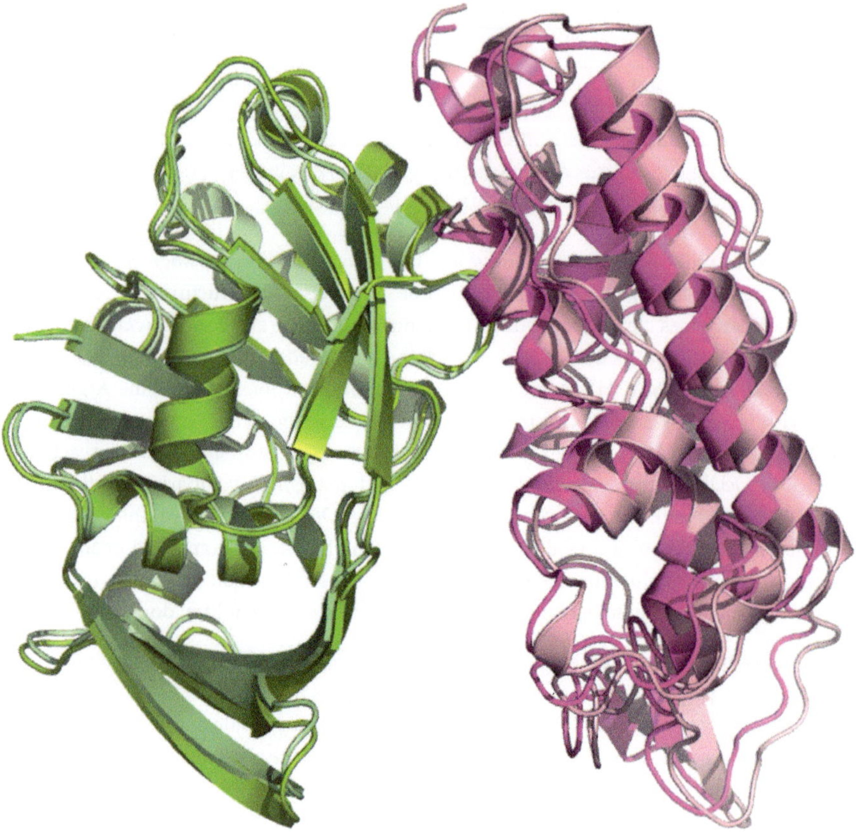

Figure 6. Overlay of the OcDH-NADH binary complex with the OcDH-NADH/L-arginine ternary complex C_I. As seen in the superposition the binding of L-arginine induces a conformational change. Domain II is rotated towards domain I which is thereby creating the pyruvate binding site. In the overlay the pyruvate structure is not shown due to clarity (PDB entries: 3C7A and 3C7D).

The crystal structures of the different states of OcDH, delivered snapshots elucidating for the first time the precise and very distinct binding order [35]. Unfortunally the crystals with the endproduct octopine did not diffract X-ray with a resolution and quality high enough for structure determination. The same hold true for a complex with all three substrates present at once. This is likely due to the fact that the immediate condensation occured and the product was formed.

To show how proteins can be crystallized with their enzymatic endproducts we chose another enzyme family as example and will describe the different procedures during the next paragraphs.

4. Enzymatic products in protein structures – How to crystallize this rather unfavored states

The state found to be important within an enzyme reaction cycle is supposedly the product bound state. After the reaction occurs the product is still sitting within the protein and will be released. Often these product have a low(er) **affinity** to the protein than the **substrates** and are therefor less often found to be successfully crystallized.

Examples of prosperous structure determination however are the shikimate dehydrogenase (SDH or AroE) of *Thermus thermophilus* (*Tth*SDH), *Aquifex aeolicus* (*Aae*SDH) and the recently deposited structures of the SDH of *Helicobacter pylori* (*Hpy*SDH) as well as the bifunctional dehydroquinase-shikimate dehydrogenase (*Ath*DHQ-SDH) from *Arabidopis thaliana* which were crystallized with its reaction product shikimic acid ([40-43]. Similar to that the closely related quinate dehydrogenase (QDH) of *Corynebacterium glutamicum* (*Cgl*QDH) was structurally characterized in four different states: as **apo-enzyme** and at atomic resolution with bound **cofactor** NAD^+ as well as in complex with quinic acid (QA) and the reduced cofactor or shikimic acid (SA) and NADH [44].

Shikimate-/quinate dehydrogenases belong to the superfamily of NAD(P)-dependent (nicotinamide adenine dinucleotide phosphate) oxidoreductases whereas SDHs catalyse the reversible reduction of 3-dehydroshikimate to shikimate under oxidation of NAD(P)H (reduced form of nicotinamide adenine dinucleotide phosphate) and QDHs the oxidation of quinate to 5-dehydroquinate with reduction of NAD(P), respectively. The overall fold consists of a N-terminal or **substrate** binding domain and a C-terminal or cofactor-binding domain and is highly conserved within that subfamily (schematically shown in Figure 7). Compared to other proteins, like the above-mentioned SBPs, the structural changes occurring during catalysis are less prominent and comprise a movement of the two domains against each other in a range of several Ångstrom.

4.1. Shikimate dehydrogenase from *Aquifex aeolicus*

Crystals of the native (apo-) *Aae*SDH were obtained with non-His-tagged protein, whereas the **ternary complex** crystals were obtained with His-tagged SDH. To get these complexes the protein solution was mixed with **substrate** and cofactor (i. e. with both natural products) to final concentrations of 5.0 mM shikimic acid and 5.0 mM $NADP^+$ before crystallization. The hanging-drop vapor diffusion method was used for crystallization trials. The drops were prepared by mixing 3 μl of the protein-ligand solution with 1 μl of well solution [41].

K_M values were determined to be 42.4 μM for both ligands, which means that there was a 100-fold excess in the crystallization drop. The bound products SA and $NADP^+$ in the protein could be explained by the low activity of the enzyme and the equilibrium constant favoring the formation of SA and $NADP^+$, both of which are caused by the low pH. The equilibrium constant ($[SA][NADP^+]/[DHSA][NADPH]$) was determined by Yaniv and Gilvarg (1955) to be 27.7 at pH 7 and 5.7 at pH 7.8 [45]. As of any dehydrogenase reaction, the equilibrium position of the *Aae*SDH-catalysed reaction depends on the hydrogen ion concentration of the environment. The pH of the well solution (0.2 M ammonium acetate, 30 % w/v PEG 4000, 0.1 M sodium acetate) was 4.6 and therefore the drop became more acidic during crystallization. They estimated the equilibrium constant at pH 5 to be around 3000 in favor of the formation of SA and $NADP^+$. The geometry of $NADP^+$ is not distinguishable from that of the NADPH at this resolution (2.2 Å) but the geometry of SA containing a tetrahedral (sp^3) C3 atom is distinct from that of DHSA, in which the geometry of C3 is planar (sp^2) [41].

There were eight (apo) and four (**ternary complex**) crystallographically independent *Aae*SDH molecules in the asymmetric unit of apo-*Aae*SDH or *Aae*SDH-$NADP^+$-SA, respectively. According to the structure of the **apo-protein** and the ternary complex a fully open (molecule F in apo-*Aae*SDH) and a closed conformation with bound ligands (molecule D in *Aae*SDH-$NADP^+$-SA; Figure 8) were observed as well as several intermediate states. From

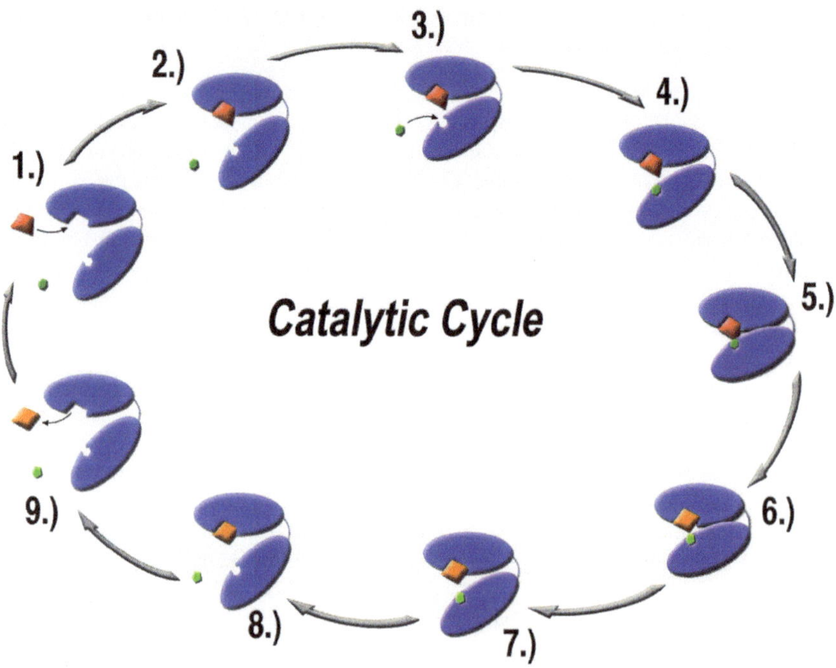

Figure 7. Schematic diagram of the conformational changes within a protein (blue ellipses) during the catalyzed reaction. 1.) Before a substrate (red trapezium) is bound the proteins exhibits an open conformation. 2.) – 4.) Binding of the substrate induces a slight domain closure before the cofactor (green hexagon) is bound. 5.) + 6.) In order to facilitate the conversion from substrate to the product (orange rhombus) both protein domains need to be in close contact. 7.) -9.) A stepwise domain opening allows the changed cofactor (light green pentagon) to leave the protein domain, followed by the product. The protein itself is not modified at all during the whole reaction and mostly all steps are reversible.

the fully open to the closed form there is a movement of three loops in the catalytic domain towards the NADP-binding domain by around 5 Å sealing the active site of the enzyme. SA and $NADP^+$ are brought in close contact in that cavity: the C3-O3 bond of SA is parallel to the C4-C5 bond of the nicotine amide ring of $NADP^+$, and the distance between the two bonds is 3.5 Å. This represents a typical distance for a hydride transfer.

The open conformation therefor represents the protein structure in state 9.) (or 1.), respectively) in Figure 7, the closed conformation correlates to state 6.) in that scheme.

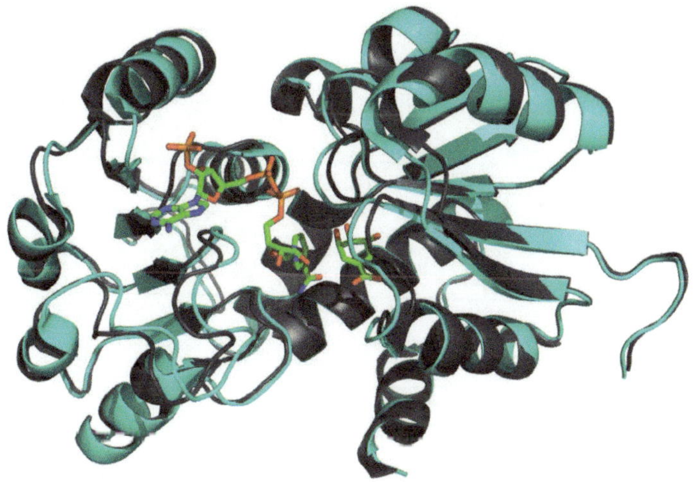

Figure 8. Shikimate dehydrogenase from *Aquifex aeolicus*. The cartoon depicted in cyan represents the open (apo) conformation of the enzyme (PDB entry: 2HK8), the structure coloured in black illustrates the closed conformation (PDB entry: 2hk9) with the bound ligands shikimic acid and $NADP^+$, shown as sticks.

4.2. Shikimate dehydrogenase from *Thermus thermophiles*

In case of *Tth*SDH, crystals of the native protein were grown in **microbatch** plates. **Co-crystallization** trials were only successful with added $NADP^+$ but failed with shikimate. To obtain complexes with bound shikimate crystals of the **apo-protein** or the SDH-$NADP^+$ complex were **soaked** for several seconds in cryosolution supplemented with shikimate. The final concentration of all added **ligands** was 5 mM. Although the kinetical parameters were not determined prior to crystallization, all K_M values of closely related SDHs are in a μM range so that there was at least a 20-fold excess of **substrate** and **cofactor** [40].

Evaluation of the complex structures revealed an open and a closed conformation of the two domains but neither the binding of shikimate nor $NADP^+$ seem to induce that conformational change. Shikimate could bind to the closed as well as to the open form, whereas $NADP^+$ was found only in closed conformation. As described for *Aae*SDH, the crystallization condition was in an acidic range of about pH 4.6, which explains that the reaction did not occur. An alignment of the three structures (apo-SDH, SDH-SA, SDH-SA-NADP$^+$) of *T. thermophilus* illustrates the domain closure while/after SA and/or NADP$^+$ binding (Figure 9). Surprisingly there seems to be no further movement of the substrate binding domain against the NADP(H) binding domain when the cofactor is bound. Thus, the apo-structure represents state 9.) (or 1.) in Figure 7 and both the **binary** and the **ternary complex** may match a state between 6.) and 7.) of that scheme.

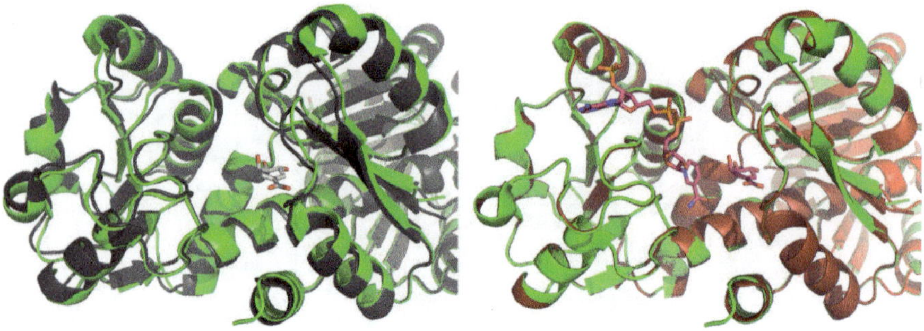

Figure 9. Shikimate dehydrogenase from *T. thermophilus*. The cartoons depicted in green (left and right side) represent the open (apo) conformation of the enzyme (PDB entry: 1WXD), the structure coloured in black illustrates the closed form with bound shikimic acid (PDB entry: 2D5C), whereas the red one corresponds to the ternary complex (PDB entry: 2EV9) with shikimic acid and NADP$^+$, shown as sticks.

4.3. Bifunctional dehydroquinase-shikimate dehydrogenase (*Ath*DHQ-SDH) from *Arabidopis thaliana*

Remarkable are the **co-crystallization** trials of Singh and Christendat with the bifunctional enzyme dehydroquinase-shikimate dehydrogenase from *Arabidopsis thaliana* (*Ath*DHQ-SDH). First crystals were obtained with the product shikimate at the SDH site and tartrate as a **substrate** analogue at the DHQ site. Later they could crystallize *Ath*DHQ-SDH with its natural products shikimate and NADP$^+$.

For the shikimate-tartrate complex crystals they used the **vapor diffusion hanging-drop** technique. Protein solution with a final concentration of 1 mM of shikimate was mixed with the reservoir solution containing 0.4 M potassium sodium tartrate tetrahydrate [42]. To obtain **ligand** bound crystals of the three different protein conditions were tested: protein only, protein with 1 mM shikimate or protein with 1 mM NADP$^+$. The protein-shikimate approach was the only one that yielded crystals (under the same conditions as mentioned above). To gain crystals of the **ternary complex** a further treatment was necessary: The above-mentioned crystals were soaked with a NADP$^+$ solution (final concentration 10 mM) for about 8 hours at pH 5.8. The K_M values were determined to be 0.6 mM for shikimic acid and 0.13 mM for NADP$^+$, respectively [42].

Not only that the closed conformation of the enzyme after binding of both products could be demonstrated (Figure 10) but also the activity of that ternary complex was proven as the oxidation of shikimate was evidenced by the generation of dehydroshikimate, – the product of the DHQ moiety – found in the DHQ site [43].

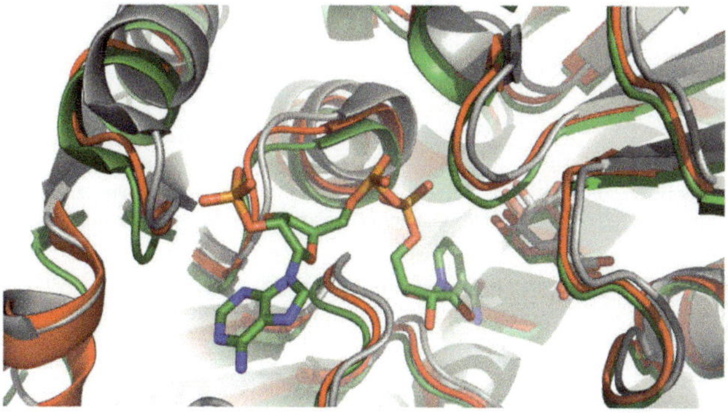

Figure 10. Bifunctional dehydroquinase-shikimate dehydrogenase (*Ath*DHQ-SDH) from *Arabidopis thaliana*. The cartoon coloured in grey reveals the binary complex (PDB entry: 2GPT) with the bound product shikimate (grey lines), the structure depicted in red shows the protein with bound substrate dehydroshikimate (red lines; PDB entry: 2O7Q), whereas the cartoon in green represents the ternary complex (PDB entry: 2O7S) with bound dehydroshikimate and the cofactor NADP(H).

The structures of the *Ath*DHQ-SDH **binary complexes** with bound product shikimate or **substrate** dehydroshikimate illustrate therefore the states 8.) or 2.), while the **ternary complex** corresponds to the transition state 5.) in Figure 7.

4.4. Shikimate dehydrogenase from *Helicobacter pylori*

Recently three different catalytic states of the *Hpy*SDH were deposited in the PDB. Unfortunately the results are not published so that detailed information about the crystallization trials are lacking. Apparently they obtained all crystals by means of the **hanging-drop** vapor diffusion method.

However, the structure is ideally suited to visualize structural changes during **cofactor** binding Figure 11.

In the **binary** structure of the *Hpy*SDH with bound shikimate there is a large loop in the C-terminal domain that obstructs the entrance to the cofactor-binding cleft and virtually acts as a lid. For cofactor binding this loop has to move away from the cleft in order to create space for NADP(H). Comparing these two structures with the overall conformation of the

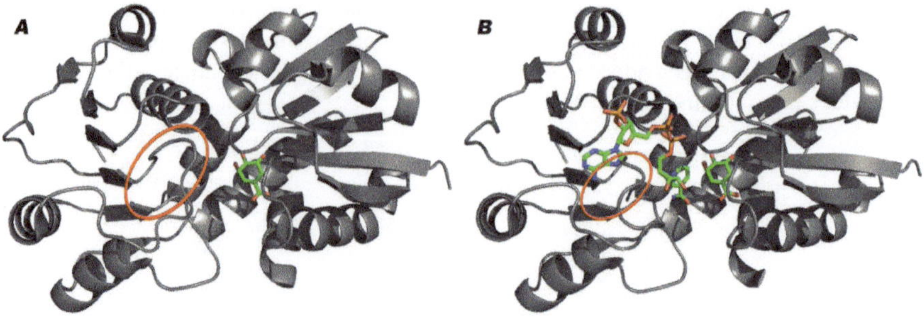

Figure 11. Binary (left; PDB entry: 3PHI) and ternary structure (right; PDB entry: 3PHH) of the shikimate dehydrogenase from *Helicobacter pylori*. The substrate dehydroshikimate and the cofactor NADP(H) are presented as sticks. The red circle indicates the loop region in the N-terminal domain which acts as a lid during cofactor binding.

apo-protein, these two conformational stages represent stages 2.)/3.) or 6./7.), in the catalytic cycle shown in Figure 7.

4.5. Quinate dehydrogenase from *Corynebacterium glutamicum*

Last but not least the bacterial quinate dehydrogenase of C. *glutamicum* could be structurally solved in four different catalytic states: apo-enzyme, with bound **cofactor** NAD^+ and in complex with quinate (QA) and the reduced cofactor or shikimate (SA) and NADH, i. e. with the natural **substrate** and the natural cofactor as product of the reaction.

For growing the crystals of the apo-form the protein solution was mixed with the reservoir solution and a NADH solution (2 µg/ml) in a drop ratio 1:1:1. The reduced cofactor could not be detected in the electron density due to the very low concentration [46].

For the **co-crystallization** trials (with the cofactor NAD^+, the substrate quinate (QA) and the reduced cofactor or shikimate (SA) and NADH) the kinetical parameters were determined first in order to get an idea of the concentrations necessary for successful **ligand** binding. The K_M values for NAD^+, QA and SA are 0.28 mM, 2.37 mM and 53,88 mM, respectively (Hoeppner et al.; publication in progress).

To obtain the **binary** and both of the **ternary complexes** the protein solution was mixed with either NAD^+ or QA plus NADH or SA plus NADH to a final concentrations of 1 mM for NAD^+ or NADH and 35 mM for QA or SA. These mixtures were incubated on ice for about 1 hour prior to crystallization. All substrates and the cofactor were bound during co-crystallization experiments by means of the **sitting drop method** with drop size of 2-4 µl in 1:1 ratio of protein and reservoir solution.

Crystals of the binary and ternary complexes were different in shape compared to the crystals of the **apo-enzyme** and grew under diffenrent conditions (Figure 12), which was a hint to (structural) changes within the protein molecules.

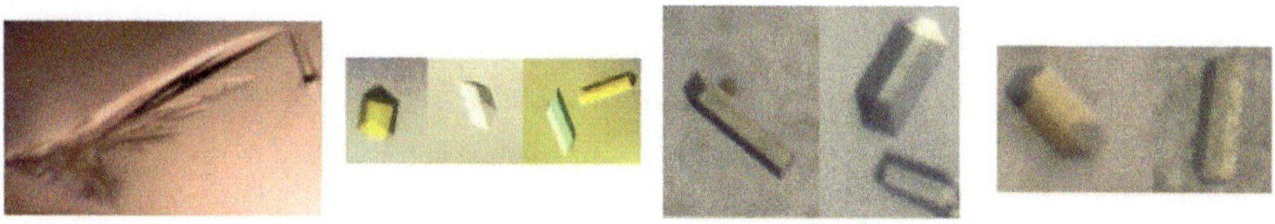

Apo *Cgl*QDH	*Cgl*QDH-NAD+	*Cgl*QDH-QA-NADH	*Cgl*QDH-SA-NADH
100 mM sodium acetate pH 4.6, 200 mM NaCl, 20 % (v/v) 2-methyl-2,4-pentanediol (MPD)	1.6 M sodium citrate tribasic dihydrate pH 6.9, plus up to 62 mM $CoCl_2$	24 % (w/v) PEG 6000, 360 to 400 mM $CaCl_2$, 100 mM Tris-HCl pH 8.0 to 9.5	

Figure 12. Comparison of the crystal shapes of the four different catalytical states of the CglQDH and the corresponding crystallization conditions.

The crystals of all three *Cgl*QDH complexes diffracted to atomic resolution and allowed us to assign the position of all **ligand** atoms unambiguously within the electron density (Figure 13).

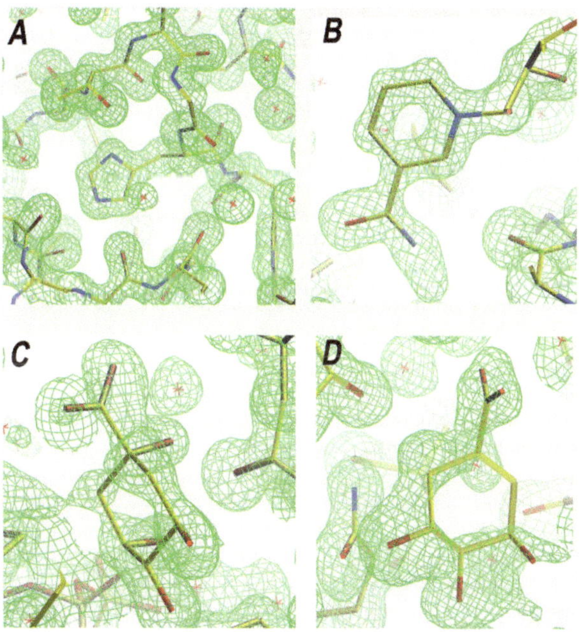

Figure 13. Representative sections of electron density maps of the *Cgl*QDH complexes at 1.0 Å (*Cgl*QDH-NAD⁺) or 1.16 Å (*Cgl*QDH-QA-NADH and *Cgl*QDH-SA-NADH) resolution. A) electron density defining protein side chains, B) density around the nicotinamide ring of the cofactor NAD(H), C) bound substrate quinate, D) bound substrate shikimate (electron density maps in A)-C) contoured at 1 σ and in D) 0.7 σ).

By comparing the overall structures of all these states an open, a semi-open and a closed conformation of the enzyme (Figure 14) was observed. Surprisingly, the apo-structure of the *Cgl*QDH exhibits the closed form although one would intuitively expect the open conformation. But it is possible that these findings were a crystallization artifact since the reservoir solution was quite acidic (pH 4.6) compared to the *Cgl*QDH pH optimum, which is 9.0-9.5 for quinate and 10.0-10.5 for shikimate (Hoeppner et al.; publication in progress).

Within the **cofactor** binding domain of *Cgl*QDH the glycine rich loop, which is highly conserved within SDH proteins and represents a classical Rossmann fold, is flapped down towards the cofactor binding cleft in the apo-structure, but moved away when the cofactor is bound. With regard to the overall arrangement of the apo-state compared to the NAD^+-bound state there is a clearly visible opening of the two domains. After forming the **ternary complex** the two domains are brought closer to each other, if only more slightly compared to the apo-conformation and thus adopt a semi-closed conformation (Hoeppner et al.; publication in progress).

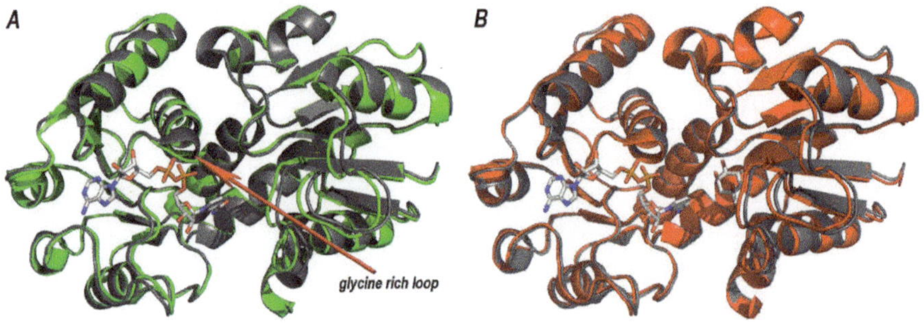

Figure 14. Structural alignments of the binary structure of *Cgl*QDH with bound NAD^+ (black; PDB entry: 3JYO) and A) the apo-protein (green; PDB entry: 2NLO) or B) the ternary structure with bound quinic acid and NADH (red; PDB entry: 3JYP). The substrate and the cofactor NAD(H) are presented as sticks. The red arrow indicates the conformational changes within the glycine rich loop.

4.6. Insights into the structural changes during catalysis and elucidation of substrate and cofactor specificity, using the example of *Cgl*QDH

4.6.1. Structure overview of C. glutamicum QDH

All *Cgl*QDH structures presented here are determined from crystals that were nearly isomorphous and belong to the same space group C2. The unit-cell parameters are very similar with one monomer per asymmetric unit.

The 282 residues in the QDH molecule form two structural domains (Figure 15): the N-terminal or catalytic domain (residues 1 to 113 and 256 to 283), which binds the **substrate** molecule, and the C-terminal or nucleotide binding domain (residues 114 to 255). The catalytic domain forms an open α/β sandwich, which is characteristic for enzymes of the S/QDH family but different from all other known proteins. The domain consists of a six-stranded, mainly parallel β sheet (strand order β2, β1, β3, β5, β6 and β4, where β5 is antiparallel. This β sheet is flanked by helices $\alpha 1$ and $\alpha 11$ at one side and $\alpha 2$, one 3_{10}-helix and $\alpha 4$ at the other. The C-terminal domain contains a six-stranded parallel β sheet (strand order β9, β8, β7, β10, β11 and β12) sandwiched by three helices ($\alpha 7$, $\alpha 6$, $\alpha 5$) on one face and by helices $\alpha 8$, $\alpha 10$ and a 3_{10}-helix on the other. The nucleotide binding domain exhibits a glycine rich loop with the sequence motive GXGGXG. The overall fold of this functional domain is very similar to that observed for other SDH proteins [47, 48] and represents the classical Rossmann fold. Both domains are linked together by helices $\alpha 5$ and $\alpha 11$. The arrangement of these two domains creates a deep active site groove in which **cofactor** and substrate are located.

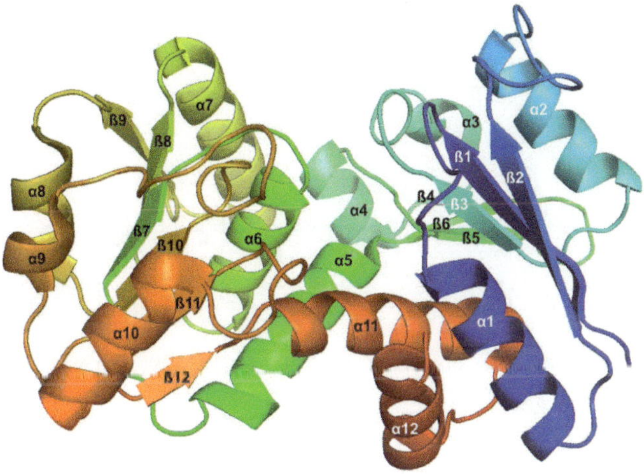

Figure 15. Schematical overview of the *Cgl*QDH fold

4.6.2. Description and analysis of QDH active site

Cofactor Binding Site: The electron densities for NAD(H) were of high quality and allowed us to assign the position of these **ligand** unambiguously at 1.0 Å. *Cgl*QDH crystallizes in the presence of NAD^+ in the same space group with similar unit cell dimensions, but under different crystallization conditions compared to the **apo-enzyme**. With regard to the overall structure we found that the catalytic domain moves away from the nucleotide-binding domain after cofactor binding making the interdomain cavity larger. Concerning the steric configuration of the residues there are only little but fundamental variances, especially in the glycine rich loop. In comparison to the QDH apo-enzyme (PDB entry 2NLO) the residues of the loop (Gly136-Val138) move out of the cavity after cofactor binding and therefore clear space for the NAD(H) molecule (Figure 14). Cofactor binding occurs in an extended groove between the N-terminal and C-terminal domain, whereas most of the molecular interactions result from the C-terminal domain. The adenine part of the adenosine moiety form hydrogen bonds only to some water molecules, while the ribose is bound by the side chains of Asp158 and Arg163. The phosphate moiety contacts the glycine rich loop and forms hydrogen bonds to Arg163 and the backbone nitrogen atom of Val138. The following ribose moiety again interacts only with water molecules, whereas the nicotinamide moiety is cramped by the backbone nitrogen of Ala255 and backbone oxygens of Val228 and Gly251, respectively. Gly251 and Ala255 are the only residues of the N-terminal domain involved in cofactor binding (Figure 16). The nucleotide-binding motive GXGGXG comprises the residues Gly134-Ala135-Gly136-Gly137-Val138-Gly139 (Hoeppner et al.; publication in progress).

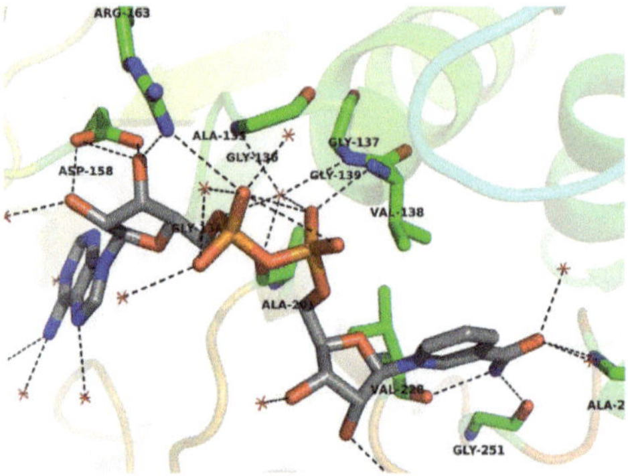

Figure 16. Interactions between the cofactor NAD(H) and *Cgl*QDH. Residues involved in hydrogen bonds (dotted lines) and bound ligand are shown as sticks, water molecules are depicted as red stars.

The strict specificity for NAD(H) is determined by the negatively charged aspartate residue 158, the neutral Leu159 and the bulk side chain of Arg163, which would result in steric hindrance with the additional phosphate group in the NADP(H) molecule.

Substrate Binding Site: We examined the **substrate** binding site of *Cgl*QDH by analysis of the two different **ternary complexes** QDH-QA-NADH and QDH-SA-NADH. The substrate binding site is located in the N-terminal domain, close to the nicotinamide ring of the **cofactor**, and is characterized by a number of highly conserved residues.

After quinate binding a slight closure of the N- and C-terminal domain of *Cgl*QDH so that the crevice becomes closer by about 0.5 Å was observed. The substrate quinate is anchored by numerous key interactions with these residues: the carbonyl group of quinate is bound by the hydroxyl groups of Ser17 and Thr19; the hydroxyl groups of the C1 and C3 atom of the substrate form hydrogen bonds to side chain of Thr69, whereas the nitrogen atom of Lys73 binds to the hydroxyl groups of C3 and C4, the latter furthermore interacts with the side chains of Asn94 and Asp110; the fourth hydroxyl group at C5 forms hydrogen bonds to the amide group of Asp110 and the oxygen atom of the Gln258 side chain, respectively. A total of eleven hydrogen bonds cause a forcipate anchorage of the substrate molecule (Figure 17 B).

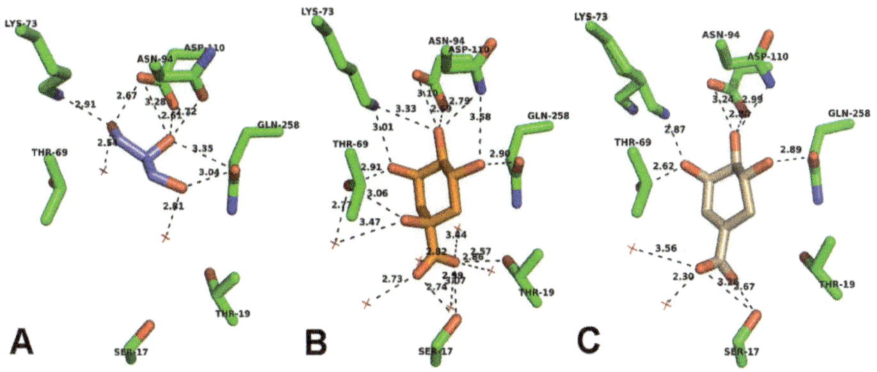

Figure 17. Active site residues of *Cgl*QDH. A) Apo-*Cgl*QDH (PDB entry: 2NLO) with bound glycerol (purple), B) ternary complex (PDB entry: 3JYP) with bound quinate (orange), C) ternary complex (PDB entry: 3JYQ) with shikimate (wheat). Residues involved in hydrogen bonds (dotted lines) and bound ligand are shown as sticks, water molecules are depicted as red stars.

In comparison to the **apo-enzyme** QDH (Figure 17 A) it is noteworthy that the side chain of Lys73 exhibits a sprawled conformation after quinate binding, which is required for interaction with the C3 and C4 hydroxyl groups of the substrate. For the hydride ion transfer from C3 of quinate to C4 of NAD^+ a particular distance between these atoms is very important. In the crystal structure the nicotinamide ring is located in a suitable orientation for the H^- transfer. After quinate binding and resulting closure of the domains the **cofactor** approaches to the substrate-binding site, whereby the distance of interest amounts to 4.27 Å.

In the case of shikimate binding a somewhat different situation was observed. In principle the above mentioned residues except Thr19 are involved in shikimate binding (Figure 17 C), but only eight polar interactions are achieved (compared to eleven when QA is bound), from which some are furthermore weaker pronounced: Thr19 is not involved in polar contacts to SA, Thr69 has contact only to the hydrogen group of C3, Asn94 is about 0.2 Å farer apart from the hydrogen atom of C4 and has no contact to the OH-group of C5. Remarkable is the appearance of an alternative side chain conformation of Lys73, as evidenced by the excellent electron density in this region. The first conformation of the Lys73 side chain in the crystal exhibits the sprawled conformation as found for the quinate binding; the second conformation reveals an angled rotamer as it occurs in apo-*Cgl*QDH. The latter conformation makes hydrophobic interactions with the shikimate molecule impossible (Hoeppner et al.; publication in progress). Furthermore the shikimate molecule exists in a half-chair conformation, whereas the quinate molecule adopts a chair conformation. Hence the distance of the C4 atom of the cofactor and the C3 atom of the substrate increases to 4.67 Å. All residues involved in cofactor and substrate binding identified here are consistent with these of further reported structures (i. e. *Tth*SDH, *Aae*SDH, *AthDHQ*-SDH).

4.6.3. Substrate and cofactor specificity and discrimination

All results of the structural analysis are also in excellent agreement with the findings of the kinetical assays. The higher **affinity** of *Cgl*QDH to the **substrate** quinate (as obvious by means of the clearly unequal K_M values) arises from the major quantity of hydrogen bonds between protein and the substrate quinate. Considering the known structures of shikimate dehydrogenases as SDH from *T. thermophilus* [40] SDH from *A. aeolicus* [41] or the SDH domain of *A. thaliana* [42, 43] they all possess at least eleven hydrogen bonds to the substrate molecule shikimate, comparable to the quinate binding in *Cgl*QDH. In contrast there are only eight polar interactions present between the enzyme and the shikimate molecule, because shikimate offers a somewhat different conformation (half-chair instead of chair) and exhibits no hydroxyl group at the C1 atom. Furthermore the formed hydrogen bonds between shikimate and the enzyme are accomplished weaker. The higher catalytic efficiency of *Cgl*QDH regarding to quinate (as obvious on the basis of significantly higher k_{cat}/K_M values) possibly results from the slightly lower distance between the C4 atom of the **cofactor** NAD(H) and the C3 atom of the quinate molecule (4.27 Å versus 4.67 Å) and the improved orientation of the substrate quinate.

A further occasion for the lower affinity and catalytic efficiency regarding shikimate results from the appearance of an alternative conformation of the Lys73 side chain (Figure 17 C), which leads to a loss of an important hydrogen bond. At last we compared the substrate binding residues of *Cgl*QDH with those of *Aae*QDH, *AthDHQ*-SDH and *Tth*SDH, which convert shikimate. We detected two differences possibly jointly responsible for quinate binding: in all the above-mentioned structures a tyrosine residue is involved in shikimate binding but not in *Cgl*QDH (Tyr230). Furthermore the second serine, which forms a hydrogen bond to the carbonyl group of shikimate, is replaced by a threonine residue in *Cgl*QDH (Thr19). Since the carbonyl group of shikimate bound in *Cgl*QDH is twisted about 90° compared to the situation of the aforesaid enzymes, Thr19 cannot take part in polar interaction with the substrate shikimate. Concerning the usage of the cofactors NAD(H) and NADP(H) the classical dinucleotid fold were identified in the past. Characteristic for all nucleotide binding proteins is the glycine rich loop with the common sequence GXGXXG, in which the number of glycine residues changes [49]. Enzymes using NAD or FAD possess a well conserved negatively charged amino acid at the C-terminus of the second β-strand of the nucleotide binding βαβ unit, mostly aspartate or glutamate. This residue interacts with the 2′-hydroxyl group of the ribose. In the majority of the NADP binding proteins this negatively charged residue is absent since the additional 2′-phosphate group is located at this position. Moreover, various NADP dependent proteins exhibit a charged amino acid (like arginine) in the position of the 2′-phosphate group which stabilizes the cofactor molecule. The *Cgl*QDH described here offers a negative charged residue (Asp158) forming a hydrogen bond to 2′-hydroxyl group of the ribose, followed by a neutral amino acid (Leu159) which is unable to interact with an additional phosphate group as present in NADP. Furthermore the bulk side chain of Arg163 constrict the cofactor binding site, which would result in steric hindrance with the additional phosphate group in a NADP(H) molecule. Due to all results of the kinetical and structural analysis we conclude that *Cgl*QDH is strictly NAD(H) specific and not able to bound NADP(H).

5. Conclusions

Crystal structures of proteins and enzymes are important to fully understand the mechanism and mode of action. Although the crystallization of a large number of proteins was successful and delivered valuable information the goal must be to fully understand the function. When crystallizing a protein a snapshot of the protein in a certain conformation is observed in the electron density. It is known that proteins are flexible and can obtain several states in solution.

Within that book chapter we explained the importance to acquire structural information of different catalytical states of proteins or enzymes, to fully understand how the protein behaves during catalysis or how the substrate bound state differs from the apo-enzyme.

The open and closed structures of the substrate binding protein ProX as apo-protein or with different substrates bound revealed enormous conformational changes during ligand binding and clearly visualzes how flexible a protein can be and elucidates the side chain movements within the substrate site upon ligand binding.

All described crystallization trials of the different transition states of the OcDH showed impressively that protein crystallization is a trial and error approach and that knowledge of the protein (especially the kinetical parameters beside others) is the essential thing to be successful. At best and as recompenses for ones effort one will achieve important insights that clearly explains the catalytic mechanism.

Last but not least the different structural information of the enzymes of the shikimate dehydrogenase family could bring to light how substrate and cofactor specificity and discrimination can be achieved throught detailed analysis of apo-, binary and ternary structure information about involved amino acids in substrate and cofactor binding.

So with these three examples the difficulties in crystallization on one hand and on the other hand the beauty of looking at proteins at work is shown.

PDB entries used

Protein	PDB Code	Title
ChoX	2RF1	Crystal structure of ChoX in an unliganded closed conformation
	3HCQ	Structural analysis of the choline binding protein ChoX in a semi-closed and ligand-free conformation
	2REJ	ABC-transporter choline binding protein in unliganded semi-closed conformation
	2RIN	ABC-transporter choline binding protein in complex with acetylcholine
	2REG	ABC-transporter choline binding protein in complex with choline

Protein	PDB Code	Title
ProX	1SW1	Crystal structure of ProX from *Archeoglobus fulgidus* in complex with proline betaine
	1SW4	Crystal structure of ProX from *Archeoglobus fulgidus* in complex with trimethyl ammonium
	1SW2	Crystal structure of ProX from *Archeoglobus fulgidus* in complex with glycine betaine
	3MAM	A molecular switch changes the low to the high affinity state in the substrate binding protein *Af*ProX
	1SW5	Crystal structure of ProX from *Archeoglobus fulgidus* in the ligand free form
CenDH	1BG6	Crystal structure of the N-(1-D-carboxylethyl)-L-norvaline dehydrogenase from *Arthrobacter sp.* strain 1C
OcDH	3C7C	A structural basis for substrate and stereo selectivity in octopine dehydrogenase (OcDH-NADH-L-arginine)
	3C7D	A structural basis for substrate and stereo selectivity in octopine dehydrogenase (OcDH-NADH-pyruvate)
	3C7A	A structural basis for substrate and stereo selectivity in octopine dehydrogenase (OcDH-NADH)
AroE	2HK8	Crystal structure of shikimate dehydrogenase from *Aquifex aeolicus*
	2HK9	Crystal structure of shikimate dehydrogenase from *Aquifex aeolicus* in complex with shikimate and $NADP^+$
	1WXD	Crystal structure of shikimate 5-dehydrogenase (AroE) from *Thermus thermophilus* HB8
	2D5C	Crystal structure of shikimate 5-dehydrogenase (AroE) from *Thermus thermophilus* HB8 in complex with shikimate
	2EV9	Crystal structure of shikimate 5-dehydrogenase (AroE) from *Thermus thermophilus* HB8 in complex with NADP(H) and shikimate
	3PHG	Crystal structure of the shikimate 5-dehydrogenase (AroE) from *Helicobacter pylori*
	3PHH	Crystal structure of the shikimate 5-dehydrogenase (AroE) from *Helicobacter pylori* in complex with shikimate and NADP(H)
	3PHI	Crystal structure of thesShikimate 5-dehydrogenase (AroE) from *Helicobacter pylori* in complex with dehydroshikimate
DHQ-SDH	2GPT	Crystal structure of *Arabidopsis* dehydroquinate dehydratase-shikimate dehydrogenase in complex with tartrate and shikimate
	2O7Q	Crystal structure of the *A. thaliana* DHQ-dehydroshikimate-SDH in complex with dehydroshikimate
	2O7S	Crystal structure of the *A. thaliana* DHQ-dehydroshikimate-SDH in complex with dehydroshikimate, NADP(H) and tartrate
QDH	3JYO	Quinate dehydrogenase from *Corynebacterium glutamicum* in complex with NAD
	3JYP	Quinate dehydrogenase from *Corynebacterium glutamicum* in complex with quinate and NADH
	3JYQ	Quinate dehydrogenase from *Corynebacterium glutamicum* in complex with shikimate and NADH
	2NLO	Crystal structure of the quinate dehydrogenase from *Corynebacterium glutamicum*

Glossary

Apo-protein/apo-enzyme
Enzymes that require a cofactor but do not have one bound are called *apo-enzymes* or *apo-proteins*. An apo-enzyme together with its cofactor(s) is called a *holoenzyme*.
Affinity
The dissociation constant is commonly used to describe the **affinity** between a ligand and a protein, i.e. how tightly a ligand binds to a particular protein. Ligand-protein affinities are influenced by non-covalent intermolecular interactions between the two molecules such as hydrogen bonding, electrostatic interactions, hydrophobic and Van der Waals forces. They can also be affected by high concentrations of other macromolecules, which causes macromolecular crowding. The smaller the dissociation constant K_d, the more tightly bound the ligand is, or the higher the affinity between ligand and protein.
Binary complex
A binary complex refers to a protein complex containing two different molecules which are bound together. In structural biology, the term binary complex can be used to describe a crystal containing a protein with one small molecule bound, for example the cofactor or the substrate; or a complex formed between two proteins.
Co-crystallization
Co-crystallization means that the protein solution is mixed with one or more ligand prior to the crystallization. Often the protein-ligand mixture is preincubated before setting up the crystallization drops.
Cofactor
A cofactor is a non-protein chemical compound that is bound to a protein and is required for the protein's biological activity. These proteins are commonly enzymes, and cofactors can be considered "helper molecules" that assist in biochemical transformations. Cofactors are either organic or inorganic. They can also be classified depending on how tightly they bind to an enzyme, with loosely bound cofactors termed coenzymes and tightly-bound cofactors termed prosthetic groups. Examples of widespread cofactors are ATP, coenzyme A, FAD, and NAD^+, vitamins or metal ions.
K_d
In chemistry, biochemistry, and pharmacology, a dissociation constant K_d is a specific type of equilibrium constant that measures the propensity of a larger object to separate (dissociate) reversibly into smaller components, as when a complex falls apart into its component molecules, or when a salt splits up into its component ions.
K_M
In biochemistry, Michaelis-Menten kinetics is one of the simplest and best-known models of enzyme kinetics. The Michaelis constant K_M is the substrate concentration at which the reaction rate is half of V_{max} (which represents the maximum rate achieved by the system, at maximum (saturating) substrate concentrations).
Intrinsic tryptophan fluorescence
Binding of ligands to proteins frequently causes changes to their three-dimensional structure. Exampes of this include the binding of substrates, inhibitors, cofactors or allosteric modulators to enzymes or of hormons to receptors. If this structural change has an effect on the environment of an intrinsic or extrinsic fluorophore in the protein, this can result in measurable changes in the fluorescence spectrum. Provided that the fluorophore has a unique location in the protein, such changes of fluorescence at a particular wavelength can be used to determine the dissociation constant (K_d) of the protein for the ligand where K_d is a measure of the affinity of the protein for the ligand [50].
Isothermal Titration Calorimetry (ITC)
This technique is useful for protein concentrations in the range of mg/ml. A typical experiment involves measurement of heat change as a function of addition of small quantities of a reagent to the calorimeter cell containing other components of the system under investigation. For example, this reagent could be a protein ligand or substrate/ inhibitor of an enzyme. At the beginning of the experiment, there is a large excess of protein compared to ligand. This means that ΔH values associated with

each aliquot can be individually measured. Initially, these values are large but, as aliquots are progressively added, eventually decrease to values similar to the ΔH of dilution of ligand into the solution in the calorimeter cell. The ΔH measured is the total enthalpy change which includes heat associated with processes such as formation of noncovalent bonds between interacting molecules and with other equilibria in the system such as conformational changes, ionization of polar groups (e.g. deprotonation) and changes due to interactions with solvent. ITC provides a useful method for studying binding processes such as those involving a protein and a ligand. It allows estimation of both the binding constant (K_b) and of the dissociation constant (K_d) [50].

Ligand

In biochemistry a ligand is a substance (usually a small molecule), that forms a complex with a biomolecule to serve a biological purpose. In the context of this chapter ligand is used as a more general expression for substrate, product or cofactor.

Ligand soaking

Ligand soaking means the addition of ligands into the mother liquid with preformed crystals. The idea is that the ligand diffuses into the crystals and binds at the active site. This technique was initially used for the incorporation of heavy atoms into protein crystals for phasing purposes.

Macro and Micro Seeding

During Macro Seeding the protein crystal is replaced into a freshly made mother liquid which allows the further enlargement of the crystals size. In Micro Seeding a suspension of microcrystals is prepared by either resuspending or crushing a protein crystal cluster or single crystals. These seeds are then used (often streaked through a new droplet of precipitant and fresh protein) to serve a crystallization starting point.

Microbatch

Microbatch is a method in which the molecule to be crystallized is mixed with the crystallizing agents at the start of the experiment. The concentration of the ingredients is such that supersaturation is achieved immediately upon mixing, thus the composition and the volume of a trial remain constant and crystals will only form if the precise conditions have been correctly chosen.

Occupancy

Occupancy means the degree of protein molecules in solution or in a crystal with bound ligand. If every second protein has attached a ligand the occupancy is 50 %.

Substrate

In biochemistry, a substrate is a molecule upon which an enzyme acts. Enzymes catalyze chemical reactions involving the substrate(s). In the case of a single substrate, the substrate binds with the enzyme active site, and an enzyme-substrate complex is formed. The substrate is transformed into one or more products, which are then released from the active site. The active site is now free to accept another substrate molecule. In the case of more than one substrate, these may bind in a particular order to the active site, before reacting together to produce products.

Surface Plasmon Resonance (SPR)

SPR is an optical technique which depends on changes in refractive index or mass changes near metal surfaces. When two surfaces, one a metal and the other a dielectric material are exposed to a beam of plane-polarized light of wavelength, λ, a longitudinal charge density wave (a surface plasmon) is propagated along the interface between them. This only happens when one of the surfaces is a metal and works best with silver, gold, copper and aluminium. This is because metals contain free oscillating electrons called plasmons. When light traveling through an optically dense medium such as glass arrives at an interface with a lower optical density (e.g. liquid), it is reflected back into the more optically dense medium, a phenomenon called total internal reflectance. Any process altering n_s (the refractive index of the dielectric medium) can be sensitively detected by SPR so the technique has found applications in the study of kinetics and thermodynamics of binding processes (e.g. protein-ligand, protein-protein) [50].

Ternary complex
A ternary complex refers to a protein complex containing three different molecules which are bound together. In structural biology ternary complex can be used to describe a crystal containing a protein with two small molecules bound, for example cofactor and substrate; or a complex formed between two proteins and a single substrate.
Vapor diffusion Hanging or Sitting drop
Two of the most commonly used methods for protein crystallization fall under the category of vapor diffusion. These are known as the hanging drop and sitting drop methods. Both entail a droplet containing purified protein, buffer, and precipitant being allowed to equilibrate with a larger reservoir containing similar buffers and precipitants in higher concentrations. Initially, the droplet of protein solution contains an insufficient concentration of precipitant for crystallization, but as water vaporizes from the drop and transfers to the reservoir, the precipitant concentration increases to a level optimal for crystallization. Since the system is in equilibrium, these optimum conditions are maintained until the crystallization is complete [51].

Author details

Astrid Hoeppner[1], Lutz Schmitt[2] and Sander H.J. Smits[2]

*Address all correspondence to: sander.smits@hhu.de

1 X-Ray Facility and Crystal Farm, Heinrich Heine University, Duesseldorf, Germany

2 Institute of Biochemistry, Heinrich Heine University, Duesseldorf, Germany

References

[1] Abts A, Schwarz CK, Tschapek B, Smits SH, Schmitt L. Rational and Irrational Approaches to Convince a Protein to Crystallize, Modern Aspects of Bulk Crystal and Thin Film Preparation, 2012 Nikolai Kolesnikov and Elena Borisenko (Ed.), ISBN: 978-953-307-610-2, InTech

[2] Berntsson RP, Smits SH, Schmitt L, Slotboom DJ, Poolman B. A structural classification of substrate-binding proteins. FEBS Lett. 2010 Jun 18;584(12):2606-17.

[3] Wilkinson J, Verschueren KHG. Crystal structures of periplasmic solute-binding proteins in ABC transport complexes illuminate their function. In: Holland IB, Cole SPC, Kuchler K, Higgins CF, editors. ABC proteins: from bacteria to man. London: Academic Press (Elsevier Science); 2003. p. 187-208.

[4] Quiocho FA, Ledvina PS. Atomic structure and specificity of bacterial periplasmic receptors for active transport and chemotaxis: variation of common themes. Mol Microbiol. 1996 Apr;20(1):17-25.

[5] Mao B, Pear MR, McCammon JA, Quiocho FA. Hinge-bending in L-arabinose-binding protein. The "Venus's-flytrap" model. J Biol Chem. 1982 Feb 10;257(3):1131-3.

[6] Sack JS, Saper MA, Quiocho FA. Periplasmic binding protein structure and function. J Mol Biol. 1989;206:171-91.

[7] Oh BH, Pandit J, Kang CH, Nikaido K, Gokcen S, Ames GF, et al. Three-dimensional structures of the periplasmic lysine/arginine/ornithine-binding protein with and without a ligand. J Biol Chem. 1993 May 25;268(15):11348-55.

[8] Loh AP, Pawley N, Nicholson LK, Oswald RE. An increase in side chain entropy facilitates effector binding: NMR characterization of the side chain methyl group dynamics in Cdc42Hs. Biochemistry. 2001;40(15):4590-600.

[9] Davidson AL, Dassa E, Orelle C, Chen J. Structure, function, and evolution of bacterial ATP-binding cassette systems. Microbiol Mol Biol Rev. 2008 Jun;72(2):317-64, table of contents.

[10] Shilton BH. The dynamics of the MBP-MalFGK(2) interaction: a prototype for binding protein dependent ABC-transporter systems. Biochim Biophys Acta. 2008 Sep; 1778(9):1772-80.

[11] Tang C, Schwieters CD, Clore GM. Open-to-closed transition in apo maltose-binding protein observed by paramagnetic NMR. Nature. 2007 Oct 25;449(7165):1078-82.

[12] Sharff AJ, Rodseth LE, Quiocho FA. Refined 1.8-A structure reveals the mode of binding of beta-cyclodextrin to the maltodextrin binding protein. Biochemistry. 1993 Oct 12;32(40):10553-9.

[13] Spurlino JC, Lu GY, Quiocho FA. The 2.3-A resolution structure of the maltose- or maltodextrin-binding protein, a primary receptor of bacterial active transport and chemotaxis. J Biol Chem. 1991 Mar 15;266(8):5202-19.

[14] Oswald C, Smits SH, Hoing M, Bremer E, Schmitt L. Structural analysis of the choline-binding protein ChoX in a semi-closed and ligand-free conformation. Biol Chem. 2009 Nov;390(11):1163-70.

[15] Oswald C, Smits SH, Hoing M, Sohn-Bosser L, Dupont L, Le Rudulier D, et al. Crystal structures of the choline/acetylcholine substrate-binding protein ChoX from *Sinorhizobium meliloti* in the liganded and unliganded-closed states. J Biol Chem. 2008 Nov 21;283(47):32848-59.

[16] Pittelkow M, Tschapek B, Smits SH, Schmitt L, Bremer E. The Crystal Structure of the Substrate-Binding Protein OpuBC from *Bacillus subtilis* in Complex with Choline. J Mol Biol. 2011 Aug 5;411(1):53-67.

[17] Bermejo GA, Strub MP, Ho C, Tjandra N. Ligand-free open-closed transitions of periplasmic binding proteins: the case of glutamine-binding protein. Biochemistry. 2010 Mar 9;49(9):1893-902.

[18] Borths EL, Locher KP, Lee AT, Rees DC. The structure of Escherichia coli BtuF and binding to its cognate ATP binding cassette transporter. Proc Natl Acad Sci U S A. 2002 Dec 24;99(26):16642-7.

[19] Linke C, Caradoc-Davies TT, Young PG, Proft T, Baker EN. The laminin-binding protein Lbp from *Streptococcus pyogenes* is a zinc receptor. J Bacteriol. 2009 Sep;191(18): 5814-23.

[20] Zou JY, Flocco MM, Mowbray SL. The 1.7 A refined X-ray structure of the periplasmic glucose/galactose receptor from *Salmonella typhimurium*. J Mol Biol. 1993 Oct 20;233(4):739-52.

[21] Quiocho FA, Spurlino JC, Rodseth LE. Extensive features of tight oligosaccharide binding revealed in high-resolution structures of the maltodextrin transport/chemosensory receptor. Structure. 1997 Aug 15;5(8):997-1015.

[22] Berntsson RP, Doeven MK, Fusetti F, Duurkens RH, Sengupta D, Marrink SJ, et al. The structural basis for peptide selection by the transport receptor OppA. EMBO J. 2009 May 6;28(9):1332-40.

[23] Schiefner A, Holtmann G, Diederichs K, Welte W, Bremer E. Structural basis for the binding of compatible solutes by ProX from the hyperthermophilic archaeon *Archaeoglobus fulgidus*. J Biol Chem. 2004 Nov 12;279(46):48270-81.

[24] Machius M, Brautigam CA, Tomchick DR, Ward P, Otwinowski Z, Blevins JS, et al. Structural and biochemical basis for polyamine binding to the Tp0655 lipoprotein of Treponema pallidum: putative role for Tp0655 (TpPotD) as a polyamine receptor. J Mol Biol. 2007 Oct 26;373(3):681-94.

[25] Muller A, Severi E, Mulligan C, Watts AG, Kelly DJ, Wilson KS, et al. Conservation of structure and mechanism in primary and secondary transporters exemplified by SiaP, a sialic acid binding virulence factor from *Haemophilus influenzae*. J Biol Chem. 2006 Aug 4;281(31):22212-22.

[26] Lecher J, Pittelkow M, Zobel S, Bursy J, Bonig T, Smits SH, et al. The crystal structure of UehA in complex with ectoine-A comparison with other TRAP-T binding proteins. J Mol Biol. 2009 May 29;389(1):58-73.

[27] Hanekop N, Hoing M, Sohn-Bosser L, Jebbar M, Schmitt L, Bremer E. Crystal structure of the ligand-binding protein EhuB from Sinorhizobium meliloti reveals substrate recognition of the compatible solutes ectoine and hydroxyectoine. J Mol Biol. 2007 Dec 14;374(5):1237-50.

[28] Oswald C, Smits SH, Bremer E, Schmitt L. Microseeding - a powerful tool for crystallizing proteins complexed with hydrolyzable substrates. Int J Mol Sci. 2008 Jun;9(7): 1131-41.

[29] Tschapek B, Pittelkow M, Sohn-Bosser L, Holtmann G, Smits SH, Gohlke H, et al. Arg149 Is Involved in Switching the Low Affinity, Open State of the Binding Protein AfProX into Its High Affinity, Closed State. J Mol Biol. 2011 Aug 5;411(1):36-52.

[30] Zaitseva J, Oswald C, Jumpertz T, Jenewein S, Wiedenmann A, Holland IB, et al. A structural analysis of asymmetry required for catalytic activity of an ABC-ATPase domain dimer. EMBO J. 2006 Jul 26;25(14):3432-43.

[31] Karpowich NK, Huang HH, Smith PC, Hunt JF. Crystal structures of the BtuF periplasmic-binding protein for vitamin B12 suggest a functionally important reduction in protein mobility upon ligand binding. J Biol Chem. 2003 Mar 7;278(10):8429-34.

[32] Chen J, Lu G, Lin J, Davidson AL, Quiocho FA. A tweezers-like motion of the ATP-binding cassette dimer in an ABC transport cycle. Mol Cell. 2003 Sep;12(3):651-61.

[33] Muller A, Janssen F, Grieshaber MK. Putative reaction mechanism of heterologously expressed octopine dehydrogenase from the great scallop, *Pecten maximus* (L). Febs J. 2007 Dec 7;274(24):6329-39.

[34] Smits SH, Mueller A, Grieshaber MK, Schmitt L. Coenzyme- and His-tag-induced crystallization of octopine dehydrogenase. Acta Crystallogr Sect F Struct Biol Cryst Commun. 2008 Sep 1;64(Pt 9):836-9.

[35] Smits SH, Mueller A, Schmitt L, Grieshaber MK. A structural basis for substrate selectivity and stereoselectivity in octopine dehydrogenase from *Pecten maximus.* J Mol Biol. 2008 Aug 1;381(1):200-11.

[36] Britton KL, Asano Y, Rice DW. Crystal structure and active site location of N-(1-D-carboxylethyl)-L-norvaline dehydrogenase. Nat Struct Biol. 1998 Jul;5(7):593-601.

[37] Asano Y, Yamaguchi K, Kondo K. A new NAD+-dependent opine dehydrogenase from Arthrobacter sp. strain 1C. J Bacteriol. 1989 Aug;171(8):4466-71.

[38] van Os N, Smits SH, Schmitt L, Grieshaber MK. Control of D-octopine formation in scallop adductor muscle as revealed through thermodynamic studies of octopine dehydrogenase. J Exp Biol. 2012 May 1;215(Pt 9):1515-22.

[39] Smits SH, Meyer T, Mueller A, van Os N, Stoldt M, Willbold D, et al. Insights into the mechanism of ligand binding to octopine dehydrogenase from *Pecten maximus* by NMR and crystallography. PLoS One. 2010;5(8):e12312.

[40] Bagautdinov B, Kunishima N. Crystal structures of shikimate dehydrogenase AroE from *Thermus thermophilus* HB8 and its cofactor and substrate complexes: insights into the enzymatic mechanism. J Mol Biol. 2007 Oct 19;373(2):424-38.

[41] Gan J, Wu Y, Prabakaran P, Gu Y, Li Y, Andrykovitch M, et al. Structural and biochemical analyses of shikimate dehydrogenase AroE from *Aquifex aeolicus*: implications for the catalytic mechanism. Biochemistry. 2007 Aug 21;46(33):9513-22.

[42] Singh SA, Christendat D. Structure of Arabidopsis dehydroquinate dehydratase-shikimate dehydrogenase and implications for metabolic channeling in the shikimate pathway. Biochemistry. 2006 Jun 27;45(25):7787-96.

[43] Singh SA, Christendat D. The DHQ-dehydroshikimate-SDH-shikimate-NADP(H) Complex: Insights into Metabolite Transfer in the Shikimate Pathway. Cryst Growth Des. 2007;7(11):2153-60.

[44] Schoepe J, Niefind K, Chatterjee S, Schomburg D. Cloning, expression, purification and preliminary crystallographic characterization of a shikimate dehydrogenase from *Corynebacterium glutamicum*. Acta Crystallogr Sect F Struct Biol Cryst Commun. 2006 Jul 1;62(Pt 7):635-7.

[45] Yaniv H, Gilvarg C. Aromatic biosynthesis. XIV. 5-Dehydroshikimic reductase. J Biol Chem. 1955 Apr;213(2):787-95.

[46] Schoepe J, Niefind K, Schomburg D. 1.6 angstroms structure of an NAD+-dependent quinate dehydrogenase from *Corynebacterium glutamicum*. Acta Crystallogr D Biol Crystallogr. 2008 Jul;D64(Pt 7):803-9.

[47] Singh S, Korolev S, Koroleva O, Zarembinski T, Collart F, Joachimiak A, et al. Crystal structure of a novel shikimate dehydrogenase from *Haemophilus influenzae*. J Biol Chem. 2005 Apr 29;280(17):17101-8.

[48] Michel G, Roszak AW, Sauve V, Maclean J, Matte A, Coggins JR, et al. Structures of shikimate dehydrogenase AroE and its Paralog YdiB. A common structural framework for different activities. J Biol Chem. 2003 May 23;278(21):19463-72.

[49] Wierenga RK, De Maeyer MCH, Hol WGJ. Interaction of Pyrophosphate Moieties with a-Helixes in Dinucleotide Binding Proteins. Biochemistry. 1985;24:1346-57.

[50] Sheenan D. Physical Biochemistry. Principles and Applications: Wiley; 2009.

[51] McRee DE. Practical Protein Crystallography. San Diego: Academic Press; 1993.

Chapter 4

Purification of Erythromycin by Antisolvent Crystallization or Azeotropic Evaporative Crystallization

Kui Chen, Li-Jun Ji and Yan-Yang Wu

Additional information is available at the end of the chapter

1. Introduction

Crystallization plays an important role in separation and purification of the antibiotics. And it is also an indispensable step in preparation of pharmaceuticals with biological activities and specific crystal form. As the last step in purification, crystallization determines the purity, crystal habit, granularity and its distribution as well as pharmacologic effect, biologic activity and product stability [1], which are actually dependent on specific mechanism for its processes and operational conditions. So it's necessary to study thermodynamics, kinetics and conditions of crystallization process, which helps increase the yield and reduce cost.

As a representative of macrolide antibiotics, erythromycin has been widely used since its introduction in 1952 [2]. As erythromycin derivatives, clarithomycin and azithromycin have exhibited remarkable improvement on stability in acid solutions and metabolism dynamics [3, 4]. A lot of researches have been done on new derivatives with features of combating drug resistance [5, 6]. In the meanwhile, high-purity erythromycin, as the raw material, is fundamental to produce its pharmaceutical derivatives.

Erythromycin is obtained from microbial fermentation in industry. Subsequent separation and purification involve multiple unit operations, such as extraction, absorption, chromatography and crystallization. Different process involves different combinations of unit operations [7].

Among them, solvent extraction accompanied with intermediate precipitation is widely used, in which butyl acetate is commonly adopted to extract erythromycin from the fermentation filtration. It is followed by reactive crystallization to form an intermediate prior to conversion into erythromycin alkaline and dissolving in acetone. Lastly, erythromycin is purified by antisolvent crystallization [8]. That is to say, both reactive crystallization and antisolvent crystallization have to be employed in this separation process.

In contrast, the technological process with membrane separation and resin absorption [9] is drawing more attention compared with the traditional solvent extraction in the above [10]. This process usually goes as follows: firstly, microfiltration is used to remove solid impurities from the fermentation broth, and the filtrate is purified by macroporous absorption resin, and then the adsorbed erythromycin is eluted with butyl acetate. Finally, either evaporative crystallization or reactive crystallization can be used to obtain the final product [11].

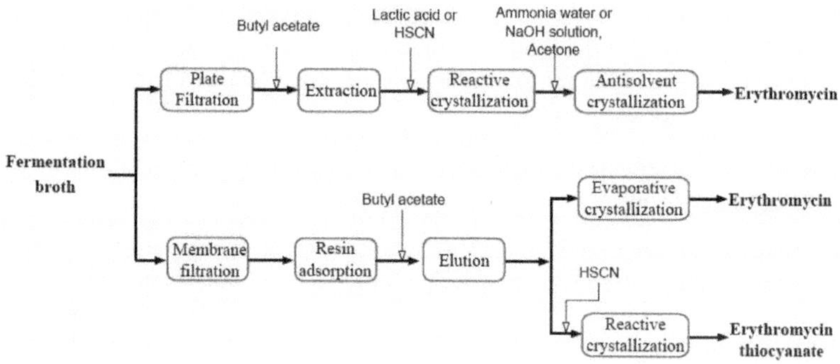

Figure 1. Schematic diagram for the purification erythromycin

The flowsheet of these two technological processes is demonstrated in Figure 1. It can be seen that crystallization is the final step to prepare erythromycin no matter which one is adopted. Different crystallization method has been used for different pretreatment.

Crystallization is a complex process involving mass transfer, heat transfer and surface reactions, which includes the formation of a supersaturated solution, nucleation and crystal growth. The operating parameters of crystallization process, such as temperature, agitation intensity and seed crystals, can affect the generation rate and scale of the supersaturation. The structure of the crystallizers and stirrer will influence the fluid mechanics properties of the crystallization system. Thus it can be seen that all these factors profoundly influence crystal nucleation and growth [12]. Over a long period of time, the crystallization processes have been carried out on according to experiences rather than theoretical researches due to the little study on thermodynamics and kinetics. Not surprisingly, it's hardly to obtain erythromycin with high purity, complete crystal form, narrow distribution of crystal size, less crystal bonding, which are very important for the stability and bioavailability of the drug.

In this paper, two crystallization processes of erythromycin in different systems, which include the antisolvent crystallization for mixed solvents of acetone and water and the azeotropic evaporative crystallization for butyl acetate-water system, are described in details. The thermodynamics and kinetics of the crystallization of erythromycin, which help to thoroughly understand the effect of a variety of factors on the nucleation, crystal growth and crystal habit, are summarized systematically. On the basis of these fundamental studies, effective control techniques are proposed to improve the quality of erythromycin product.

2. Purification of erythromycin by antisolvent crystallization

In the solvent extraction process for purification of erythromycin, erythromycin alkaline was converted from erythromycin thiocyanate by adding ammonia or NaOH solution and dissolving in acetone. Then erythromycin product was prepared by antisolvent crystallization, in which water was served as antisolvent.

The traditional crystallization process, which was too simple, only involved modulating two indicators including antisolvent quality and crystallization temperature.

Water was poured into erythromycin acetone solution at room temperature, and then the product was obtained by filtration after standing for a period of time. Such operation made obvious differences of supersaturation, nucleation rate and crystal growth rate and then caused the discrepancy in product quality for different batch.

2.1. Solubility and metastable limit of erythromycin

As we know, the phase equilibrium between solid and its solution is fundamental to choose crystallization method and also determines the maximum yield of solution crystallization [12]. Erythromycin is soluble in acetone, but insoluble in water [13]. Thus, erythromycin can be precipitated by adding water into erythromycin acetone solution.

2.1.1. Solubility

The solubility of erythromycin in acetone increases with the increasing temperature, whereas it becomes less soluble with the higher temperature in water. So, the solubility of erythromycin in acetone-water binary solvent system is influenced by the solvent composition and temperature.

Some data on solubility of erythromycin in acetone-water solution was reported in literatures [14,15]. In this paper, the solubility above 303.15K has been measured. As can be seen in Figure 2, the solubility of erythromycin in the medley acetone-water solution increased with increasing acetone concentration and increasing temperature, respectively. In the same range of acetone content, the slope of the solubility curve increased with increasing temperature, which meant the rate of increase of erythromycin solubility increased.

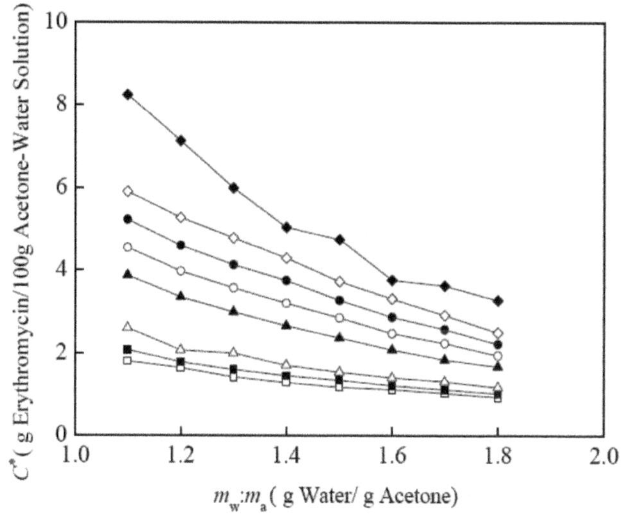

Figure 2. Solubility of erythromycin in acetone-water solution at different temperatures; -□-: 293.15K; -■-: 298.25K; -△-: 303.15K; -▲-: 308.15K; -○-: 310.15K; -●-: 312.15K; -◇-: 314.15K; -◆-: 323.15K

The impact of acetone on the solubility of erythromycin increased as the mass fraction of acetone increasing. It was not hard to infer that the difference of the solubility at different temperatures tended to decrease with the mass fraction of water increase.

An empirical model was proposed to relate the experimental data of the solubility of erythromycin and the parameters was obtained by fitting. The empirical equation for the solubility of erythromycin in mixed solvents of acetone and water was expressed as below:

$$C^* = \frac{1.02 \times e^{0.0491T}}{0.395 - 0.00383T + x^{(0.0138T+1.298)}} \tag{1}$$

where C^* was solubility (g Erythromycin/100g Acetone-Water Solution) and x was the mass proportion of water to acetone ($x=m_w:m_a$). Equation (1) was practicable in the range of $293.15\text{K} \leq T \leq 323.15\text{K}$, $1.0 \leq x \leq 2.0$.

Equation (1) could be used to calculate erythromycin solubility C^*_{Cal} under various experimental conditions. The calculated solubility C^*_{Cal} and the experimental solubility C^*_{Exp} were shown in Figure 3. It was indicated that Equation (1) was appropriate to predict the solubility of erythromycin within the range of temperature and acetone concentration for the equation.

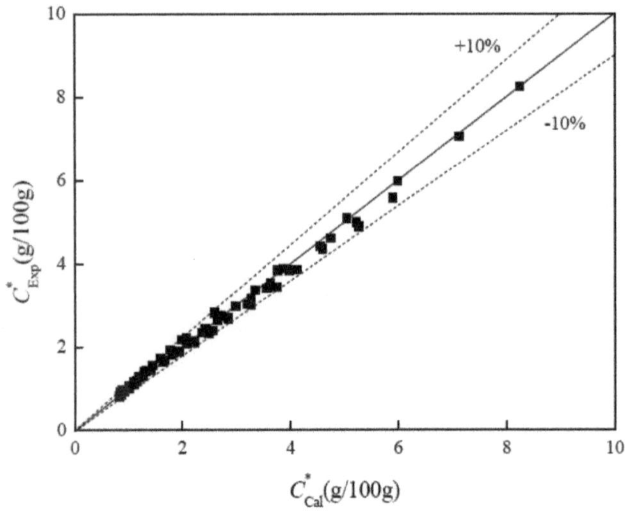

Figure 3. Comparison of simulated value and experimental data of solubility

2.1.2. Metastable zone

Metastable zone width is fundamental to choose suitable supersaturation of crystallization. It is also used as a restrictive operating condition to avoid crystallization system going to unstable zone [16] that results in the worse product.

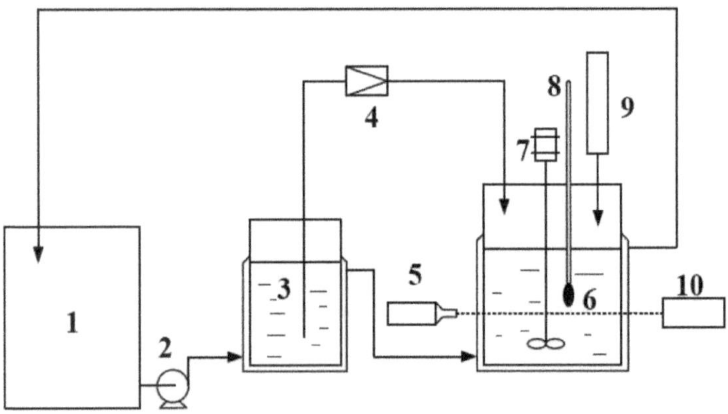

Figure 4. Apparatus for antisolvent crystallization of erythromycin; 1. Thermostat Bath; 2. Circulating Pump; 3. Water Storage Tank; 4. Peristaltic Pump; 5. Laser Generator; 6. Double-Wall Crystallizer; 7. Stirrer; 8. Thermometer; 9. Condenser; 10. Laser Power Detector

Supersolubility of erythromycin was measured by the method of laser scattering [15]. As shown in Figure 4, the experimental device consisted of crystallizer, mixing system, feeding system, temperature control system and detection system. Wherein, the crystallizer was a double-wall kettle with internal diameter 75mm and height 130mm. Stirrer with four inclined propellers was driven by variable speed motor, the propeller diameter was 12mm, and the stirring shaft diameter was 5mm. The peristaltic pump continuously pumped antisolvent water at a fixed temperature into crystallizer. The detection system consisted of He-Ne laser generator and laser power detector. He-Ne laser generator outputted 632.8nm red line, scattering and diffraction occurred when monochrome laser beam encountered with body of similar length of wavelength, the laser intensity received by detector was drastically reduced, thus the nucleation could be detected.

The relationship between metastable zone width ΔC of erythromycin and solvent composition at 323.15K was shown in Figure 5. It could be seen form the figure that metastable zone width decreased gradually with the increase of the quality of water in solution. In m_w:m_a range of 1.0 to1.8, the supersolubility presented apparent downward trend. After m_w:m_a reached 1.8, the change of the metastable zone width weakened, but the metastable zone width of this region was already narrow and was not suitable for crystallization operation.

The equation was obtained by correlating the metastable zone width and solvent composition, which was listed as follows:

$$\Delta C = 3.09 \times x^{-2.49} \tag{2}$$

where x was the mass ratio of water to acetone, $x = m_w : m_a$

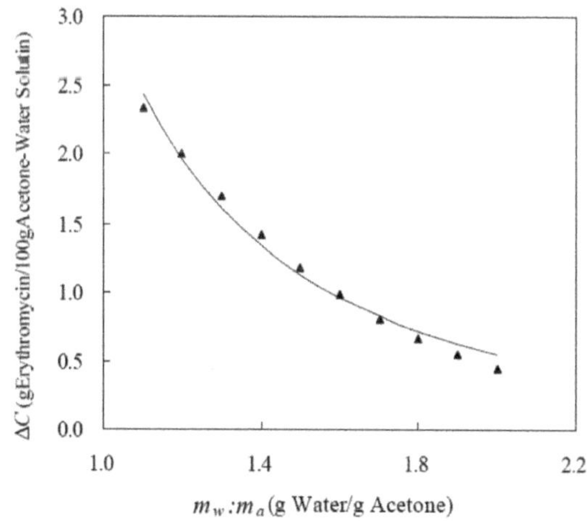

Figure 5. Effect of solvent composition on metastable zone width of erythromycin at 323.15K

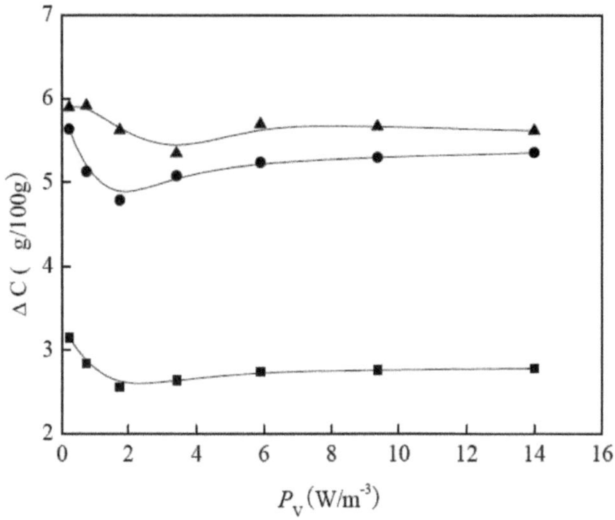

Figure 6. The effect of stirring intensity on supersaturation of erythromycin at different temperatures; –▲–: 308.15K; –●–: 313.15K; –■–: 323.15K

It could be seen form Figure 5, the calculated value was in good agreement with experimental data. Similar results could be obtained at other temperatures.

As shown in Figure 6, the metastable zone width of erythromycin decreased with the increase of temperature. The metastable zone width was similar at 308.15K and 313.15K, while it was quite different at 313.15K and 323.15K, which indicated metastable zone width was temperature sensitive in the range of 313.15K to 323.15K. The variation of metastable zone width with agitation power presented a consistency at different temperatures. The metasable zone width was wider under the same agitation power at lower temperature.

2.2. Antisolvent crystallization kinetics of erythromycin

In this paper, the intermittent dynamic method [17] was used to study the kinetics of erythromycin antisolvent crystallization under different conditions. With the empirical models deduced from the Larson-Randolph population balance equation [18,19], the model parameters were obtained from the experimental data through the matrix convertion and the method of linear squares regression. Thus, the equations of nucleation and crystal growth of antisolvent crystallization of erythromycin were established to help find the suitable operation parameters.

The experimental apparatus were shown in Figure 4. Firstly, at the start of recording the time, antisolvent water at set temperature was poured into the erythromycin-acetone solution in the crystallizer. Once the nucleation appeared in the solution, water was stopped importing and the time was recorded. Then the agitation rate and temperature were maintained constant. It was sampled at different interval of time. The indexes of each sample, such as magma density, degree of supersaturation and crystal size distribution (CSD) of production, were measured respectively.

2.2.1. Crystal size correlation of crystal growth

The crystal nucleation and growth kinetics were solved according to the size-independent model [16], using a set of the experimental data of magma density and CSD at 323.15K. The calculated value was in good agreement with the experimental data, as shown in Figure 7. In the crystal size (L_i) range, erythromycin crystal population density logarithm ($\ln n_i$) was basically a straight line. At the same time, the results of matrix convertion also showed that the use of size-dependent model to describe the crystal growth was of large error. Therefore, erythromycin crystal growth was size-independent.

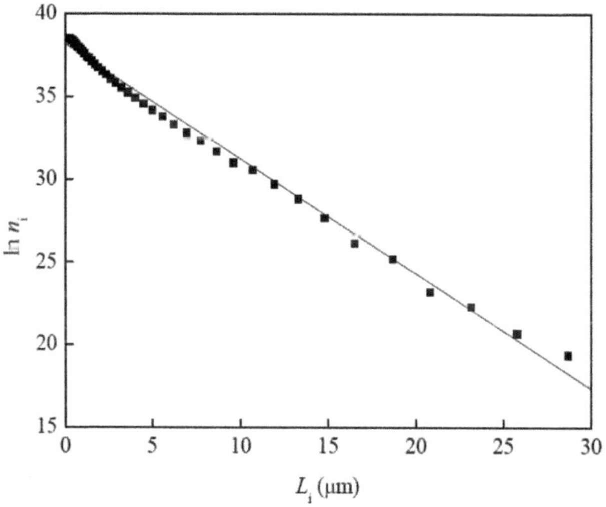

Figure 7. Typical population density distribution of erythromycin

2.2.2. *Kinetics model*

On the basis of the above, the size-independent model was adopted to describe the crystal growth rate of erythromycin. According to the study on the effects of temperature, agitation and dosing rate of antisolvent on nucleation rate and crystal growth rate, the corresponding equations for nucleation rate and crystal growth rate were shown as follows:

The nucleation equation

$$\Delta BG = 3.23\times 10^{14}\exp(\frac{2343}{T})P_V^{0.378}M_T{}^{0.317} \qquad {}^{3.303} \tag{3}$$

The crystal growth equation

$$\Delta GG = 1.18\times 10^{-5}\exp(\frac{4539}{RT})P_V^{0.102} \qquad {}^{3.053} \tag{4}$$

where M_T was magma density (kg/m^3), P_V was unit volume of stirring power (W/m^3).

In the antisolvent crystallization of erythromycin, slurry density had less effect on the nucleation rate than supersaturation did. The influence of stirring intensity and supersaturation on nucleation rate was greater than those on crystal growth rate. The supersaturation series 3.303 in the nucleation equation was much smaller than the primary nucleation kinetics series [12]. So the mechanism of nucleation of antisolvent crystallization of erythromycin was secondary nucleation.

2.2.3. *Online study of crystallization process*

In order to further reveal the intrinsic principles of the antisolvent crystallization process of erythromycin, the Focused Beam Reflectance Measurement (FBRM) technique was adopted to monitor in situ the variation of crystal quantity and crystal size distribution in this paper.

The total number and the chord length distribution (CLD) of crystal particles were measured by using the equipment and method shown in literature [20]. A mathematical procedure based on Monte Carlo simulation was established to transform the CLD into CSD.

The change of the number of crystals and CSD of erythromycin antisolvent crystallization were studied under different temperature and feeding rate of antisolvent. The results indicated that the faster water was fed, the earlier new crystals came into being, the faster the crystal grew at the initial stage. The total number of crystals at the stable stage tended to decrease as temperature increased [20].

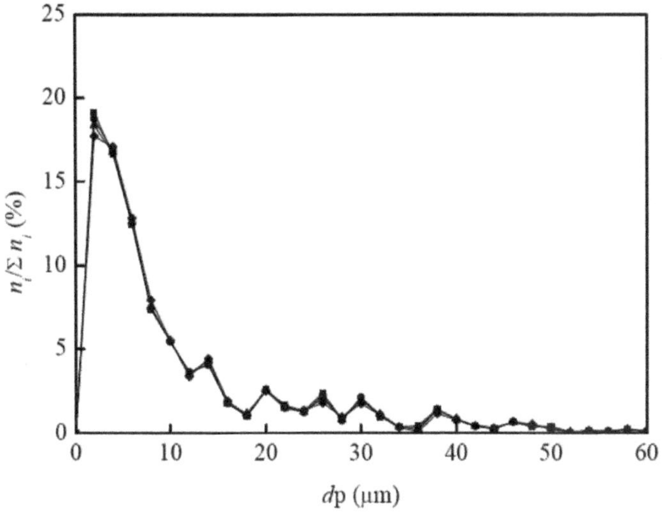

Figure 8. CSD after peak value of overall crystal number count at 308.15K; -■-: 0min; -●-: 30min; -▲-: 90min; -◆-: 180min

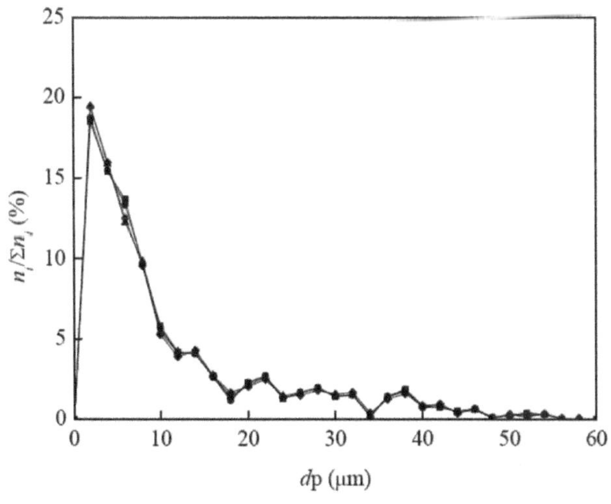

Figure 9. CSD after peak value of overall crystal number count at 314.15K; -■-: 0min; -●-: 30min; -▲-: 90min; -◆-: 180min

The proportion of particles of different size was the grain size frequency distribution. Figure 8 and Figure 9 showed the size frequency distribution curve after nucleation at 308.15K and 314.15K respectively. As could be seen from those, the curves were similar at different temperatures, which were sharp and steep. Particle size which was less than 20μm accounted for the vast majority and the peak of the curves was close to 20%.

It could be found from Figure 8 and Figure 9 that the number of both small size crystal and large size crystal hardly change with time. It meant that particles with small size were constantly dissolving, while saturated solute of erythromycin was precipitated to form new crystal, or the existing crystal grew larger in volume. The dissolution and precipitation of erythromycin reached equilibrium.

In order to properly characterize the crystal growth, volume mean diameter D_V (also known as D_{43}) which was the equivalent diameter of the particles with same volume (or mass), was used to investigate the changes of crystal size with time at different temperatures. As shown from Figure 10, erythromycin D_V monotonically decreased with the increasing temperature at the same crystallization time.

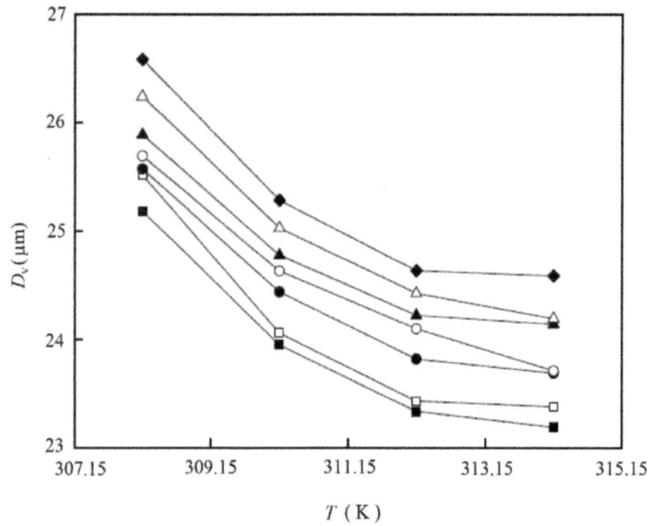

Figure 10. The effect of temperature on D_V of erythromycin crystal; -■-: 0min; -□-: 30min; -●-: 60min; -○-: 90min; -▲-: 120min; -△-: 150min; -◆-: 180min

2.3. The antisolvent crystallization technique of erythromycin

The thermodynamics and kinetics of the antisolvent crystallization of erythromycin were summarized systematically to understand thoroughly the effect of a variety of factors on the nucleation, crystal growth and crystal habit. On the basis of these fundamental studies, appropriate technological parameters were explored to develop the efficient industrialized crystallization process of erythromycin.

2.3.1. *Technological parameters*

Crystal quality, such as crystal purity, crystal habit, crystal size, and CSD, was related closely to the crystallization conditions. Accordingly, the effect of the dosing rate of antisolvent, crystallization time, stirring intensity and crystallization temperature on CSD of erythromycin was studied in details in this paper.

Dosing rate of antisolvent For antisolvent crystallization of erythromycin, the dosing rate of antisolvent determined the generation rate of supersaturation, and also affected the rate of nucleation and crystal growth.

The definition of dosing rate of antisolvent was the importing water volume of per unit time and per unit volume of erythromycin-acetone solution.

$$v_d = \frac{V_w}{V} \tag{5}$$

where v_d was dosing rate of antisolvent (min^{-1}), V_w was the volume rate of importing water (mL/min), V was the erythromycin-acetone volume (mL).

Figure 11 showed the relationship between the dosing rate of antisolvent v_d and erythromycin CSD, where d_p was the crystal diameter and R_v was the cumulative volume fraction. It could be seen from the figure that the proportion of crystals with large size increased with the increasing dosing rate of water, but the CSD tended to disperse. While the CSD of crystals obtained in lower water dosing rate was more concentrated.

Therefore, in process of the crystallization, an appropriate increase in generation rate of supersaturation could speed up the crystallization rate and improve the capability of the crystallizer. However, the rapid generation of crystals will increase the chance of crystal breakage and secondary nucleation and make the CSD disperse.

Crystallization time The cumulative volume distribution at different crystallization time was shown in Figure 12, where d_p was the crystal diameter and R_v was the cumulative volume fraction. As could be seen from the figure, the increase of the crystallization time was conducive to crystal growth, while the CSD did not tend to concentrate. The crystal growth needed some time, however, long time crystallization couldn't promise CSD being more consistent.

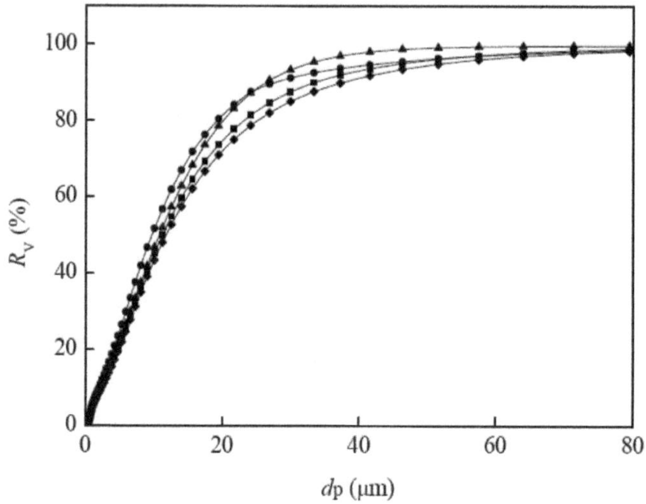

Figure 11. CSD based on cumulative volume of erythromycin at different water-pumping velocities; -▲-: 0.0138min^{-1}; -●-: 0.188 min^{-1}; -■-: 0.024 min^{-1}; -◆-: 0.0389 min^{-1}

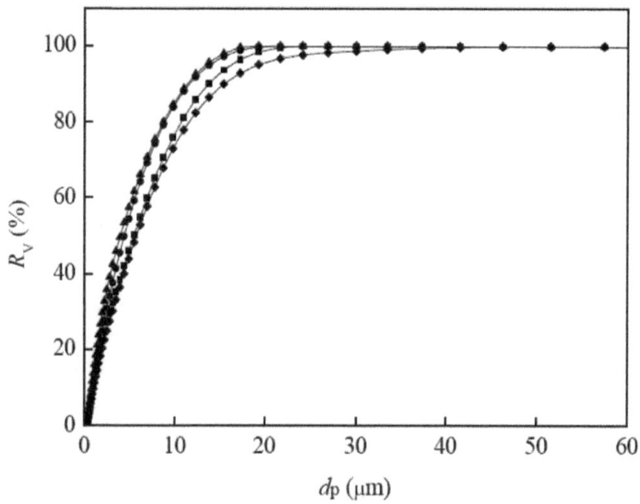

Figure 12. CSD of erythromycin based on cumulative volume at different crystallization time; -▲-: 40min; -●-: 50min; -■-: 70min: -◆-: 100min

Agitation power The CSD was the result of the interaction of primary nucleation, secondary nucleation, and crystal growth. Meanwhile, agitation power had a significant impact on all the above. Figure 13 showed the particle volume distribution of erythromycin at different stirring intensity, where x_v was the particle volume distribution.

It could be seen from Figure 13 that the erythromycin product had the widest CSD and the highest proportion of small size crystals when the stirring power was 13.99 W/m^3, and the distribution curve had smearing phenomenon in the range of large particle size. While the crystal had the narrowest CSD and the lowest proportion of small size crystals when the stirring power was 1.749 W/m^3, and the distribution curve had no smearing. The energy imported by stirring was conducive to nucleation and crystal growth. In the meanwhile, crystal breakage could easily occur with too strong stirring, while the obvious differences of supersaturation would occur with too weak stirring and then caused variation of rate of nucleation and crystal growth.

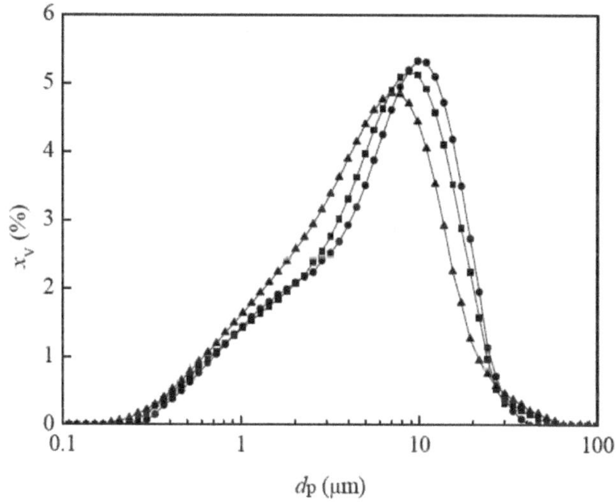

Figure 13. CSD based on volume of erythromycin at different agitation power; -■-: 0.02179W/m^3; -●-: 1.749 W/m^3; -▲-: 13.99 W/m^3

Crystallization temperature Figure 14 showed the variation of volume mean diameter (D_V) of erythromycin at different crystallization temperature, D_V decreased with the increasing of temperature. The previous thermodynamic study showed that the metastable zone width of erythromycin reduced with the increase of temperature. The intensified thermal motion of molecule caused by the increasing temperature accelerated the frequency of contact and collision of crystals, and then promoted the formation of tiny crystals, and decreased the supersaturation required for nucleation. On the other hand, the driving force of crystallization decreased with the narrowing metastable zone width, so did the rate of crystal growth. Therefore the volume mean diameter of the crystals decreased as the temperature increased.

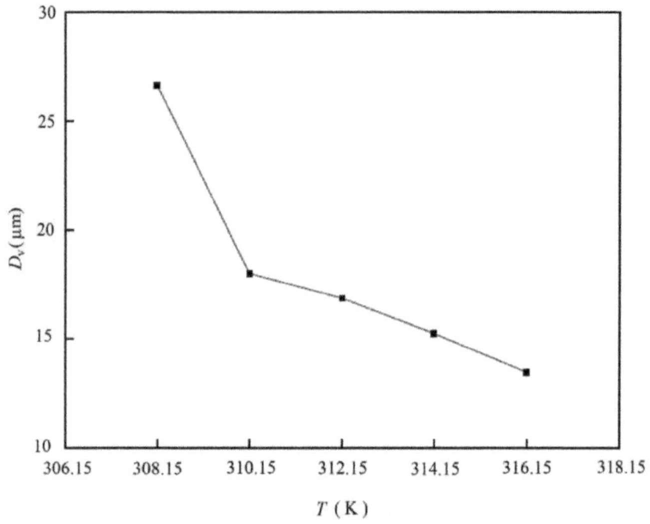

Figure 14. Volume mean diameter of erythromycin at different temperatures

2.3.2. *The novel technique of antisolvent crystallization of erythromycin*

For the traditional antisolvent crystallization, water was poured into erythromycin acetone solution at room temperature. Then after standing for a period of time, the erythromycin alkaline product was obtained by filtration.

It was not difficult to find the shortages of this crystallization method. Firstly, the dosing rate of antisolvent was too fast. When the antisolvent water was fed rapidly, the supersaturation formed suddenly and leaded to the outbreak of the nucleation. Nucleation was active and occupied the dominant position of the crystallization process. Meanwhile, the impurities easily accompanied with crystals by precipitation in the fast crystallization process. Secondly, stirrer and stirring intensity were inappropriate. Poor mixing effect made uneven distribution of supersaturation, so it was hard to obtain erythromycin with complete crystal form and narrow distribution of crystal size [12,21]. Thirdly, crystallization temperature was uncontrolled. Then the differences of solubility between erythromycin and impurities in acetone-water solution could not be fully explored to improve the separation efficiency.

The operation of the crystallization mentioned above lacked of crystallization process control and could not play a good role in purification of erythromycin by crystallization. Then the erythromycin product would be highly influenced by fermentation broth and pre-purification. That was to say, the quality of erythromycin was restricted by erythromycin thiocyanate. So it was hard to obtain the erythromycin product with stable and high quality and yield.

There were some other studies [8,22] on the improvement of erythromycin crystallization method by adding seed crystals.

On the basis of thorough research on the antisolvent crystallization process of erythromycin, a novel technique for antisolvent crystallization of erythromycin by dynamic control of temperature and stirring power was proposed in this paper, which was listed as follows.

1. Dosing the antisolvent. The polarity of mixed solvents was changed gradually when the antisolvent was imported into erythromycin acetone solution slowly. In the meantime, the solubility of erythromycin decreased gradually until crystal nucleus formed. The supersaturation could be controlled within the thermodynamic metastable zone by dosing antisolvent continuously, the crystal growth was moderated and in order, and the CSD of erythromycin tended to be narrow.

2. Appropriate stirring intensity. The suitable stirring power could be conducive to maintaining uniform supersaturation and crystallization rate. Meanwhile, stirring could promote dynamic balance of crystallization and dissolution, and reduce the crystal bonding, and then improve the purity of the crystal.

3. Increasing nucleation temperature. Substance usually had higher solubility at a higher temperature, so did the impurities. Increasing nucleation temperature could reduce the chance of impurities precipitation and improve the purity of erythromycin.

4. Cooling crystallization and aging with lower stirring intensity. After nucleation at high temperature, the stirring power should be reduced to avoid excessive shear force on the crystal collision and maintain a uniform concentration distribution in the slurry at the same time. Then, the supersaturation produced by cooling maintained crystal growth at a steady rate after dosing all antisolvent. Lastly, aging with lower stirring power at lower terminal temperature could improve the quality and yield of product.

On the basis of the above, the key operation parameters which affect the quality of crystal, such as temperature, dosing rate of antisolvent and stirring intensity, were determined by measuring the crystal shape, titer and yield [23]. Then the novel technique of erythromycin antisolvent crystallization was established in this paper, which was characteristic of dynamic control of temperature and stirring intensity [24].

Figure 15 and Figure 16 showed the crystal shape and CSD of industrial erythromycin products obtained by the traditional method (a) and novel technique (b), respectively. For the crystal shape, product (b) had a more regular and bigger size than product (a) did. For the CSD, product (b) was narrower. For titer, product (b) was 935.6 U/mg, while product (a) was 920 U/mg. Those meant that the quality of erythromycin had been improved by the novel technique of antisolvent crystallization [23].

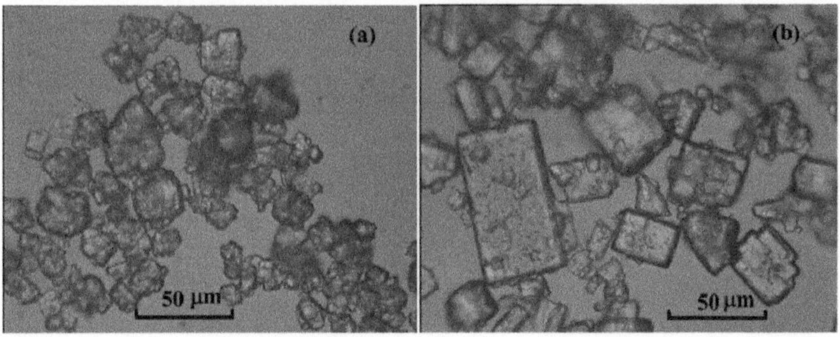

(a) traditional process (b) novel process

Figure 15. Crystal shape of erythromycin from different antisolvent crystallization processes [23]; (a) traditional process; (b) novel process

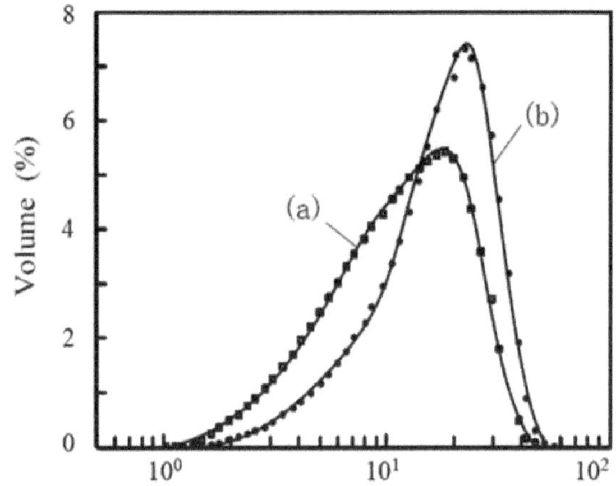

Figure 16. CSD of erythromycin from two antisolvent crystallization processes [23]; (a) traditional process; (b) novel process

In the commercial use of the antisolvent crystallization process, erythromycin with high specific activity was obtained at high yield. Over 90% of the products met the demands per year, which was much higher than the 53% with the traditional crystallization process.

3. Purification erythromycin by azeotropic evaporative crystallizaion

The development of the crystallization technique of erythromycin is limited to some extent by the extraction and purification prior to the crystallization. Taking the production of erythromycin as an example, the widely used process is frame filtration of fermentation broth - solvent extraction - salting-out crystallization – alkalization - antisolvent crystallization. Due to the limited interception capability for fine particles and macromolecules impurities such as proteins by frame filtration, and the low selectivity of the object over pigment and the small un-ionized organic molecules by solvent extraction, the impurity content is high in the organic phase. Therefore, the object should be further purified by coupling two crystallization methods in the subsequent refining process.

In recent years, a different technological process by membrane separation and resin absorption is gradually introduced into industrial application [25,11]. The process consists of several steps including membrane separation, resin absorption, elution and crystallization. Firstly, microfiltration is used to remove mycelium, a variety of fine suspension particles and some protein from fermentation broth, then pigment and small un-ionized organics are removed by resin absorption and the elution with butyl acetate. An improvement of the purity of erythromycin butyl acetate solution is obtained by using this pretreatment. And it makes crystallization preparation of erythromycin alkaline from butyl acetate elution become possible.

For the preparation of erythromycin alkaline from erythromycin butyl acetate solution, the product yield is low due to the high solubility of erythromycin. So the urgent task is to increase the yield. To remove butyl acetate is feasible, while high temperature for solvent evaporation may cause the destruction of erythromycin. Although erythromycin has better thermal stability than some other sorts of antibiotics, there is no precedent on the separation and purification of erythromycin with temperature being above 323.15K in industrial application till now. Thus, azeotropic evaporative crystallization of erythromycin is proposed in this paper. The method takes erythromycin, butyl acetate and water as crystallization system. Then butyl acetate-water azeotrope is removed by vacuum azeotropic evaporation to make erythromycin precipitate and disperse into water. Excessive water is added to the erythromycin butyl acetate solution for azeotropic evaporation, which can also play a role of washing crystals. The solubility of butyl acetate in water is quite small, so the azeotrope is easy to split into two phases at room temperature. The schematic diagram of azeotropic evaporative crystallization of erythromycin is demonstrated in Figure 17.

The solubility of erythromycin in butyl acetate-water saturated solution (solution A) and in water-butyl acetate saturated solution (solution B) was detected, respectively. The result indicated that the solubility of erythromycin in solution A was quite low and had little change with temperature. So for the azeotropic evaporative crystallization of erythromycin, the proportion of water was based on its effect on operation, such as the viscosity of the solution and crystal dispersion, as well as the utilization of equipment and the efficiency of production, rather than on the yield of crystallization.

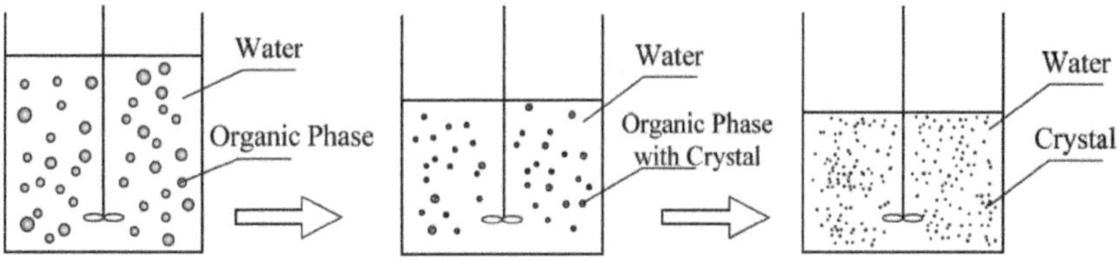

Figure 17. Principle illustration for azeotropic evaporative crystallization process of erythromycin

3.1. Technological parameters

Once the azeotropic evaporative crystallization of erythromycin was established, the optimization of parameters was directed by the quality and yield of crystal. The process parameters related to the crystal shape, crystal size and CSD were shown as follow: firstly, the supersaturation, which was related to the quantity of butyl acetate removed by azeotropic evaporation; secondly, the generation rate of supersaturation, which was dependent on the azeotropic evaporation rate and the cooling rate; thirdly, the crystallization temperature, which was bound up with vacuum of system and the cooling rate; fourthly, the stirring intensity, and etc,. The yield of erythromycin was determined by the evaporation quantity of butyl acetate and the terminal crystallization temperature.

System vacuum and operation temperature The butyl acetate-water azeotrope was removed from the crystallization system by vacuum evaporation, and the azeotropic temperature varied with the pressure. Then the high temperature leading to the damage of erythromycin could be avoided by adjusting the pressure.

According to the phase equilibrium data reported in the literatures [26,27], the azeotropic temperature and composition under different vacuum was calculated by using Pro II simulation software and NRTL thermodynamic model. When the system vacuum was controlled above 0.084MPa, the crystallization temperature was below 323.15K.

Supersaturation The supersaturation of erythromycin increased with the increasing volume of butyl acetate evaporated. The supersaturation varied with temperature and evaporation volume of butyl acetate, which affected CSD of the product.

Figure 18 showed the relationship between the cumulative volume distribution and supersaturation at 316.15K, where d_p was the crystal diameter and R_v was the cumulative volume fraction. As can be seen from the figure, the crystal size of erythromycin increased with the increase of supersaturation. However, it was necessary to choose the supersaturation range carefully due to the variation of solution viscosity and the difficulties of the crystal dispersion.

Cooling rate The cooling crystallization started after evaporating some amount of butyl acetate. The supersaturation caused by cooling made the crystallization process proceed continuously. It could improve the quality yield of the product by reducing the terminal crystallization temperature.

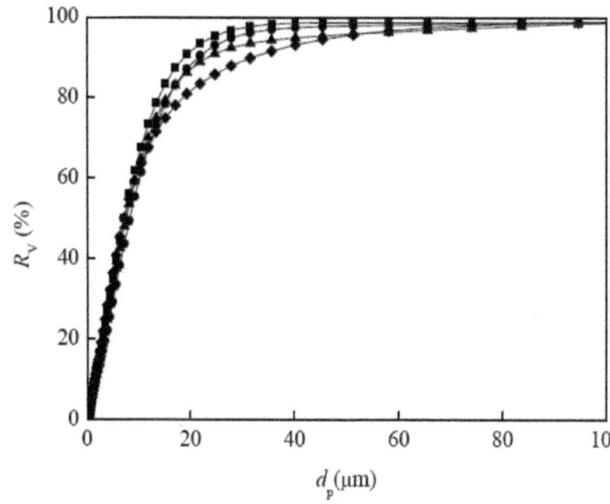

Figure 18. CSD based on cumulative volume of erythromycin at different supersaturation -■-: 39.91 g/100g; -▲-: 40.76 g/100g; -●-: 41.26 g/100g; -◆-: 41.60 g/100g

Figure 19 showed the relationship between the cumulative volume distribution of erythromycin and cooling rate, where d_p was the crystal diameter and R_v was the cumulative volume fraction of the crystal. It could be seen form the figure that speeding up the cooling rate was not conducive to the growth of crystal and made the crystal size decrease.

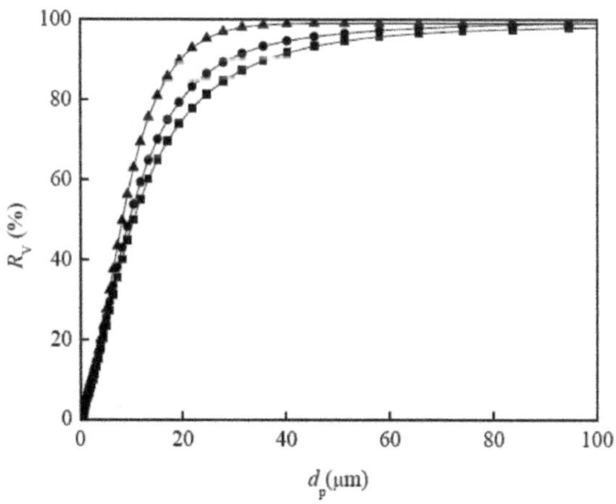

Figure 19. CSD based on cumulative volume of erythromycin at different cooling rate -■-: 273.17K/min; -●-: 273.20K/min; -▲-: 273.28K/min

3.2. The technique of erythromycin azeotropic evaporative crystallization

On the basis of the studies above, the crystallization technique combining the azeotropic evaporation with cooling crystallization was established to prepare the erythromycin from erythromycin butyl acetate solution directly. This process included mainly the following steps: firstly, introduction of entrainer. Adding entrainer (water) to erythromycin butyl acetate solution could form azeotropic crystallization system and decrease the evaporation temperature; secondly, vacuum evaporation. Adjusting the vacuum could promise the azeotropic evaporation temperature of butyl acetate and water was low enough to avoid the destruction of erythromycin; thirdly, appropriate evaporation quantity of butyl acetate. The supersaturation could be maintained within the thermodynamic metastable zone by adjusting the evaporation quantity of butyl acetate; fourthly, modulating cooling rate. The rate of crystallization could be regulated by adjusting cooling rate, so the supersaturation produced by cooling also could be maintained within the thermodynamic metastable zone to promise crystal growth; finally, the agitation power should be adjusted with the variation of crystallization stages.

There was an application for erythromycin purification by azeotropic evaporative crystallization. The technological conditions were listed as follows, the raw material of erythromycin was provided by a pharmaceutical company, the volume of water in the crystallization system was three times the volume of butyl acetate, the supersaturation was about 45g erythromycin/100g butyl acetate, cooling rate was 273.22K/min, the terminal crystallization temperature was 303.15K. With the conditions above and the technology in this paper, the purity of erythromycin A in the product was 95.87% and the yield in mass was 75.7%, which was higher than the yield 64.6% of erythromycin product by traditional antisolvent crystallization process using the same batch of raw materials.

4. Conclusion

In this paper, the thermodynamics, crystallization kinetics and operating conditions were studied systematically for the antisolvent crystallization of erythromycin. A brand-new technique with dynamic control of temperature and agitation intensity was henceforth presented. This process included nucleation at high temperature (313.15K~323.15K), regulation of temperature and agitation power according to the different stage of nucleation, crystal growth and crystal aging. It made the operation parameters of crystallization process more reasonable, and the erythromycin with high specific activity had high yield. The commercial use of the antisolvent crystallization technique had been successful.

Meanwhile, a novel purification method of erythromycin by azeotropic evaporative crystallization was also put forward. With this method, erythromycin could be produced from erythromycin butyl acetate solution directly. By the introduction of water, the evaporation temperature of azeotrope of butyl acetate and water was decreased and the supersaturation was induced. Then, crystallization nucleation and crystal growth were controlled by the regulation of cooling rate. With the azeotropic evaporative crystallization, qualified erythromycin product could be obtained without recrystallization, which leaded to less solvent consumption, simplified purification process and crystal product with narrow size distribution and perfect crystal shape.

Acknowledgments

The authors are grateful to the Jiawen Zhu and Bin Wu for helpful discussions. Thanks are extended to Qing Zhang, Bin Cao, Wenjian Zheng and Peixue Mao for technical assistance.

Author details

Kui Chen*, Li-Jun Ji and Yan-Yang Wu

*Address all correspondence to: chenkui@ecust.edu.cn

Chemical Engineering Research Center, East China University of Science and Technology, Shanghai, China

References

[1] Han, J.Y.; Yan, Y.C.; Chang, H.Y.; Wang, H. Progress in drug crystallization technology. Chemical Industry and Engineering Process 2002; 21(12) 945-948.

[2] McGuire, J.M.; Bunch, R.L.; Andersen, R.C. et al. Erythromycin a new antibiotic. Antibiotics and Chemotherapy 1952; 2(2) 281-283.

[3] Morimoto, S; Takahashi, Y.; Watanabe, Y.; Omura, S. Chemical modification of erythromycin. I. Synthesis and antibacterial activity of 6-O-methylerythromycin A. Journal of Antibiotics 1984; 37(2) 187-189.

[4] Retsema, J.; Girard, A.; Schelkly, W.; Manousos, M.; Anderson, M.; Bright, G. et al. Spectrum and mode of action of azithromycin (CP62,993), a new 15-membered-ring macrolide with improved potency against gram-negative organisms. Antimicrobial Agents and Chemotherapy 1987; (31) 1939-1947.

[5] Chu, D.T.; Plattner, J.J.; Katz, L. New direction in antibacterial research. Journal of Medicinal Chemistry 1996; 39(20) 3853-3874.

[6] Strigl S, Roblin P M, Reznik T, Hammerschlag M R. In vitro activity of ABT 773, a new ketolide antibiotic, against Chlamydia pneumoniae[J]. Antimicrobial Agents and Chemotherapy 2000; 44(4) 1112-1113.

[7] Fujita S., Takatsu A., Shibuya K. Process for purifying erythromycin 1971; US Patent 3629233.

[8] Zhao, Q,; Gao, D.W.; Yu, S.J.; Li, G.J. Study on improvement of erythromycin lactate alkalization process. Chinese Journal of Antibiotics 1998; 23(1) 14-16.

[9] Alves, A.M.B.; Morao, A.; Cardoso, J.P. Isolation of antibiotics from industrial fermentation broths using membrane technology. Desalination 2002; (148) 181-186.

[10] Sun, Y.; Zhu, J.W.; Chen, K.; Xu, J. Modeling erythromycin adsorption to the macroporous resin Sepabead SP825. Chemical Engineering Communications 2009; (196) 1-11.

[11] Zhu, J.W.; Sun, Y.; Chen, K.; Zhu, S.; Xu, J.; et al. A method of purification of erythromycin A 2011; China Patent 101367855B.

[12] Mullin J W. Crystallization. Oxford: Butterworth; 2001.

[13] Fukumori, Y.; Fukuda, T.; Yamamoto, Y.; Shigitani, Y.; Hanuy, Y.; et al. Physical characterization of erythromycin dihydrate, anhydrate and amorphous solid and their dissolution properties. Chemical and Pharmaceutical Bulletin 1983; 31(11) 4029-4039.

[14] Hang, A.G; Wu, Y.Y. Solubility of erythromycin in acetone-water solution. Chinese Journal of Antibiotics 1999; 24(6) 415-416.

[15] Chen, K.; Zhu, J.W.; Ji L.J.; Wu, B. The Metastable Characteristic of Erythromycin Slovent-out Crystallization Process. Journal of Chemical Engineering of Chinese Universities 2006; 20(5) 847-851.

[16] Allan S. Myerson. Handbook of Industrial Crystallization. Oxford: Butterworth; 2002.

[17] Bennema, P. Crystal growth from solution-Theory and experiment. Journal of Crystal Growth 1944; 24(25) 76-83.

[18] Randolph, A.D.; Larson, M.A. Theory of Particulate Processes. New York: Academic Press; 1988.

[19] Randolph, A.D.; Larson, M.A. Transient and Steady state size distributions in Continuous mixed suspension crystallizers. AIChEJ 1962; (8) 639-645.

[20] Chen, K.; Cao, B.; Zhu, J.W.; Wu, B.; Ji, L.J. In-situ measurement of erythromycin crystal size distribution by focused beam reflective measurement technology. Asia-pacific Journal of Chemical Engineering 2009; (4) 832-836.

[21] Kim, K.J.; Ryu, S.K. Nucleation of thiourea adduct crystals with cyclohexane-methylcyclopentane system. Chemical Engineering Communications 1997; 159(1) 51-66.

[22] Gao, D.W.; Zhao, Q.; Li, G.J. The mechanism and enhancing methods of solventing-out crystallization. Journal of South China University of Technology 1998; 26(11) 16-23.

[23] Chen, K.; Zhu, J.W.; Ji, L.J.; Wu, B. Dynamic Solventing-out Crystallization Process of Erythromycin. Journal of East China University of Science and Technology (Natural Science Edition) 2006; 32(8) 897-901.

[24] Chen, K.; Zhu, J.W.; Wu, B.; Ji, L.J. The dynamic control of preparation of erythromycin from erythromycin lactate 2005; China Patent 1219788c.

[25] Song, Y.H.; Zhu, J.W.; Chen, K.; Wu, B. Adsorption of erythromycin on macroporous resins and thermodynamic analysis. Journal of Chemical Industry and Engineering (China) 2006; 57(4) 715-718.

[26] Gmehling, J.; Onken, U. Vapor-liquid equilibrium data collection 1. Frankfurt: DECHEMA; 1977.

[27] Gmehling, J.; Onken, U.; Arlt, W. Vapor-liquid equilibrium data collection 1a. Frankfurt: DECHEMA; 1981.

Chapter 5

Characterizations of Functions of Biological Materials Having Controlling-Ability Against Ice Crystal Growth

Hidehisa Kawahara

Additional information is available at the end of the chapter

1. Introduction

Subzero winter temperatures pose a significant challenge to the survival of organisms in temperate and polar regions. Many organisms living in these areas have evolved a number of strategies for surviving in extreme environments such as subzero temperature [1-5]. Some strategies in a given organism use a mechanism based on freezing point depression trough accumulation of cryoprotectants such as sugars and polyhydric alcohol [6]. Other strategies use a mechanism based in physical damage avoidance through production of antifreeze material and ice nucleators [7-9]. Overwintering strategies based in freeze tolerance and freeze avoidance play an important role in adaptation promoting cold hardiness. Freeze-tolerant organisms survive the formation of extracellular ice but typically do not survive intracellular freezing [3]. In contrast, freeze-avoiding organisms must avoid freezing or death will result. These two alternative overwintering strategies share many of the same physiological adaptations, such as the accumulation of polyhydric alcohols, antifreeze protein and/or glycoprotein during cold acclimation [3, 10].

In subzero conditions, all organisms are exposed to conditions that necessitate the partial removal of water from the intracellular space in order to maintain the structure and function of the cell. Any significant deviation in the accessibility of water due to dehydration, desiccation or alteration of water's physical state, that is, from the aqueous phase to an ice crystal, will pose a severe threat to the normal function and survival of an organism [11]. Some bacteria among various organisms can counteract or minimize the deleterious effect of ice crystal formation in the intracellular and extracellular spaces [12]. As shown in Figure 1 [12], ice crystal- controlling proteins and other materials were related to the phenomenon of three steps in the formation and growth of ice. Ice nuclei can be formed by homogeneous (no particle present) or heterogeneous (particle-induced) nucleation in the first step. The formation of ice nuclei through heterogeneous ice nucleation is promoted by foreign particles that act as ice nucleation activator. Various types of ice nucleation activators of biogenic origin are known to exist in plant bacteria, fungi, insects, plants and lichens. Inhibitors of heterogeneous ice nucleation, which can favour supercooling, have been found in various organisms. These inhibitors can minimize the threats of intra- and extracellular ice formation. These inhibitors are known to exist in the xylem parenchyma cells of Katsura trees (*Cercidiphyllum japonicum*) [13]. Other ice crystal-controlling materials, which can play a crucial role in the second step of ice formation, are antifreeze proteins, antifreeze glycoproteins and antifreeze glycolipids. The function of AFP is to inhibit ice formation and ice crystal growth by suppressing the binding of water molecules to the ice crystal surface [14].

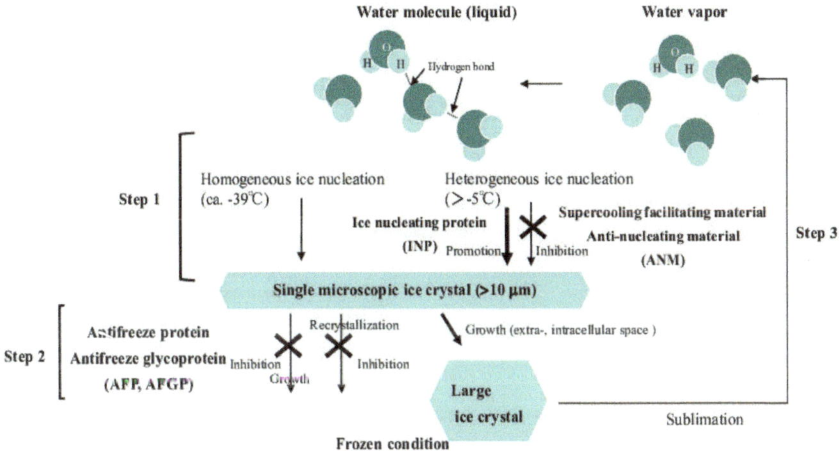

Figure 1. The representative functions on various ice crystal-controlling materials

In this chapter, we pay particular attention to the steps of ice crystal formation and growth along with the biogenic ice crystal- controlling materials. Among biogenic ice crystal-controlling materials, ice nucleation protein having the ability to promote ice nuclei formation, supercooling-facilitating materials having the ability to inhibit ice nucleation, and antifreeze materials having the ability to inhibit ice crystal growth and ice recrystallization are each explained as their structures, functions, and applications. Also, we mention the assay systems for each activity to seek these materials from various organisms and food wastes.

2. The mechanism of ice crystal formation

When pure liquid water is cooled at atmospheric pressure, it does not freeze spontaneously at 0ºC. Due to density fluctuations in liquid water, water molecules form clusters that have the same water molecular arrangement (Figure 2) as ice crystals but remain in a liquid state due to the fluctuation of energy. This state is called supercooling. A drop of pure water without perfectly foreign particles can display a supercooling temperature or freezing temperature at -39ºC [15]. This process has been called 'homogeneous ice nucleation' (Figure 1, Step 1). However, impurities or foreign particles present in water can attach water molecules onto their surfaces. As water molecules may be oriented in a way such as to resemble an ice nucleus, these become compatible with the critical dimension of ice nucleation. Franks reported that the deciding factors for the formation of ice nuclei by materials included the following three conditions: similarity to the crystal lattice, paucity of surface charge, and high hydrophobicity of the ice nuclei [16]. This process is called 'heterogeneous ice nucleation', and occurred at a temperature between -2ºC and -15ºC. The formed ice crystal nuclei may become 'ice crystals' by starting crystal growth (Figure 1, Step 2). This type of ice crystal growth exhibits three different mechanisms [17]. The first mechanism of ice crystal growth is growth from a perfect crystal side, and the growth rate at the interface of an ice crystal serves as the controlled surface nucleation rate. The second mechanism of ice crystal growth is growth by screw dislocation. The ice crystal growth rate is related to the degree of interface supercooling. The third mechanism of ice crystal growth is called continuous growth with large driving energy of crystal growth. In this case, the nucleation obstacle, which should be overcome in the case of crystal growth, does not exist, but the crystal growth rate is proportional to the degree of interface supercooling. This growth rate is affected by freezing temperatures. As shown in Figure 3, the maximum ice crystal generation temperature region is from 0ºC to -7ºC. This temperature region is important for ice crystal structure formation. When the time to pass through this temperature region is short, a detailed ice crystal is formed, and when the time is long, a large and rough ice crystal is formed. Difference in the shape of this formed ice crystal could affect the nature of the physical damage to the cells and organs during freezing. The differences in how quickly this temperature region is passed through influences the survival rate of cells and organisms after freezing and thawing. In the case of this passage time with one of late and slow freezing in the realm of nature, all organisms have acquired high freezing tolerance through production of various ice crystal-controlling materials.

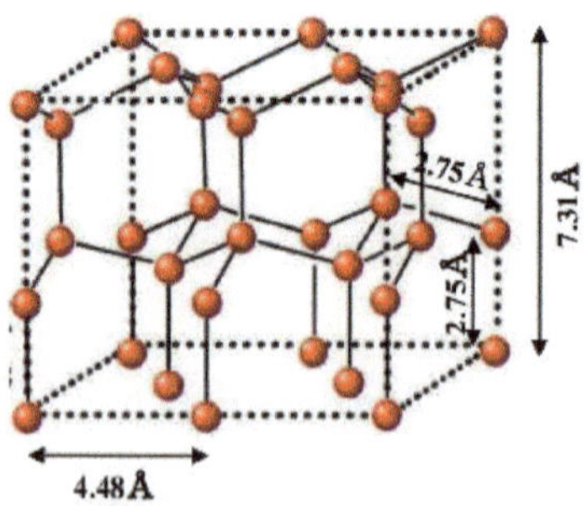

Figure 2. The structure of Ice crystal Ih.

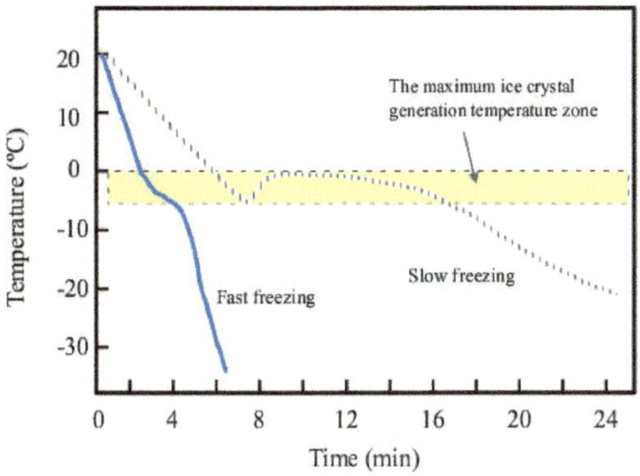

Figure 3. The maximum ice crystal generation temperature zone.

3. Structure and function of ice nucleation proteins from various organisms

As shown in Figure 1 Step1, the process called heterogeneous ice nucleation always occurs at a temperature higher than homogeneous ice nucleation. Ice nucleation proteins (INP) are integral components of various types of ice nucleation activators (INA) of biogenic origin INAs are present in a variety of plant bacteria [18], insects [19], intertidal invertebrates [10], plants [20], and lichen [21-23]. The INA found in a species of frost-resistant frog, *Rana sylvatica*, has also been shown to be composed of proteins [24]. This protein was present in *Rana sylvatica* plasma collected in the autumn and spring.

Various Gram-negative epiphytic bacteria, which have been called ice-nucleating bacteria, have been known to produce INA at temperatures higher that -3ºC. These bacteria belong to genera *Pseudomonas, Erwinia, Pantoea* and *Xanthomonas*. Six species of ice-nucleating bacteria have been found and various INPs from these bacteria have been analyzed to determine their amino acid sequences [25-30]. Also, some strains of *Fusarium acuminatum* and *F. avenaceum* are active in ice nucleation at a temperature of -2.5ºC [31]. These substances have different properties when compared to those of bacterial and fungal INPs. These differences might be caused by the different components of each ice nucleation material which contain ice nucleation protein as the active center. Extracellular ice-nucleating material secreted into the culture broth and localized on the surface of cell wall was found to be composed of lipid, protein, saccharide, and polyamine as the minor component [32, 33]. This localization was caused by the formation of large homoaggregates on the surface of the outer membrane. Genes conferring ice-nucleating activity have encoded INPs (120- 150 kDa) with similar primary structures. All INPs are composed of a highly repetitive central domain flanked by nonrepetitive N- and C-terminal domains (Figure 4). The tandem consensus octapeptide,

Ala-Gly-Tyr-Gly-Ser-Thr-Leu-Thr, of the central domain is hypothesized to form a β-helical fold secondary structure. This structure can bind water molecules in a configuration similar to an ice lattice [34]. This β-helical fold's secondary structure plays an important role in structures resembling ice lattices. Furthermore, the conserved glycine residues involved in chain bending are located at every turn of the proposed R-domain structure while the high Ser and Tyr residues are only present in the middle of β-strands, allowing them to act as an ice-like template. They are involved in the aggregation of each INP, thereby increasing the INP's hydrophobicity [35]. Based on their ice-binding abilities, it was suggested that INPs may have a similar β-helical fold and may interact with water through a repetitive TXT motif [36] (Figure 4). Large INPs having a molecular mass of 120-150 kDa could express high supercooling temperatures through both different tertiary structures of the R-domain. The N-domain is at least responsible for the binding of phosphatidylinositol as a lipid, saccharide (mannan) and INP [37]. Also, the C-terminal domain is rich in basic amino acid residues and is very hydrophilic. Among C-terminal amino acid residues, Tyr27 in this domain is important for ice nucleation, although not exclusively required, since nucleation was lost to a great extent when this residue was replaced by Gly or Ala but to a much lesser extent when it was replaced by Leu [38]. These results point to the important of the secondary and/or tertiary structure of the C-domain region for the ice nucleation with the hydroxyl group in the surface of its protein, which may interact with water molecule. Based on the structure and component of ice nucleation materials, we could predict that each domain has the following important role for the nucleation: N-domain through association with lipid and saccharide thereby increasing hydrophobicity: R-domain through structuralization of ice lattice-resembling protein: and C-domain through stabilization of tertiary structure of the complex.

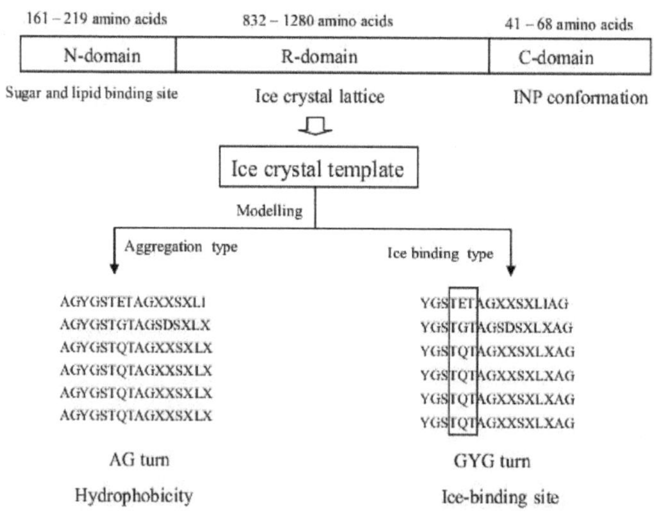

Figure 4. The structure of ice-nucleating protein and both models of different properties.

The wood frog (*Rana sylvatica*) is able to tolerate freezing of its body tissue. This tolerance is promoted by the initiation of ice formation at high subzero temperatures which allows ice to form gradually [39]. Also, some ice-nucleating bacteria, including *Pseudomonas putida, P. fluorescens,* and *Pantoea* (*Enterobacter*) *agglomerans* were isolated from the gut of frogs collected in the field. The maximum nucleation temperature of an aqueous suspension of *P. putida* cells ranged from -1.6 to -3.0ºC [40]. These ice-nucleating bacteria may play a role in enhancing winter survival by promoting ice nucleation at high subzero temperature.

Frost-sensitive plant species have a limited ability to tolerate ice formation in their tissues [41]. Alternatively, some plants can supercool to some extent below 0ºC and avoid damaging ice formation [42]. The temperature to which a given plant can supercool varies by plant species and is influenced by the presence of ice-nucleating agents that may be of plant origin [43]. Ice nuclei active at approximately -2ºC and intrinsic to woody tissues of *Prunus* sp. were shown to have properties distinct from bacterial ice nuclei [20]. Development of ice nuclei in immature peach buds and sweet cherry stems did not occur until midsummer and their formation was essentially complete by late seasonal changes in growth. The apparent physiological function of the ice nuclei in promoting cold hardiness of woody plants illustrates the importance of supercooling and endogenously-controlled ice nucleation during dormancy and deacclimation [20].

Then, how is this ice nucleation activity measured? Ice-nucleating activity of bacterial cells was measured with a freezing nucleus spectrometer (thermoelectric plate, Mitsuwa model K-1), as described by Vali [44]. Thirty drops, 10 μl each, were placed on a controlled-temperature surface and the temperature was slowly lowered from ambient to -20ºC at a rate of 1ºC per min. The ice- nucleating spectra were obtained by the droplet-freezing method as modified by Lindow et al. [45]. After examining the shapes of these cumulative spectra, it was suggested that the sample nuclei could be separated into three classes: type I, II and III, with respective threshold temperature ranges of -5ºC or warmer, -5ºC to -8ºC, and -10ºC or colder [46]. Another simple procedure is to measure the highest threshold temperature of the INA in the sample using a glass capillary [47]. However, this method does not assay for less active nucleators and is best suited for cases where INA does not exhibit activity for screening of the ice nucleator.

The most representative application of INP is its use as the template of artificial snow. The sterilized and freeze-dried cell powder of the ice-active bacterium, *Pseudomonas syringae,* was used for the Calgary Winter Olympics in 1988 as an artificial snow agent. However, these highly active ice-nucleating bacteria were almost epiphytic bacteria causing frost damage. *Xanthomonas campestris,* which are known as a species of xanthun gum-producing bacterium, was isolated from frost-damaged tea leaves [48]. This strain, INXC-1, can be easily sterilized by a high-pressure treatment at low temperature without decreasing ice nucleation activity [49]. This cell preparation has been used for various processed foods, such as freeze concentration. Watanabe et al. have succeeded in applying freeze concentration to soy sauce, removing about half of the salt as eutectic crystals and leaving behind the flavor substances [50]. Soy sauce was frozen in the presence of this cell preparation at -25ºC. After removing the ice and eutectic crystals of salt and water, the product retained well its original aroma and taste substances at 1.6 times concentration. However, this pressurized-cell preparation has not yet been permitted for food use by the Japanese Ministry of Health and Welfare.

4. Structure and function of supercooling-facilitating material (anti nucleating material) in various organisms and chemicals

Ice-nucleating inhibitors have the ability to lower the supercooling point of water. This activity is termed either 'supercooling-facilitating activity' or 'anti-nucleating activity'. An enzyme-modified gelatin (EMG-12) has been reported as an ice-nucleating inhibitor of silver iodine, AgI, a well-known ice-nucleating agent [51]. Also, there are some reports regarding anti-ice nucleation substances that enhance the supercooling of water as shown in Table 1. Antifreeze proteins from insects [52], antifreeze proteins and antifreeze glycoproteins from fish [53], anti-nucleating proteins from bacteria [54], and polysaccharides from bacteria [55] all exhibit anti-ice nucleation activity toward water droplets. As substances originating from plants, hinokitiol from the leaves of Taiwan yellow cypress [56] and eugenol from cloves both reduce the ice-nucleation activity of water [57]. Crude extracts from the seeds of woody plants and supernatant liquids from germinating legume seeds exhibit very high anti-ice nucleation activity toward water droplets, although the causative substances for supercooling in these plant extracts were not identified [58]. As chemical substances, polyvinyl alcohol and polyglycerol enhance supercooling of aqueous solutions [59, 60]. Recently, it was reported that deep supercooling xylem parenchyma cells (XPCs) of the katsura tree (*Cercidiphyllum japonicum*) contain four kinds of flavonol glycosides with high anti- nucleating activities. These flavonol glycosides have very similar structures, but their activities are very different [61]. It was clear that the combination of the position of attachment of the glycosyl moiety, the kind of attached glycosyl moiety and the structure of aglycone determined the magnitude of this activity [62] (Figure 5). We have also purified an anti-nucleating protein from *Acinetobacter calcoaceticus* KINI-1, which was isolated from the camphor leaf [54]. This anti-nucleating protein has a molecular mass of 550 kDa. It exhibits a broad specificity with the capacity to lower the nucleating activity of a wide range of ice nucleators, including some bacterial components and AgI (Table 1). However, the expression mechanism of its anti-nucleating activity remains unknown. Also, the xylem extract of the katsura tree exhibited anti-nucleating activity against a wide range of ice nucleators. The anti-nucleating activities (ºC) of this extract at the same concentration against cell suspensions of *P. fluorecsens, E. ananas* and *X. campestris* were found to be 0.7, 1.9, and 1.3, respectively. This activity against AgI was 1.8. After isolating each active compounds, the main active compounds in the xylem extract found to be four flavonol glycosides; kaempferol 7-O-β-D-glucopyranoside, kaempferol 3-O-β-D- glucopyranoside, 8-methoxykaempferol 3-O-β-D- glucopyranoside, and querdcetin 3-O-β-D- glucopyranoside [61]. It is clear that the activity of the flavonol glycosides are controlled by a combination of the position of attachment of the glycosyl moiety, the kind of attached glycosyl moiety and the structure of aglycone. Although the features of the structures in flavonol glycosides that clearly affect this activity were not found, judging from the active low molecular compounds in regards to the flavonol related compound, hinokitiol [56] and eugenol [57], the functional group in polyphenol may interact with the active site Tyr residue in the C-domain of ice nucleation protein.

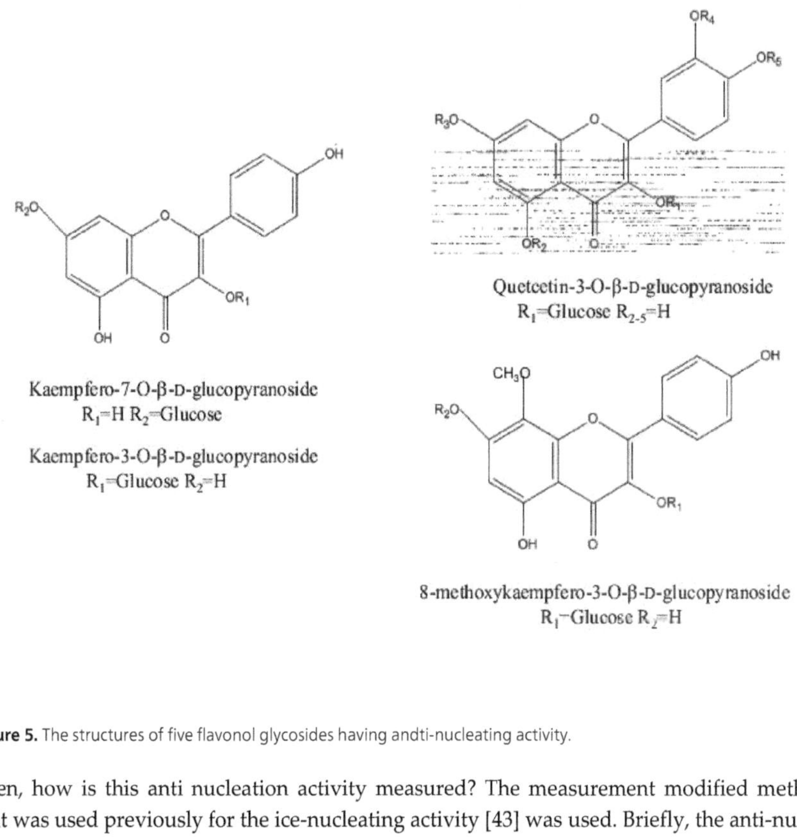

Figure 5. The structures of five flavonol glycosides having andti-nucleating activity.

Then, how is this anti nucleation activity measured? The measurement modified method that was used previously for the ice-nucleating activity [43] was used. Briefly, the anti-nucleating activity was measured as follows. A sample solution (270 μl) and a suspension (30 μl) containing lyophilized cells of various ice-nucleating bacteria in a potassium phosphate buffer to an absorbance at 0.1 of 660 nm (50 mM, pH 7.0) were mixed and incubated in ice for 10 min. The ice-nucleating temperature of this mixture solution was measured. A mixture solution including 270 μl of 50 mM potassium phosphate buffer (pH 7.0) was measured as a control. Also, a mixture solution of the sample solution (270 μl) and the AgI (1mg/ml) suspension (30 μl) was examined. The difference between the ice-nucleating temperature of the sample and the control was defined as the anti-nucleating activity or supercooling-facilitating activity (ºC).

Materials	Concentration	Ice nucleator	Anti ice nucleating activity (°C)	Condition
Xylem extract				
Castonea crenata	100 mosmol/kg	*Erwinia ananas*	1.2	1
Fagus crenata	100 mosmol/kg	*E. ananas*	1.3	1
Cercidipkyllum japonicum	100 mosmol/kg	*E. ananas*	1.9	1
	100 mosmol/kg	*Pseudmonas syringae*	0.7	1
	100 mosmol/kg	Silver iodide	1.8	1
55-kDa protein *A. calcoaceticus*	10 μg/ml	*Erwinia uredovora*	2.2	2
130-kDa polysaccharide *B. thuringiensis*	50 μg/ml	*Pntoea ananas*	2.3	2
	50 μg/ml	Silver iodide	4.2	2
Hinokitiol	10 mM	*P. fluorescens*	2.1	2
Eugenol	1 mg/ml	*P. fluorescens*	1.9	2

*This list was modified table 1 in reference [61]

Condition 1: Volume of droplets was 2 μl and cooling rate was 0.2 °C/min

Condition 2: Volume of droplets was 10 μl and cooling rate was 1.0 °C/min

Table 1. A list of anti-nucleating materials*

Only a few studies have been performed on the practical application of supercooling facilitating material. Organ cryopreservation is hindered by ice-inflicted damages and non-freezing preservation of livers at subzero temperature over -5ºC might offer advantages over the current method of preservation. A solution containing bacterial anti-nucleating protein (20 μg/ml) [52] and ascorbic acid 2-glucoside (100 μg/ml) as an antioxidant was used as a subzero non-freezing storage method (SZNF) for rat liver graft [63]. When liver grafts were kept for 24 h at SNZF storage (-3.0ºC), apoptotic cells were greatly diminished. Also, ATP concentrations in grafted liver tissues preserved with SNZF were significantly higher than those that underwent normal storage at 4ºC for 24 h. In the case of flavonol glycoside, the supplemental addition of kaempferol 7-O-β-D-glucopyranoside to diluted vitrification solution, which consists of 2.0 M glycerol, 0.4 M sucrose and 4% dimethylsulfoxide (Me_2SO, w/v) in basal culture medium was examined [64]. The addition of 0.5 mg/ml kaempferol 7-O-β-D-glucopyranoside to the diluted plant vitrification solution 2, which consists of 30% glycerol (w/v), 15% ethylene glycol (w/v) and 15%Me_2SO (w/v) in basal culture medium containing 0.4 M sucrose (pH 5.2). resulted in significantly higher regrowth rates after cryopreservation.

5. Structure and function of antifreeze protein (AFP) and AFP related material from various organisms

In the late 1960s, DeVries and Wohlschlag reported that a carbohydrate-containing protein (antifreeze glycoprotein; AFGP) that was isolated from the blood plasma of an Antarctic notothenioid fish accounted for a freezing point depression of -1.31ºC [65]. This discovery provided a biophysical explanation for how such organisms escape lethal freezing events despite continual contact with -1.9ºC sea water. Many mechanisms containing the production of AFP have been utilized by various species. Other than these adaptive mechanisms, other mechanisms include seasonal migration, hibernation, supercooling, synthesis of small cryoprotectant molecules such as glycerol, trehalose, mannitol and others. Almost all AFPs identified in various organisms were orders of magnitude more active than that which could be explained by colligative properties. AFPs excepting some AFP-related materials, had thermal hysteresis (TH) activity without change in the melting point and recrystallization inhibition (RI) activity [66]. Ice can exist in several crystalline polymorphic structures and also in an amorphous or vitreous state of rather uncertain structure. Of these, only ordinary or hexagonal ice (I_h) is stable under normal pressure at 0ºC (Figure 2). This ice structure, I_h, grows along the a and c axis (Figure 6 a). The plane growing along the *a* axis is called the prism face, and the plane growing along the c axis is called the basal face.

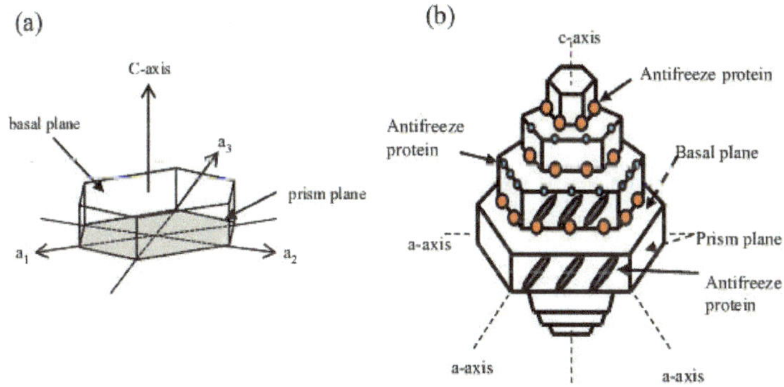

Figure 6. The strucutire of hexagonal ice crystal and its binding sites of antifreeze proteins. (a) Hexagonal ice crystal (b) Various binding site of antifreeze protein

Flat AFP peptides and the flat sides of AFP tertiary structure contact sides could bind to ice lattices (Figure 6) and interfere with crystal growth along the *a*- axis by making it thermodynamically unfavorable for water molecules to join the ice surface [67]. Therefore, AFPs appeared to inhibit the normal growth direction of ice by preferentially adsorbing to the prism faces of ice crystals, thereby forming needle–shape crystals (Figure 7 (a)). Several models have been proposed to describe how molecular binding between the peptide and ice occurs

[68], but the specific nature of this interaction is still not well-understood. Not all AFPs have an effect on freezing point depression, that is, Thermal hysteresis activity (TH activity), and have only recrystallization inhibiting activity (RI activity) [69]. Those ice-binding proteins having ice affinity are often referred to as ice-active or ice-structuring proteins [70] and ice recrystallization inhibiting proteins. Various AFPs have been isolated from fishes, plants, insects, fungi, and bacteria [71]. Among AFPs from various organisms, each AFP of different origins could be divided into groups based on its structure. As shown in Table 2, fish AFPs and AFGP could be divided into five groups. Four groups of fish AFP having different structures and molecular weights each had different TH activity. For instance, type I AFPs were defined as small (3 ~ 4 kDa), Ala-rich (~60% Ala) α-helices [72]. These AFPs were isolated from three taxonomic orders; pleuronectiforme such as the winter flounder [73], scorpaeniforme such as sculpins [74] and perciforme such as cunner [75]. Typically, type I AFPs form amphipathic helices with a well-conserved Ala-rich surface opposite a less conserved, more hydrophilic helix side [76]. Although AFP types I, II, III and IV, as well as AFGP, produce ~1ºC of thermal hysteresis at high concentrations (10 ~ 40 mg/ml), hyperactive AFP type I, which provides ~1.1ºC of TH activity at a concentration of 0.1 mg/ml, was isolated from winter founder [77]. The structure of hyperactive AFP type I was two extended 195-amino acid α-helices forming an amphipathic homodimer with a series of linked Ala- and Thr-rich patches on the surface of the dimer [78]. As with the discovery of hyperactive AFP, further study may be in progress to find and characterize new type of fish AFPs.

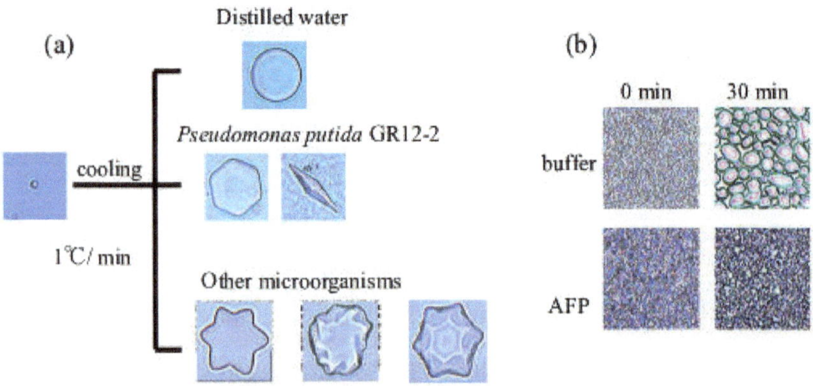

Figure 7. Ice crystal regulation by some antifreeze inhibition. (a) Ice crystal morphology (b) Ice crystal recrystallization inhibition.

Animal AFPs exhibit significant differences in the levels of TH, ranging from 1 to 2ºC in fishes and 5 to 10ºC in insects [79]. In contrast, plant AFPs, which characteristically have low levels of TH activity (0.1 ~ 0.6ºC) [80], were divided into two groups based on structure. In winter rye, six AFPs ranging in size from 15 to 35 kDa have been identified from the apoplastic fraction. These AFPs are similar to pathogenesis-related proteins containing chitinase, β-glucanase and thomathin-like proteins [81]. An AFP with higher RI activity and lower TH activity compared with other AFPs was isolated from the perennial ryegrass *Lolium prerenne* [82]. In carrots, an LT-up-regulated AFP shows a significant similarity (50-65%) to the polygalacturonase inhibitor family of plant leucine-rich repeat (LRR) proteins [83].

Characteristics	AFGP	Type I AFP	Type II AFP	Type III AFP	Type IV AFP
Mass (Da)	2600-33000	3300-4500	11000-24000	6500	12000
Key Properties	AAT repeat Disaccharide	Alanine-rich α-helix	Disulfide bonded	β-sandwich	Alanine rich herical bundle
Representative Structure	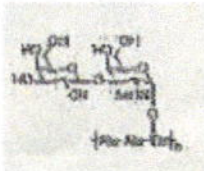				
Natural Sources	Antarctic-nototheniods	Right-eyed, flouders	Sea raven, Smelt, herring	Ocean pout, Wolfish, eel pout	Longhorn, sculpin

Table 2. Structure and properties of AFP and AFGP from various fishes.

Based on the presence of TH activity in the extract of various plants, some grains like winter and spring rye, some vegetables including cabbage and carrot, *Ammopiptanthus mongolicus, Solonum dulcamara, Lolium perenne* and tobacco have been chosen to investigate AFPs [84, 85] as shown in Figure 8. In Japan, many vegetables are harvested during the winter. The Japanese radish particularly is one of the typical winter vegetables and the most productive vegetable with the largest amount in Japan. However, AFPs in the Japanese radish leaf and tuber were found to accumulate in the apoplastic spaces of vegetables harvested until April. We examined the effect of cold acclimation time on the AFP production by measurement of protein amount and TH activity [86]. The protein amount and TH were almost constant during 2 weeks of acclimation time. Each maximum value (46.5 μg/ml and 0.20°C, respectively) was attained after 4 weeks of cold acclimation time. The TH was almost constant (0.18 - 0.20°C) until 7 weeks of cold acclimation time had elapsed. When the Japanese radish tuber was stored at 4°C for 7 weeks, the protein amount in its apoplastic space diminished remarkably (22 μg/ml) (Figure 9). Some proteins in this apoplastic fraction reacted with the anti-glucanase-like protein (GLP) antiserum and anti-chitinase-like protein (CLP) antiserum produced against isolated winter rye AFPs. Also, these prepared proteins exhibited chitinase and β-1,3-glucanase activities. The structure of the chitinase-type AFP and glucanase-type AFP from winter rye leaf were elucidated by sequencing the gene of each AFP [87], but the binding sites in these AFPs were unclear. Also, the structures of these proteins from Japanese radish remain unknown.

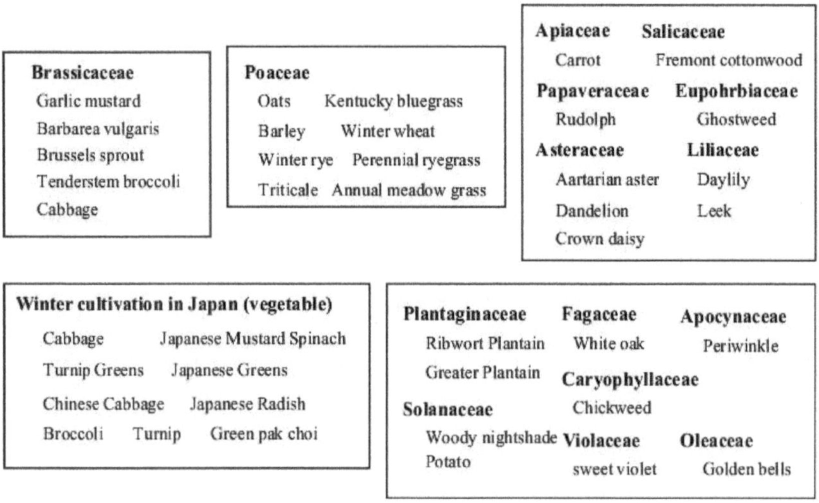

This list is modified in table from reference No. 83

Figure 8. List of some plants having antefreze actibity (TH activity).

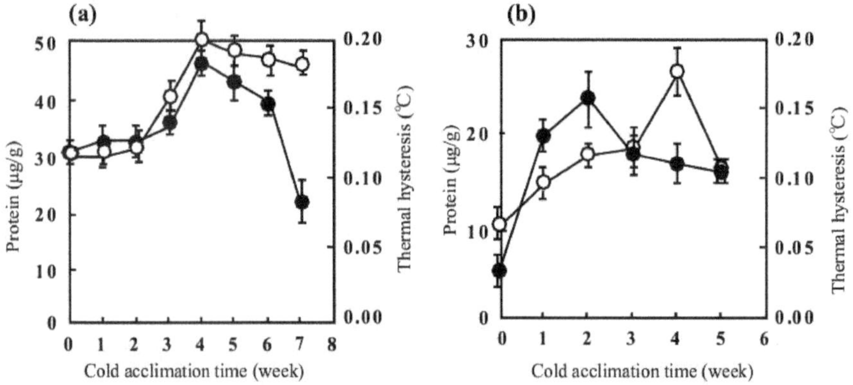

Each value is the mean ± SD (n=4). Values obtained from different cold acclimation times are significaly different at ρ<0.05 ●, protein; ○, Thermal hysteresis

Figure 9. Effects of cold acclimation times on the apoplastic protein and thermal hysteresis. (a) tuber (b) leaf

Other than AFP having high TH activity (0.1 – 0.2ºC), some proteins having high RI activity were isolated from various plants. Two related genes encoding ice recrystallization-inhibiting protein from wheat were identified and characterized by assay of IR activity. Two proteins share homology with two subsets of proteins: their N-terminal parts are similar to the Leucine rich repeat-containing regions present in the receptor domain of receptor-like kinases, while their C-terminuses are homologous to the RI domain of AFPs [88]. This C-terminal part is homologous to LpAFP, a partial gene coding for an AFP from the rye grass *Lollium perenne*. The encoded, incomplete 118 residue LpAFP has an RI activity higher than that of other AFPs [24]. This primary structure shows a series of highly conserved repeated motifs with regularly spaced serine and threonine residues that may form hydrogen bonds with the ice surface [89]. One model predicted that two ice binding sides were each on side of the AFP molecule, with xxNxVxG and xxNxVx consensus motifs [90]. This consensus motif is different from the ice-binding motif (xTxTx) for TH activity in some beetle AFPs [91]. It was predicted that this unusual duplication of putative-binding sites on opposite sides of the protein could be responsible for the high RI activity. Homology search within this RI domain (113 amino acids of C-terminal region) has shown that this region exists only in four species that are known to be cold–tolerant cereals from the *Pooideae* subfamily of the *Poaceae* (*Gramineae*) family: wheat, *L. perenne*, rye, and barley. The consensus sequence of some AFPs may be found from some organisms in the same subfamily.

Then, how is this antifreeze activity measured? Both activities, that is, TH activity and RI activity, were measured by different methods. TH activity was measured using a nanoliter osmometer (for example, Otago osmometers). The osomometer was calibrated using deionized water (Milli-Q) and osmolarity standards. Droplets of the test sample were transferred to the sample wells of the osmometer., which were filled with oil. The droplet was frozen by rapidly cooling it to about -30ºC and then was observed under a dissecting microscope. The temperature was then raised rapidly until close to the expected melting temperature, at which the last ice crystal melted, and the melting temperature was determined and the osmolality calculated. Then the temperature was decreased to refreeze the sample and increased to melt the sample back to a single small ice crystal. In the use of nanoliter osmometer, the temperature was lowered by 0.02ºC/min until discernable growth of the ice crystal occurred. This was taken as the hysteresis freezing point and from this the mount of thermal hysteresis (TH) was calculated. The shape of the ice crystal upon growth at the hysteresis freezing point was noted using the microscope [69]. Other the method using a micro-osmometer, a microscope with temperature–controlled freezing stage (Model THM 600, Linkham Scientific Instruments, Surrey, UK) was used to measure TH activity [83] (Figure 10). One microliter of protein sample was applied to the center of a temperature-controlled freezing stage on a circular glass cover. The freezing stage was fitted onto the stage of a conventional microscope and was connected to a pressurized air supply cooled by liquid N_2. The stage temperature was controlled by a programming unit (Model TMS 90, Linkham Scientific Instruments, Surrey, UK). After sample application, the stage was heated to 20ºC, then cooled to -40ºC at the rate of 100ºC/min to freeze the sample, after which it was heated at the same rate to -5ºC. The warming was slowed to 5ºC/min to thaw the sample until only a single ice crystal was present. Subsequently, the temperature was slowly (1ºC/min) lowered to observe ice crystal growth. The time (s) at which the ice crystal growth started, were measured and the TH value (ºC) was calculated this time 60^{-1}. Under these conditions, high levels of antifreeze activity were indicated by the multi-faceted or bipyramidal shape of the ice crystal, whereas low levels of antifreeze activity were indicated by the flat, hexagonal shape of the growing ice crystal. In the absence of AFPs, the ice crystals were round and flat. Measurement of RI activity was performed with various methods. The assay for the inhibition of ice recrystallization was performed using the method described by Smallwood et al. [92]. This method was called the 'sucrose sandwich method'. Each sample contained a known dilution of the AFP preparation and 30% sucrose in water. This mixture (1.5 μl) was 'sandwiched' between two labeled, 13 mm diameter circular glass cover slips. The sandwich was cooled to -80°C using the programming unit and then maintained at -6ºC. The sandwich was observed using a phase-contrast microscope with a 10X objective and with a temperature-controlled freezing stage (Figure 11). Visual assessment of any recrystallization was made by comparison of the test sample with a control sample (30% sucrose solution without AFP) after 30 min. The presence of RI activity in the sample was compared with crystal sizes in photographs of both the sample and positive control, that is, a fish AFP sample after 30 min of annealing. Before the development of the sucrose-sandwich method, RI activity was measured by a technique known as the 'splat cooling assay' [93]. In this method, a small volume of sample liquid (10 μl) is dropped from a height of about 2 m onto a polished metal block precooled in solid carbon dioxide. The polished block is usually at a temperature of about -78ºC. The resulting ice splat is about 1 cm in diameter and is a thin disc of polycrystalline ice. This is then transferred to a cold stage on a microscope where it is maintained at -6 to -9ºC and is observed between crossed photographs to determine changes in the average size of the ice crystals over time. Also, to compare each ice crystal size in serial diluted samples, a capillary method using 10 μl glass capillaries was developed [94]. Serial dilutions of sample are prepared to determine the concentration below which RI activity was no longer detected, termed the RI endpoint. All of the diluted samples can be assayed and evaluated simultaneously. Also, Warlton et al. have developed a new technique to measure and qualify levels of crystallization in a sample solution using an optical recrystallometer [95]. However, these assay methods for RI activity were not defined in terms of the unit of RI activity, such as the enzyme unit for various enzyme activities. As shown in Figure 11, photographs of both the test sample and control sample after 30 min at -6ºC were analyzed by average area (in pixels) of one crystal using Image Factory (Ruka International Co., Japan). The value of RI was calculated as the relative rate of the average area of one crystal from both the control and test samples. On high RI activity this high activity eqaul to the decrease of this RI value (Figure 12). Also, one unit of RI activity was defined as activity with the relative ratio of RI=0.5. To confirm the presence of AFP in crude extract from various organisms, the best method is the separation of those proteins having the affinity ability against ice crystal surface. The cold finger method [96] can purify ice-binding protein like AFP from a crude mixture and confirm the presence of ice-binding protein.

Many companies around the world have been expecting to apply AFPs to frozen food. The representative application of AFP is quality preservation in various processed frozen foods. Unilever Group developed AFP type III HPLC 12 preparations produced by recombinant baker's yeast, which is used commercially for the quality preservation of commercial ice cream. They established the safety of this recombinant AFP based on a set of *in vitro* and *in vivo* genotoxicity assays (bacterial mutation, chromosome aberration, mammalian cell gene mutation and rat bone marrow micronucleus) as well as a 3-month repeat-dose gavage study on rats [97]. This recombinant protein, that is, ice structuring protein (ISP), is used as a food additive and ice cream containing ISP is sold in North America. Also, concentrated carrot protein (CCP) containing 15.4% (w/w) carrot (*Daucus carota*) AFP was found to potentially improve the fermentation capacity of frozen dough [98]. The effects of pre-slaughter administration of AFGP to lambs were assessed on lamb meat quality after thawing [99]. Meats were vacuum packed and stored frozen at -20°C for 2-16 weeks. Upon thawing, meats were assessed for drip loss and sensory properties (foreign flavour, storage flavour, texture, tenderness, juiciness and overall acceptability). Injection of AFGP at either 1 or 24 h before slaughter reduced drip loss and ice crystal size. Ice crystals were smallest in the lambs injected with a final concentration of 0.01 μg/kg. AFGP, particularly when injected 24 h before slaughter. These results suggest that the addition of AFGP could reduce damage owing to frozen storage of meat. Perhaps the qualities of various frozen meats and fishes could be improved by the injection of AFP and AFGP. However, the applications of non-recombinant AFPs for these applications are marred by high cost due to low yield. Applications for frozen food may require future technology in which highly active extracts can be manufactured inexpensively.

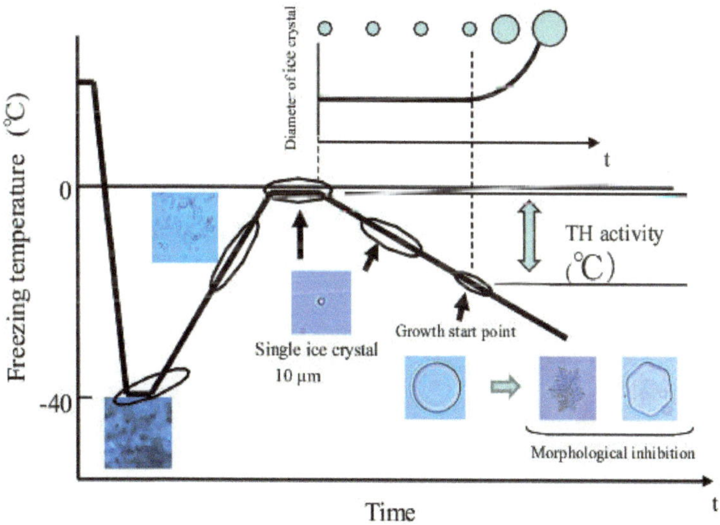

Figure 10. Scheme of freezing curve and ice morphology of the TH activity using microscope with temperature-controlled sample stage.

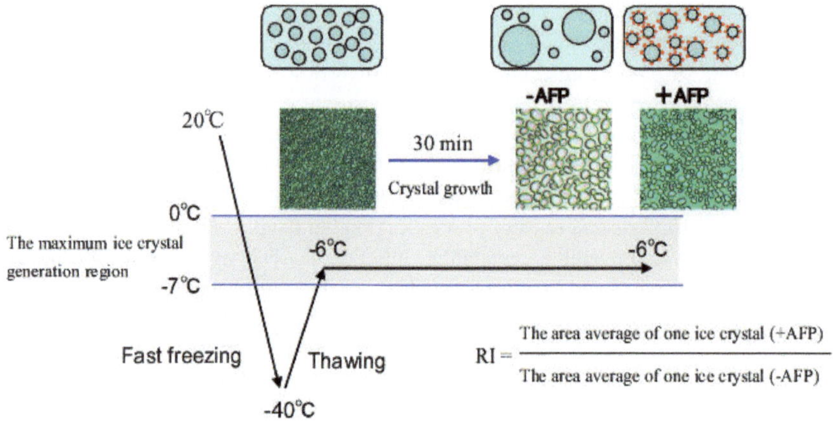

Figure 11. Scheme of freezing curve and ice morphology of sucrose sandwich assay using microscope with temperature-controlled sample stage.

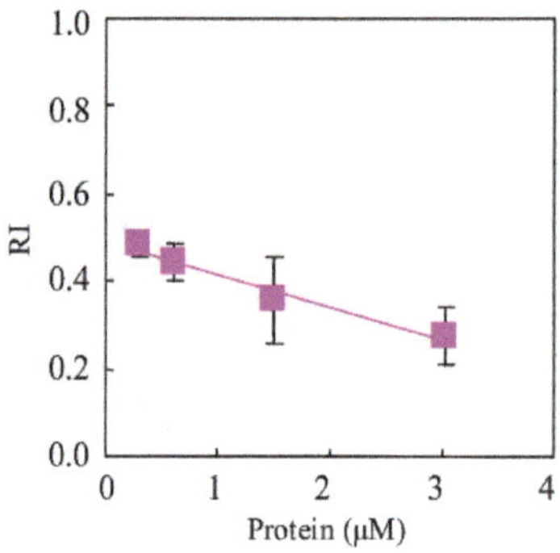

Each value is the mean ±SD (n=3)

Figure 12. Effect of proteins concentration of RI value in the purified fish type II AFP.

6. Conclusion

All steps related to the processes from nucleus generation as shown in Figure 1 without sublimation have correlated with the survivals of various organisms under subzero temperature. However, the sublimation of ice take place under frozen conditions (-10 °C) during a long term over 1 month. The sublimation-controlling ability must be the materials with the ice-binding ability. These organisms have developed the ability to tolerant freezing conditions by producing ice crystal-controlling materials in their extracellular or intracellular spaces. Although these materials have different functions, all of them seem to have ice-binding sites and therefore a high affinity for ice lattice surface. Among all materials, some AFP groups had consensus sequence or consensus tertiary structure. Also, both INPs and AFPs exhibit their each opposite functions in the case of special regions containing ice binding sites on all amino acid sequence [100] or aggregations of each molecule [101]. This phenomenon is an important factor in consideration of AFP's potentially harmful effect on the quality of frozen food processing and the viability of cultured cells after cryopreserving owing to excess concentration. However, these materials can be advantageous in various industries concerned with freezing and preservation through use of more diluted concentrations. Among these materials, sublimation-inhibiting or -facilitating materials remain unexplored. As AFPs and INPs had ice binding site on each molecule, both proteins may have sublimation-inhibiting ability. From now we will try to confirm an assay system for sublimating- inhibiting activity. In the future, this inhibitor will likely be discovered and applied to various processed frozen food.

Author details

Hidehisa Kawahara

Department of Life Science and Biotechnology, Kansai University, Yamate-cho, Suita-shi, Osaka, Japan

References

[1] Denlinger DL, Lee RE. Low temperature biology of insects. Cambridge University Press ; 2010.

[2] Robinson CH. Cold adaptation in Arctic and Antarctic fungi. New Phytology 2001; 151 341-353.

[3] Storey KB and Storey JM. Natural freezing survival in animals. Annu. Rev. Ecol. Syst. 1996; 27 365-386.

[4] Thomashow MF. Plant cold acclimation: freezing tolerance genes and regulatory mechanisms. Annu. Rev. Plant Physiol. Plant Mol. Biol. 1999; 50 571-599.

[5] Voituron Y, Joly P, Eugène M, Barrè H. Freezing tolerance of the European water frogs: the good, the bad, and the ugly. Am. J. Physiol. Regul. Integr. Comp. Physiol. 2005; 1563-1570.

[6] Storey KB. Metabolism and bound water in overwintering insects. Cryobiology 1983; 20 365-379.

[7] DeVries AL. The role of antifreeze glycopeptide and peptides in the freezing avoidance of Antarctic fishes. Comp. Biochem. And Physiol. B 1988; 90 611-621.

[8] Duman JG, Olesen TM. Thermal hysteresis protein activity in bacteria, fungi, and phylogenetically diverse plants. Cryobiology 1993; 30 322-328.

[9] Zacchariasen KE, Kristiansen E. Ice nucleation and antinucelation in nature. Cryobiology 2000; 41 257-259.

[10] Duman JG. Antifreeze and ice nucleator proteins in terrestrial arthropods. Annu. Rev. Physiol. 2001; 63 327-357.

[11] Beall PT. States of water in biological systems. Int. J Biochem. Cell Biol. 1983; 20 324-334.

[12] Kawahara H. The structure and function of ice crystal-controlling proteins from bacteria. J Biosci. Bioeng. 2002; 94 492-496.

[13] Kasuga J, Mizuno K, Arakawa K, Fujikawa S. Anti-ice nucleation activity in xylem extracts from trees that contain deep supercooling xylem parenchyma cells. Cryobiology 2007; 55 305-314.

[14] Davies PL, Baardsnes J, Kuiper MJ, Walker VK. Structure and function of antifreeze proteins. Phil. Trans. R. Soc. Lond. B. 2002; 357 927-935.

[15] Biggs EK. The supercooling of water. Proc. Phy. Soc., B66, 688-703 (1953).

[16] Franks F. in "Biophysics and Biochemistry at Low Temperatures", Japan UNI Agency Inc., 1985 pp. 36-37.

[17] Fletcher NH. in "The Chemical Physics of ice", Cambridge University Press, 1970 pp. 271

[18] Wolber PK. Bacterial ice nucleation. Adv. Microb. Physiol. 1993; 34 203-237.

[19] Murase Y, Ruike M, Matsunaga N, Hayakawa M, Kaneko Y, Ono Y. Spider silk has an ice nucleation activity. Naturwissenschaften 2001; 88 117-118.

[20] Gross DC, Proebsting EL, Maccrindle-Zimmenrman H. Development, distribution, and characteristics of intrinsic, nonbacterial ice nuclei in prunus wood. Plant Physiol. 1988; 88 915-922.

[21] Kieft TL. Ice nucleation activity in lichen. Appl. Environ. Microbiol. 1988; 54 1678-1681.

[22] Kieft TL, Ruscetti T. Characterization of biological ice nuclei from a lichen. J. Bacteriol. 1990; 172 3519-3523.

[23] Obata H, Shiga T, Takemura T, Kawahara H, Yamamoto Y. Properties of Cell-Free Ice Nuclei from a Novel Ice Nucleation- Active *Heterodermia obscurata* (Nyl.) *Trevis.* Lichenology 2006; 5 37-44.

[24] Storey KB, Baust JG, Wolanczyk JP. Biochemical modification of plasma ice nucleating activity in a freeze-tolerant frog. Cryobiology 1992; 29 374-384.

[25] Warren G, Corotto L, Wolber P. Conserved repeats in diverged ice nucleation structural gene from two species of *Pseudomonas*. Nucleic. Acids Res. 1986; 14 8047-8060.

[26] Green R. Warren G. Physical and functional repetition in a bacterial ice nucleation gene. Nature 1985; 317 645-648.

[27] Abe K, Watabe S, Emori Y, Watanabe M, Arai S. An ice nucleation active gene of *Erwinia ananas*. FEBS Lett 1989; 258 297-300.

[28] Michigami Y, Watabe S, Abe K, Obata H, Arai S. Cloning and Sequencing of an ice nucleation active gene of *Erwinia uredovoa*. Biosci. Biotechnol. Biochem. 1994; 58 762-764.

[29] Warren G, Corotto L. The consensus sequence of ice nucleation protein from *Erwinia herbicola, Pseudomonas fluorescens,* and *Pseudomonas syringae*. Gene 1989; 85 241-244.

[30] Zhao J, Orser CS. Conserved repetition in the ice nucleation gene inaX from *Xanthomonas campestris* pv. *translucens*. Mol. Gen. Genet. 1990; 223 163-166.

[31] Pouleur S, Richard C, Martin JG, Antoun H. Ice nucleation activity in *Fusarium acuminatum* and *Fusarium avenaceum*. Appl. Environ. Microbiol. 1992; 58 2960-2964.

[32] Kawahara H, Mano Y, Obata H. Purification and characterization of extracellular ice-nucleating matter from *Erwinia uredovora* KUIN-3. Biosci. Biotechnol. Biochem. 1993; 57 1429-1432.

[33] Hasegawa Y, Ishihara Y, Tokuyama T. Characteristics of ice-nucleation activity in *Fusarium avenaceum* IFO7158. Biosci. Biotechnol. Biochem. 1994; 58 2273-2274.

[34] Gurian-Sherman D, Lindow SE. Bacterial ice nucleation: significance and molecular basis. FASEB J. 1993; 7 1338-1343.

[35] Kajaba AV, Lindow SE. A model of the three-dimensional structure of ice nucleation proteins. J. Mol. Biol. 1993; 232 709-7176.

[36] Graether SP, Jia Z. Modeling *Pseudomonas syringae* ice-nucleation protein as a β-helical protein. Biophys J. 2001; 80 1169-1173.

[37] Kozloff LM, Turner MA, Arellano F, Lute M. Phophatydylinositol, a phospholipid of ice-nucleating bacteria. J Bacteriol. 1991; 65 2053-2060.

[38] Michigami Y, Abe K, Obata H, Arai S. Significance of the C-terminal domain of Erwinia uredovora ice nucleation-active protein (inaU). J Biochem (Tokyo) 1995; 118 1279-1284.

[39] Schmid WD. Survival of frogs in low temperature. Science 1982; 215 697-698.

[40] Lee MR, Lee RE Jr, Strong-Gunderson JM, Mings SR. Isolation of ice-nucleating active bacteria from the freeze-tolerant frog, *Rana sylvatica*. Cryobiology 1995; 358-365.

[41] Burke MJ, Gusta LA, Quamme HA, Weiser CJ, Li PH. Freezing and injury to plants. Annu. Rev. Plant Physiol. 1976; 27 507-528.

[42] Proebsting EL Jr, Andrews PK, Gross D. Supercooling young developing fruit and floral buds in deciduous orchards. Hortic Sci. 1982; 17 67-68.

[43] Andrews PK, Proebsting EL Gross DC. Ice nucleation and supercooling in freeze-sensitive peach and sweet cherry tissues. J Am. Soc. HOrtic Sci. 1986; 111 232-236.

[44] Vali G. Quantitative evaluation of experimental results on the heterogeneous freezing nucleation of supercooled liquids. J. Atoms Sci. 1971; 28 402-409.

[45] Lindow SE, Arny DC, Upper CD. Bacterial ice nucleation: a factor in frost injury to plants. Plant Physiol. 1982; 70 1084-1089.

[46] Yankofsky SA, Levin Z, Bertold T, Sandlrman N. Some basic characteristics of bacterial freezing nuclei. J Appl Meteorol 1983: 20 1013–1019.

[47] Wharton DA, Mutch JS, Wilson PW, Marshall CJ, Lim M. A simple ice nucleation spectrometer. Cryo Letters 2004: 25 335-340.

[48] Watanabe M, Watanabe J, Makino T, Homma K, Kumeno K, Arai S. Isolation and cultivation of a novel ice nucleation-active strain of *Xanthomonas campestris*. Biosci. Biotechnol. Biochem. 1993; 57 994-995.

[49] Homma K, Makino T, Kumeno K, Watanabe M. High-pressure sterilization of ice nucleation-active *Xanthomonas campestris* and its application to egg processing. Biosci. Biotechnol. Biochem. 1993; 57 1091-1094.

[50] Watanabe M, Tesaki S, Arai S. Production of low-salt soy sauce with enriched flavor by freeze concentration using bacterial ice nucleation activity. Biosci. Biotechnol. Biochem. 1996; 60 1519-1521.

[51] Arai S, Watanabe M. The protein that prevent the formation of ice nuclei. Kagaku To Seibutsu 1985; 23 363-367 (in Japanese).

[52] Duman JG. The inhibition of ice nucleators by insect antifreeze proteins is enhanced by glycerol and citrate. J Comp. Physiol. B 2002; 172 163-168.

[53] Parody-Morreale A, Murpy KP, Di Cera E, Fall R, DeVries AL, Gill SJ. Inhibition of bacterial ice nucleators by fish antifreeze glycoproteins. Nature 1988; 333 782-783.

[54] Kawahara H, Nagae I, Obata H. Purification and characterization of a new anti-nucleating protein isolated from *Acinetobacter calcoaceticus* KINI-1. Biocotrol Sci. 1996; 1 11-17.

[55] Yamashita Y, Kawahara H, Obata H. Identification of a novel anti-ice-nucelating polysaccharide from *Bacillus thuringiensis* YY529. Biosci. Biotechnol. Biochem. 2002; 66 948-954.

[56] Kawahara H, Masuda K, Obata H. Identification of a compound in *Chamaecyparis taiwabebsis* inhibiting the ice-nucleating activity of *Pseudomonas fluorescens* KUIN-1. Biosci. Biotechnol. Biochem. 2000; 64 2651-2656.

[57] Kawahara H, Obata H. Identification of a compound in spices inhibiting the ice-nucleating activity of *Erwinia uredovora* KUIN-3. J. Antibact. Antifungi. Jpn. 1996; 24 95-100.

[58] Caple G, Layton RG, McCurdy SN, Dunn C, Culbertson L. Biogenic effects in heterogeneous ice nucleation. CryoLetters 1983; 4 59-64.

[59] Holt CB. The effect of antifreeze proteins and poly (vinyl alcohol) on the nucleation of ice: a preliminary study. CryoLetters 2003; 24 323-330.

[60] Wowk B, Fahy GM. Inhibition of bacterial ice nucleation by polyglycerol polymer. Cryobiology 2002; 44 14-23.

[61] Kasuga J, Hashidoko Y, Nishioka A, Yoshida M, Arakawa K, Fujikawa S. Deep supercooling xylem parenchyma cells of Katsura tree (*Cercidiphyllum japonicum*) contain flavonol glycosides exhibiting high anti-ice nucleating activity. Plant Cell Environ. 2008; 31 1335-1348.

[62] Kasuga J, Fukushi Y, Kuwabara C, Wang D, Nishioka A, Fujikawa E, Arakawa K, Fujikawa S. Analysis of supercooling-facilitating (anti-ice nucleation) activity of flavonol glycosides. Cryobiology 2010; 60 240-243.

[63] Matsukawa H, Yagi Y, Matsuda H, Kawahara H, Yamamoto I, Matsuoka J, Tanaka N. Ascorbic acid 2-glucoside prevents sinusoidal endothelial endothelial cell apoptosis in supercooled preserved grafts in rat liver transplantation. Transplant. Proc. 2000; 32 313-317.

[64] Kami D, Kasuga J, Arakawa K, Fujikawa S. Improved cryopreservation by diluted vitrification solution with supercooling- facilitating flavonol glycoside. Cryobioloigy 2008; 57 242-245.

[65] DeVries AL, Wohlschlag DE. Freezing resistance in some Antarctic fishes. Science 1969; 163 1073-1075.

[66] Knight CA, DeVries AL, Oolman LD. Fish antifreeze protein and the freezing and recrystallization of ice. Nature 1984; 308 295-296.

[67] Raymond JA, DeVries AL. Adsorption inhibition as a mechanism of freezing resistance in polar fishes. Proc. Natl. Acad. Sci. USA 1977; 74 2589-2593.

[68] Davies PL, Baardsnes J, Kuiper MJ, Walker VK. Structure and function of antifreeze proteins. Phil Trans R Soc Lond 2002; 357 927-935.

[69] Wharton DA, Barrett J, Goodall G, Marshall CJ, Ramløv H. Ice-active proteins from the Antarctic nematode Panagrolaimus davidi. Cryobiology 2005; 51 198-207.

[70] Clarke CJ, Buckley SL, Linder N. Ice structuring proteins- a new name for antifreezing proteins. Cryo. Letters 2002; 23 89-92.

[71] Barrett J. Thermal hysteresis proteins. Int. J. Biochem. Cell Biol. 2001; 33 105-117.

[72] Harding MM, Ward LG, Haymet AD. Type I 'antifreeze' proteins. Structure-activity studies and mechanisms of ice growth inhibition. Eur. J. Biochem. 1999; 264 653-665.

[73] Duman JG, and DeVries AL. Freezing resistance in winter flounder. Nature 1974; 274 237-238.

[74] Hew CL, Joshi S, Wang NC, Kao MH, Ananthanarayanan VS. Structures of shorthorn sculpin antifreeze polypeptides. Eur. J Biochem. 1985; 151 161-172.

[75] Evans RP, Fletcher GL. Isolation and purification of antifreeze proteins from skin tissues of snailfish, cunner and sea raven. Biochim Biophys Acta 2004; 1700 209-217.

[76] Baardsnes J, Kondejewski LH, Hodges RS, Chao H, Kay C, Davies PL. New ice-binding face for type I antifreeze protein. FEBS Lett. 1999; 463 87-91.

[77] Marshall CB, Fletcher GL, Davies PL. Hyperactive antifreeze protein in a fish. Nature 2004; 429 153.

[78] Graham LA, Marshall CB, Lin F-H, Campbell RL, Davies PL. Hyperactive antifreeze protein from fish contains multiple ice-binding sites. Biochemistry 2008; 47 2051-2063.

[79] Jia Z, Davies PL. Antifreeze proteins: an unusual receptor-ligand interaction. Trends Biochem Sci 2002; 27 101-106.

[80] Worrall D, Elias L, Ashford D, Smallwood M, Sidebottom C, Lillford P, Telford J, Holt C, Bowles D. A carrot leucine-rich-repeat protein that inhibits ice recrystallization. Science 1998; 282 115-117.

[81] Yu XM, Griffith M. Winter rye antifreeze activity increases in response to cold and drought, but not abscisic acid. Physiol Plant 2001; 112 78-86.

[82] Purdy PD, Buckley SL, Sidebottom CM, Twigg SN, Sevilla MP, Holt CB, Roper D, Telford JH, McArthur AJ, Lillford PJ. The physico-chemical caharacteriztaion of a boiling stable antifreeze protein froma pereninial grass (Lolium perenne). Arch Biochem Biophys. 2003; 410 238-245.

[83] Meyer K, Keil M, Naldrett MJ. A leucine-rich repeat protein of carrot that exhibits antifreeze activity. FEBS Lett. 1999; 447 171-178.

[84] Atici O, Nalbantoglu B. Antifreeze proteins in higher plant. Phytochemistry 2003; 64 1187-1196.

[85] Urrutia ME, Duman JG, Knight CA. Plant thermal hysteresis proteins. Biophys Biochem Acta 1992; 1121 199-206.

[86] Kawahara H, Fujii A, Inoue M, Kitao S, Fukuoka J, Obata H. Antifreeze activity of cold acclimated Japanese radish and purification of antifreeze peptide. Cryo Letters. 2008;

[87] Yeh S, Moffatt BA, Griffith M, Xiong F, Yang DS, Wiesman SB, Sarhan F, Danyluk J, Xue YQ, Hew CL. Chitinase genes responsive to cold encode antifreeze proteins in winter cereals. Plant Physiol 2000; 124 1251-1264.

[88] Tremblay K, Ouellet F, Fournier J, Danyluk J, Sarhan F. Molecular characterization and origin of novel bipartite cold-regulated ice recrystallization inhibition proteins from cereals. Plant Cell Physiol. 2003; 46 884-891.

[89] Sidebottom C, Buckley S, Pudney P, Twigg S, Jarman C, Holt C, Telford J, McArther A, Worrall D, Hubbard R, Lillford P. Heatstable antifreeze protein from grass. Nature 2000; 406 256.

[90] Kuiper MJ, Davies PL, Walker VK. A theoretical model of a plant antifreeze protein from *Lollium perenne*. Biophys J. 2001; 81 3560-3565.

[91] Graham LA, Qin W, Lougheed SC, Davies PL, Walker VK. Evolution of hyperactive, repetitive antifreeze proteins in beetles. J. Mol. Evol. 2007; 64 387-398.

[92] Smallwood M, Worrall D, Byass L, Elias L, Ashford D, Doucet CJ, Holt C, Telford J, Lillford P, Bowles DJ. Isolation and characterization of a novel antifreeze protein from carrot (*Daucus carota*). Biochem. J. 1999; 340 385-391.

[93] Knight CA, Hallett J, DeVries AL. Solute effects on ice recrystallization: an assessment technique Cryobiology 1988; 25 55-60.

[94] Tomczak MM, Marshall CB, Gilbert JA, Davies PL. A facile method for determining ice recrystallization inhibition by antifreeze proteins. Biochem. Biophys. Res Commun. 2003; 311 1041-1046.

[95] Wharton DA, Wilson PW, Mutch JS, Marshall CJ, Lim M. Recrystallization inhibition assessed by splat cooling and optical recrystallometry. Cryo.Letters 2007; 28 61-68.

[96] Kuiper MJ, Lankin C, Gauthier SY, Walker VK, Davies PL. Purification of antifreeze proteins by adsorption to ice. Biochem. Biophys. Res Commun. 2003; 645-648.

[97] Hall-Manning T, Spurgeon M, Wolfreys AM, Baldrick AP. Safety evaluation of ice-structuring protein (ISP) type III HPLC 12 preparation. Lack of genotoxicity and subchronic toxicity. Food Chem. Tocycol. 2004; 42 321-333.

[98] Zhang C, Zang H, Wang L. Effect of carrot (*Daucus carota*) antifreeze proteins on the fermentation capacity of frozen dough. Food Res Int. 2007; 40 763-769.

[99] Payne SR, Young OA. Effects of pre-slaughter administration of antifreeze protein on frozen meat quality. Meat Sci. 1995: 41 147-155.

[100] Kobashigawa Y, Nishiyama Y, Miura Y, Ohgiya S, Miura A, Tsuda S. A part of ice nucleation protein exhibits the ice-binding ability. FEBS Lett. 2005; 579 1493-1497.

[101] Wilson PW, Osterday KE, Heneghan AF, Haymet AD. Type I antifreeze proteins enhance ice nucleation above certain concentrations. J. Biol. Chem. 2010; 285 34741-34745.

Chapter 6

The Mineralization of Bone and Its Analogies with Other Hard Tissues

Ermanno Bonucci

Additional information is available at the end of the chapter

1. Introduction

Bone is one of the many biological tissues characterized by the still poorly defined process of mineralization, that is, by the deposition of inorganic substance in their organic matrix, and is one that, chiefly due to its important biological functions, but also because of its widespread presence in primates and its easy availability, has been investigated for hundred of years by thousand of investigators (see reviews by [1-3]). Its study, however, is a demanding task, because of the complexity and variability of the tissue – factors that derive mainly from differences in its histological types, biochemical composition, phases of maturation, and degree of mineralization (the terms 'mineralization', 'biomineralization', and 'calcification' are treated as equivalents in this paper; see Bonucci [2]). These differences, although recently emphasized [4 5], have not always received due recognition. In this connection, revealing indicative examples include, for instance, the differences between the compact, lamellar bone and spongy, woven bone [4], or between the dense rostrum bone of toothed whales [6] and the medullary bone of birds [7]. These differences may lead to incorrect interpretations of apparently contrasting results, but may also lead to new insights if they are recognized to derive from distinctive traits of tissues belonging to the same broad type, which is what they are. The concepts that follow, which chiefly focus on the local mechanism of matrix calcification, mainly refer to woven bone, which raises fewer technical issues than compact bone, but they are valid for, and can be extended to, all other types of bone and hard tissue.

2. Historical notes

So much research has been carried out on bone mineralization that it is hard to provide an inclusive summary; moreover, research of this kind is often interlaced with that on other calcifying tissues, especially cartilage, enamel and mollusk shells, so that, even if references are limited to the main studies closely connected with the subject of this review, i.e., the local mechanisms of matrix calcification, their selection may appear subjective, and omissions are hard to avoid. With these limitations in mind, the first important investigations and theories on the calcification of bone and other hard tissues date back to the first quarter of the twentieth century. In 1923, Robison [8], considering that the concentration of phosphate ions in the bone matrix is too low for allowing calcium phosphate precipitation, suggested that it could be locally increased by the hydrolysis of phosphate esters promoted by an enzyme, alkaline phosphatase. This theory was later challenged, chiefly because the local amounts of phosphate esters are insufficient to produce concentrations of phosphate ions high enough to exceed the solubility product of calcium phosphate, nor could the problem be solved by resorting to the consideration of phosphate sources such as glycogen or adenosine triphosphate (ATP). It did, however, have the great merit of drawing attention to an enzyme that is actually fundamental to calcification [9], while stressing the primary role of organic molecules in biological calcifications. This was also the important indication given by Freudenberg and György [10] in the same year, 1923. They noted that cartilage calcification occurs in vitro if the tissue is first treated with calcium chloride and then with phosphate, but not the reverse, and concluded that this was due to the binding of calcium to a colloid, with the subsequent binding of phosphate and formation of calcium-phosphoprotein complexes. In line with this concept, Robison and Rosenheim [11] suggested that an enzymatic 'second mechanism' should constitute a 'local factor' that could induce the precipitation of calcium and phosphate ions even when their concentration is too low. On the basis of the correlation between the degree of matrix metachromasia and the calcification process, Sobel [12] suggested that chondroitin sulfate, or the collagen-chondroitin sulfate complex, were possible constituents of the local factor, which was, in any case, supposed to be a component of the organic matrix. Again early in the twentieth century, DeJong [13] first observed that bone powder gives a crystalline X-ray diffraction pattern similar to that of inorganic apatite, an observation which is fundamental to an understanding of the organization of the calcium phosphate that forms the bone mineral, but is, at the same time, the basis for a number of misinterpretations of its crystalline nature, as discussed below (the word 'crystals' or 'crystallites' used in this paper in referring to the bone mineral has been retained for historical reasons, even if it may lead to misconceptions about their true structure and nature). On the basis of the coincidence of their X-ray diffractograms, Caglioti [14] suggested that the inorganic substance and components of the organic matrix were closely connected, and Dawson [15], noticing that trypsin digestion increases the sharpness of the diffractograms, supposed that this was due to the binding of the inorganic substance to the organic matrix. Dallemagne and Melon [16] and Ascenzi [17], on the basis of bone birefringence, hypothesized the existence of organic-inorganic bonds.

The concepts summarized above prompted a number of studies. The histochemical demonstration that in bone and cartilage the matrix that calcifies contains glycoproteins and acid proteoglycans (i.e., is PAS-positive and metachromatic; Pritchard [18]) drew attention to these substances and supported the suggestion that the local factor could be chondroitin sulfate [12]). At almost the same time, Di Stefano et al. [19] suggested that the formation of crystals in the matrix was a process catalyzed by a template, and Neuman and Neuman [20] discussed the possibility that it might be the result of the stereotactic properties of a component of the organic matrix: in their opinion, calcium and phosphate ions could bind to an organic template according to the space relationships in the apatite lattice, so giving rise to apatite nuclei that would grow into definitive crystals by further ion aggregation. Specific steric relationships between amino-acid side-chains groups in the collagen fibrils were then considered the possible nucleation centers within the fibrils [21] (see below). A close relationship between the inorganic substance and the collagen fibrils was shown and repeatedly confirmed by electron microscopy, starting with the pioneering studies of Robinson and Watson [22] and Ascenzi and Chiozzotto [23]. Another fundamental observation was obtained by microradiography: Amprino and Engström [24] first showed that the degree of bone mineralization is not constant, because that of primary bone is always higher than that of secondary bone; moreover, they found that osteon calcification occurs in two stages, of which the first is characterized by quick deposition of about 70% of the final mineral content, and the second by the slow completion of this process.

3. Bone organization

Although, as reported above, the structure and composition of bone vary with its type, it is organized according to well-defined characteristics that allow an easy recognition of the nature and type of the tissue (reviewed by [1]). In this connection, and with reference to the local mechanism of bone mineralization, three main components must be considered, two of which, the collagen fibrils and the interfibrillar, non-collagenous proteins, constitute the bone organic matrix, while the third is the inorganic substance. Obviously, a fourth component consists of the whole complex of osteogenic and osteoclastic cells, as well as osteocytes and bone marrow cells, and cellular products – including matrix vesicles – which play a primary role in bone formation, maintenance and metabolism; they are only considered indirectly in this review, and reference is made to them when needed. In particular, the matrix vesicles, in spite of their fundamental role, are not considered in detail because the complexity and importance of the topic would excessively broaden the text and because excellent reviews are available [25-27].

3.1. Collagen fibrils

Collagen fibrils are the most representative bone constituents, amounting to 18.64% by weight [28] and about 90% [29] of the organic matrix of compact bone. They are type I collagen fibrils, i.e., fibrils about 78 nm in diameter characterized under the electron microscope by an axial periodic structure consisting of the repetition of two consecutive 'bands': an electron-dense band about 0.4D in length and a less electron-dense band about 0.6D in length, the D value being 68-70 nm [30,31]. This ultrastructural feature depends on the spatial arrangement of the collagen molecules, which are rich in glycine (33% of the chain), proline and hydroxyproline (22% of the chains), are 280-300 nm long and 1.4 nm thick, and are formed by three polypeptide chains aggregated in a left-handed helical configuration and twisted around a common axis to form a supercoil [32,33]. The three polypeptide chains are characteristic of type I collagen: two of them have the same amino acid sequence and are called $\alpha1(I)$, one has a different amino acid sequence and is called $\alpha2(I)$ [34]. The way molecules are assembled into fibrils is still uncertain. Hodge and Petruska [35] have suggested that the parallel alignment of the molecules occurs in such a way that they are staggered by approximately a quarter of their length (depicted in Fig. 1), so generating regions where the molecules alternately overlap (dense zones) and are separated by gaps (hole zones) that correspond to the interval between the 'tail' and the 'hand' of successive molecules in the same plane. The fibril structure is stabilized by intra- and intermolecular cross-links [36]. These are also responsible for the low degree of solubility of the bone collagen, which is essentially insoluble in reagents such as NaCl and acetic acid under conditions that solubilize collagen from a wide variety of soft tissues [37]. The three-dimensional assembly of the collagen fibrils is still uncertain; the various theories on this topic have been discussed in several reviews [32,38].

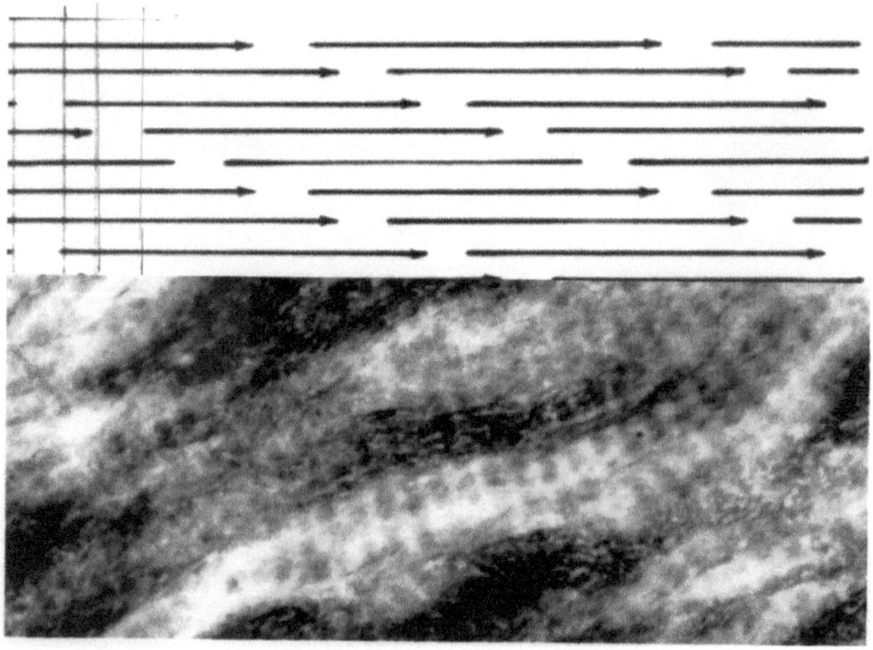

Figure 1. Above: diagram depicting the suggested arrangement of collagen molecules in a fibril: they are shifted in such a way as to generate overlapping (d) and holes (h) zones. Below: in low calcified bone matrix, the inorganic substance is collected in electron-dense bands that cross the collagen fibrils and emphasize their periodic banding. Unstained, x 60,000.

The amount and arrangement of collagen fibrils in bone are not constant and depend on the type of bone (reviewed by [1]). The highest collagen concentration is found in the compact bone of the diaphysis of long bones, where the fibrils are closely packed in lateral register and give rise to the so-called lamellar or parallel-fibered bone. The interfibrillar spaces are reduced to a minimum in this type of bone and the non-collagenous components of the matrix are also at a minimum. A loose arrangement of fibrils is found in woven bone, where irregular interfibrillar spaces contain abundant but variable amounts of non-collagenous proteins [4]. This arrangement is particularly evident in embryonic bone [39] and in the medullary bone of pigeons and other birds [7].

Although type I collagen fibrils are the most abundant component of the bone matrix, and are intimately connected with the inorganic substance (see below), they are not components of all the tissues that calcify and do not appear, therefore, to be essential for the biomineralization process to take place [40]. Thus, the matrix of the calcifying epiphyseal cartilage contains collagen type II, which is thinner than type I and lacks a clearly recognizable periodic pattern; a number of tissues that calcify do not contain collagen fibers at all, as typically occurs in the case of dental enamel. This does not, of course, exclude the possibility that collagen fibrils can induce the calcification process, especially if in combination with phosphoproteins, the collagen-phosphoproteins complex facilitating calcification better than collagen alone [41].

3.2. Non-collagenous components

Besides collagen fibrils, the bone matrix contains a mixture of substances, most of which are permanent components of its structure, while others are transitory cellular products (alkaline phosphatase, growth factors) and molecules present in the circulating fluids (albumin, α_2HS-glycoprotein) [42]. As already mentioned, these substances, roughly indicated as 'non-collagenous components' or 'non-collagenous proteins', vary qualitatively and quantitatively, depending on the type of bone (reviewed by [1,2,43]). Qualitatively, they include proteoglycans, so-called Gla-proteins, phosphoproteins and phospholipids. Quantitatively, their local concentration correlates directly with the speed of formation of the bone tissue and indirectly with the packing density of its collagen fibrils [44].

Proteoglycans. As indicated by their name, these molecules consist of a core protein and a variable number of glycosaminoglycan chains, their properties depending on the composition and arrangement of both these components [45]. There are plenty of them in the cartilage matrix but they are much less frequent in the bone matrix, where their concentration reflects the relatively low amounts of non-collagenous components. They are, in any case, better represented in the uncalcified, osteoid tissue, which is PAS-positive and metachromatic [46], than in the calcified matrix [47], suggesting that they are partly lost during calcification. This has been confirmed by the biochemical analyses of isolated osteons at different degrees of calcification carried out by Pugliarello et al. [48], who found values for exosamines of 0.61% of dry weight in the osteoid tissue, 0.31% in osteons at the lowest degree of calcification, and 0.28% in osteons at the highest degree of calcification (see below for further discussion of this important topic).

The proteoglycan molecules have been divided into three groups on the basis of their properties – the small leucine-rich proteoglycans, the modular proteoglycans, and the cell-surface proteoglycans [49] – the first two groups being components of the bone matrix.

The **small leucine-rich proteoglycans** (SLRPs) are characterized by leucine-rich repeats in their proteic core; they have N-terminal cysteine clusters and at least one GAG chain. Divided into 5 classes [50], they are able to modulate cell-matrix interactions and cell functions, and for this reason are also known as 'matricellular proteins' [51]. The most representative SLRP in bone appears to be decorin (DCN), which modulates collagen matrix assembly and mineralization. It may have an inhibitory role, as suggested by the observation that a high or a low expression of DCN by osteoblasts causes delay or acceleration, respectively, of the calcification process. Studies by Hoshi et al. [52] are in agreement with this possibility: using ultrastructural histochemistry, they have shown that DCN is localized near the collagen fibrils of the osteoid border and that it is removed whenever the fibril fusion occurs at the onset of calcification.

Another member of the SLRP family is biglycan. Its role in bone calcification is not known exactly: however, biglycan deficiency, probably by affecting BMP-4 signal transduction and reducing the Cbfa1 transcription factor [53], causes abnormalities in collagen fibrils [54], induces delayed reparative osteogenesis and leads to an osteoporosis-like phenotype [55].

Other SRLPs (asporin, fibromodulin, lumican, osteoaherin) have been isolated from bone. Although they seem to be involved in collagen fibrillogenesis, their precise function is poorly known.

The **modular proteoglycans** (also called lecticans) form a heterogeneous group of large, often highly glycosylated molecules divided into hyalectans, which bind hyaluronan (hyaluronic acid, HA) and non-hyalectans [49]. The former include aggregan, versican, neurocan and brevican; perlecan is the most representative molecule of the latter. These molecules are poorly represented in bone matrix and their role in calcification, if any, is uncertain.

Aggrecan, which consists of a core protein and many glycosaminoglycans, especially chondroitin sulfate (reviewed by [56]), is abundant in cartilage, where it contributes to the regulation of the mechanical properties. It has been found in the matrix of normal and ectopic bone, where its concentration falls with tissue maturation [57].

Gla-proteins. This name refers to substances whose molecules contain the amino acid γ-carboxyglutamic acid (Gla). They include the bone Gla-protein and the matrix Gla protein.

Bone Gla-protein (BGP), also known as osteocalcin (OC), is contained in the calcified bone matrix, whereas its concentration is very low in the uncalcified osteoid tissue [58]. Its role in bone mineralization is uncertain. When in solution, it causes a delay in calcium phosphate precipitation, an effect that disappears if it is immobilized on sepharose beads [59]; inhibition by warfarin of vitamin K, a cofactor in the carboxylation of glutamate residues, induces a reduction in bone osteocalcin content without altering the bone structure of the rat [60]. Warfarin does, however, disrupt the assembly of the calcification nodules in the rat [61] and osteocalcin-deficient mice show a fall in crystal size and perfection [62]. Krueger et al. [63], in reviewing the topic, stress the role of osteocalcin as a regulator of the mineralization process and suggest that osteocalcin deficiency due to prolonged warfarin treatment may contribute to bone demineralization and vascular calcification (known as the calcification paradox).

Matrix Gla protein (MGP; reviewed by [64]) is found not only in bone, but in all soft tissues too. It probably plays an inhibitory role in calcification. MPG-deficient mice show chondrocyte metaplasia of vascular smooth muscle cells and widespread vascular calcification.

Glycoproteins (phosphoproteins). The bone matrix contains acidic molecules that are rich in glutamic, aspartic and sialic acids, but mostly display *o*-phosphoserine and *o*-phosphothreonine, which goes to show they are phosphoproteins, and are covalently bound to collagen. Those that appear to be most involved in bone mineralization are osteonectin, acidic glycoprotein-75, bone sialoprotein, osteopontin, dentin matrix protein 1, and matrix extracellular phosphoglycoprotein. Apart from the first two, they are grouped under the acronym SIBLING (Small Integrin-Binding Ligand, N-linked Glycoprotein) which represents a family of genetically related proteins clustered on human chromosome 4 [65]. They can also be grouped as intrinsically disordered proteins (IDP), that is, proteins with an irregular, extended conformation and acidic character that facilitate interaction with counter ions and biomineralization [66].

Osteonectin (ON), also called SPARC (Secreted Protein, Acid and Rich in Cysteine) or BM-40 protein, is a glycoprotein expressed in all connective tissues and a number of other soft tissues. Its concentration in bone appears to be inversely correlated with the degree of calcification, with a maximum immunohistochemical reactivity in osteoid tissue [67]. ON-null mice show a higher mineral content and degree of crystallinity than the age-matched wildtype controls [68]. Crystal growth appears to be inhibited by ON in vitro [69], whereas even high concentrations of ON seem to lack nucleating activity. The cDNA sequence of ON reveals potential binding regions for calcium and hydroxyapatite; as it is a matricellular protein, ON may have several functions, such as the regulation of calcium-mediated processes, cell-matrix interactions, and the regulation of bone remodeling [51]. However, its role in bone calcification remains elusive.

Acidic glycoprotein-75 (BAG-75) is a sialic acid-rich phosphoglycoprotein that has been found as 75 and/or 50 kDa forms in mineralized phases of bone and epiphyseal cartilage and in serum [70]. In bone its localization predicts the limits of subsequent mineralization; it delineates condensed mesenchyme regions that accumulate bone sialoprotein and nucleate hydroxyapatite, forming macromolecular complexes that have the potential to sequester phosphate ions [71]. A specific proteolytic processing of bone sialoprotein and bone acidic glycoprotein-75 in these complexes by an osteoblast-derived serine protease seems to be a prerequisite for their mineralization [72].

Bone sialoprotein (BSP) is a glycosylated, sulfated, phosphorylated, sialic acid-rich protein that can bind both hydroxyapatite and cell-surface integrins through the Arg-Gly-Asp motif (reviewed by [73]). Mainly localized in the bone matrix [44], where its concentration varies with the bone type and the degree of calcification, BSP is also expressed by osteoclasts, fetal epiphyseal chondrocytes and the trophoblastic cells of the placenta [74]. It has a close relationship with the collagen fibrils [75] and with the calcification nodules at the calcification front [76]. Its overexpression enhances osteoblast differentiation, calcium incorporation and the formation of calcification nodules by osteoblast cultures [77]. BSP-null mice show abnormal bone growth and defective mineralization that lead to reduced bone formation and the delayed repair of cortical defects [78], whereas BSP overexpression leads to osteopenia and mild dwarfism in mice with an increase in bone resorption and reduction in osteoblast numbers [79]. These results point to a role of BSP in the early stages of calcification, a possibility supported by the demonstration of its capacity for nucleating hydroxyapatite [69].

Osteopontin (OPN), like BSP, is a glycosylated, phosphorylated, sulfated sialoprotein (reviewed by [80]). It contains the Arg-Gly-Asp motif and the amino acid sequence that allows its binding to cell surface and hydroxyapatite. Like BSP, it is contained in the calcified matrix and the calcification nodules, and is found at the bone surface of osteoclasts and other cells; it can, however, be found in many soft tissues, as well. Moreover, unlike BSP, it inhibits apatite nucleation and crystal formation, so that its deficiency increases mineral content and mineral crystallinity in mouse bone [81]. OPN appears to be a multifunctional cytokine that plays a role not only in the regulation of bone formation and resorption, but in many other processes too, such as tissue inflammation and repair, wound healing, angiogenesis, and immunological reactions [82].

Dentin matrix protein 1 (DMP1) is not an exclusive feature of dentin, as its name suggests, but is also found in bone, where it is mainly expressed by osteocytes, and in soft tissues [83]. Its molecule is highly phosphorylated and is cleaved into three distinct segments, two of which are promoters, while the third inhibits the calcification process [84]. DMP1 can, in fact, nucleate hydroxyapatite when immobilized on collagen fibrils. On the basis of electron microscope studies showing that DMP1 induces in vitro the formation of parallel arrays of crystallites similar to those found in the DMP1-rich, collagen-devoid peritubular dentin, Beniash et al. [85] concluded that DMP1 controls the mineral organization outside the collagen fibrils. Studies in vitro by Tartaix et al. [86] suggest that the native form of DMP1 inhibits calcification, but becomes a promoter of the process when cleaved or dephosphorylated. These results, and the observation that its deficiency results in hypomineralized matrix, support the conclusion that DMP1 is a key regulator of mineralized matrix formation [87].

Matrix extracellular phosphoglycoprotein (MEPE; also known as osteoblast/osteocyte factor 45; OF45; Osteoregulin) is highly expressed in osteocytes in human bone [88]. It is a phosphate-regulating factor (phosphatonin) that induces dose-dependent hyperphosphaturia and hypophosphatemia in mice and is a product of the cells of the tumors that cause osteomalacia. Its specific activity is regulated by posttranslational modifications: the phosphorylated intact protein is an effective promoter of mineralization in the gelatin gel diffusion system, while the associated ASARM peptide (acidic serine-aspartate-rich MEPE-associated motif) is an effective inhibitor, and neither MEPE nor ASARM have any effect on mineralization once they have been dephosphorylated [89]. Moreover, dentonin (or AC-100), a synthetic 23-amino-acid peptide derived from MEPE, can stimulate stem cell proliferation in dental pulp [90] and can therefore be active in osteogenesis.

α2-HS glycoprotein/Fetuin, also known as α2-Heremans-Schmid glycoprotein (AHSG), is a prominent, non-collagenous component of the bone matrix, although it is a serum protein synthesized in the liver and is only secondarily accumulated in bone. It plays an inhibitory role on calcification by forming fetuin-mineral complexes, corresponding to high molecular mass complexes of calcium phosphate mineral, fetuin and matrix Gla protein [91]. According to Heiss et al. [92], these fetuin-mineral aggregates give rise to 'calciprotein particles', i.e., colloidal spheres, 30-150 nm in diameter, which are initially amorphous and soluble, but then become progressively more crystalline and insoluble. Their inhibitory effect on calcification is confirmed by studies in cell culture and in the test tube [93] and by the diffuse ectopic calcifications that develop in AHSG-deficient mice on a mineral and vitamin D rich diet [94]. This inhibitory effect might also be effective in regulating collagen mineralization: in vitro studies by Price et al. [95] have shown that the homogeneous nucleation of calcium phosphate occurs within the collagen fibrils in the presence of fetuin, whereas without it mineral grows outside the fibrils. On the basis of these and other results, they have advanced the hypothesis that calcification occurs by inhibitor exclusion ('mineralization by inhibitor exclusion'), that is, the selective mineralization of a matrix using a macromolecular inhibitor of mineral growth.

The function of the enzyme **alkaline phosphatase** (AP) in mineralization has been discussed by Orimo [96] in a recent review, to which the reader can refer. AP is a glycoprotein with calcium-binding properties [97]. It is a component of the bone matrix [98], although it is mainly distributed on the cell membrane and in matrix vesicles. Its tissue isoenzyme TNSALP (tissue non-specific alkaline phosphatase) is critical for mineralization, as is shown by the skeletal rickets like changes that develop in congenital hypophosphatasia and in TNSALP-knockout mice [99]. Interestingly, even in these conditions, the matrix vesicles give rise to the formation of apatite crystals; these, however, do not spread through the surrounding matrix [100], suggesting that TNSALP removes an inhibitor of crystal diffusion, a process that might involve the matrix inorganic pyrophosphate and the expression of the plasma cell membrane glycoprotein-1 (PC-1) that is needed for its synthesis [101]. The Ca-binding property of alkaline phosphatase suggests that another function is feasible, that is, the formation of organic-inorganic hybrids that would constitute the first step in crystal formation (see below).

Phospholipids. Observing that calcification areas are stained by Sudan black B after hot pyridine extraction, Irving [102] first suggested that areas of early calcification contain lipidic material. He subsequently reported that this material was very resistant to extraction before decalcification and consisted predominantly of phosphatidylserine and phosphatidylinositol [103]. Lipids are, in fact, intrinsic components of bone, a calcium-phospholipid-phosphate complex has been extracted from bone, and lipids can be demonstrated histochemically and immunohistochemically [104] in calcification nodules and in matrix vesicles [105]. Lipid involvement in bone calcification has been supported by the results of several studies (reviewed by [106]).

4. The inorganic substance

Composition. It has long been known that the inorganic substance of bone is a calcium phosphate with about 5% by weight of carbonate and traces of other elements (reviewed by [107]), a composition that makes uncertain the true nature of the bone mineral, even if it has been regarded as an hydroxyapatite with the formula $Ca_{10}(PO_4)_6(OH)_2$ [108]. This uncertainty is increased by the variability of values for the Ca/P molar ratio, which are reported to range between 1.57 and 1.71 [109] and, for the weight ratio, between 2.09 and 2.25 in adult human bone and between 1.82 and 1.98 in fetal human bone [110]. In reality, the size and composition of bone apatite changes with age, so that the Ca/P molar ratio varies too. It increases, in fact, from a mean value of 1.35 in the calcification nodules found in the osteoid tissue (i.e., the earliest mineral deposits) to 1.60 in the fully calcified areas [111], or from 1.60-1.70 to 1.81-1.97, respectively [112].

The problem of the nature and composition of bone mineral has been only partly solved by X-ray and electron-diffraction. The early studies by deJong [13]) had shown that the X-ray diffractograms of the compact bone are similar to those of a polycrystalline hydroxyapatite, a conclusion that has been repeatedly confirmed (see reviews by [107,108,113]). This observation has greatly helped to clear up the nature of the bone mineral but, at the same time, has often led to considering it from a purely mineralogical point of view. Actually, the concept that the bone mineral is polycrystalline, and that the crystals are the same as those of the natural hydroxyapatite, has been challenged in several studies. Arnott and Pautard [114] were the first to stress that inorganic particles in bone are defined 'crystals' without any proof that any portion of the mineralized area actually consists of crystals. Arnold et al. [115], by energy-filtering transmission electron microscopy in the selected area electron diffraction mode of dentine, bone, enamel and inorganic apatite mineral, found that the early formed crystallites have a paracrystalline character comparable to biopolymers and that, with the maturation of, and the consequent fall in, the organic proportion in the matrix, the lattice fluctuations of the crystallites diminish, so allowing them to acquire a typical (para)crystalline character. Wheeler and Lewis [116] showed that the crystalline apatite content of untreated mature cortical bovine bone has, in fact, a paracrystalline structure (i.e., no long-range order) and calculated the paracrystalline mean distance fluctuations. Several diffraction studies have shown that the structure, composition and crystallinity of the bone mineral and of the first apatite crystals formed in solution are not constant, and that crystallinity, in agreement with the Ca/P changes reported above, rises with age and degree of maturation [117-121]. In this connection, Landis and Glimcher [112] found that no electron diffraction pattern of poorly crystalline hydroxyapatite, of the type shown by the diffractograms of heavily calcified regions, was generated from the early bone mineral deposits, that the absence of diffraction rings was not due to an insufficient mass of the solid phase, and that there was a progressive change in this pattern towards crystalline diffractograms as the matrix became fully calcified. Moreover, neutron spectroscopic studies by Loong et al. [122] have shown that the bone mineral differs significantly from hydroxyapatite, not only due to the presence of labile and stable CO3 and HPO4 groups, but also because of the predominant, if not total, absence of OH groups from the specific crystallographic lattice sites which they occupy in pure HA. The question is further complicated by the possibility that apatite crystal formation may be preceded by that of an unstable precursor such as amorphous calcium phosphate (for a discussion of this topic see [2]). In conclusion, bone hydroxyapatite appears to differ from natural apatite, and its structure and composition vary with its age and its relationship with organic matrix components.

Morphology. Morphological studies have not provided a definitive answer to the problem of the morphology, size and organization of the inorganic substance in bone. This mainly depends on the fact that electron microscope techniques, the best for visualization of inorganic particles in bone, face major issues in investigating a tissue whose hardness makes difficult to handle at the ultramicrotomic level, so that unexpected changes (for instance, decalcification) can easily be introduced in ultrastructural components [123]. On the other hand, the other physical techniques available for the study of inorganic nanostructures have a poor degree of discrimination and can hardly distinguish different ultrastructures within the same segment of bone (for instance, in areas at different degrees of calcification), because their characteristics are masked by those of the volumetrically prevailing structures. The microscopic dissection, or special physical devices and techniques, can be used to avoid this handicap but they are time-consuming, technically demanding and hard to implement.

One disagreement of great importance for the implications it may have on the understanding of the mechanism of calcification concerns the shape and the size of the mineral particles and their relationship with the components of the organic matrix. The earliest electron microscope studies had led to the conclusion that bone crystals have a platelet-like shape [22,124,125]. This finding was then repeatedly reported in bone and other calcified tissues (dentin, tendons, cartilage) whether using the electron microscope or other methods of investigation [126 139] and the dimensions of the platelets were calculated (mean values reviewed by [2]).

The concept that bone mineral particles are structured as platelets clashes with earlier results obtained from polarized light studies: as early as 1933, Schmidt [140] had shown that bone has an intrinsic birefringence deriving both from its organic and inorganic components, and that the latter has a 'form birefringence' which complies with Wiener's law for rod-like composite bodies. That bone crystals may have a rod-like shape was then confirmed by using the polarizing microscope [16,141,142] and X-ray diffraction [143-148]). It was the electron microscope that definitively showed that bone crystals, at least in part, have a rod-, needle-, or filament-like shape [149-153]), like that found in the crystals of calcified cartilage and other hard tissues (Fig. 2).

The results of the goniometric tilting under the electron microscope of single crystals that had been isolated after bone dissociation have led to the suggestion that the needle-shaped crystals could be no more than a side view of small, thin platelets [126-128]. Obviously, the fact that thin platelets may appear as needles when viewed from one side is not a proof that all needle-like structures observed in bone are due to sideways views of platelets. On the other hand, others suggested that platelet-like and rod-like crystals could be found together in the same bone areas and are different aspects of the same mineral [154] connected to different organic components of the matrix (discussed below). The question is made still more complex by recent atomic force microscopy observations that isolated mineral particles (named 'mineralites') extracted from bovine bone measure 9 x 6 x 2 nm [155,156], values that are much smaller than those obtained with other methods of investigation. In any case, depending on the compactness of bone, crystals may appear as elongated structures shaped as straight, rigid, needle-like or rod-like structures; more often, however, they appear as bent or irregular, filament-like structures (Fig. 2). For this reason, 'filament-like crystals' will be the term used preferentially in the following pages.

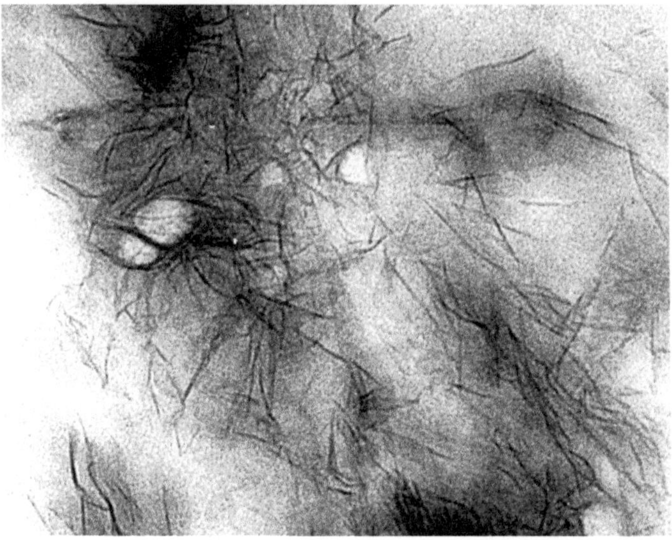

Figure 2. Detail of calcification nodules: crystals appears as elongated, bent, needle- and filament-like structures. Unstained, x 312,000.

Relationships with matrix organic components. Electron microscope studies have clearly shown that the bone mineral is only partly contained in the collagen fibrils, while it is mostly (about 75% of the total according to Pidaparti et al. [157]) located in the extrafibrillar space. This appears to be in contrast with the observation, originally reported in the earliest ultrastructural investigations on bone, that the mineral substance is closely related to the periodic banding of collagen [22,23,158], a relationship later repeatedly confirmed (reviewed by [2,3,38,159,160]. The mineral, in fact, because of its intrinsic electron density, produces a reinforcement of the collagen periodic banding (Figs. 1,3), as if it was a 'negative stain' [161]. Actually, this kind of picture is due to the location of inorganic particles within the collagen fibrils; more exactly, there is a general consensus that the mineral is contained in the 'holes' of the fibril gap-zones (reviewed by [2]. Moreover, because collagen fibrils aggregate side by side in lateral register during mineralization, their hole zones come to be in register, too, and give rise to transverse, 'electron-dense bands' that cross the collagen bundles (Figs. 1,3). One possibility is that these bands might also be located in pores, channels or grooves resulting from collagen fibril aggregation [132,162]. This ultrastructural pattern is above all expressed in compact bone, especially if incompletely mineralized (bone at the initial stage of mineralization, pathologically hypomineralized bone), whereas it is poorly expressed, or completely lacking, in woven bone or in bones with loose collagen fibrils (Fig. 4). Characteristically, in secondary osteons, which calcify in two phases, it is more clearly recognizable during the initial phase of formation than during the later phase of mineral completion [151]. It has been suggested that the 'dense bands', which consist of nanogranules, might give rise to platelet-like fragments if removed from the fibrils, and this might explain why platelet-like crystals have chiefly been described in studies based on bone dissolution and crystal isolation (reviewed by [2]).

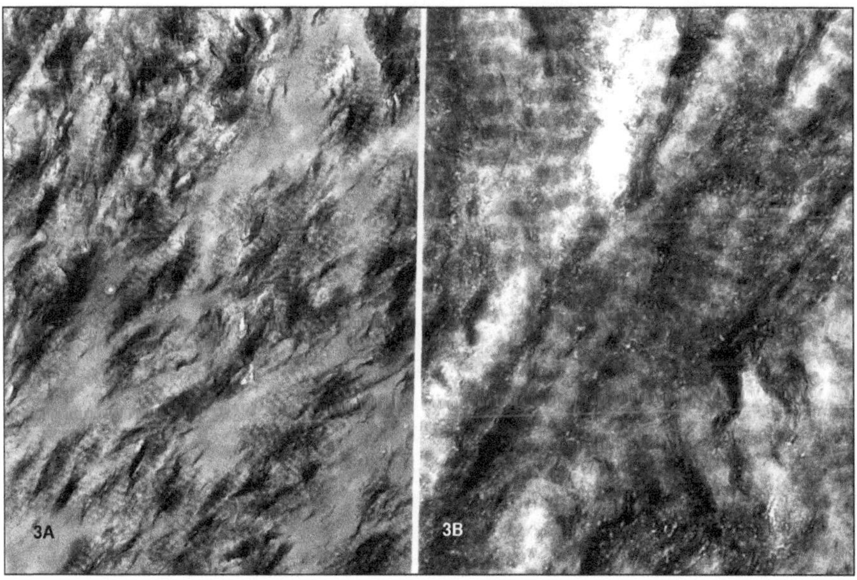

Figure 3. A) Area of initial calcification in compact bone, and B) osteon at the initial stage of formation: the relationship between inorganic substance and collagen periodic banding is clearly recognizable; the apparently empty areas correspond to zones of still uncalcified matrix. Unstained, x 70,000.

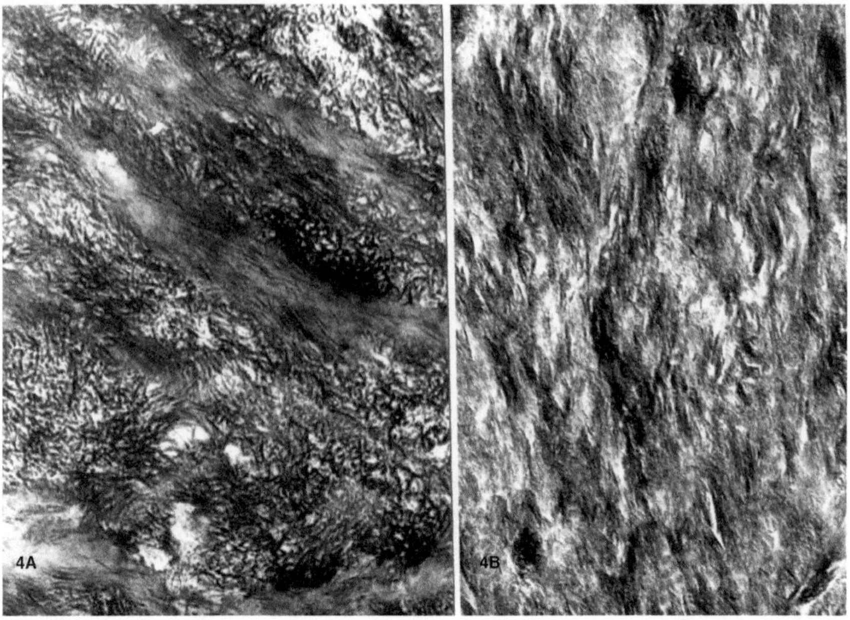

Figure 4. A) Embryonic avian bone: most of the inorganic substance correspond to filament-like crystals; the mineral in bands is practically unrecognizable. B) Osteon at a final stage of calcification: the inorganic substance in bands is poorly visible. Unstained, A x 90,000 and B x 75,000.

The relationship of filament-like crystals with collagen fibrils is much less clear than that of the 'dense bands', both because of the intrinsic difficulty of the problem, and because the crystal arrangement seems to change during mineralization (reviewed by [2,3,159]. At the outset, the elongated crystals form roundish (calcification nodules) or elongated (calcification islands) aggregates in the context of the still uncalcified osteoid borders (Fig. 5). The crystals of the calcification nodules very probably arise in matrix vesicles (Fig. 5, inset), then increase in numbers, acquire a roughly radial arrangement and spread from the nodules (i.e., the fully calcified matrix vesicles) into the surrounding matrix. These crystals are located in the interfibrillar spaces and fail to show any direct relationship with the collagen fibrils, which have no preferential arrangement and are still uncalcified. At this mineralization stage, the calcification nodules closely resemble those visible in the early stages of cartilage calcification; their numbers change with the type of bone, and are inversely related to the compactness of the collagen fibrils. Plenty of them can, in fact, be found in the medullary bone of birds [7]) and in embryonic bone [39], whereas they are much rare in compact bone [151].

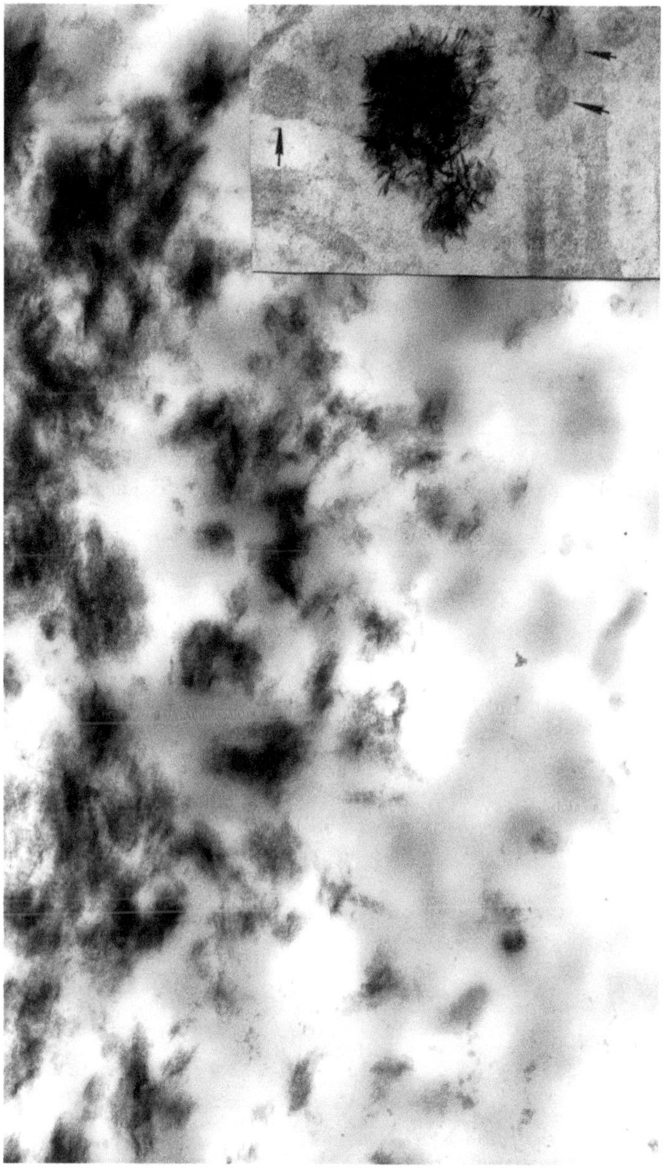

Figure 5. Calcification frontof spongy bone: most of inorganic substance is collected in calcification nodules; no relationship between inorganic substance and the collagen periodic banding is recognizable. Unstained, x 30,000. Inset: Detail of calcification front: a calcification nodule and a few matrix vesicles (arrows) are recognizable. Uranyl acetate and lead citrate, x 60.000.

The calcification islands are characterized by small bundles of needle-like crystals that are arranged side by side, in direct contact with collagen fibrils, and are oriented parallel to them. These crystals, in spite of their adhesion to the fibril surface, appear to be entirely located in the extrafibrillar space, as is also shown by the electron microscope observation that the cross-sectioned collagen fibrils in areas of initial calcification appear as electron-lucent circular spaces surrounded by crystals [6,138,163]. No relationships are found between the crystals of the calcification islands and the collagen period.

There seems to be little doubt that the elongated, filament-like crystals (whether in the calcification nodules or in calcification islands) are located between the collagen fibrils and on their surface, that is, in the extrafibrillar space. The problem arises, therefore, of knowing if they have a relationship with some specific organic component of the bone matrix. The problem is that the already noted intrinsic electron density of the inorganic substance masks the underlying organic structures, if any, so making any direct electron microscope study impossible; moreover, their ultrastructural resolution cannot be improved by removing the mineral through decalcification, because of the extraction artifacts that are implicit in this technique. The only solution available is that of resorting to special decalcification and staining techniques (post-embedding decalcification and staining; cationic dye stabilization) that do not affect the organic components.

The Post-Embedding Decalcification and Staining (PEDS) method relies on the decalcification of the bone matrix after embedding it in an epoxy resin (by floating ultrathin sections on the calcifying solution); the method is based on the observation that the embedding process prevents the distortion and solubilization of the organic components without blocking the removal of the mineral [164]. This has been clearly shown in epiphyseal cartilage (Fig. 6); the results recorded for this tissue are paradigmatic for the organic-inorganic relationship during the early phases of calcification in other tissues and in bone itself. As epiphyseal cartilage is easier to handle than these tissues, it is often used instead of them.

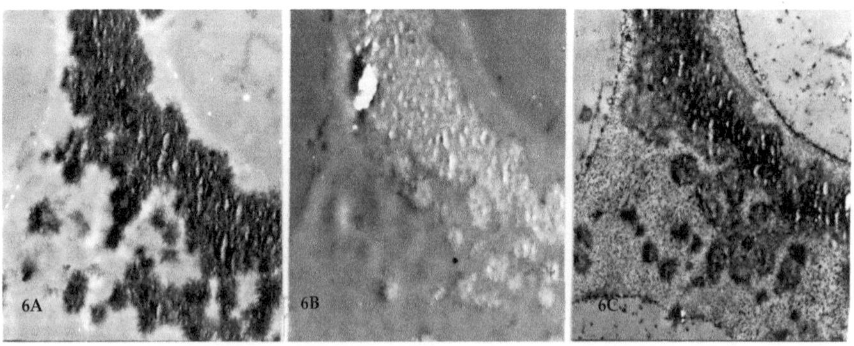

Figure 6. Series of sections from the epiphyseal cartilage: A) unstained; B) decalcified and unstained; C) decalcified and stained with uranyl acetate and lead citrate. PEDS method, x 18,000.

If the PEDS method is interrupted after its first phase, that is, after floating sections on the decalcifying solution, the decalcified areas appear empty and electron-lucent, so showing that all the mineral substance has been removed (Fig. 6B). If the method is completed, that is, the decalcified sections are treated with a heavy metal (usually uranyl acetate, lead citrate, or phosphotungstic acid), the decalcified areas appear to consist of aggregates of filament-like structures (Fig. 6C) whose ultrastructure is very similar to that of untreated crystals (Fig. 7) and for this reason are called 'crystal ghosts' [165]. Crystals and crystal ghosts show such close similarity that the latter were initially considered to be crystals left undecalcified in sections. In reality, they are organic structures. It is crucially important that they are clearly recognizable where the calcification process is active, as in the smallest calcification nodules and at the border of the large but still developing ones; conversely, they disappear in the fully calcified areas (Fig. 8). They have been found in other calcifying tissues, in particular in cartilage, dentin, enamel, calcifying tendons and some pathological tissues [166-174].

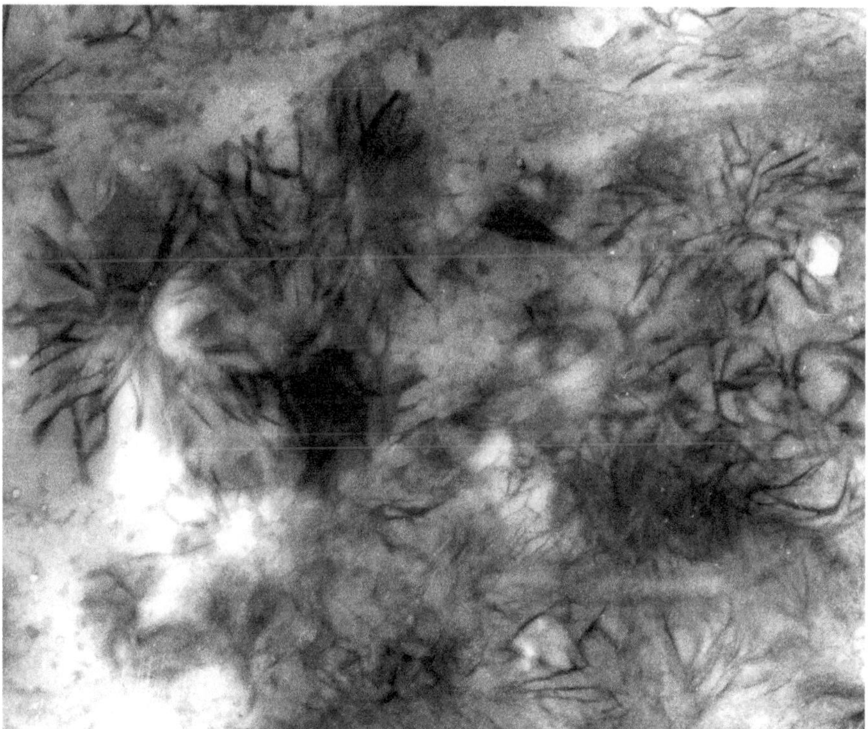

Figure 7. Front of calcification: detail of crystal ghosts. PEDS method (formic acid, uranyl acetate and lead citrate), x 120,000.

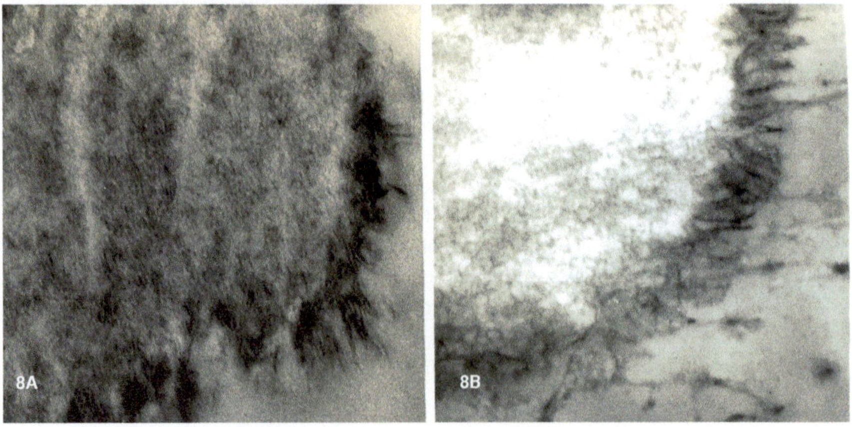

Figure 8. Two serial sections from rat epiphyseal cartilage. A) untreated; B) PEDS method (decalcified with formic acid and stained with uranyl acetate and lead citrate): crystal ghosts are only recognizable at the border of the calcified matrix, where the calcification process is still active, and not in the central, fully calcified area. x 70,000.

Almost the same ultrastructural pictures can be obtained using the cationic dye stabilization method (CDS). This is based on the fact that, if stabilized by cationic substances (usually cationic dyes), acidic molecules resist the extraction produced by decalcifying solutions [175,176]. When this method is used, crystal ghosts are less sharply outlined than those shown by the PEDS method; even so, they are highly reminiscent of the aggregates of untreated filament-like crystals (Fig. 9).

5. The mechanism of mineralization

Although our understanding of bone and other hard tissues has risen sharply in recent years (reviewed by [2]), the mechanism of mineralization is still uncertain. On the basis of the close relationship between the inorganic substance and the hole zones of the collagen period, Glimcher [21] proposed the theory that specific atomic groups located in the hole zones of collagen fibrils are arranged in such a way as to induce heterogeneous nucleation of hydroxyapatite, and that the nuclei subsequently grow by addition of further inorganic ions, so giving rise to crystals (reviewed by [113]. Most of the subsequent investigations on biomineralization, either of bone or other hard tissues, have been carried out on the basis of this theory and important improvements have been obtained. They cannot, however, be considered conclusive because of a number of experimental faults and incongruities. Several electron microscope studies have shown, for instance, that the filament-like crystals are longer than the collagen period [118,130,138,151,177-179] and are not confined to the gap zones of the fibrils but extend into the overlap zones [123,180-182]. This means that the crystals grow outside the holes and, to allow this to happen, they must pierce the intermolecular spaces and enlarge them to find the space required for them to grow. It follows that they should break the intermolecular cross-links – a change that should raise the solubility of bone collagen, which actually decreases during mineralization [37], and should modify the fibril periodic banding as well, which, conversely, is not appreciably affected. Moreover, according to Bachra [183], the many side-chains of the amino acid residues that point laterally at a distance of about 0.286 nm along the axis of the collagen molecules would prevent the growth of the nuclei into crystals. On the basis of neutron diffraction studies of fully mineralized ox bone, it has also been reported that less space is available within collagen fibrils than was previously assumed, and that most of mineral is therefore located outside the fibrils [184]. In line with these findings, Lees and Prostak [153] point out that the size of the crystallites is greater than the intermolecular spacing of collagen fibrils, so that crystallites could not fit in them. They report, in addition, that in fish dentin (which is similar to bone) at the initial stage of calcification the filament-like crystallites form dense strips between the collagen fibrils and practically none of them can be found within fibrils. The same authors calculate that the ratio of the weight of mineral contained in the hole zones to the total mineral content is not greater than 20% in loose bones like that of deer antlers, and this percentage falls to less than 5% for hyperdense bone like porpoise petrosal.

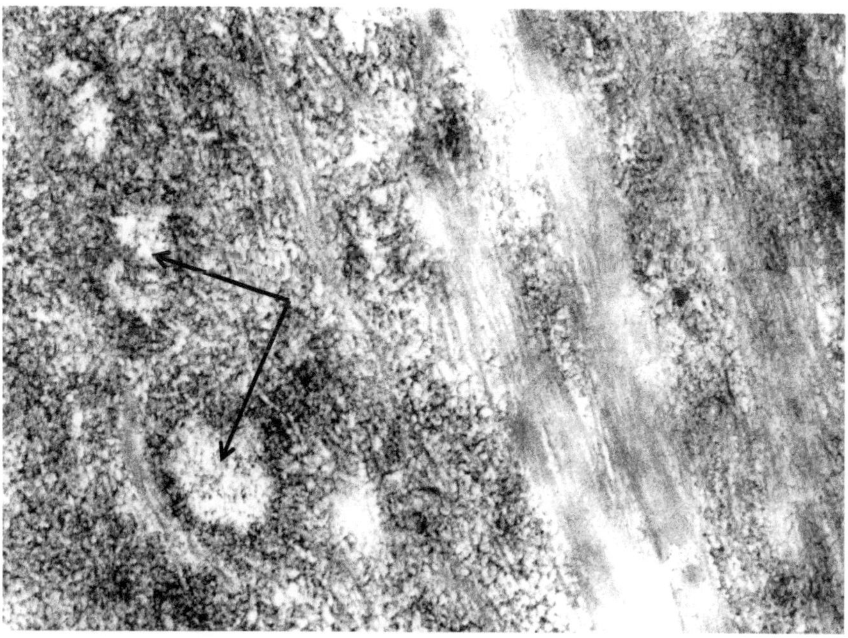

Figure 9. Front of calcification; section from a specimen fixed with glutaraldehyde-acridine orange, decalcified with EDTA and treated with ruthenium red; structures resembling crystal ghosts are recognizable (arrows point to the empty, central zones of the calcification nodules). x 18,000.

These observations converge in suggesting that the filament-like crystals cannot be nucleated in the holes found in collagen fibrils, and that the theory of heterogeneous nucleation must therefore be reconsidered. The electron microscope results leave little doubt that the calcified matrix of bone displays two types of mineral: electron-dense bands related to the periodic banding of collagen, and filament-shaped crystals. Probably because of the suggestion of the theory of heterogeneous nucleation, the latter have been considered to derive from the growth and development of the former. This conclusion is actually arbitrary: a number of electron microscope results have clearly shown that these two types of structure, although both consisting of hydroxyapatite, are separate entities. The aggregates of filament-like crystals, known as calcification nodules, arise in matrix vesicles and spread through the surrounding matrix; they only have secondary contact with collagen fibrils. The crystals of the calcification islands, on the other hand, seem to be in contact with the surface of the collagen fibrils, but are not contained within them. Only the electron-dense, granular bands are related to the collagen period, and are contained in the hole zones. The fact that they become less recognizable as the matrix mineralization proceeds is not due, as often arbitrarily believed, to their transformation into filament-like crystals, but to the masking effect arising from the increase in numbers of the latter.

6. Crystal ghosts and calcification

The obvious conclusion to be drawn from the results discussed above is that the filament-like crystals are in contact with non-collagenous organic structures contained in the interfibrillar spaces and that, as a result, their formation is probably regulated by these same structures. The finding of crystal ghosts, as described above, now takes on a new prominence, because it leads logically into a plausible explanation for the calcification mechanism (discussed by [2,3,159]). This is based on the very close ultrastructural similarity between crystal ghosts and untreated filament-like crystals – a fact that strongly suggests that they are components of the same organic-inorganic, crystal-like structures that are formed through a chemical process in which the crystal ghosts operate as templates. During the early stages of calcification, the structures called crystals would not develop through a process of heterogeneous nucleation but through an epitaxial process implying a link between inorganic cations and organic molecules (crystal ghosts) with the formation of organic- inorganic hybrids whose filament-like shape would simply derive from, and reflect, the filament-like shape of the template. It is also possible that during this initial phase of calcification crystal ghosts stabilize amorphous calcium phosphate, whose existence in bone is still, however, a controversial issue (discussed by [2]). The organic-inorganic composition of the early crystals would explain why the calcification nodules give amorphous diffraction rings at the beginning of their formation and why their crystals have a paracrystalline character and show crystal lattice distortions, as discussed above.

This possibility has been challenged because the location of organic material within the crystals appears to conflict with mineralogical laws [185]. This criticism is based on the concept that the inorganic substance of bone consists of crystals in a mineralogical sense. In reality, as reported above, the early crystals give amorphous diffractograms and the term "crystal" is used only as a matter of habit. If the early crystals are organic-inorganic hybrids, each of them must consist of a mixture of, and imply mutual penetration by, the two components (organic and inorganic), and it is not possible to differentiate between an inner and an outer constituent, especially considering that the filament-like crystals of bone measure less than 10 nm (from 1 to 7.7 nm; see [2]) in thickness. But even allowing the hypothesis that the organic component is contained in the inorganic one, examples of incorporation of organic material in crystal-like structures (biominerals) are quite numerous [186-199]. The organic content can induce lattice distortions [200], but at the same time it enhances the mechanical properties of crystals [192]).

At this point the need for adequate knowledge about the nature of crystal ghosts becomes a priority. So far, ultrastructural histochemical investigations, which might permit the acquisition of this knowledge, have mainly been carried out in cartilage. In this tissue, crystal ghosts are stained by acidic phosphotungstic acid, periodic acid-silver methenamine, periodic acid thiosemicarbazide-osmium; they are reactive with cations and with colloidal iron at pH 2.0 (Fig. 10), but are unreactive after methylation and saponification [201]. These results show that they correspond to acid proteoglycans, a conclusion supported by the finding of chondroitin sulfate in their molecule [202]) and by other histochemical results [167,169,203,204]. That acid proteoglycans can be involved in calcification, either promoting or inhibiting it, has long been known [12,175], and the possibility that hydroxyapatite nanocrystals can self-assemble on chondroitin sulfate templates has been put forward [205]. Of course, other types of the many non-collagenous molecules located in the bone matrix could, alternatively, constitute the crystal ghosts found in bone. In this connection, polyanionic molecules (especially phosphoglycoproteins), often containing the repetitive sequence of aspartic acid, have been described in all calcified tissues [71,206-210] and could well have the same role as that of acid proteoglycans in cartilage; anyway, independently of their nature, crystal ghosts cannot be ignored and any proposal to account for the mechanism of calcification should include their role.

In any case, the formation of crystals appears to be more complex than an epitaxial process alone. It has been stressed above that crystal ghosts disappear from the central zone of developing calcification nodules (i.e., they disappear as the degree of calcification rises; Figs. 8, 9, 10) and that at the same time the electron diffractograms, which are of amorphous type in the calcification nodules at the initial stage of calcification, change into crystalline and acquire the characteristics of hydroxyapatite. These findings show that, as calcification proceeds, the initial organic-inorganic hybrids gradually lose their organic component and change into almost pure inorganic crystals. It may be speculated that an enzymatic mechanism removes the organic component, and that at the same time the residual inorganic backbone acquires further inorganic ions and 'matures', so turning into a hydroxyapatite crystal. This hypothesis is in agreement with the well-known fact that in cartilage the calcification process is characterized by changes in the organic matrix that chiefly imply the loss of acid proteoglycans [211-224]. In this connection, it is known that the calcification process is inhibited in vitro by proteoglycans and is increased by their degradation or removal [219,225], so that their controlled enzymatic breakdown could be considered a possible mechanism by which the calcification of growth plate cartilage might be allowed to advance in vivo [214]. In discussing this problem, Bonucci and Gomez [226] reported the observation that aggregates of filamentous structures, similar to crystal ghosts, can be found in uncalcified cartilage after staining with lanthanum chloride [227]. These structures are focal concentrations of proteoglycans that, due to a modification of their molecule, have acquired high La-binding capacity (in a broader perspective, high Ca-binding capacity) and behave as pre-crystal ghosts.

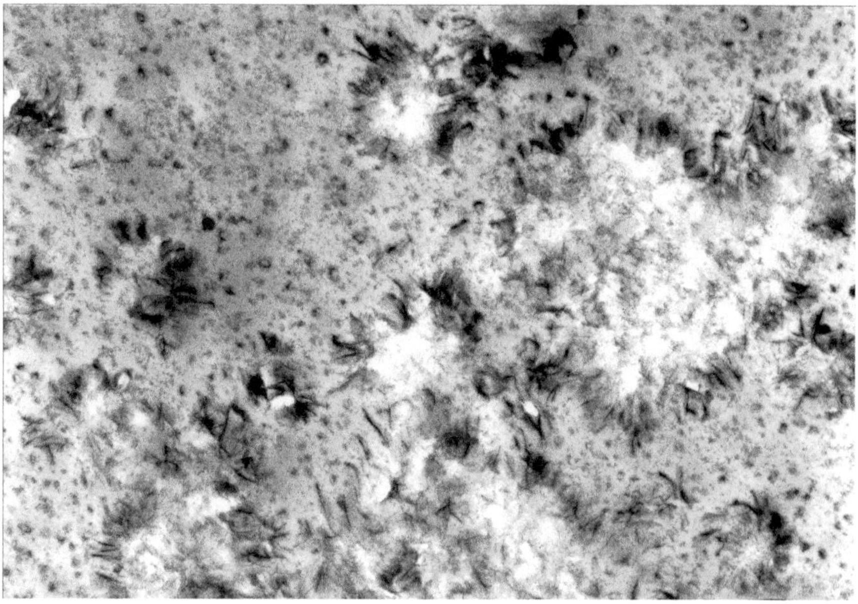

Figure 10. Cartilage section from a calf scapula: PEDS method (decalcified with formic acid and treated with colloidal iron at pH 2.8). Reactive crystal ghosts are recognizable at the periphery of the calcification nodules, whose central, fully calcified areas are negative. The granular structures scattered through the uncalcified matrix correspond to proteoglycan aggregates. X 15,000.

The enzymatic removal of organic material before calcification has been repeatedly demonstrated in enamel, where it is a prerequisite to calcification (reviewed by [228]) and in bone [47,48,229]. Hoshi et al. [52] have reported that bone calcification follows the removal of decorin and the fusion of collagen fibrils. Obviously, loss of other matrix, non-collagenous components may occur too.

Another issue to be resolved is whether this mechanism may also be effective in the case of the intrafibrillar calcification that gives rise to the mineral in bands. The PEDS method shows that lightly stained bands replace the inorganic electron-dense bands. In his case, too, there is a superimposition of organic and inorganic ultrastructures, as if the inorganic material reacts with, and is occluded by, an organic material contained in the hole zone of collagen fibrils, a possibility already suggested by Veis [38].

7. Analogies with other hard tissues

There can be no doubt that the primary role played by the organic components in bone calcification is operative in inducing and regulating the calcification process in other hard tissues. This certainly occurs in tissues, such as dentin, cementum, and calcified tendons, whose matrix, like that of bone, consists of a meshwork of type I collagen fibrils enclosing interfibrillar non-collagenous proteins; but a number of studies shows that it also occurs in tissues that, like epiphyseal cartilage, have a different type of collagen [226,230] or no collagen at all, like tooth enamel [228], mollusk shells [231], spicules of echinoderms [232], crustacean cuticle [233], eggshell [234], and other tissues and organisms. In all these circumstances, an active role in calcification is credited to proteic material, with special reference to acid molecules. In this connection, it is worth mentioning what occurs in enamel, where the organic-inorganic relationships closely resemble those described in bone [170-172,235]: organic septa are intrinsic components of crystals during the early phase of enamel formation, only to be degraded and to disappear as enamel matures. On this topic, it is mandatory to distinguish the early from the final arrangement of the calcified areas. Technical problems often make this distinction impossible to visualize, but, whenever possible, it has shown that the early mineral consists of organic-inorganic hybrids, within which the organic phase is occluded by the inorganic one. This finding has been reported not only in vertebrate calcified tissues, but also in invertebrates and calcified non-skeletal tissues, such as eggshell and several unicellular organisms, as well as in pathologically calcified tissues (reviewed by [2]), suggesting that the initial stabilization of inorganic material by organic structures may be a process shared by all calcified tissues. This possibility is strengthened by what is known about silicification, the process that leads to the hardness of integumental hard tissues in diatom walls, sponge spicules and radiolarian micro-skeletons [236]. As in vertebrate calcification, the silicification process is characterized by a close association of silica with organic components (silicateins) which have both catalytic, enzymatic and templating functions and lead to the formation of organic-inorganic hybrids [237-239]).

8. Conclusions

Without considering the role of the matrix vesicles and alkaline phosphatase, which can be looked up in excellent recent reviews [27,96], this article has examined the possible role in calcification of the organic components of the bone matrix. Most remains to be discovered. The role of collagen fibrils in this tissue appears to be less necessary than previously thought: they are chiefly stromal structures and certainly their participation is not an absolute priority for calcification, which can occur in tissues where they are not found [40]; their hole zones seem to be penetrated by inorganic substance without any change in their molecular arrangement, probably through the same type of mechanism that leads to extrafibrillar calcification. Most of the mineral substance is actually contained in the interfibrillar spaces (about 60% according to [153]), where it is associated with the non-collagenous components of the matrix. There are many of these, and although their individual function is known with a good degree of confidence, their mutual role is still uncertain. The electron microscope clearly shows, however, that there is an intimate contact between the inorganic substance and organic frameworks pertaining to one or several of these non-collagenous molecules. These are named 'crystal ghosts', a term justified by the close similarity between them and the structures that are called 'crystals' or 'crystallites'; their histochemical properties justify their recognition as acidic proteins, which in cartilage are mainly acid proteoglycans; and their evolution as calcification develops justifies considering them as organic templates that link inorganic ions, thus giving rise to organic-inorganic hybrids that gradually acquire a crystalline, hydroxyapatite character as the organic component is eliminated. A number of questions must still be answered: chiefly, the nature of crystal ghosts; how they acquire calcium-binding properties; how they are removed; and their relationship with matrix vesicles and collagen fibrils. They are, in any case, actively involved in the early phases of the calcification process and cannot be ignored. They may prove to be the key to opening up and entering a still partly obscure biological mechanism (biomineralization), so providing insights into the biomimetic construction of biologically inspired materials.

Author details

Ermanno Bonucci

Department of Experimental Medicine, La Sapienza University, Rome, Italy

References

[1] Bonucci E. Basic composition and structure of bone. In: An YH, Draughn RA (eds). *Mechanical testing of bone and the bone-implant interface*. Boca Raton, CRC Press, 2000, p3-21

[2] Bonucci E. *Biological calcifications. Normal and pathological processes in the early stages*. Heidelberg Berlin, Springer, 2007

[3] Bonucci E. Bone mineralization. Front Biosci 2012; 17, 100-128

[4] Gorski JP. Is all bone the same? Distinctive distributions and properties of non-collagenous matrix proteins in lamellar vs. woven bone imply the existence of different underlying osteogenic mechanisms. *Crit Rev Oral Biol Med* 1998; 9, 201-223

[5] Midura RJ, Midura SB, Su X, Gorski JP. Separation of newly formed bone from older compact bone reveals clear compositional differences in bone matrix. *Bone* 2011; 49, 1365-1374

[6] Zylberberg L, Traub W, De Buffrenil V, Allizard F, Arad T, Weiner S. Rostrum of a toothed whale: ultrastructural study of a very dense bone. *Bone* 1998; 23, 241-247

[7] Bonucci E, Gherardi G. Histochemical and electron microscope investigations on medullary bone. *Cell Tiss Res* 1975; 163, 81-97

[8] Robison R. The possible significance of hexosephosphoric esters in ossification. *Biochem J* 1923; 17, 286-293

[9] Posen S, Cornish C, Kleerekoper M. Alkaline phosphatase and metabolic bone disorders. In: Avioli LV, Krane SM (eds). *Metabolic bone disease*. New York, Academic Press, 1977, p141-181

[10] Freudenberg E, György P. Der Verkalkungsvorgang bei der Entwicklung des Knochens. *Ergebn inn Med* 1923; 24, 17-28

[11] Robison R, Rosenheim AH. Calcification of hypertrophic cartilage in vitro. *Biochem J* 1934; 28, 684-698

[12] Sobel AE. Local factors in the mechanism of calcification. *Ann N Y Acad Sci* 1955; 60, 713-731

[13] deJong WF. La substance minérale dans les os. *Rec Trav Chim* 1926; 45, 445-446

[14] Caglioti V. Sulla struttura delle ossa. *Atti V Congr Naz Chimica Pura Applicata*. Rome, Associazione Italiana di Chimica, 1935, p320-331

[15] Dawson JM. X-ray diffraction pattern of bone: evidence of reflexions due to the organic constituent. *Nature* 1946; 157, 660-661

[16] Dallemagne MJ, Melon J. Nouvelles recherches relatives aux propriétés optique de l'os: la biréfringence de l'os minéralisé; relations entre les fractions organiques et inorganique de l'os. *J Washington Acad Sci* 1946; 36, 181-195

[17] Ascenzi A. On the existence of bonds between ossein and inorganic bone fraction. *Science (Lancaster, Paris)* 1950; 112, 84-86

[18] Pritchard JJ. A cytological and histochemical study of bone and cartilage formation in the rat. *J Anat* 1952; 86, 259-277

[19] DiStefano V, Neuman WF, Rouser G. The isolation of a phosphate ester from calcifiable cartilage. *Arch Biochem Biophys* 1953; 53, 218-220

[20] Neuman WF, Neuman MW. The nature of the mineral phase of bone. *Chem Rev* 1953; 53, 1-45

[21] Glimcher MJ. Molecular biology of mineralized tissues with particular reference to bone. *Rev Modern Phys* 1959; 31, 359-393

[22] Robinson RA, Watson ML. Collagen-crystal relationships in bone as seen in the electron microscope. *Anat Record* 1952; 114, 383-409

[23] Ascenzi A, Chiozzotto A. Electron microscopy of the bone ground substance using the pseudo-replica technique. *Experientia* 1955; 11, 140

[24] Amprino R, Engstrom A: Studies on X ray absorption and diffraction of bone tissue. *Acta Anat* 1952; 15, 1-22

[25] Anderson HC, Garimella R, Tague SE. The role of matrix vesicles in growth plate development and biomineralization. *Front Biosci* 2005; 10, 822-837

[26] Golub EE. Role of matrix vesicles in calcification. *Biochim Biophys Acta* 2009; 1790, 1592-1598

[27] Wuthier RE, Lipscomb GF. Matrix vesicles: structure, composition, formation and function in calcification. *Front Biosci* 2011; 17, 2812-2902

[28] Eastoe JE, Eastoe B. The organic constituents of mammalian compact bone. *Biochem J* 1954; 57, 453-459

[29] McLean FC, Urist MR. *Bone*. Chicago, The University of Chicago Press, 1968

[30] Olsen BR. Electron microscope studies on collagen I. Native collagen fibrils. *Z Zellforsch* 1963: 59, 184-198

[31] Bairati A, Petruccioli MG, Torri Tarelli L. Studies on the ultrastructure of collagen fibrils 1. Morphological evaluation of the periodic structure. *J Submicr Cytol* 1969; 1, 113-141

[32] Hulmes DJS. Building collagen molecules, fibrils, and suprafibrillar structures. *J Struct Biol* 2002; 137, 2-10

[33] Ramachandran GN, Kartha G. Structure of collagen. *Nature* 1954; 174, 269-279

[34] van der Rest M. The collagens of bone. In: Hall BK (ed). *Bone. Vol. 3: Bone matrix and bone specific products*. Boca Raton, CRC Press, 1991, p187-237

[35] Hodge AJ, Petruska JA. Recent studies with the electron microscope on ordered aggregates of the tropocollagen molecules. In: Ramachandran GN (ed). *Aspects of protein structure*. London, Academic Press, 1963, p289-300

[36] Knott L, Bailey AJ. Collagen cross-links in mineralizing tissues: a review of their chemistry, function, and clinical relevance. *Bone* 1998; 22, 181-187

[37] Glimcher MJ, Katz EP. The organization of collagen in bone: the role of noncovalent bonds in the relative insolubility of bone collagen. J Ultrastruct Res 1965; 12, 705-729

[38] Veis A. Mineralization in organic matrix frameworks. *Rev Mineral Geochem* 2003; 54, 249-289

[39] Bonucci E, Silvestrini G. Ultrastructure of the organic matrix of embryonic avian bone after en bloc reaction with various electron-dense 'stains'. *Acta Anat* 1996; 156, 22-33

[40] Hosseini MM, Peel SAF, Davies JE. Collagen fibres are not required for initial matrix mineralization by bone cells. *Cells and Materials* 1996; 6, 233-250

[41] Glimcher MJ. Mechanism of calcification: role of collagen fibrils and collagen-phosphoprotein complexes in vitro and in vivo. Anat Record 1989; 224, 139-153

[42] Quelch KJ, Cole WG, Melick RA. Noncollagenous proteins in normal and pathological human bone. *Calcif Tissue Int* 1984; 36, 545-549

[43] Gorski JP. Biomineralization of bone: a fresh view of the roles of non-collagenous proteins. *Front Biosci* 2011; 17, 2598-2621

[44] Nanci A: Content and distribution of noncollagenous matrix proteins in bone and cementum: relationship to speed of formation and collagen packing density. *J Struct Biol* 1999; 126, 256-269

[45] Iozzo RV. Matrix proteoglycans: from molecular design to cellular function. *Ann Rev Biochem* 1998; 67, 609-652

[46] Caretto L. Contributo allo studio della matrice ossea neodeposta. *Arch Putti Chir Org Mov* 1958; 10, 211-230

[47] Baylink D, Wergedal J, Thompson E.: Loss of proteinpolysaccharides at sites where bone mineralization is initiated. *J Histochem Cytochem* 1972; 20, 279-292

[48] Pugliarello MC, Vittur F, de Bernard B, Bonucci E, Ascenzi A. Chemical modifications in osteones during calcification. *Calcif Tissue Res* 1970; 5, 108-114

[49] Schaefer L, Schaefer RM. Proteoglycans: from structural compounds to signaling molecules. *Cell Tissue Res* 2010; 339, 237-246

[50] Schaefer L, Iozzo RV. Biological functions of the small leucine-rich proteoglycans: from genetics to signal transduction. *J Biol Chem* 2008; 283, 21305-21309

[51] Bornstein P, Sage EH. Matricellular proteins: extracellular modulators of cell function. *Curr Opin Cell Biol* 2009; 14, 608-616

[52] Hoshi K, Kemmotsu S, Takeuchi Y, Amizuka N, Ozawa H. The primary calcification in bones follows removal of decorin and fusion of collagen fibrils. *J Bone Miner Res* 1999; 14, 273-280

[53] Parisuthiman D, Mochida Y, Duarte WR, Yamauchi M. Biglycan modulates osteoblast differentiation and matrix mineralization. *J Bone Miner Res* 2005; 20, 1878-1886

[54] Corsi A, Xu T, Chen X-D, Boyde A, Liang J, Mankani M, Sommer B, Iozzo RV, Eichstetter J, Robey PG, Bianco P, Young MF. Phenotypic effects of biglycan deficiency are linked to collagen fibril abnormalities, are synergized by decorin deficiency, and

mimic Ehlers-Danlos-like changes in bone and other connective tissues. *J Bone Miner Res* 2002; 17, 1180-1189

[55] Xu T, Bianco P, Fisher LW, Longenecker G, Smith E, Goldstein S, Bonadio J, Boskey A, Heegaard A.-M, Sommer B, Satomura K, Dominguez P, Zhao C, Kulkarni AB, Gehron Robey P, Young MF. Targeted disruption of the biglycan gene leads to an osteoporosis-like phenotype in mice. *Nature Genet* 1998; 20, 78-82

[56] Kiani C, Chen L, Wu YJ, Yee AJ, Yang BB. Structure and function of aggrecan. *Cell Res* 2002; 12, 19-32

[57] Mania VM, Kallivokas AG, Malavaki C, Asimakopoulou AP, Kanakis J, Theocharis AD, Klironomos G, Gatzounis G, Mouzaki A, Panagiotopoulos E, Karamanos NK. A comparative biochemical analysis of glycosaminoglycans and proteoglycans in human orthotopic and heterotopic bone. *IUBMB Life* 2009; 61, 447-452

[58] Bianco P, Hayashi Y, Silvestrini G, Termine JD, Bonucci E. Osteonectin and Gla-protein in calf bone: ultrastructural immunohistochemical localization using the protein A-gold method. *Calcif Tissue Int* 1985; 37, 684-686

[59] Doi Y, Horiguchi T, Kim S-H, Moriwaki Y, Wakamatsu N, Adachi M, Shigeta H, Sasaki S, Shimokawa H. Immobilized DPP and other proteins modify OCP formation. *Calcif Tissue Int* 1993; 52, 139-145

[60] Price PA, Williamson MK. Effects of warfarin on bone. Studies on the vitamin K-dependent protein of rat bone. *J Biol Chem* 1981; 256, 12754-12759

[61] Amizuka N., Li M, Hara K, Kobayashi M., de Freitas PH, Ubaidus S, Oda K, AkiyamaY. Warfarin administration disrupts the assembly of mineralized nodules in the osteoid. J Electron Microsc (Tokyo) 2009; 58, 55-65

[62] Boskey AL, Gadaleta S, Gundberg C, Doty SB, Ducy P, Karsenty G. Fourier transform infrared microspectroscopic analysis of bones of osteocalcin-deficient mice provides insight into the function of osteocalcin. *Bone* 1998; 23, 187-196

[63] Krueger T, Westenfeld R, Schurgers L, Brandenburg V. Coagulation meets calcification: the vitamin K system. *Int J Artif Org* 2009; 32, 67-74

[64] Price PA. Gla-containing proteins of bone. *Connect Tissue Res* 1989; 21, 51-69

[65] Fisher LW, Fedarko NS. Six genes expressed in bones and teeth encode the current members of the SIBLING family of proteins. *Connect Tissue Res* 2003; 44 (Suppl. 1), 33-40

[66] Wojtas M, Dobryszycki P, Ozyhar A. Intrinsically disordered proteins in biomineralization. In: Seto J (ed). *Advanced topics in biomineralization*. Rijeka, InTech 2012, p3-32

[67] Bianco P, Silvestrini G, Termine JD, Bonucci E. Immunohistochemical localization of osteonectin in developing human and calf bone using monoclonal antibodies. Calcif Tissue Int 1988; 43, 155-161

[68] Boskey AL, Moore DJ, Amling M, Canalis E, Delany AM. Infrared analysis of the mineral and matrix in bones of osteonectin-null mice and their wildtype controls. *J Bone Miner Res* 2003; 18, 1005-1011

[69] Hunter GK, Hauschka PV, Poole AR, Rosenberg LC, Goldberg HA. Nucleation and inhibition of hydroxyapatite formation by mineralized tissue proteins. *Biochem J* 1996; 317, 59-64

[70] Gorski JP, Griffin D, Dudley G, Stanford C, Thomas R, Huang C, Lai EL, Karr B, Solursh M. Bone acidic glycoprotein-75 is a major synthetic product of osteoblastic cells and localized as 75- and/or 50-kDa forms in mineralized phases of bone and growth plate and in serum. *J Biol Chem* 1990: 265, 14956-14963

[71] Gorski JP, Kremer EA, Chen Y, Ryan S, Fullenkamp C, Delviscio J, Jensen K, McKee MD. Bone acidic glycoprotein-75 self-associates to form macromolecular complexes in vitro and in vivo with the potential to sequester phosphate ions. *J Cell Biochem* 1997; 64, 547-564

[72] Huffman NT, Keightley JA, Chaoying C, Midura RJ, Lovitch D, Veno PA, Dallas SL, Gorski JP. Association of specific proteolytic processing of bone sialoprotein and bone acidic glycoprotein-75 with mineralization within biomineralization foci. *J Biol Chem* 2007; 282, 26002-26013

[73] Ganns A, Kim RH, Sodek J. Bone sialoprotein. *Crit Rev Oral Biol Med* 1999; 10, 79-98

[74] Bianco P, Fisher LW, Young MF, Termine JD, Gehron Robey P. Expression of bone sialoprotein (BSP) in developing human tissues. *Calcif Tissue Int* 1991; 49, 421-426

[75] Baht GS, Hunter GK, Goldberg HA. Bone sialoprotein-collagen interaction promotes hydroxyapatite nucleation. *Matrix Biol* 2008; 27, 600-608

[76] Nefussi JR, Brami G, Modrowski D, Oboeuf M, Forest N. Sequential expression of bone matrix proteins during rat calvaria osteoblast differentiation and bone nodule formation in vitro. *J Histochem Cytochem* 1997; 45, 493-503

[77] Gordon JA, Tye CE, Sampaio AV, Underhill TM, Hunter GK, Goldberg HA. Bone sialoprotein expression enhances osteoblast differentiation and matrix mineralization in vitro. *Bone* 2007; 41, 462-473

[78] Monfoulet L, Malaval L, Aubin JE, Rittling SR, Gadeau AP, Fricain JC, Chassande O. Bone sialoprotein, but not osteopontin, deficiency impairs the mineralization of regenerating bone during cortical defect healing. *Bone* 2010; 46, 447-452

[79] Valverde P, Zhang J, Fix A, Zhu J, Ma W, Tu Q, Chen J. Overexpression of bone sialoprotein leads to an uncoupling of bone formation and bone resorption in mice. *J Bone Miner Res* 2008; 23, 1775-1788

[80] McKee MD, Nanci A. Osteopontin: an interfacial extracellular matrix protein in mineralized tissues. *Connect Tissue Res* 1996; 35, 197-205

[81] Boskey AL, Spevak L, Paschalis E, Doty SB, McKee M.D. Osteopontin deficiency increases mineral content and mineral crystallinity in mouse bone. *Calcif Tissue Int* 2002; 71, 145-154

[82] Scatena M, Liaw L, Giachelli CM. Osteopontin. A multifunctional molecule regulating chronic inflammation and vascular disease. *Arterioscler Thromb Vasc Biol* 2007; 27, 2302-2309

[83] Terasawa M, Shimokawa R, Terashima T, Ohy K, Takagi Y. Expression of dentin matrix protein 1 (DMP1) in nonmineralized tissues. *J Bone Miner Metab* 2004; 22, 430-438

[84] Gericke A, Qin C, Sun Y, Redfern R, Fujimoto Y, Taleb H, Butler WT, Boskey AL. Different forms of DMP1 play distinct roles in mineralization. *J Dent Res* 2010; 89, 355-359

[85] Beniash E, Deshpande AS, Fang PA, Lieb NS, Zhang X, Sfeir CS. Possible role of DMP1 in dentin mineralization. *J Struct Biol* 2011; 174, 100-106

[86] Tartaix PH, Doulaverakis M, George A, Fisher LW, Butler WT, Qin C, Salih E, Tan M, Fujimoto Y, Spevak L, Boskey AL. *In vitro* effects of dentin matrix protein-1 on hydroxyapatite formation provide insights into *in vivo* functions. *J Biol Chem* 2004; 279, 18115-18120

[87] George A, Ramachandran A, Albazzaz M, Raviandran S. DMP1- a key regulator in mineralized matrix formation. *J Muscoloskel Neuron Interact* 2007; 7, 308

[88] Nampei A, Hashimoto J, Hayashida K, Tsuboi H, Shi K, Tsuji I, Miyashita H, Yamada T, Matsukawa N, Matsumoto S, Ogihara T, Ochi T, Yoshikawa H. Matrix extracellular phosphoglycoprotein (MEPE) is highly expressed in osteocytes in human bone. *J Bone Miner Metab* 2004; 22, 176-184

[89] Boskey AL, Chiang P, Fermanis A, Brown J, Taleb H, David V, Rowe PS. MEPE's diverse effects on mineralization. *Calcif Tissue Int* 2010; 86, 42-46

[90] Liu H, Li W, Gao C, Kumagai Y, Blacher RW, DenBesten PK. Dentonin, a fragment of MEPE, enhanced dental pulp stem cell proliferation. *J Dent Res* 2004; 83, 496-499

[91] Price PA, Nguyen TM, Williamson MK. Biochemical characterization of the serum fetuin-mineral complex. *J Biol Chem* 2003; 278, 22153-22160

[92] Heiss A, DuChesne A, Denecke B, Grötzinger J, Yamamoto K, Renné T, Jahnen-Dechent W. Structural basis of calcification. Inhibition by α_{2}-HS glycoprotein/fetuin-A. Formation of colloidal calciprotein particles. *J Biol Chem* 2003; 278, 13333-13341

[93] Schinke T, Amendt C, Trindl A, Poschke O, Muller-Esterl W, Jahnen-Dechent W. The serum protein α2-HS glycoprotein/fetuin inhibits apatite formation *in vitro* and in mineralizing calvaria cells. A possible role in mineralization and calcium homeostasis. *J Biol Chem* 1996; 271, 20789-20796

[94] Schäfer C, Heiss A, Schwarz A, Westenfeld R, Ketteler M, Floege J, Müller-Esterl W, Schinke T, Jahnen-Dechent W. The serum protein α_2-Heremans-Schmid glycopro-

tein/fetuin-A is a systematically acting inhibitor of ectopic calcification. *J Clin Invest* 2003; 112, 357-366

[95] Price PA, Toroian D, Lim JE. Mineralization by inhibitor exclusion: the calcification of collagen with fetuin. *J Biol Chem* 2009; 284, 17092-17101

[96] Orimo H. The mechanism of mineralization and the role of alkaline phosphatase in health and disease. *J Nippon Med Sch* 2010; 77, 4-12

[97] de Bernard B, Bianco P, Bonucci E, Costantini M, Lunazzi GC, Martinuzzi P, Modricky C, Moro L, Panfili E, Pollesello P, Stagni N, Vittur F. Biochemical and immunohistochemical evidence that in cartilage an alkaline phosphatase is a Ca^{2+}-binding glycoprotein. *J Cell Biol* 1986; 103, 1615-1623

[98] Bonucci E, Silvestrini G, Bianco P. Extracellular alkaline phosphatase activity in mineralizing matrices of cartilage and bone: ultrastructural localization using a cerium-based method. *Histochemistry* 1992; 97, 323-327

[99] Fedde KN, Blair L, Silverstein J, Coburn SP, Ryan LM, Weinstein RS, Waymire K, Narisawa S, Millan JL, MacGregor GR, Whyte MP. Alkaline phosphatase knock-out mice recapitulate the metabolic and skeletal defects of infantile hypophosphatasia. *J Bone Miner Res* 1999; 14, 2015-2026

[100] Anderson HC, Sipe JB, Hessle L, Dhamyamraju R, Atti E, Camacho NP, Millàn JL. Impaired calcification around matrix vesicles of growth plate and bone in alkaline phosphatase-deficient mice. *Am J Pathol* 2004; 164, 841-847

[101] Johnson KA, Hessle L, Vaingankar S, Wennberg C, Mauro S, Narisawa S, Goding JW, Sano K, Millan JL, Terkeltaub R. Osteoblast tissue-nonspecific alkaline phosphatase antagonizes and regulates PC-1. *Am J Physiol Regul Integr Comp Physiol* 2000; 279, R1365-R1377

[102] Irving JT. A histological staining method for site of calcification in teeth and bone. *Arch Oral Biol* 1959; 1, 89-96

[103] Irving JT. Interrelations of matrix lipids, vesicles, and calcification. *Fed Proc* 1976; 35, 109-111

[104] Bonucci,E., Silvestrini G, Mocetti P. MC22-33F monoclonal antibody shows unmasked polar head groups of choline-containing phospholipids in cartilage and bone. *Eur J Histochem* 1997; 41, 177-190

[105] Wuthier RE. Lipids of matrix vesicles. *Feder Proc* 1976; 35, 117-121

[106] Dziak R. Role of lipids in osteogenesis: cell signaling and matrix calcification. In: Bonucci E (ed). *Calcification in biological systems*. Boca Raton, CRC Press, 1992, p59-71

[107] Eanes ED, Posner AS. Structure and chemistry of bone mineral. In: Schraer H (ed). *Biological calcification*. New York, Appleton-Century-Crofts, 1970, p1-26

[108] Posner AS. Bone mineral and the mineralization process: In: Peck WA: *Bone and mineral research 5*. Amsterdam, Elsevier Science Publisher, 1987, p65-116

[109] Woodard HQ. The elementary composition of human cortical bone. *Health Phys* 1962; 8, 513-517

[110] Zipkin I. The inorganic composition of bones and teeth. In: Schraer H (ed). *Biological calcification: cellular and molecular aspects*. New York, Appleton-Century-Crofts, 1970, p69-103

[111] Wergedal JE, Baylink DJ. Electron microprobe measurements of bone mineralization rate in vivo. *Am J Physiol* 1974; 226, 345-352

[112] Landis WJ, Glimcher MJ. Electron diffraction and electron probe microanalysis of the mineral phase of bone tissue prepared by anhydrous techniques. *J Ultrastruct Res* 1978; 63, 188-223

[113] Glimcher MJ. The nature of the mineral component of bone and the mechanism of calcification. In: Coe FL, Favus MJ (EDS). *Disorders of bone and mineral metabolism*. New York, Raven Press, 1992, p265-286

[114] Arnott HJ, Pautard FGE. Osteoblast function and fine structure. *Israel J Med Sci* 1967; 3, 657-670

[115] Arnold S, Plate U, Wiesmann H-P, Straatmann U, Kohl H, Höhling H-J. Quantitative analyses of the biomineralization of different hard tissues. *J Microsc* 2001; 202, 488-494

[116] Wheeler EJ, Lewis D. An X-ray study of the paracrystalline nature of bone apatite. *Calcif Tissue Res* 1977; 24, 243-248

[117] Smith CB, Smith DA. An X-ray diffraction investigation of age-related changes in the crystal structure of bone apatite. *Calcif Tissue Res* 1976; 22, 219-226

[118] Ascenzi A, Bonucci E, Ripamonti A, Roveri N. X-ray diffraction and electron microscope study of osteons during calcification. *Calcif Tissue Res* 1978; 25, 133-143

[119] Burnell JM, Teubner EJ, Miller AG. Normal maturation changes in bone matrix, mineral, and crystal size in the rat. *Calcif Tissue Int* 1980; 31, 13-19

[120] Bonar LC, Roufosse AH, Sabine WK, Grynpas MD, Glimcher MJ. X-ray diffraction studies of the crystallinity of bone mineral in newly synthesized and density fractionated bone. *Calcif Tissue Int* 1983; 35, 202-209

[121] Matsushima N, Hikichi K. Age changes in the crystallinity of bone mineral and in the disorder of its crystal. *Biochim Biophys Acta* 1989; 992, 155-159

[122] Loong C-K, Rey C, Kuhn LT, Combes C, Wu Y, Chen SH, Glimcher MJ. Evidence of hydroxyl-ion deficiency in bone apatites: an inelastic neutron-scattering study. *Bone* 2000; 26, 599-602

[123] Arsenault AL. A comparative electron microscopic study of apatite crystals in collagen fibrils of rat bone, dentin and calcified turkey leg tendons. *Bone and Mineral* 1989; 6, 165-177

[124] Robinson RA. An electron-microscopic study of the crystalline inorganic component of bone and its relationship to the organic matrix. *J Bone Joint Surg* 1952; 34-A, 389-434

[125] Johansen E, Parks HF. Electron microscopic observations on the three dimensional morphology of apatite crystallites of human dentine and bone. *J Biophys Biochem Cytol* 1960; 7, 743-746

[126] Bocciarelli DS. Morphology of crystallites in bone. *Calcif Tissue Res* 1970; 5, 261-269

[127] Boothroyd B. Observations on embryonic chick-bone crystals by high resolution transmission electron microscopy. *Clin Orthop* 1975; 106, 290-310

[128] [128 Landis WJ, Hauschka BT, Rogerson CA, Glimcher MJ. Electron microscopic observations of bone tissue prepared by ultracryomicrotomy. *J Ultrastruct Res* 1977; 59, 185-206

[129] Jackson SA, Cartwright AG, Lewis D. The morphology of bone mineral crystals. *Calcif Tissue Res* 1978; 25, 217-222

[130] Weiner S, Price PA. Disaggregation of bone into crystals. *Calcif Tissue Int* 1986; 39, 365-375

[131] Weiner S, Traub W. Organization of hydroxyapatite crystals within collagen fibrils. *Feder Eur Bioch Soc* 1986; 206, 262-266

[132] Heywood BR, Sparks NH, Shellis RP, Weiner S, Mann S. Ultrastructure, morphology and crystal growth of biogenic and synthetic apatites. *Connect Tissue Res* 1990; 25, 103-119

[133] Christoffersen J, Landis WJ. A contribution with review to the description of mineralization of bone and other calcified tissues in vivo. *Anat Record* 1991; 230, 435-450

[134] Moradian-Oldak J, Weiner S, Addadi L, Landis WJ, Traub W. Electron imaging and diffraction study of individual crystals of bone, mineralized tendon and synthetic carbonate apatite. *Connect Tissue Res* 1991; 25, 219-228

[135] Kim H-M, Rey C, Glimcher MJ. X-ray diffraction, electron microscopy, and Fourier transform infrared spectroscopy of apatite crystals isolated from chicken and bovine calcified cartilage. *Calcif Tissue Int* 1996; 59, 58-63

[136] Siperko LM, Landis WJ. Aspects of mineral structure in normally calcifying avian tendon. *J Struct Biol* 2001; 135, 313-320

[137] Rubin MA, Jasiuk I, Taylor J, Rubin J, Ganey T, Apkarian RP. TEM analysis of the nanostructure of normal and osteoporotic human trabecular bone. *Bone* 2003; 33, 270-282

[138] Su X, Sun K, Cui FZ, Landis WJ. Organization of apatite crystals in human woven bone. *Bone* 2003; 32, 150-162

[139] Mahamid J, Aichmayer B, Shimoni E, Ziblat R, Li C, Siegel S, Paris O, Fratzl P, Weiner S, Addadi L. Mapping amorphous calcium phosphate transformation into crystalline mineral from the cell to the bone in zebrafish fin rays. *Proc Natl Acad Sci USA* 2010; 107, 6316-6321

[140] Schmidt WJ. Der Feinbau der anorganischen Grundmasse der Knochengewebes. *Ber Oberhess Ges Natur u Heilk, naturwiss Abt* 1933; 15, 219-247

[141] Ascenzi A. Quantitative researches on the optical properties of human bone. *Nature (London)* 1949; 163, 604

[142] Ascenzi, A, Bonucci E. A quantitative investigation of the birefringence of the osteon. *Acta Anat* 1964; 44, 236-262

[143] Clark JH. A study of tendons, bones, and other forms of connective tissue by means of x-ray diffraction patterns. *Am J Physiol* 1931; 98, 328-337

[144] Finean JB, Engström A. The low-angle scatter of x-rays from bone tissue. *Biochim Biophys Acta* 1953; 11, 178-189

[145] Carlström D, Finean JB. X-ray diffraction studies on the ultrastructure of bone. *Biochim Biophys Acta* 1954; 13, 183-191

[146] Fernández-Morán H, Engström A. : Electron microscopy and X-ray diffraction of bone. *Biochim Biophys Acta* 1957; 23, 260-264

[147] Carlström D, Glas J-E. The size and shape of the apatite crystallites in bone as determined from line-broadening measurements on oriented specimens. *Biochim Biophys Acta* 1959; 35, 46-53

[148] Matsushima N, Akiyama M, Terayama Y, Izumi Y, Miyake Y. The morphology of bone mineral as revealed by small-angle X-ray scattering. *Biochim Biophys Acta* 1984; 801, 298-305

[149] Knese K-H, Knoop A-M. Elektronenoptische Untersuchungen über die periostale Osteogenese. *Z Zellforsch* 1958; 48, 455-478

[150] Aho AJ, Isomäki AM. Electron microscopic observations on experimental callus formation in rats. *Acta Path Microbiol Scand* 1962; 154 (suppl.), 103-105.

[151] Ascenzi A, Bonucci E, Steve Bocciarelli D. An electron microscope study of osteon calcification. *J Ultrastruct Res* 1965; 12, 287-303

[152] Ascenzi A, Bonucci E , Steve Bocciarelli D. An electron microscope study on primary periosteal bone. *J Ultrastruct Res* 1967; 18, 605-618

[153] Lees S, Prostak K. The locus of mineral crystallites in bone. *Connect Tissue Res* 1988; 18, 41-54

[154] Fratzl P, Groschner M, Vogl G, Plenk HJr, Eschberger J, Fratzl-Zelman N, Koller K, Klaushofer K. Mineral crystals in calcified tissues: a comparative study by SAXS. *J Bone Miner Res* 1992; 7, 329-334

[155] Eppell SJ, Tong W, Katz JL, Kuhn L, Glimcher MJ. Shape and size of isolated bone mineralites measured using atomic force microscopy. *J Orthop Res* 2001; 19, 1027-1034

[156] Tong W, Glimcher MJ, Katz JL, Kuhn L, Eppell SJ. Size and shape of mineralites in young bovine bone measured by atomic force microscopy. *Calcif Tissue Int* 2003; 72, 592-598

[157] Pidaparti RMV, Chandran A, Takano Y, Turner CH. Bone mineral lies mainly outside collagen fibrils: prediction of a composite model for osteonal bone. *J Biomechanics* 1996; 29, 909-916

[158] Robinson RA, Cameron DA. Electron microscopy of cartilage and bone matrix at the distal epiphyseal line of the femur in the newborn infant. *J Biophys Biochem Cytol* 1956; 2, 253-263

[159] Bonucci E. Crystal ghosts and biological mineralization: fancy spectres in an old castle, or neglected structures worthy of belief? *J Bone Miner Metab* 2002; 20, 249-265

[160] Murshed M, McKee MD. Molecular determinants of extracellular matrix mineralization in bone and blood vessels. *Curr Opin Nephrol Hypertens* 2010; 19, 359-365

[161] Glimcher MJ, Krane SM. The organization and structure of bone, and the mechanism of calcification. In: Gould BS (ed) *Biology of collagen*. London, Academic Press, 1968, p67-251

[162] Landis WJ, Hodgens KJ, Arena J, Song MJ, McEwen BF. Structural relations between collagen and mineral in bone as determined by high voltage electron microscopic tomography. *Microsc Res Techn* 1996; 33, 192-202

[163] Zylberberg L, Géraudie J, Meunier F, Sire J-Y. Biomineralization in the integumental skeleton of the living lower vertebrates. In: Hall BK. *Bone, volume 4: Bone metabolism and mineralization*. Boca Raton, CRC Press, 1992, p171-224

[164] Bonucci E, Reurink J. The fine structure of decalcified cartilage and bone: a comparison between decalcification procedures performed before and after embedding. *Calcif Tissue Res* 1978; 25, 179-190

[165] Bonucci E. Fine structure of early cartilage calcification. *J Ultrastruct Res* 1967; 20, 33-50

[166] Bonucci E. Further investigation on the organic/inorganic relationships in calcifying cartilage. *Calcif Tissue Res* 1969; 3, 38-54

[167] Appleton J. Ultrastructural observations on the inorganic/organic relationships in early cartilage calcification. *Calcif Tissue Res* 1971; 7, 307-317

[168] Goldberg M, Noblot MM, Septier D. Effets de deux méthodes de démineralisation sur la préservation des glycoprotéines et des protéoglycanes dans les dentines intercanaliculaires et péricanaliculaires chez le cheval. Jour Biol Buccale 1980; 8, 315-330

[169] Davis WL, Jones RG, Hagler HK. An electron microscopic histochemical and analytical X-ray microprobe study of calcification in Bruch's membrane from human eyes. *J Histochem Cytochem* 1981; 29, 601-608

[170] Hayashi Y, Bianco P, Shimokawa H, Termine JD, Bonucci E. Organic-inorganic relationships, and immunohistochemical localization of amelogenins and enamelins in developing enamel. *Basic Appl Histochem* 1986; 30, 291-299

[171] Kallenbach E. Crystal-associated matrix components in rat incisor enamel. An electron-microscopic study. *Cell Tissue Res* 1986; 246, 455-461

[172] Hayashi,Y.: Ultrastructural demonstration of the carbohydrate in developing rat enamel using soybean agglutinin-gold complexes. *Archs Oral Biol* 1989; 34, 517-522

[173] Bonucci E, Lozupone E, Silvestrini G, Favia A, Mocetti P. Morphological studies of hypomineralized enamel of rat pups on calcium-deficient diet, and of its changes after return to normal diet. *Anat Record* 1994; 239, 379-395

[174] Prostak KS, Lees S. Visualization of crystal-matrix structure. *In situ* demineralization of mineralized turkey leg tendon and bone. *Calcif Tissue Int* 1996; 59, 474-479

[175] Takagi M. Ultrastructural cytochemistry of cartilage proteoglycans and their relation to the calcification process. In: Bonucci E, Motta PM (eds). *Ultrastructure of skeletal tissues*. Boston, Kluwer Academic Publishers, 1990, p111-127

[176] Shepard N. Role of proteoglycans in calcification. In: Bonucci E (ed). *Calcification in biological systems*. Boca Raton, CRC Press, 1992, p41-58

[177] Ascenzi A, Bonucci E, Generali P, Ripamonti A, Roveri N. Orientation of apatite in single osteon samples as studied by pole figures. *Calcif Tissue Int* 1979; 29, 101-105

[178] Höhling HJ, Barckhaus RH, Krefting E-R, Althoff J, Quint P. Collagen mineralization: aspects of the structural relationship between collagen and the apatitic crystallites. In: Bonucci E, Motta PM (eds). *Ultrastructure of skeletal tissues*. Boston, Kluwer Academic Publishers, 1990, p41-62

[179] Landis WJ, Song MJ, Leith A, McEwen L, McEwen BF. Mineral and organic matrix interaction in normally calcifying tendon visualized in three dimensions by high-voltage electron microscopic tomography and graphic image reconstruction. *J Struct Biol* 1993; 110, 39-53

[180] Arsenault,A.L.: The ultrastructure of calcified tissues: methods and technical problems. In: Bonucci E, Motta PM (eds) *Ultrastructure of skeletal tissues*. Boston, Kluwer Academic Publishers, 1990. p1-18

[181] Arsenault AL. Image analysis of collagen-associated mineral distribution in cryogenically prepared turkey leg tendons. *Calcif Tissue Int* 1991; 48, 56-62

[182] Ziv V, Weiner S. Bone crystal sizes: a comparison of transmission electron microscopic and X-ray diffraction line width broadening techniques. *Connect Tissue Res* 1994; 30, 165-175

[183] Bachra BN. Some molecular aspects of tissue calcification. *Clin Orthop* 1967; 51, 199-222

[184] Bonar LC, Lees S, Mook HA. Neutron diffraction studies of collagen in fully mineralized bone. *J Mol Biol* 1985; 181, 265-270

[185] Warshawsky H. Organization of crystals in enamel. *Anat Record* 1989; 224, 242-262

[186] Watabe N. Decalcification of thin sections for electron microscope studies of crystal-matrix relationships in mollusk shells. *J Cell Biol* 1963; 18, 701-703

[187] Mutvei H. Ultrastructure of the mineral and organic components of molluscan nacreous layers. *Biomineralization* 1970; 2, 48-72

[188] Meenakshi VR, Hare PE, Wilbur KM. Amino acids of the organic matrix of neogastropod shells. *Comp Biochem Physiol* 1971; 40B, 1037-1043

[189] Crenshaw MA. The soluble matrix from *Mercenaria mercenaria* shell. *Biomineralization* 1972; 6, 6-11

[190] Wong V, Saleuddin ASM. Fine structure of normal and regenerated shell of *Helisoma duryi duryi*. *Can J Zool* 1972; 50, 1563-1568

[191] Benson S, Jones EM, Crise-Benson N, Wilt F. Morphology of the organic matrix of the spicule of the sea urchin larva. *Exp Cell Res* 1983; 148, 249-253

[192] Berman A, Addadi L, Kvick Å, Leiserowitz L, Nelson M, Weiner S: Intercalation of sea urchin proteins in calcite: study of a crystalline composite material. *Science* 1990; 250, 661-667

[193] Albeck S, Addadi L, Weiner S. Regulation of calcite crystal morphology by intracrystalline acidic proteins and glycoproteins. *Connect Tissue Res* 1996; 35, 365-370

[194] Cho JW, Partin JS, Lennarz WJ. A technique for detecting matrix proteins in the crystalline spicule of the sea urchin embryo. *Proc Natl Acad Sci USA* 1996; 93, 1282-1286

[195] Miyashita T, Takagi R, Okushima M, Nakano S, Miyamoto H, Nishikawa E, Matsushiro A. Complementary DNA cloning and characterization of Pearlin, a new class of matrix protein in the nacreous layer of oyster pearls. *Mar Biotechnol* 2000; 2, 409-418

[196] Dauphin Y: Comparison of the soluble matrices of the calcitic prismatic layer of *Pinna nobilis* (Mollusca, Bivalvia, Pteriomorpha). *Comp Biochem Physiol A Mol Integr Physiol* 2002; 132, 577-590

[197] Tong H, Hu J, Ma W, Zhong G, Yao S, Cao N. In situ analysis of the organic framework in the prismatic layer of mollusc shell. *Biomaterials* 2002; 23, 2593-2598

[198] Marin F, Pokroy B, Luquet G, Layrolle P, De Groot K. Protein mapping of calcium carbonate biominerals by immunogold. *Biomaterials* 2007; 28, 2368-2377

[199] Kim YY, Ganesan K, Yang P, Kulak AN, Borukhin S, Pechook S, Ribeiro L, Kröger R, Eichhom SJ, Armes SP, Pokroy B, Meldrum FC. An artificial biomineral formed by incorporation of copolymer micelles in calcite crystals. *Nat Mater* 2011; 10, 890-896

[200] Pokroy B, Fitch AN, Lee PL, Quintana JP, Caspi EN, Zolotoyabko E. Anisotropic lattice distortions in the mollusk-made aragonite: a widespread phenomenon. *J Struct Biol* 2006; 153, 145-150

[201] Bonucci E, Silvestrini,G., Di Grezia,R.: Histochemical properties of the "crystal ghosts" of calcifying epiphyseal cartilage. *Connect Tissue Res* 1989; 22, 43-50

[202] Bonucci E, Silvestrini G. Immunohistochemical investigation on the presence of chondroitin sulfate in calcification nodules of epiphyseal cartilage. *Eur J Histochem* 1992; 36, 407-422

[203] Sundström B, Takuma S. A further contribution on the ultrastructure of calcifying cartilage. *J Ultrastruct Res* 1971; 36, 419-424

[204] Davis WL, Jones RG, Knight JP, Hagler HK. Cartilage calcification: an ultrastructural, histochemical, and analytical X-ray microprobe study of the zone of calcification in normal avian epiphyseal growth plate. *J Histochem Cytochem* 1982; 30, 221-234

[205] Rhee SH, Tanaka J. Self-assembly phenomenon of hydroxyapatite nanocrystals on chondroitin sulfate. *J Mater Sci Mater Med* 2002; 13, 597-600

[206] Weiner S. Aspartic acid-rich proteins: major components of the soluble organic matrix of mollusk shells. *Calcif Tissue Int* 1979; 29, 163-167

[207] Butler WT. Matrix macromolecules of bone and dentin. *Collagen Rel Res* 1984; 4, 297-307

[208] George A, Sabsay B, Simonian PA, Veis A. Characterization of a novel dentin matrix acidic phosphoprotein. Implications for induction of biomineralization. *J Biol Chem* 1993; 268, 12624-12630

[209] Gotliv BA, Addadi L, Weiner S. Mollusk shell acidic proteins: in search of individual functions. *Chembiochem* 2003; 4, 522-529

[210] Rahman MA, Oomori T. In vitro regulation of CaCO(3) crystal growth by the highly acidic proteins of calcitic sclerites in soft coral, Sinularia Polydactyla. *Connect Tissue Res* 2010; 50, 285-293

[211] Campo RD, Dziewiatkowski DD. Turnover of the organic matrix of cartilage and bone as visualized by autoradiography. *J Cell Biol* 1963; 18, 19-29

[212] Hirschman A, Dziewiatkowski DD. Protein-polysaccharide loss during endochondral ossification: immunochemical evidence. Science 1966; 154, 393-395

[213] Matukas, VJ, Krikos GA. Evidence for changes in protein polysaccharide associated with the onset of calcification in cartilage. *J Cell Biol* 1968; 39, 43-48

[214] Silbermann M, Frommer J. Dynamic changes in acid mucopolysaccharides during mineralization of the mandibular condylar cartilage. *Histochemie* 1973; 36, 185-192

[215] Lohmander S, Hjerpe A. Proteoglycans of mineralizing rib and epiphyseal cartilage. *Biochim Biophys Acta* 1975; 404, 93-109

[216] Althoff J, Quint P, Krefting E-R, Höhling HJ. Morphological studies on the epiphyseal growth plate combined with biochemical and X-ray microprobe analyses. *Histochemistry* 1982; 74, 541-552

[217] Mitchell N, Shepard N, Harrod J. The measurement of proteoglycan in the mineralizing region of the rat growth plate. An electron microscopic and X-ray microanalytical study. *J Bone Joint Surg* 1982; 64-A, 32-38

[218] Takagi M, Parmley RT, Denys FR. Ultrastructural cytochemistry and immunocytochemistry of proteoglycans associated with epiphyseal cartilage calcification. *J Histochem Cytochem* 1983; 31, 1089-1100

[219] Chen C-C, Boskey AL, Rosenberg LC. The inhibitory effect of cartilage proteoglycans on hydroxyapatite growth. *Calcif Tissue Int* 1984; 36, 285-290

[220] Campo RD, Romano JE. Changes in cartilage proteoglycans associated with calcification. *Calcif Tissue Int* 1986; 39, 175-184

[221] Buckwalter JA, Rosenberg LC, Ungar R. Changes in proteoglycan aggregates during cartilage mineralization. *Calcif Tissue Int* 1987; 41, 228-236

[222] Hagiwara H, Aoki T, Yoshimi T. Immunoelectron microscopic analysis of chondroitin sulfates during calcification in the rat growth cartilage. *Histochemistry* 1995; 103, 213-220

[223] Akisaka T, Nakayama M, Yoshida H, Inoue M. Ultrastructural modifications of the extracellular matrix upon calcification of growth plate cartilage as revealed by quick-freeze deep etching technique. *Calcif Tissue Int* 1998; 63, 47-56

[224] Hoshi K, Ejiri S, Ozawa H. Localizational alterations of calcium, phosphorus, and calcification-related organics such as proteoglycans and alkaline phosphatase during bone calcification. *J Bone Miner Res* 2001; 16, 289-298

[225] Boskey AL, Stiner D, Binderman I, Doty SB. Effects of proteoglycan modification on mineral formation in a differentiating chick limb-bud mesenchymal cell culture system. *J Cell Biochem* 1997; 64, 632-643

[226] Bonucci E, Gomez S. Cartilage calcification. In: Seto J (ed) *Advanced topics in biomineralization*. Rijeka, InTech, 2012, p85-110

[227] Gomez S, Lopez-Cepero JM, Silvestrini G, Mocetti P, Bonucci E. Matrix vesicles and focal proteoglycan aggregates are the nucleation sites revealed by the lanthanum in-

cubation method: a correlated study on the hypertrophic zone of the rat epiphyseal cartilage. *Calcif Tissue Int* 1996; 58, 273-282

[228] Moradian-Oldak J. Protein-mediated enamel mineralization. *Front Biosci* 2012; 17, 1996-2023

[229] Prince CW, Rahemtulla F, Butler WT. Metabolism of rat bone proteoglycans *in vivo*. *Biochem J* 1983; 216, 589-596

[230] Amizuka N, Hasegawa T, Oda K, Luiz de Freitas PH, Hoshi K, Li M, Ozawa H. Histology of epiphyseal cartilage calcification and endochondral ossification. *Front Biosci (Elite Ed.)* 2012; 4, 2085-2100

[231] Marin F, Le Roy N, Marie B. The formation and mineralization of mollusk shell. *Front Biosci* (Schol Ed) 2012; 4, 1099-1125

[232] Veis A. Organic matrix-related mineralization of sea urchin spicules, spines, test and teeth. *Front Biosci* 2011; 17, 2540-2560

[233] Nagasawa H. The crustacean cuticle: structure, composition and mineralization. *Front Biosci (Elite Ed)* 2012; 4, 711-720

[234] Hincke MT, Nys Y, Gautron J, Mann K, Rodriguez-Navarro AB, McKee MD. The eggshell: structure, composition and mineralization. *Front Biosci* 2012; 17, 1266-1280

[235] Smales FC. Structural subunit in prisms of immature rat enamel. *Nature* 1975; 258, 772-774

[236] Bonucci E. Calcification and silicification: a comparative survey of the early stages of biomineralization. *J Bone Miner Metab* 2009; 27, 255-264

[237] Wang X, Müller WE. Complex structures - smart solutions: Formation of siliceous spicules. *Commun Integr Biol* 2011; 4, 684-688

[238] Wang X, Schlossmacher U, Wiens M, Batel R, Schröder HC, Müller WE. Silicateins, silicatein interactors, and cellular interplay in sponge skeletogenesis: Formation of glass fiber-like spicules. *FEBS J* 2012; 279, 1721-1736 (doi: 10.1111/j.1742-4658.2012.08533.x)

[239] Andre R, Tahir MN, Natalio F, Tremel W. Bioinspired synthesis of multifunctional inorganic and bio-organic hybrid materials. *FEBS J* 2012; 279, 1737-1749 (doi: 10.1111/j.1742-4658.2012.08584.x)

Chapter 7

Crystallization: From the Conformer to the Crystal

J.S. Redinha, A.J. Lopes Jesus, A.A.C.C. Pais and
J. A. S. Almeida

Additional information is available at the end of the chapter

1. Introduction

Crystallization, commonly defined as a process of formation of a crystalline solid from a supersaturated solution, melt or vapor phase, is an old technique widely used in laboratory and in industrial processes to separate and purify substances. In various modern industries, crystalline forms with a certain habit, size and structure, constitute the basic materials for the production of highly sophisticated materials [1, 2]. Integrated circuits as well as piezoelectric and optical materials are just a few examples of devices whose properties are dependent on the crystal structure. Also, in organic chemistry, molecular crystals with determined characteristics are now-a-days of utmost importance for the production of pharmaceuticals, dyestuffs, pigments, foodstuffs, chemicals, cosmetics, etc. For all these reasons, crystal growth has become an important and attractive research field.

Crystallization from solution is one of the preferred methods to obtain a crystal since it can be carried out under different experimental conditions and provides a wide variety of products. In fact, it can occur by lowering the temperature of a supersaturated solution, partial evaporation of the solvent, precipitation by adding an anti-solvent or vapor diffusion of a gas into the solution. Furthermore, solvents with different properties can be used. Such a diversity of experimental conditions has great influence in the output, *i.e.*, in the type of crystalline phase that can be obtained. Therefore, any aprioristic selection of the conditions leading to a desired final product is a fundamental challenge but it is also an almost unfeasible task [3-5]. On respect to the harvest, crystallization is, in a certain extension, a trial-and-error operation.

The various steps of crystallization from solvents are summarized in the following scheme:

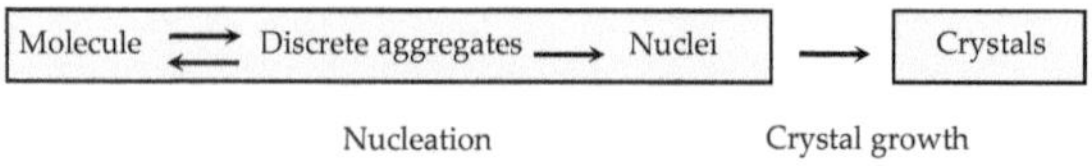

Scheme 1. Main crystallization pathways from solution

The competition between the solute-solvent and solute-solute interactions to form the nuclei, and between the specific aggregation forces and packing in order to minimize repulsive interactions, determine the structure of the new crystalline phase. During this process the molecular conformation can change drastically in an unpredicted way. Various computational approaches have been employed to predict the crystalline structure (unit cell, space group and atomic coordinates) formed by a given molecule [6-8], most of them relying on the assumption that the most probable forms are those with lowest lattice energy. Despite the recent success of these methods in the prediction of crystalline structures of small rigid molecules [9], their applicability to high size flexible molecules is not yet satisfactory [10, 11]. The main drawbacks lie in the difficultly of describing accurately the high complexity of inter and intramolecular interactions (e.g. hydrogen bonds and van der Waals interactions) and of selecting, among the energetically feasible polymorphs, that or those that really exist [3]. In truth, crystal prediction is still a very hard task so far.

Nevertheless, computational calculation can be a powerful tool to get information about the different steps of crystallization. The main goal of the present chapter is included in this research field. Quantum-chemical calculations or molecular dynamics simulations, adapted to the molecular complexity can be used for this purpose. In molecular terms, a compound with a certain conformational flexibility exists in a supersaturated aqueous solution as a mixture of conformers with different populations. The knowledge of the conformational equilibrium in this medium is a way of identifying the conformers that are likely to be involved in the early stages of nucleation and to interpret the mechanism of the crystallization process. In this chapter, these topics are addressed by presenting the results obtained for erythritol and *L*-threitol, two isomeric alditols with four carbon atoms, and glutamic acid. These compounds were chosen because of the role played by their conformational features on crystallization. Furthermore, the two polyols exhibit a different crystallization behavior despite being chemically similar. The energies of the different conformers were obtained through theoretical calculations at the DFT level of theory [12, 13] using the popular B3LYP (Becke, 3-parameter, Lee-Yang-Parr) exchange-correlation functional [14-16] combined with the aug-cc-pVDZ [17, 18] or 6-311+ +G(d,p) [19] basis set. Solvent effects have been accounted by using the well-established conductor-like polarizable continuum model (CPCM) [20-22].

As crystallization proceeds, the self-association of the solute molecules gives rise to molecular aggregates which play an important role in the resulting crystalline structure [23-25]. Data from the application of molecular dynamics simulations in the study of aqueous solutions of erythitol and *L*-threitol near the saturation concentration are reported in an attempt to give an insight into the type of supramolecular species formed in solution, and hence a step forward to elucidate about their crystallization behavior.

As stated before, depending on the experimental conditions and intrinsic molecular features of the molecules, different crystalline phases or polymorphs can be obtained by crystallization. The polymorphs differ from one another in the arrangement (packing) and/or conformation of the molecules in the crystal lattice [26-29]. They are commonly called packing polymorphs in the first case and conformational polymorphs in the second one. Polymorphism has significant implications in a wide range of areas and for this reason it has been matter of extensive investigation. Some typical examples are here reported with the objective of pointing out the diversity of structures that can obtained by crystallization.

2. Nucleation and crystal growth

Fluctuations of the order of a solution, for example density, are accompanied by the formation of solute molecular clusters. The driving force to bring the molecules close together is given by the following chemical potential difference:

$$\mu - \mu^{*} = kT \ln \frac{C}{C^{*}} \tag{1}$$

In this equation, μ and μ^{*} are the chemical potentials of the solute at the actual (C) and saturation (C^{*}) concentrations, respectively. The Gibbs energy change per molecule (Δg) corresponding to the formation of a spherical cluster of radius r is [30-32]

$$\Delta g = \frac{4/3\pi r^{3}}{v}\Delta\mu + 4\pi r^{2}\sigma \tag{2}$$

where v is the molecular volume and σ the interfacial energy per surface. The first term of the right hand equation gives the work produced by the intermolecular attractive forces (Δg_b) while the second one corresponds to the work required to increase the surface area in 1cm^2 (Δg_s). Owing to the meaning of both terms, equation (2) can be rewritten as

$$\Delta g = \Delta g_b + \Delta g_s \tag{3}$$

With the increase of concentration, the amplitude of the fluctuations also increase and so do the size of the clusters. For smaller size clusters, Δg_s is the dominant contribution and $\Delta g/dr > 0$, *i.e.*, the clusters are unstable. Conversely, for larger size clusters, Δg_b is dominant and $\Delta g/dr < 0$, meaning that the clusters are stable solid particles. The curve representing the variation of Δg with the radius, shown in Figure 1, passes through a maximum at $r = r_c$ ($d\Delta g/dr = 0$), with r_c representing the limiting size for stable clusters. A cluster of this size is the embryo of the solid form nucleus. The following relationships involving r_c can be established:

$$\begin{aligned} r_c &= -\frac{2\upsilon\sigma}{\Delta\mu} && \text{a} \\ r_c &= -\frac{2\sigma}{\Delta g_b} && \text{b} \\ \Delta g_c &= \frac{4\pi r_c^2\sigma}{3} && \text{c} \end{aligned} \tag{4}$$

where Δg_c corresponds to the value of Δg at r_c. The supersaturation concentration (S) can be related with the clusters size by the Gibbs-Thomson equation:

$$\ln\frac{C}{C^*} = \ln S = \frac{2\upsilon\sigma}{kTr_c} \tag{5}$$

Combining (4c) and (5), the following expression is obtained for Δg_c as function of the supersaturation:

$$\Delta g_c = \frac{16\pi\sigma^3\upsilon^2}{3kT\ln S^2} \tag{6}$$

Δg_c represents the energy barrier for the nucleation process and the rate of nucleation (J) will be given by:

$$J = Ae^{\left(\frac{-\Delta g_c}{kT}\right)} \tag{7}$$

This equation shows that the rate of nucleation is strongly dependent on the supersaturation and tends to a finite value for a given supersaturation.

Organic crystals are supramolecules, *i.e*, an ensemble of molecules, bonded by non-covalent forces involving ions, polar or non-polar groups. Such interactions, whatever their strength, do not change the molecular individuality. Hydrogen bonding occupies a special place among the intermolecular interactions occurring in molecular crystals. Its energy ranges from 160 kJ mol^{-1} (strongly covalent) down to 16 kJ mol^{-1} (mostly electrostatic) [33] and it is highly dependent on the orientation of the hydrogen donor group (D–H) relative to the acceptor atom (A). The maximum strength occurs when the D–H···A angle is approximately 180°, though values greater than 110° are acceptable for a hydrogen bond [34].

It is obvious that a crystal packing in which the specific interactions between the functional groups (molecular recognition) are more favorable is, in principle, that leading to the lowest energy. Thus, the lowest energy crystalline structure would correspond to the stronger interaction between a group and its complementar in nature. For example, to a positive charge there would be a negative one, or to a H-bond donor a H-bond acceptor. The molecular packing bringing the molecules close together also gives rise to steric repulsion. Hence, the equilibrium structure is the result between the attraction force and the steric repulsion. This compromise makes the molecule to adopt a conformation corresponding to the lowest Gibbs energy.

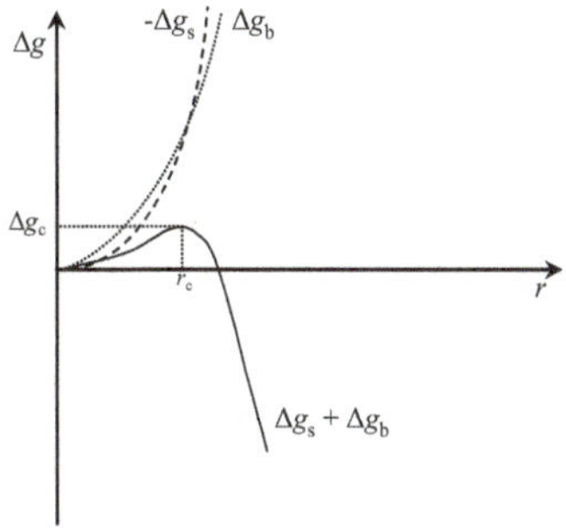

Figure 1. Variation of Δg and its components with the radius of the molecular clusters.

Let us consider the conformational variation from the gas phase to the solid for a simple molecule as paracetamol (acetaminophen). This is a drug of great commercial interest in the pharmaceutical industry owing to its wide use as antipyretic and analgesic agent. This molecule exhibits two relevant conformers differing from one another in the orientation of the OH relatively to the carbonyl group [35]. In one of them the dihedral angle formed by the H-O(1)-C-O(2) atoms is close to 180° (*trans* conformation, Figure 2), while in the other the angle is approximately 0° (*cis* conformation). Geometry optimization of the two conformers followed by vibrational frequencies calculation at the B3LYP/aug cc pVDZ level shows that in the gas phase the population (%) of *cis* relative to *trans* is 57:43. Including water effects by applying the CPCM model, both conformers are practically isoenergetic and so their relative population is 50:50. However, in the crystalline phase only the *trans* conformer is presented. This means that the lattice energy yielded by the *trans* conformer is more negative than that of the *cis* conformer. Along this chapter some examples will be given concerning the relationship between the conformers' population in the three states of matter.

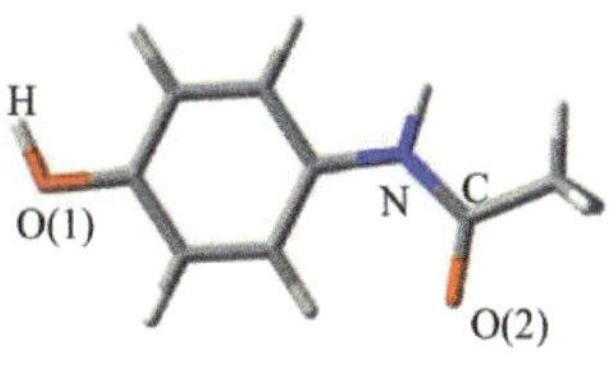

Figure 2. Optimized geometry of the *trans* conformer of paracetamol at the B3LYP/aug-cc-pVDZ level.

3. Polymorphism

According to McCrone [36], polymorphism of a compound corresponds to its ability to crystallize into more than one crystalline structure. From the thermodynamic point of view, only the lowest Gibbs energy form exists. However, higher energy polymorphs can remain as metastable forms for a period of time long enough to be used for practical purposes, providing the height of the energy barrier separating these polymorphs and the most stable one is sufficiently high [28]. Since different solid-state modifications exhibit distinct physicochemical properties, such for example the melting point, solubility, dissolution rate and density, polymorphism has a great impact in the pharmaceutical, food and dyes technologies, among others. For example, in the pharmaceutical industry, the desired polymorph for a given active pharmaceutical ingredient (API) would be the one with highest bioavailability and structural stability during shelf life [37].

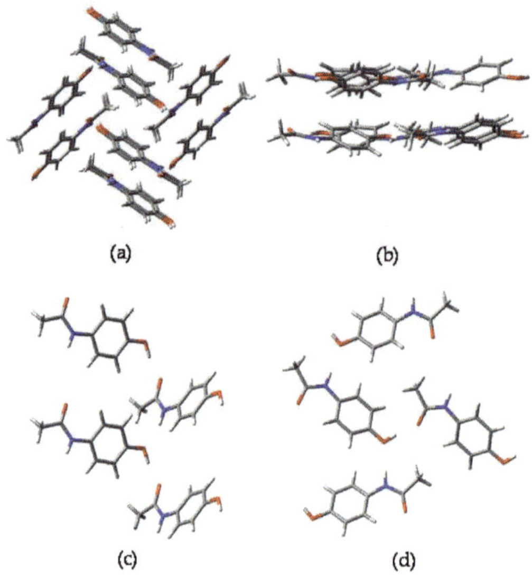

Figure 3. Views of the crystalline structures of forms I (a) and II (b) of paracetamol and detail of the H-bonds formed in both polymorphs [(c) and (d)].

A typical example of a drug exhibiting polymorphism is paracetamol. Three polymorphs, labeled as I, II and III, have been identified for this compound [38-40]. The first (form I) is the thermodynamically stable form while the second (form II) is metastable but exists long enough to be experimentally studied. It is usually prepared from cooled melt [41]. Form III is very unstable [41] and thus it is not possible to investigate its structure and properties. Form I crystallizes in the monoclinic system (space group $P2_1/a$) [42, 43]. As shown in Figure 3, the molecules in the crystalline structure are linked through O–H O=C and N–H O–H hydrogen bonds forming chains which in turn give rise to pleated layers [frames (a) and (a')].

Regarding form II, it crystallizes in the orthorhombic system (space group *Pcab*) [44, 45]. The H-bonding system is identical to that existing in form I but the layers are plane interconnected by van der waals forces [frames (b) and (b′)].

The geometric parameters of the H-bonds in forms I and II are summarized in Table 1. In both forms the O–H···O=C hydrogen bond is stronger than the N–H···O–H one. In addition, the hydrogen bonds in the stable polymorph are stronger than in the metastable one.

The two polymorphs were also characterized by infrared spectroscopy [46, 47]. The values of the stretching vibration frequencies for the groups involved in hydrogen bonds are given in Table 2. To work as reference, the frequencies of the "free" groups, obtained from a spectrum of paracetamol in a $CDCl_3$ solution were also included in the Table. From the frequency shift ($\Delta\nu$) of the NH and OH groups one can estimate the enthalpy involved in the two hydrogen bonds by applying the empirical relationship proposed by Iogansen: $\Delta H^2 = 1.92(\Delta\nu - 40)$ [48]. The value of ΔH estimated for the N–H···O–H H-bond was found to be 12 kJ mol^{-1} for both polymorphs, while that estimated for the O–H···O=C one was found to be 28 kJ mol^{-1} for form I and 19 kJ mol^{-1} for form II.

Form	O–H···O=C			N–H···O–H		
	H···O/Å	O···O/Å	O–H···O/°	H···O/Å	N···O/Å	N–H···O/°
I	1.719	2.653	167.5	2.032	2.904	161.6
II	1.838	2.709	170.6	2.069	2.943	163.8

[a] Values taken from ref. [43, 45]

Table 1. Distances and angles of the hydrogen bonds in the paracetamol polymorphs.[a]

	νNH	νOH	νCO
form I	3324	3160	1656
form II	3326	3205	1666 ; 1656
$CDCl_3$ solution	3438	3600	1683

[a] Values taken from ref. [46, 47]

Table 2. Vibrational frequencies (ν/cm^{-1}) corresponding to the NH, OH and CO stretching vibrations of paracetamol in the two crystalline phases and in a $CDCl_3$ solution.[a]

The higher stability of form I is also confirmed by the thermodynamic properties obtained for sublimation, fusion and vaporization of the two polymorphic forms [38]. The value of the packing density, determined by flotation or X-ray (or neutron) diffraction [49] is higher in form II than in form I. In fact, at 298K the density of the former is 1.336 g cm^{-3} while that of the latter is 1.297 g cm^{-3}. That is, the higher stability of form I results from a stronger hydrogen bond interaction rather than from a more favorable packing. In this compound, the differences between specific interactions resulting from the conformational recognition overcome that due to the crystal packing.

The polymorphism of paracetamol assumes great importance from the practical point of view in so far the commercialized polymorph is not the most suitable solid for formulation [45, 50, 51]. Despite it is easily obtained from various solvents, it has the inconvenient to require a binding agent to make tablets for compression. During this process it gives a brittle solid, consequence of its rough molecular layers. Conversely, form II is constituted by thin plates that glide on pressing. Its plasticity allows the formulation into tablets by direct compression without the need of incorporating any binder agent [45]. The main difficulty to obtain this form comes up against the existence of an adequate method for their preparation at an industrial scale [41].

Aspirin (acetylsalicylic acid) is a drug commonly used as analgesic, antipyretic and anti-inflammatory. It is also known to act as anticoagulant by reducing the blood platelet aggregation [52]. About 35,000 tons of aspirin are taken a day throughout the world.

Using X-ray diffraction, the structure of aspirin has been characterized for the first time in 1964 as a monoclinical crystal, space group P21/c [53]. In 2004, the existence of a second polymorph (form II) was predicted computationally [54] and observed experimentally one year later [55]. However, in view of the slight differences between both structural modifications, it was speculated that this new form might result from uncertainties of the methods used in the structural characterization [56, 57]. Just recently it has been settled the existence of a second polymorph for this compound [58]. In both forms, the carboxyl groups of two neighbor molecules are connected by O–H···O=C H-bonds giving rise to centrosymmetric dimers as illustrated in Figure 4. These dimers are arranged into stacked layers, which are identical in the two polymorphs. The acetyl groups of molecules belonging to different layers are in turns interconnected through C–H···O interactions. In Form I, these interactions have centrosymmetric geometry, *i.e.*, the CH_3 and CO groups of one molecule are bonded to the complementary groups of a neighbor molecule. The structure is a sequence of AAAA layers. Unlike, the structure of form II is a sequence of ABAB layers [59]. The difference between the B and A layers lies in a relative translation of the lattice repetition. This means that in the latter form the CH_3 and CO groups of one molecule are connected to the complementary groups of two neighbor molecules, yielding catemers.

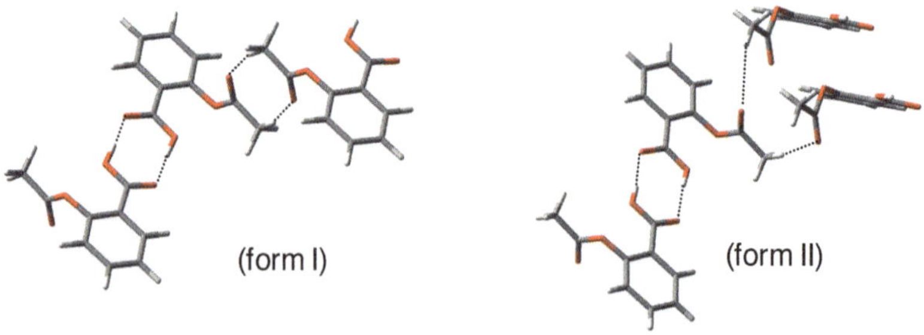

Figure 4. H-bonds in the crystalline structures of forms I and II of aspirin.

An interesting feature of aspirin single crystals is that they exhibit simultaneously domains of forms I and II of variable relative size. Deran *et al.* [58] explain this type of behavior as an accidental degeneracy. According to these authors the two polymorphs are isoenergetic and during crystallization an intergrowth of both forms occurs. The more favorable internal conformation in form I is compensated by an enhanced cooperativity of the catameric H-bonding network in form II. These results raise the question of whether or not these two forms of aspirin can be called polymorphs. According to the definition given by McCrone [36], they do not fit into the concept.

Besides polymorphism, further crystalline structure complexity arises with the existence of disorder in crystals, situation quite common in many organic crystals [60-64]. This occurs when portions of the structure under analysis occupy two or more positions. For example, molecular fragments of organic molecules not involved in intermolecular interactions can be free enough to acquire different conformations. Such solids are designated by the conflicting denomination of disordered crystals. An example of a compound exhibiting crystal disorder is betaxolol hydrochloride, 1-{4-[2-(cyclopropylmethoxy)ethyl]phenoxy}-3-isopropyl-amino-2-propanol hydrochloride (Figure 5). It is a cardio selective β adrenergic drug that crystallizes into the triclinic system with P_1 symmetry [65].

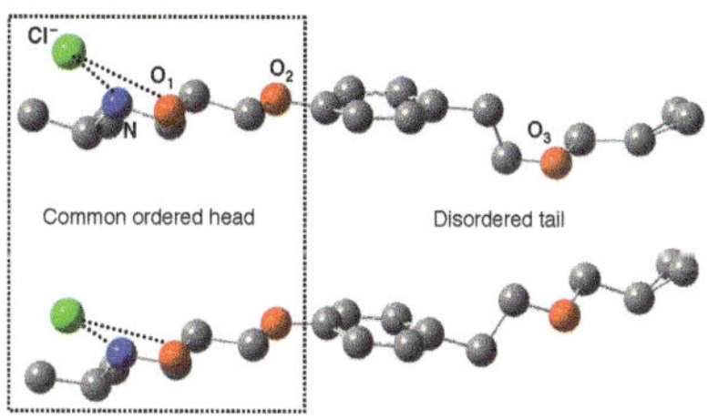

Figure 5. Conformations of betaxolol hydrochloride in the crystalline structure. Hydrogen bonds between the amino and hydroxyl groups with the chloride ion of the same molecule are represented by dotted lines. For better visualization the hydrogen atoms are not shown.

The intermolecular arrangement in the crystalline structure is as follows: the NH_2^+ and the OH groups of the isopropylaminoethanol moiety form H-bonds with the chloride ion of the same molecule as shown in Figure 5. In addition, one of the hydrogens of the NH_2^+ group is also H-bonded to oxygen atom of the OH group of a second molecule while the other hydrogen of the same group is linked to the chloride ion of a third molecule [63, 64]. That is, the betaxolol molecules are connected each other just through the isopropylaminoethanol fragment, leaving a long molecular chain free to move. The X-ray diffraction data shows the existence of an ordered head from the isopropyl group up to O_2 and a disordered tail from this atom to the end. By refinement, the crystalline structure can be satisfactory interpreted as being constituted by two conformations which are displayed in Figure 5. Since we have a no well defined unit cell, the structure of betaxolol hydrochloride can not be included in the polymorphism or isomorphism definitions.

4. Erythritol and threitol: identical chemical structure, different crystalline assembling

Erythritol and *L*-threitol (see Figure 8) are two diastereomers of 1,2,3,4-butanetetrol. The first is a *meso* compound while the second is optically active, existing as *D*- or *L*-threitol. An important difference between both alditols lies in their crystallization ability. Whilst erythritol is easily crystallized from aqueous solution or melt, *L*-threitol is much more difficult to crystallize [66, 67]. As illustrated in the DSC curves shown in Figure 6, molten erythritol transforms readily into a crystalline phase providing it is cooled at scanning rates lower than or equal to 10°C/min. On the contrary, no crystallization peak is observed for *L*-threitol, even when the melt is cooled very slowly, say 2°C/min. Since the two compounds are chemically similar, their different crystallization tendency is likely to have a conformational origin [68].

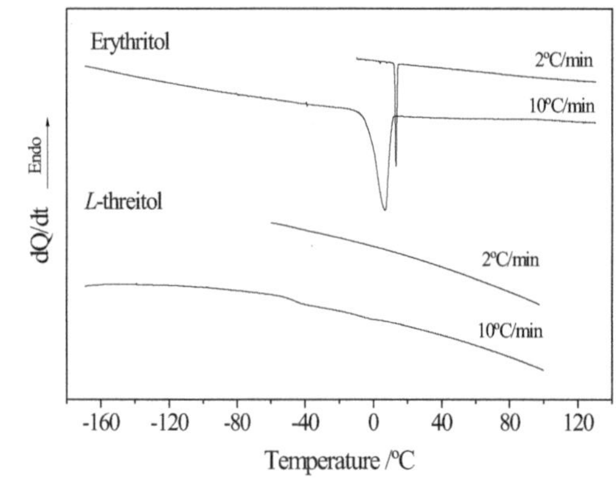

Figure 6. DSC cooling curves of molten erythritol (m.p = 118°C) and *L*-threitol (m.p= 89°C) obtained at different cooling rates

Conformationally, both diastereomers are highly flexible. In fact, the existence of seven independent dihedral angles, each of them assuming three standard orientations: gauche$^+$ (60°), gauche$^-$ (-60°) and trans (180°), originate a total of 2187 possible conformers. Such a large number of structures make their systematic exploration by means of *ab initio* calculations unfeasible. Therefore, a previous conformational search method using random generation and subsequent molecular mechanics energy minimization has been employed to sampling the most relevant conformations. All generated structures falling within an energy window of 20 kJ mol^{-1} above the global minimum have been then optimized at DFT level using the B3LYP functional and the 6-311++G(d,p) basis set. In order to simulate solvent effects, their energy was also evaluated by single-point calculations in aqueous solution using the CPCM continuum model. Details of the computational calculations are given elsewhere [69, 70].

The Boltzmann populations of the erythritol and *L*-threitol conformers in gas phase and aqueous solution are displayed in Figure 7. The geometries of the relevant conformers in both media, as well as the conformations exiting in the crystalline structures, are depicted in Figure 8. For the sake of simplicity the conformers were grouped according to their backbone family, being labeled with three italic letters: *g* (gauche$^+$), *g′* (gauche$^-$) and *t* (trans), to specify the orientation of the O(1)-C(1)-C(2)-O(2), C(1)-C(2)-C(3)-C(4) and O(3)-C(3)-C(4)-O(4) dihedrals, respectively.

As shown in Figure 7, isolated erythritol exists preferentially in the *tgg* and *ggg* bent conformations (both make up 38% of the conformational mixture at 298.15K). Their stability results from the formation of intramolecular hydrogen bonds which are particularly strong in the *tgg* conformers due to the participation of the terminal OH groups. In the presence of the solvent, the population of these conformers decreases significantly (11%). Conversely, two higher energy conformers in gas phase (*g′tg* and *gtg*), both with a distended carbon backbone, emerge as the most stable forms in solution (49%). This population change can be explained by the stronger hydration of conformers with the OH groups less internally H-bonded and therefore more available to interact with water. In the crystal, the neutron diffraction data available for this compound indicates that the molecule exhibits two conformations, labeled as A and B, differing each other by the positions of the H[O(2)] and H[O(3)] hydrogen atoms. At 22.6 K the occupancy of conformations A and B are 85% and 15%, respectively [71]. In terms of backbone, both have a *g′tg* conformation which is the same as that exhibited by the preferred conformers of erythritol in aqueous solution. As shown in Figure 8, the geometry of the most stable conformer in solution matches that of the crystal conformation having the highest occupancy percentage.

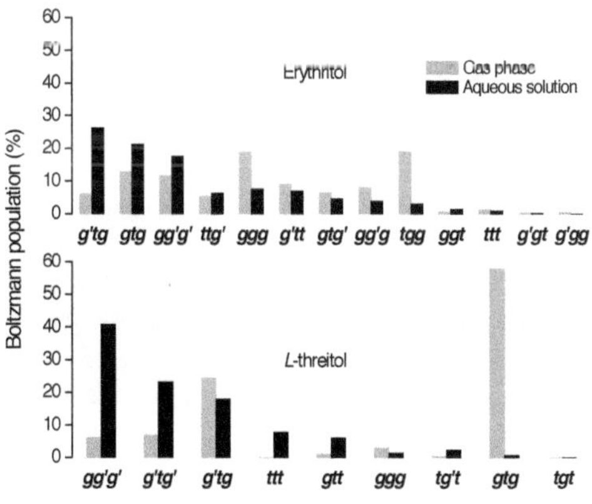

Figure 7. Conformational distribution of erythritol and *L*-threitol in gas phase and aqueous solution. Boltzmann populations were estimated from the Gibbs energies of the conformers [B3LYP/6-311++G(d,p)] in both media.

Regarding *L*-threitol, its gas phase conformational mixture is largely dominated by the *gtg* conformers which are characterized by a cyclic, symmetric and cooperative hydrogen bonding network. Since this OH groups' arrangement is very unfavorable to interact with water, they practically disappear in solution and are replaced by the *gg'g'*, *g'tg'* or *g'tg* conformations which are not so highly intramolecularly bonded. Unlike erythritol, none of these conformations has been identified in the crystalline *L*-threitol [72]. Here, the molecules adopt a distended conformation (*ttt*) which has a relatively low weight in solution (8%) and does not exist in the vacuum.

The results just presented provide an important basis to understand the different crystallization ability of erythritol and *L*-threitol. In the *meso* form, this process is facilitated due to the resemblance between the backbone conformation of the molecules existing in the crystal lattice and that characteristic of the predominant forms in solution. The fact that such resemblance is not found to exist in *L*-threitol may contribute to its lower crystallization tendency. This behavior is in agreement with the results obtained for other alditols [68, 73]

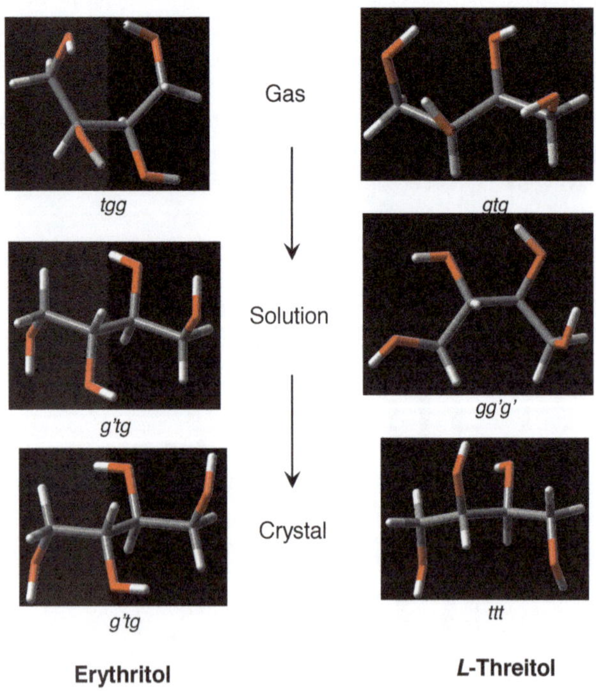

Figure 8. Most stable conformers of erythritol and *L*-threitol in gas phase and aqueous solution, as well as the respective crystal conformations. The crystal conformation of erythritol corresponds to that having highest occupancy in the crystal lattice (Conformation A).

5. Conformational variation during the molecular incorporation into the crystal: Glutamic acid

Glutamic acid, 2-aminopentanedioic acid ($C_5H_9NO_4$), is another example illustrating the role played by conformation on crystallization. Two conformational polymorphs have been identified for this compound, labeled as α and β, with the latter being the thermodynamically stable form over all temperature range [74-77]. The crystals have different morphologies: the metastable form exhibits a prismatic shape while the stable one has a needle-like shape. Both belong to the ortorrombic space group with four molecules in the unit cell (Z=4) [78, 79]. In the crystal lattice the molecules are in the zwitterionic state and all functional groups participate in a complex intermolecular hydrogen bonding network [63].

The conformations adopted by the *L*-glutamic acid molecules in the two polymorphs are displayed in Figure 9. Their main difference lies in the orientation of the two C-C-C-C dihedral angles. In the crystal lattice the C(1)-C(2)-C(3)-C(4) and C(2)-C(3)-C(4)-C(5) dihedrals assume, respectively, values of 59.2°, 68.3° in α and -171.1°, -73.1° in β. Using the same dihedral labeling scheme previously adopted for the polyols, the conformation of glutamic acid in the α and β-crystals is *gg* and *tg'*, respectively. Depending on the experimental conditions, namely the temperature, both polymorphs can be crystallized from aqueous solution. Cooling a supersaturated solution to temperatures below 25°C originates almost pure α-crystals, whereas as the temperature raises the proportion of form β in the precipitated crystals increases. For example at 45°C, the fraction of β-crystals is 45% [74, 75].

The conformational behavior of zwitterionic glutamic acid is aqueous solution is a useful starting point to understand the crystallization of this compound in molecular terms. This has been done theoretically by performing full geometry optimizations using the CPCM continuum solvation model and the B3LYP/aug-cc-pVDZ model chemistry, both implemented in the Gaussian 03 program. The cavity was built with the Bondii radii which have been found to yield accurate results for the hydration of similar molecules in the zwitterion ic state, such for example glycine [80]. Nine starting geometries were built by assuming the three standard orientations (g, g' and t) for each one of the C-C-C-C dihedrals. The remaining dihedrals were kept in their preferred orientations. The optimized structures were further submitted to a vibrational frequency calculation at the same level to ensure that they correspond to minima on the aqueous potential energy surface and also to calculate the thermal corrections at 298.15K.

The Gibbs energies of the conformers in aqueous solution at 298.15 K (G_{sol}) can be expressed as: $G_{sol} = E_{int} + \Delta G_{thermal} + \Delta G_{hyd}$, with E_{int} representing the intrinsic (abbreviated as "int") electronic energy of the conformer at 0 K, *i.e.*, excluding solvent effects, $\Delta G_{thermal}$ the thermal correction to the Gibbs energy from 0 to 298K and ΔG_{hyd} the Gibbs energy of hydration at 298.15K. The values of E_{int}, G_{sol}, ΔG_{hyd}, as well as the equilibrium populations at 298.15 K, are given in Table 3. In this table are also included the values of the C(1)-C(2)-C(3)-C(3) and C(2)-C(3)-C(4)-C(5) dihedrals, abbreviated as D(1) and D(2), respectively. The relative stability of the conformers in solution is governed by a balance between two effects: interaction with the solvent (quantified by ΔG_{hyd}) and intramolecular interaction (quantified by E_{int}). In general, conformers with a favorable intramolecular interaction are weakly hydrated and *vice-versa*. The most abundant conformer in solution, *tg* (Pop. = 67%), is the one having a less negative ΔG_{hyd}; however, the intrinsic energy decrease arising from the formation of an internal N^{+}–H···O hydrogen bond (see Figure 9), turns it into the lowest energy conformer. Unlike, the second most stable conformer (*tt*) has a fully distended backbone that hampers the formation of the just referred intramolecular hydrogen bond and so it has a relatively high intrinsic energy. Its stabilization results essentially from a favorable interaction with the solvent. Both conformers represent *ca.* 81% of the conformational composition in aqueous solution at 298.15K.

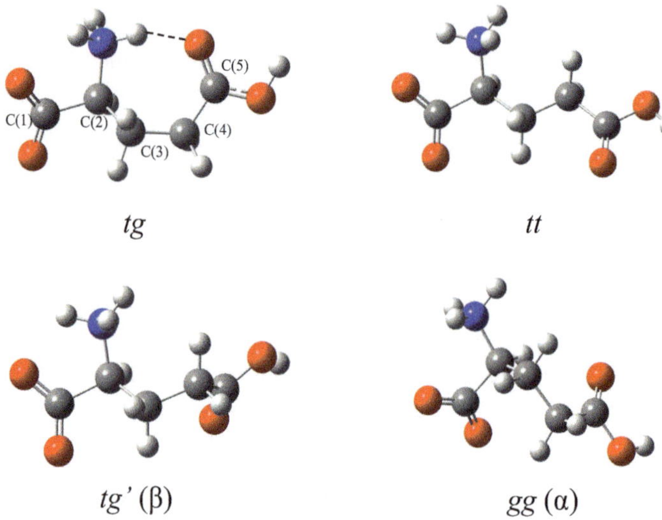

Figure 9. CPCM/B3LYP/aug-cc-pVDZ optimized geometries of relevant conformers of the zwitterionic glutamic acid in aqueous solution. Dashed line represents an intramolecular hydrogen bond. Atom numbering scheme is shown for the most stable conformer.

Regarding the two crystal conformations, *tg*' (β) and *gg* (α), they are stable forms in aqueous solution with the values of the C-C-C-C dihedral angles in solution not differing significantly from those in the crystal. However, somehow surprisingly, their estimated equilibrium populations in solution are too low: 2.3 and 0.2% for conformers β and α, respectively. Glutamic acid thus constitutes an interesting example of a molecule where the conformers experimentally observed in the crystalline structure are much diluted in aqueous solution. Two immediate conclusions can be drawn from this result: (1) the conformational composition in aqueous solution does not determine the conformations exhibited by this compound in the final crystalline state; (2) crystallization is accompanied by a conformational change. The second conclusion implies that conformers *tg* and *tt*, namely the first, are interconverted into the *gg* and *tg*'. These interconversions can be characterized by the energy barriers ($\Delta G^{\ddagger}$) separating these conformers. For this purpose, we have used the synchronous transit-guided quasi-Newton method (STQN) [81] in its QST2 variety at the CPCM/B3LYP/aug-cc-pVDZ level. The activation energies corresponding to the interconversion of the most stable conformer into conformers α and β were calculated to be $\Delta G^{\ddagger} = 18$ and 29 kJ mol^{-1}, respectively. This barrier height difference may help to explain the preferential crystallization of form α at temperatures below 25°C and the increase of the percentage of form β in the crystals obtained at higher temperatures. Apparently, the dependence of the crystallization behavior of glutamic acid on temperature is not due to a change on the relative abundance of the crystallizing conformers in solution with temperature [75], but rather to the value of the energy barriers for the interconversion of the most stable conformer in solution into the α and β conformers.

Conf.	Dihedral angles/°		E_{int}/ kJ mol^{-1}	-ΔG_{hyd}/ kJ mol^{-1}	G_{sol}/ kJ mol^{-1}	Pop. (%)
	D(1)	D(2)				
tg	178.5	75.3	13.1	184.8	0.0	67.1
tt	168.9	-177.2	56.9	223.7	3.9	13.6
g'g'	-52.5	-79.9	0.0	168.2	4.8	9.8
gt	63.1	178.8	36.4	196.1	7.4	3.5
g´t	-56.5	178.7	52.1	213.7	7.7	3.0
tg'(β)	172.5	-69.8	59.6	221.9	8.4	2.3
gg'	62.4	-85.1	49.8	208.1	12.5	0.4
gg(α)	66.1	61.1	54.7	210.5	14.0	0.2
g´g	-61.7	81.4	71.1	225.3	18.3	0.0

Table 3. Dihedral angles, intrinsic electronic energies at 0K (E_{int}), Gibbs energies in aqueous solution (G_{sol}), Gibbs energies of hydration (ΔG_{hyd}) and Boltzmann populations (Pop.) at 298.15K of the zwitterionic glutamic acid conformers, calculated at the CPCM/B3LYP/aug-cc-pVDZ level.

6. Molecular dynamics study of pre-nucleation clusters

The molecular aggregates individualized in the early stages of nucleation (pre-nucleation) are windows to follow the crystallization process. Data on the self-assembly of erythritol and *L*-threitol in supersaturated aqueous solutions are reported through molecular dynamics (MD) simulations. MD is one of the most comprehensive computer simulation tools, with enormous application in the study of large systems and molecular phenomena requiring longer observation times. Relevant molecular phenomena are often adequately described by classical mechanics, providing reliable force-fields are available for a variety of systems. The results of the MD simulations presented in this chapter were carried out in the NpT ensemble and under periodic boundary conditions, resorting to the GROMACS package, version 4.5.4 [82] and GROMOS 96 43a1 force field [83]. Long-range electrostatics was computed using the particle mesh Ewald (PME) method, while for Lennard-Jones energies a cut-off of 1.4

nm was applied. Temperature (298K) and pressure (1bar) were coupled to Berendsen external baths, with coupling constants of 0.1 and 0.5 ps, respectively. Each system was firstly subjected to an energy minimization step, and then left to evolve up to 80ns, using in both parts a standard time step of 2 fs. The last 40 ns of production runs were subsequently subjected to standard analysis, such as radial distribution functions [RDF, g(*r*)].

Useful data about the type of aggregates formed in solution can be given by the RDFs for the different pairs of solute OH groups (OH–OH) which are depicted in Figure 10. They were obtained from the simulation of an aqueous solution containing 11 solute molecules and 4000 solvent molecules, corresponding to a concentration of 75g /100 cm^3, which is near to the saturation concentration [84].

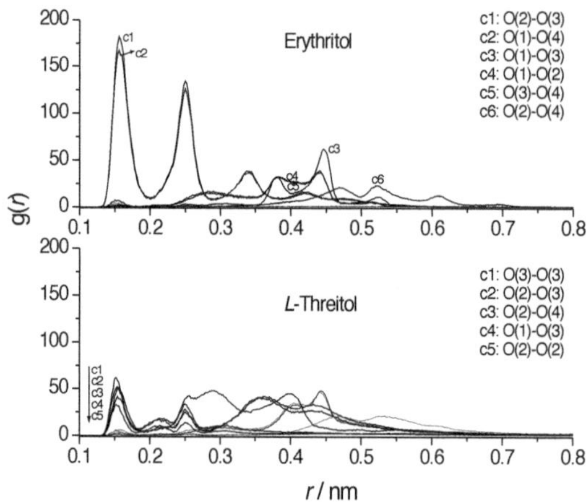

Figure 10. OH–OH RDFs for erythritol and *L*-threitol in aqueous solution. For better visualization the most relevant curves are numbered as c1, c2, etc.

In erythritol, the most prominent RDFs are found for the O(2)H–O(3)H and O(1)H–O(4)H pairs. Both RDFs, labeled in Figure 10 as c1 and c2, respectively, exhibit sharp and intense peaks at *r* (nearest distance between groups) = 0.16 and 0.25 nm. This indicates that the self-assembly of erythritol in solution yields well organized clusters formed by hydrogen bonds involving preferentially the O(1)H and O(4)H groups or the O(2)H and O(3)H groups. Analysis of the MD trajectory shows that these clusters are mostly dimers with some of them involving simultaneously the two just referred H-bonds (cyclic dimers). Interestingly, the cyclic dimer formed by these two H-bonds is the one existing in the crystalline structure. Some illustrative snapshots of these dimers are displayed in Figure 11. From the other OH–OH RDFs shown in Figure 10 one can conclude that the formation of intermolecular hydrogen bonds involving the remaining groups is very unlikely to occur. In fact, the integration of all RDFs up to 0.28 nm (cut-off distance for the formation of an H-bond [85]) reveals that their probability of formation is lower than 15%.

Regarding *L*-threitol, the RDF profiles are substantially different from those obtained for erythritol. In fact, the most relevant RDFs, labeled in Figure 10 as c1 to c5 and corresponding to the O(3)H–O(3)H, O(2)H–O(3)H, O(2)H–O(4)H, O(1)H–O(3)H and O(2)H–O(2)H interactions, show relatively well defined first peaks centered at 0.15 nm. The first peaks heights are much lower than in erythritol which means that the organized part of the *L*-threitol clusters is much smaller as compared to that in erythitol. The remaining part of the clusters has a disordered structure as evidenced by the broad peaks located at larger distances. This result is an argument in favor of the lower crystallization tendency of this compound, in addition to the conformational one discussed in section 4. The snapshots of *L*-threitol clusters during the simulation show the existence of various types of disordered oligomers (Figure 12).

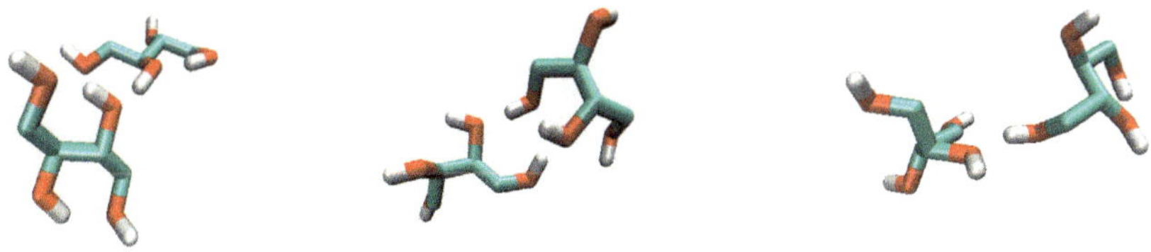

Figure 11. Snapshots of the erythritol clusters formed in aqueous solution.

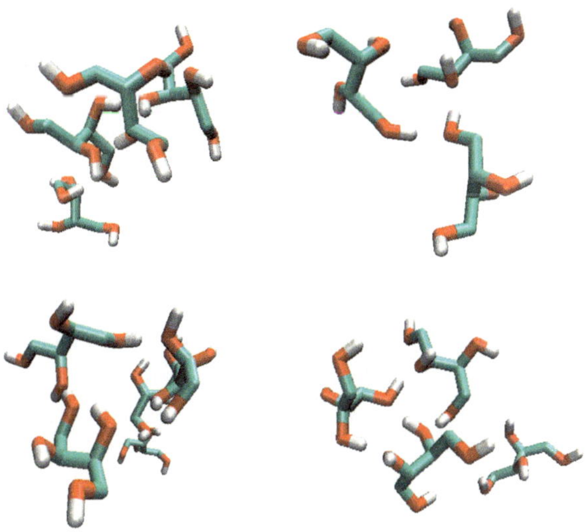

Figure 12. Snapshots of the *L*-threitol clusters formed in aqueous solution.

7. Conclusion

Structural aspects of crystallization from solvents were pointed out throughout this chapter. Particular attention was paid to the conformational variation of flexible molecules during this process. Three compounds were taken as examples: erythritol, *L*-threitol and glutamic acid. The different crystallization behavior shown by the two alditols was understood in terms of the resemblance between the solution and crystal conformers. Regarding glutamic acid, it is a quite peculiar example since the conformers existing in the two identified conformational polymorphs have a negligible weight in aqueous solution. Their selective crystallization has been interpreted on the grounds of the energy barriers separating the dominant conformer in solution and those found in both crystalline forms.

The investigation of the structure of the molecular aggregates formed in solution by molecular dynamics, here exemplified for erythritol and *L*-threitol, has proven to be a valuable contribution to better understand crystallization.

Polymorphism, an important property of the solid state structure with various practical implications, was called in the present chapter as a tangly pathway in the crystallization process. Additional crystalline structure complexity may result from crystal disorder, hardly to be included in the polymorphism concept.

Author details

J.S. Redinha[1], A.J. Lopes Jesus[1,2], A.A.C.C. Pais[1] and J. A. S. Almeida[1]

1 University of Coimbra / Department of Chemistry, Portugal

2 University of Coimbra / Faculty of Pharmacy, Portugal

References

[1] Brian Henderson and Ralph HB. Crystal-Field Engineering of Solid-State Laser Materials. New York: Cambridge University Press; 2000.

[2] Brooks JS. Organic crystals: properties, devices, functionalization and bridges to biomolecules. Chemical Society Reviews. 2010;39(7):2667-94.

[3] Day GM, Motherwell WDS, Ammon HL, Boerrigter SXM, Della Valle RG, Venuti E, et al. A third blind test of crystal structure prediction. Acta Crystallographica Section B. 2005;61(5):511-27.

[4] Motherwell WDS, Ammon HL, Dunitz JD, Dzyabchenko A, Erk P, Gavezzotti A, et al. Crystal structure prediction of small organic molecules: a second blind test. Acta Crystallographica Section B. 2002;58(4):647-61.

[5] Lommerse JPM, Motherwell WDS, Ammon HL, Dunitz JD, Gavezzotti A, Hofmann DWM, et al. A test of crystal structure prediction of small organic molecules. Acta Crystallographica Section B. 2000;56(4):697-714.

[6] Oganov AR, editor. Modern Methods of Crystal Structure Predictions. Weinheim: Wiley-VCH; 2011.

[7] Woodley SM, Catlow R. Crystal structure prediction from first principles. Nature Materials. 2008;7(12):937-46.

[8] Oganov AR, Glass CW. Crystal structure prediction using ab initio evolutionary techniques: Principles and applications. Journal of Chemical Physics. 2006;124(24): 244704-15.

[9] Day GM, Cooper TG, Cruz-Cabeza AJ, Hejczyk KE, Ammon HL, Boerrigter SXM, et al. Significant progress in predicting the crystal structures of small organic molecules - a report on the fourth blind test. Acta Crystallographica Section B. 2009;65(2): 107-25.

[10] Kim S, Orendt AM, Ferraro MB, Facelli JC. Crystal structure prediction of flexible molecules using parallel genetic algorithms with a standard force field. Journal of Computational Chemistry. 2009;30(13):1973-85.

[11] Day GM, S. Motherwell WD, Jones W. A strategy for predicting the crystal structures of flexible molecules: the polymorphism of phenobarbital. Physical Chemistry Chemical Physics. 2007;9(14):1693-704.

[12] Hohenberg P, Kohn W. Inhomogeneous Electron Gas. Physical Review. 1964;136(3B):B864-B71.

[13] Kohn W, Sham LJ. Self-Consistent Equations Including Exchange and Correlation Effects. Physical Review. 1965;140(4A):A1133-A8.

[14] Becke AD. Density-Functional Exchange-Energy Approximation with Correct Asymptotic-Behavior. Physical Review A. 1988 Sep 15;38(6):3098-100.

[15] Becke AD. Density-Functional Thermochemistry.3. The Role of Exact Exchange. Journal of Chemical Physics. 1993 Apr 1;98(7):5648-52.

[16] Lee CT, Yang WT, Parr RG. Development of the Colle-Salvetti Correlation-Energy Formula into a Functional of the Electron-Density. Physical Review B. 1988 Jan 15;37(2):785-9.

[17] Dunning J, Thom H. Gaussian basis sets for use in correlated molecular calculations. I. The atoms boron through neon and hydrogen. Journal of Chemical Physics. 1989 1989/01/15/;90(2):1007-23.

[18] Woon DE, Dunning J, Thom H. Gaussian basis sets for use in correlated molecular calculations. III. The atoms aluminum through argon. Journal of Chemical Physics. 1993 1993/01/15/;98(2):1358-71.

[19] Krishnan R, Binkley JS, Seeger R, Pople JA. Self-consistent molecular orbital methods. XX. A basis set for correlated wave functions. Journal of Chemical Physics. 1980;72(1):650-4.

[20] Cossi M, Rega N, Scalmani G, Barone V. Energies, structures, and electronic properties of molecules in solution with the C-PCM solvation model. Journal of Computational Chemistry. 2003 Apr 30;24(6):669-81.

[21] Cossi M, Barone V, Cammi R, Tomasi J. Ab initio study of solvated molecules: A new implementation of the polarizable continuum model. Chemical Physics Letters. 1996 Jun 14;255(4-6):327-35.

[22] Klamt A, Schuurmann G. Cosmo - a New Approach to Dielectric Screening in Solvents with Explicit Expressions for the Screening Energy and Its Gradient. Journal of the Chemical Society- Perkin Transactions 2. 1993(5):799-805.

[23] Weissbuch I, Kuzmenko I, Vaida M, Zait S, Leiserowitz L, Lahav M. Twinned Crystals of Enantiomorphous Morphology of Racemic Alanine Induced by Optically Resolved.alpha.-Amino Acids; A Stereochemical Probe for the Early Stages of Crystal Nucleation. Chemistry of Materials. 1994 2012/09/14;6(8):1258-68.

[24] Etter MC. Hydrogen bonds as design elements in organic chemistry. Journal of Physical Chemistry. 1991 2012/09/14;95(12):4601-10.

[25] Weissbuch I, Lahav M, Leiserowitz L. Toward Stereochemical Control, Monitoring, and Understanding of Crystal Nucleation. Crystal Growth & Design. 2003 2012/09/14;3(2):125-50.

[26] Bernstein J. Polymorphism in Molecular Crystals. Oxford: Oxford University Press; 2002.

[27] Hilfiker R, editor. Polymorphism in the Pharmaceutical Industry. Weinheim: Willey-VCH; 2006.

[28] Brittain HG, editor. Polymorphism in Pharmaceutical Solids. 2 ed. New York: Informa Healthcare USA, Inc; 2009.

[29] Nangia A. Conformational Polymorphism in Organic Crystals. Accounts of Chemical Research. 2008;41(5):595-604.

[30] J.W.Mullin. Crystallization. 4 ed. Oxford: Butterworth-Heinemann; 2001.

[31] Schmelzer JWP, editor. Nucleation Theory and Applications: WILEY-VCH; 2005.

[32] Laaksonen A, Talanquer V, Oxtoby DW. Nucleation: Measurements, Theory, and Atmospheric Applications. Annual Review of Physical Chemistry. 1995;46(1):489-524.

[33] Steiner T. The hydrogen bond in the solid state. Angewandte Chemie, International Edition in English. 2002;41(1):48-76.

[34] Elangannan Arunan, Gautam R. Desiraju, Roger A. Klein, Joanna Sadlej, Steve Scheiner, Ibon Alkorta, et al. Definition of the hydrogen bond (IUPAC Recommendations 2011). Pure and applied chemistry. 2011;83(8):1637-41.

[35] Lee SJ, Min A, Kim Y, Ahn A, Chang J, Lee SH, et al. Conformationally resolved structures of jet-cooled acetaminophen by UV-UV hole-burning spectroscopy. Physical Chemistry Chemical Physics. 2011;13(37):16537-41.

[36] McCrone WC. In: Fox D, Labes MM, Weissberger A, editors. Physics and Chemistry of the Organic Solid State. London: Interscience Publishers; 1965. p. 725-67.

[37] Singhal D, Curatolo W. Drug polymorphism and dosage form design: a practical perspective. Advanced Drug Delivery Reviews. 2004;56(3):335-47.

[38] Perlovich G, Volkova T, Bauer-Brandl A. Polymorphism of paracetamol. Journal of Thermal Analysis and Calorimetry. 2007;89(3):767-74.

[39] Kolesov BA, Mikhailenko MA, Boldyreva EV. Dynamics of the intermolecular hydrogen bonds in the polymorphs of paracetamol in relation to crystal packing and conformational transitions: a variable-temperature polarized Raman spectroscopy study. Physical Chemistry Chemical Physics. 2011;13(31):14243-53.

[40] Boldyreva E, Drebushchak V, Paukov I, Kovalevskaya Y, Drebushchak T. DSC and adiabatic calorimetry study of the polymorphs of paracetamol. Journal of Thermal Analysis and Calorimetry. 2004;77(2):607-23.

[41] Di Martino P, Conflant P, Drache M, Huvenne JP, Guyot-Hermann AM. Preparation and physical characterization of forms II and III of paracetamol. Journal of Thermal Analysis and Calorimetry. 1997;48(3):447-58.

[42] Haisa M, Kashino S, Kawai R, Maeda H. The Monoclinic Form of p-Hydroxyacetanilide. Acta Crystallographica Section B-Structural Science. 1976;32(4):1283-5.

[43] Bouhmaida N, Bonhomme F, Guillot B, Jelsch C, Ghermani NE. Charge density and electrostatic potential analyses in paracetamol. Acta Crystallographica Section B. 2009;65(3):363-74.

[44] Haisa M, Kashino S, Maeda H. The orthorhombic form of p-hydroxyacetanilide. Acta Crystallographica Section B. 1974;30(10):2510-2.

[45] Nichols G, Frampton CS. Physicochemical characterization of the orthorhombic polymorph of paracetamol crystallized from solution. Journal of Pharmaceutical Sciences. 1998;87(6):684-93.

[46] Burgina EB, Baltakhinov VP, Boldyreva EV, Shakhtschneider TP. IR Spectra of Paracetamol and Phenacetin. 1. Theoretical and Experimental Studies. Journal of Structural Chemistry. 2004;45(1):64-73.

[47] Ivanova BB. Monoclinic and orthorhombic polymorphs of paracetamol - solid state linear dichroic infrared spectral analysis. Journal of Molecular Structure. 2005;738(1-3):233-8.

[48] Iogansen AV. Direct proportionality of the hydrogen bonding energy and the intensification of the stretching v(XH) vibration in infrared spectra. Spectrochimica Acta, Part A: Molecular and Biomolecular Spectroscopy. 1999;55(7-8):1585-612.

[49] Wendy C. Duncan-Hewitt, Grant DJW. True density and thermal expansivity of pharmaceutical solids: comparison of methods and assessment of crystallinity. International Journal of Pharmaceutics. 1986;28(1):75-84.

[50] Joiris E, Martino PD, Berneron C, Guyot-Hermann A-M, Guyot J-C. Compression Behavior of Orthorhombic Paracetamol. Pharmaceutical Research. 1998;15(7):1122-30.

[51] Di Martino P, Guyot-Hermann AM, Conflant P, Drache M, Guyot JC. A new pure paracetamol for direct compression: The orthorhombic form. International Journal of Pharmaceutics. 1996;128:1-8.

[52] Roth GJ, Calverley DC. Aspirin, Platelets, and Thrombosis: Theory and Practice. Blood. 1994;83(4):885-98.

[53] Wheatley PJ. The crystal and molecular structure of aspirin. Journal of the Chemical Society (Resumed). 1964:6036-48.

[54] Ouvrard C, Price SL. Toward Crystal Structure Prediction for Conformationally Flexible Molecules: The Headaches Illustrated by Aspirin. Crystal Growth & Design. 2004;4(6):1119-27.

[55] Vishweshwar P, McMahon JA, Oliveira M, Peterson ML, Zaworotko MJ. The Predictably Elusive Form II of Aspirin. Journal of the American Chemical Society. 2005 2012/09/10;127(48):16802-3.

[56] Bond AD, Boese R, Desiraju GR. On the Polymorphism of Aspirin. Angewandte Chemie International Edition. 2007;46(4):615-7.

[57] Chang C-J, Díaz LE, Morin F, Grant DM. Solid-state 13C NMR study of drugs: Aspirin. Magnetic Resonance in Chemistry. 1986;24(9):768-71.

[58] Wen S, Beran GJO. Accidental Degeneracy in Crystalline Aspirin: New Insights from High-Level ab Initio Calculations. Crystal Growth & Design. 2012 2012/09/10;12(5): 2169-72.

[59] Bond AD, Boese R, Desiraju GR. What Is A Polymorph? Aspirin As A Case Study. American Pharmaceutical Review. 2007:1-4.

[60] Glusker JP, Lewis M, Rossi M. Crystal Structure Analysis for Chemists and Biologists (Methods in Stereochemical Analysis). New York: Wiley-VCH; 1994.

[61] Wilson CC. A basic introduction to thermal motions of atoms in crystal structures, the underlying potentials and the physical information available from their analysis. Crystallography Reviews. 2009 2012/09/11;15(1):3-56.

[62] Habgood M, Grau-Crespo R, Price SL. Substitutional and orientational disorder in organic crystals: a symmetry-adapted ensemble model. Physical Chemistry Chemical Physics. 2011;13(20):9590-600.

[63] Redinha JS, Lopes Jesus AJ. Molecular Recognition and Crystal Growth. In: McEvoy JA, editor. Molecular Recognition: Biotechnology, Chemical Engineering and Materials Applications: Novapublishers; 2011.

[64] Redinha JS, Lopes Jesus AJ. Crystal Growth of Pharmaceuticals from Melt. In: Borisenko E, editor. Crystallization and material science of modern artificial and natural crystals. Rijeka: InTech; 2012.

[65] Pascard C, Tran Huu Dau E, Manoury P, Mompon B. Betaxolol hydrochloride: 1-{4-[2-(cyclopropylmethoxy)ethyl]phenoxy}-3-isopropylamino-2-propanol hydrochloride, C18H30NO3+.Cl. Acta Crystallographica, Section C: Crystal Structure Communications. 1984;40(8):1430-2.

[66] Lopes Jesus AJ, Nunes SCC, Ramos Silva M, Matos Beja A, Redinha JS. Erythritol: Crystal growth from the melt. International Journal of Pharmaceutics. 2010;388(1-2): 129-35.

[67] Lopes Jesus AJ, Redinha JS. On the structure of erythritol and L-threitol in the solid state: An infrared spectroscopic study. Journal of Molecular Structure. 2009;938(1-3): 156-64.

[68] Yu L, Reutzel-Edens SM, Mitchell CA. Crystallization and Polymorphism of Conformationally Flexible Molecules: Problems, Patterns, and Strategies. Organic Process Research & Development. 2000 2012/09/14;4(5):396-402.

[69] Lopes Jesus AJ, Tomé LIN, Rosado MTS, Leitão MLP, Redinha JS. Conformational study of erythritol and threitol in the gas state by density functional theory calculations. Carbohydrate Research. 2005;340(2):283-91.

[70] Lopes Jesus AL, Tomé LIN, Eusébio MES, Redinha JS. Determination of the Enthalpy of Solute-Solvent Interaction from the Enthalpy of Solution: Aqueous Solutions of Erythritol and L-Threitol. Journal of Physical Chemistry B. 2006 May 11, 2006;110(18): 9280-5.

[71] Ceccarelli C, Jeffrey GA, McMullan RK. A Neutron-Diffraction Refinement of the Crystal-Structure of Erythritol at 22.6 K. Acta Crystallographica Section B. 1980;36:3079-83.

[72] Kopf J, Morf M, Zimmer B, Haupt ETK, Jarchow O, Köll P. The crystal and molecular structure of threitol. Carbohydrate Research. 1993;247:119-28.

[73] Jeffrey GA, Kim HS. Conformations of the alditols. Carbohydrate Research. 1970;14(2):207-16.

[74] Kitamura M. Polymorphism in the crystallization of L-glutamic acid. Journal of Crystal Growth. 1989;96(3):541-6.

[75] Kitamura M. Controlling factor of polymorphism in crystallization process. Journal of Crystal Growth. 2002;237-239:2205-14.

[76] Davey RJ, Blagden N, Potts GD, Docherty R. Polymorphism in Molecular Crystals: Stabilization of a Metastable Form by Conformational Mimicry. Journal of the American Chemical Society. 1997;119(7):1767-72.

[77] Ono T, ter Horst JH, Jansens PJ. Quantitative Measurement of the Polymorphic Transformation of l-Glutamic Acid Using In-Situ Raman Spectroscopy. Crystal Growth & Design. 2004;4(3):465-9.

[78] Lehmann M, Koetzle T, Hamilton W. Precision neutron diffraction structure determination of protein and nucleic acid components. VII. The crystal and molecular structure of the amino acid L-lysine monohydrochloride dihydrate. Journal of Chemical Crystallography. 1972;2(5):225-33.

[79] Lehmann MS, Nunes AC. A short hydrogen bond between near identical carboxyl groups in the [alpha]-modification of l-glutamic acid. Acta Crystallographica, Section B: Structural Science. 1980;36(7):1621-5.

[80] Kim CK, Park B-H, Lee HW, Kim CK. Comprehensive Studies on the Free Energies of Solvation and Conformers of Glycine: A Theoretical Study. Bulletin of the Korean Chemical Society. 2011;32(6):1985 - 92.

[81] Peng CY, Ayala PY, Schlegel HB, Frisch MJ. Using redundant internal coordinates to optimize equilibrium geometries and transition states. Journal of Computational Chemistry. 1996;17(1):49-56.

[82] Hess B, Kutzner C, van der Spoel D, Lindahl E. GROMACS 4: Algorithms for Highly Efficient, Load-Balanced, and Scalable Molecular Simulation. Journal of Chemical Theory and Computation. 2008;4(3):435-47.

[83] Schuttelkopf AW, van Aalten DMF. PRODRG: a tool for high-throughput crystallography of protein-ligand complexes. Acta Crystallographica Section D. 2004;60(8): 1355-63.

[84] Cohen S, Marcus Y, Migron Y, Dikstein S, Shafran A. Water sorption, binding and solubility of polyols. Journal of the Chemical Society, Faraday Transactions. 1993;89(17):3271-5.

[85] Gilli G, Gilli P. Towards an unified hydrogen-bond theory. Journal of Molecular Structure. 2000;552(1-3):1-15.

Chapter 8

Modeling Ice Crystal Formation of Water in Biological System

Zhi Zhu He and Jing Liu

Additional information is available at the end of the chapter

1. Introduction

Ice crystal formation of water plays a very important role in broadest scientific and engineering fields such as cloud evolution in atmosphere and brine pockets in marine ice. One of the most attractive attentions of this issue is about the mechanisms of cell injury induced by freezing in cryobiology [1]. During the past few decades, a series of cryogenic techniques have been developed for clinical applications in cryosurgery and cryopreservation. The former is aimed to destroy the target diseased tissues (such as tumor) while minimize pain and bleeding of the tissue as much as possible. It has been demonstrated to be a safe, minimally invasive and cost-effective treatment approach. In contrast to the cryosurgery, the task of cryopreservation is however to preserve cell, tissues or organs at a super-low temperature condition with the addition of protective agents, and ensure that they can be normally used after several re-warming stages. Clearly, no matter which process it involves, the formation of the ice crystal is ineluctable and decisive to the final fate of those cells and tissues subject to treatment.

Cryoinjury induced by ice formation is usually considered as a main killer of cells during cryosurgery or cryopreservation. A two-factor hypothesis [2], "intracellular ice formation" (IIF) at rapid cooling rates and "solution effects" at slow cooling rates, proposed by Mazur, has been serving well to explain the mechanism of freeze injury. The biological materials would subject to damage due to extracellular and intracellular ice formation during freezing. For the cryopreservation at super-cooling temperature, the occurrence of IIF is the single most important factor determining whether or not cells will survive cryopreservation [3]. So far, the ice crystal formation problems have been partially addressed in material and atmospheric sciences. However, similar researches are hard to find in cryobiology area due to complexity of the water structure in a biological system. This chapter is dedicated to summarize the typical physical and mathematical model for ice nucleation and growth in cryobiology area. In addition, the heat and solute transport, and interaction between the cell and ice during freezing are also discussed.

2. Ice nucleation

Freezing often denotes a process of new phase defined by ice creation via nucleation and subsequent crystal growth from supercooled water. This process undergoes three stages including the chemical diffusion stage for water molecule transportation, the surface kinetic stage for hydrogen bonds formation, and the final heat conduction stage for heat releasing and diffusing. It implies that the ice nucleation—whose formation determines the growth of ice crystal to a large extent—turns out to be the initial and crucial step during crystallization. In terms of whether there exist foreign bodies or not, the nucleation is usually classified as homogeneous or heterogeneous cases. By definition, homogeneous ice nucleation requests a severe condition that the liquid water is absolutely dust-free, or there exist foreign particles indeed, but whose sizes are small enough to be negligible. Since this rigid demand is hard to be satisfied, the nucleation is in fact regarded as heterogeneous in most cases and may be influenced by many external factors, such as the size of foreign particles, surface roughness and the effective size of the nucleators at different supercoolings. There exist a great many barriers to build up an integrated theory on nucleation, because of the dissimilar properties might exhibit between bulk fluid and the micro/nanoscale molecules. Up to now, the two major theoretical approaches to characterize the formation of ice can be classified as thermodynamic approximation and molecular simulations [4].

2.1. Thermodynamic model of ice nucleation

The ice nucleation does not happen necessarily if only temperature is below the freezing point. This process is determined by the free energy of system, which implies that the ice germ growth or disappearance depends on the free energy decrease or increase. For constant rate cooling protocols, the temperature of system is given by $T(t)=T_0-Bt$, where B is the cooling rate. Considering a cluster of ice denoted by β with i molecules from its mother water phase α, the driving force of nucleation is derived from the Gibbs free energy differenceΔG_i [5]

$$\Delta G_i = iv^{\beta}\Delta G_t + (36\pi)^{1/3} i^{2/3} \sigma^{\alpha\beta} (v^{\beta})^{2/3} \quad (1)$$

where ΔG_t is negative and denotes the Gibbs free-energy difference between α and β phase per volume, according to classical homogeneous nucleation theory; $v^{\beta}$is the molecular volume of phase β, and $\sigma^{\alpha\beta}$ is the interfacial free energy per unit area. According to thermodynamics, Gibbs free-energy difference between α and β could be given as

$$\Delta G_t = \frac{\Delta H_f \Delta T}{T_f} \frac{2T}{T_f + T} \tag{2}$$

where ΔH_f is the heat of fusion, T_f is the equilibrium freezing temperature, and $\Delta T = T_f - T$. The free energy of form of a cluster of ice is initially positive and goes through a maximum as the cluster size is increased. Maximizing with respect to i and assuming closely packed water molecules, the following relationships for the critical cluster could be derived:

$$r_c = -\frac{2\sigma^{\alpha\beta}}{\Delta G_t} \text{ and } \Delta G_{i_c} = \frac{16\pi}{3} \frac{\left(\sigma^{\alpha\beta}\right)^3}{\left(\Delta G_t\right)^2} \tag{3}$$

where r_c is the radius of the critical cluster. For heterogeneous nucleation theory ΔG_{i_c} is respectively modified as

$$\Delta G_{i_c,h} = f(\eta, x) \Delta G_{i_c} \tag{4}$$

where x is defined as $x = R_s / r_p$, R_s is the radius of the foreign particle, and r_p is the radius of curvature of the critical nuclei. η is approximately regarded as $\cos\theta$ where θ is the contact angle between the nucleation phase and the substrate. The factor f for convex surface of particle could read as

$$f(\eta, x) = \frac{1}{2}\left\{1 + \left(\frac{1 - \eta x}{g}\right)^3 + x^3\left[2 - 3\left(\frac{x - \eta}{g}\right) + \left(\frac{x - \eta}{g}\right)^3\right] + 3\eta x^2\left(\frac{x - \eta}{g} - 1\right)\right\} \tag{5}$$

where $g = (1 + x^2 - 2\eta x)^{1/2}$

The homogeneous nucleation rate J considering both diffusion barrier $\Delta G_{i'}$ and nucleation barrier ΔG_{i_c} could be given by

$$J = \frac{nkT}{h} \exp\left(-\frac{\Delta G_{i'}}{kT}\right) \exp\left(-\frac{\Delta G_{i_c}}{kT}\right) \tag{6}$$

where h is Planck constant, k the Boltzmann constant, and n denotes molecules per unit volume of mother phase. Thus the heterogeneous nucleation rate J_{het} could be obtained by modifying ΔG_{i_c} as $\Delta G_{i_c,h}$ in the expression of J.

2.2. Diffusion limited model of ice growth inside cell

Karlssonet al. [6] has developed a physiochemical theory of ice growth inside biological cells by coupling the water transport models, theory of ice nucleation, and theory of crystal growth. Considering a given cell with the control volume of $V_{cv}(t)$, which is composed of water-NaCl-glycerol solution. Evolution of the number of ice nuclei denoted by $N(t)$ in a given cell is a stochastic process given as

$$N(t) = \text{int}\left[\int_0^t J(t) V_{cv}(t) dt\right] \tag{7}$$

where int denote the nearest integer truncated from the ensemble average. When the crystallizing phase is of different composition than the bulk solution, such as the case with ice crystallization from aqueous solutions, the rate of crystal growth is believed to be limited by the diffusion water molecules to the ice-solution interface. Considering a spherically symmetric ice crystal under isothermal conditions, its radius at time t could be given as [6]

$$r(t,t_i) = \left[\int_{T(t_i)}^{T(t)} \varphi^2 \overline{D}\left(-\frac{dT}{B}\right) dt\right]^{1/2} \tag{8}$$

where $\overline{D}$ is the effective diffusivity of water and is calculated according to the viscosity of the intracellular solution. φ is the nondimensional crystal growth parameter and is derived from the equation obtained by curve fitting in advance based on the nondimensional supersaturation [6]. If the crystals within a given cell are small enough such that neighboring crystals and their depletion regions do not overlap, the total crystallized volume could be given as

$$V_{ice}(t) = \sum_{i=1}^{N(t)} \frac{4\pi}{3} r^3(t,t_i) \tag{9}$$

Thus the crystallized volume fraction considering the impingement appearance could be given as

$$X_{ice}(t) = 1\text{-}\exp\left(\frac{V_{ice}}{V_{cv}}\right) \tag{10}$$

According the water-transport model developed by Mazur [7] to cells containing a ternary water-NaCl-glycerol solution, one could derive the time evolution of cell volume

$$\frac{dV_{cv}}{dt} = -\frac{L_p A R_g T}{v_w}\left[\frac{\Delta H_f}{R_g}\left(\frac{1}{T_0}-\frac{1}{T}\right)-\ln\left(\frac{V_{cv}-\left(n_s v_s+n_g v_g\right)}{V_{cv}-\left(n_s v_s+n_g v_g\right)+v_w\left(n_s \nu_s+n_g\right)}\right)\right] \tag{11}$$

where L_p is the water permeability; A is the cell membrane surface area; R_g gas constant; v_w, v_s and v_g denote specific volumes of water, NaCl and the glycerol, respectively; n_s and n_g are the number of moles of NaCl and glycerol in cell. Coupling the Eq. (10) and Eq. (11), one could calculate the crystallized volume fraction during freezing. Recently more sophisticated models [8-10] of intracellular ice formation and its growth have been developed to obtain more close to the real cell environments.

2.3. Molecular simulation of ice nucleation

Molecular dynamics (MD) is a powerful computer simulation technique to track time evolution of a system of interacting particles, such as atoms, molecules and coarse-grained particles. It has been widely used in the study of the structure, dynamics, and thermodynamics of water phase change and their complexes addressing a variety of issues including ice nucleation. The basic idea of the MD simulations is to generate the atomic trajectories of a system of finite particles by numerical integration, of a set of Newton's equations of motion for all particles in the system with certain given boundary conditions, an initial set of positions and velocities. In addition, the potential energy of the particles system needs to be specified in order to describe the interaction between atoms or molecules. The detailed potential energy expression and its parameter are derived from both experimental work and high-level quantum mechanical calculations.

The dynamics of the ice–water interface determines the ice nucleation and growth such that considerable MD methods have been developed. Karim and Haymet [11, 12] have used the TIP4P water model to simulate the contact of the basal plane of ice with water and investigate dynamics of the ice–water interface. Their results show that the interface was found to be stable at 240 K. In addition, there appears a gradual change in the orientation and mobility of the molecules in the interfacial region. A very extensive MD simulation by Matsumoto et al. [13] predicted the freezing of pure water at different supercooling. They found that the occurrence of the ice nucleation was impressed extensively by the number of spontaneously developing long-lived hydrogen bonds at the same location. Handel et al. [14] had performed direct calculations of the ice Ih-water interfacial free energy for the TIP4P model by extending the cleaving method to molecular systems, which agrees with the experiments results.

MD simulations also could be used to investigate the effects of salt on ice nucleation, which is more close to biological environments. Smith et al. [15] had studied the potential of mean force of single ions as a function of the distance from the ice-water interface, and the free energy of transfer of Na^+ and Cl^- ions from bulk water into the ice and to the interface. Their calculations suggested that the interface may have a potential that is attractive for Cl^- and repulsive for Na^+. The investigation from Vrbka and Jungwirth [16], and Carignano et al. [17] indicated that the growth rate of ice was much lower in solutions with salt than in pure water.

The mechanism of ice growth inhibition by antifreeze proteins (AFP) is still poorly understood, owing to the difficulty in elucidating the molecular-scale structure of an ice–water interface, and in experimentally observing the structure and dynamics of AFP at an ice–water interface. MD is one of most popular methods that currently have the potential to clarify the mechanism of ice growth inhibition by AFP at the molecular level [18-21]. Wathen et al. [19] developed a MD which combines molecular representation and detailed energetics calculations of molecular mechanics techniques with the less-sophisticated probabilistic approach to study systems containing millions of water molecules undergoing billions of interactions. Such model could enable the 3-D shape and surface properties of the molecules to directly affect crystal formation. They have applied this technique to study the inhibitory effects of antifreeze proteins (AFPs) (from both fish and insect) on ice-crystal formation, including the replication of ice-etching patterns, ice-growth inhibition, and specific AFP-induced ice morphologies. Their work suggests that the degree of AFP activity results more from AFP ice-binding orientation than from AFP ice-binding strength. The simulation results were consistent with experimental observations very well. The effects of cryoprotectants such as ethylene glycol and glycerol solutions on ice nucleation have also been investigated [22, 23].

The electrostatic field was recently proposed to improve the output of a cryopreservation process [24-26]. As the molecular dynamics simulated [27], a DC field with magnitude of $E = 5.0 \times 10^9 \mathrm{V/m}$ could induce crystallization of supercooled water. Recently, experiments [24-26, 28] have demonstrated that the external electric field can effectively affect the water nucleation and ice growth. A lower electrostatic field strength in the range of $E = 10^3 \sim 10^5 \mathrm{V/m}$, which is easily obtained in experiments, can induce increase in the nucleation temperature of water [25] and the asymmetric growth of ice [24].

3. Ice crystal growth using phase field method

To better understand the detailed events in cell damage due to extracellular and intracellular ice formation during freezing, it is rather important to model and predict the micro-scale ice crystal formation and growth behavior, and thus characterize the effects of several core factors, such as undercooling, anisotropy, thermal noise, and external electric field etc. Among various ways ever developed, the phase field method is rather flexible in characterizing the crystal formation and has been extensively adopted to model solidification process in material science [29]. Recently some authors [30, 31] begun to pay attention to possible application of this method in cryopreservation. He and Liu [32] extended deeply this theoretical strategy to characterize the ice crystal growth under electrostatic field and describe multiple ice nucleuses' competitive growth behavior.

A dimensionless phase field model [33] is adopted to characterize the ice growth. A non-conserved phase field ϕ ($\phi=0$ is corresponding to ice and $\phi=1$ to water) describes the transition from water to ice, which is driven by undercooling. A conserved dimensionless temperature field u is used to characterize the energy variation due to temperature diffusion and phase transition. It reads as

$$\phi_t = m\nabla\cdot\left(B(\theta)\nabla\phi\right)+m\varepsilon^{-2}\phi(1-\phi)\left[\phi-0.5+30\varepsilon a S u\phi(1-\phi)\right], \tag{12}$$

$$u_t = \nabla^2 u - S^{-1}30\phi^2\left(1-\phi\right)^2\phi_t+\psi\left(x,y,t\right), \tag{13}$$

where

$$B(\theta)=\begin{pmatrix}\gamma^2(\theta) & -\gamma'(\theta)\gamma(\theta)\\ \gamma'(\theta)\gamma(\theta) & \gamma^2(\theta)\end{pmatrix} \tag{14}$$

In the above equations, the term ψ represents the thermal noise, which indicates the fluctuation of temperature, and $u=(T-T_M)/\Delta T$. $\gamma(\theta)=1+\varsigma\cos n(\theta-\theta_0)$ denotes dimensionless anisotropic surface tension and $\theta=\arctan(\psi_y/\psi_x)$, where θ_0 is the angle between reference direction and x axis. Four dimensionless parameters are defined as follow

$$m=\frac{\nu\sigma T_m}{\kappa L},\ \varepsilon=\frac{\delta}{w},\ a=\frac{\sqrt{2}w}{12d},\ S=\frac{c_p\Delta T}{L}, \tag{15}$$

where $d=c_p\sigma T_m/\rho L^2$ [33].

3.1. Ice crystal growth from pure water

The physical properties of pure water [30] used in the simulation are listed as: $\kappa=1.31\times10^{-3}cm^2/s$, $L=333.5J/cm^3$, $T_M=273.15K$, $c_p=4.212J/cm^3\cdot K$ and $\sigma=7.654\times10^{-6}J/cm^2$. The length scale w is considered as $2.7\times10^{-4}cm$. Thus the phase field model parameters for all the simulations, according to Ref. [33], can be fixed with $\varepsilon=0.005$, $m=0.035$, $a=400$ and $S=0.5$, which is corresponding to the cooling temperature $233.56K$.

An explicit time-differencing scheme is adopted to solve ϕ-equation, and implicit Crank-Nicolson algorithm is chosen for solving u-equation. All the simulations are performed in a two dimensional domain with 1500×1500 grid points ($\Delta x=10^{-2}$ and $\Delta t=10^{-4}$) under the

Neumann boundary conditions. The initial condition of the phase field for single ice nucleus is set for $\phi = \frac{1}{2}\left[\tanh\left(\frac{r - r_c}{2\sqrt{2}\varepsilon}\right) + 1\right]$. The initialization of the dimensionless temperature is $u = -\phi$ for all the simulations, which is corresponding to initial temperature T_M in solid and $T_M - \Delta T$ in liquid.

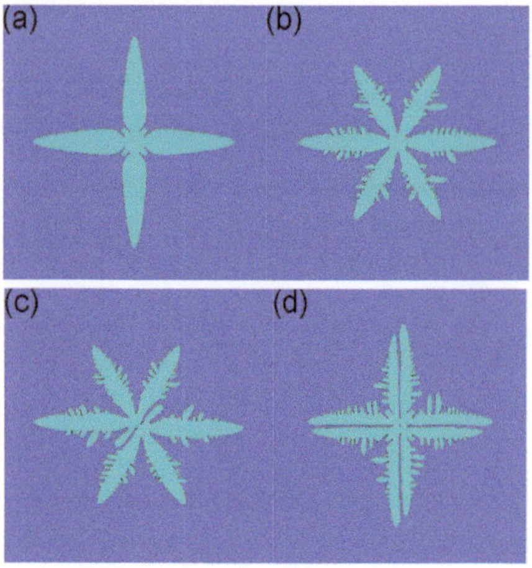

Figure 1. Micro-scale dendritic ice crystal patterns with (a) 4-fold structures without thermal noise, (b) 6-fold structures with thermal noise, (c) two seeds and (d) four seeds competitive growth. The parameters $\gamma = 0.04$ and $n = 4$ in (a), (d), and $\gamma = 0.02$ and $n = 6$ in (b), (c). The angle $\theta_0 = 0$ for (a-d). [32]

Simulations indicate that the influence of surface tension anisotropy on the shape of ice is more evident compared with that on the growth rate. The variation of the anisotropy strength does not significantly change the velocity at the dendritic tip, which is strongly affected by the local cooling strength. Four-fold and six-fold dendritic ice crystals are often observed in experiments. Fig. 1(a) shows a four-fold dendritic ice crystal evolving from an initial circle shaped seed [32]. The thermal noise, which often induces the temperature fluctuation on the interface, is considered as the main reason of sidebranching growth [34]. A six-fold dendritic ice with sidebranching is shown in Fig. 1(b). It is necessary to consider the influence of the other crystal seeds on the ice growth. Fig. 1(c), (d) represent two seeds and four seeds competitive growth, which results in the crease of growth for both collided dendrites and preferred growth in directions of unhindered grains.

3.2. The effects of external electric on Ice crystal growth

The growth rate depends on the dynamical state of the ice-water interface, such as the surface tension and the temperature at interface. In addition, the effects of the external electric field on the ice formation have been demonstrated in both theoretical and experimental research. The influence of low electrostatic field strength on water nucleation and ice growth may be attributed to dipole polarization of the water molecules by the electrostatic field. Under the electrostatic field, the phase transition must overcome excess energy $\Delta G^E = E\Delta\mu$, where $\Delta\mu = \mu_w - \mu_i$ is the difference of electrical dipole moments between water and ice. However, for the low electrostatic field strength ($E < 10^5$V/m), ΔG^E is much smaller than L and can be negligible. In other words, the water molecules with dipole moments may follow the Boltzmann distribution function $p = A\exp(\mu_w E\cos\varphi / kT)$ [25], and get in the most stable state with the maximum energy along the direction of the electrostatic field ($\varphi = 0$). Although for the low electrostatic field magnitude, the value variation of p with respect to φ is very small in bulk phase, the effects of electrostatic field may become largely strong at the interface and affect dynamical state of interface [24]. The polar water molecules (or clusters) may be torqued, rearranged, and forced to join the ice lattice via a special orientation and position under action of the electrostatic field. Thus the rate of attachment (or detachment) of water molecules on the ice interface is affected by external electrostatic field. In order to characterize such influence of the electrostatic field, the surface tension should be modified as $\exp(\lambda\cos(\theta - \varphi))\sigma$ during the ice growth, where λ is proportional to $\mu_w E / kT$.

From the above dissection, the effects of electrostatic field with low magnitude can induce change of the surface tension with the form $\sigma_E = \exp(\lambda\cos(\theta - \varphi))\sigma$. Thus in phase field model, the dimensionless anisotropic surface tension should be modified as $\gamma_E(\theta) = \exp(\lambda\cos(\theta - \varphi))\gamma(\theta)$. The value of λ is proportional to $\mu_w E / kT$, which characterizes the strength of electric field. In our simulations, the value of λ is chosen in the range 0.02~0.2 and $\varphi = 0$, which indicates that the orientation of electric field is along positive x axis.

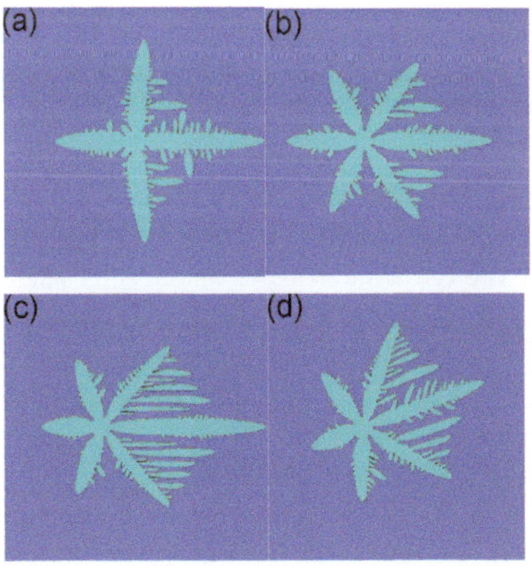

Figure 2. The influence of electric field on ice crystal growth at (a) $\gamma = 0.04$, $n = 4$ and $\lambda = 0.1$, at (b) $\gamma = 0.02$, $n = 6$ and $\lambda = 0.1$, at (c), (d) $\gamma = 0.02$, $n = 6$ and $\lambda = 0.2$. The angle $\theta_0 = 0$ for (a-c) and $\theta_0 = \pi / 9$ for (d). [32]

Under the electrostatic field, the growth of dendritic ice crystals represents asymmetry behavior. Fig 2 shows that the main branches parallel to the electrostatic field grow faster than the other branches [32]. One can also find that the growth of sidebranching is strengthened along the direction of electric field and weakened along the inverse direction. The above influence becomes clearer by increasing the value of λ (comparing Fig. 2(b) with (c)), which indicates that a more strong magnitude of electrostatic field would strengthen the asymmetric growth of ice. Fig. 2(d) shows how the reference direction affects the ice growth under electric field. The above results concerning the effects of electric field are in good accordance with the qualitative analysis in Ref. [25] and experimental observation in Ref. [24].

It is noteworthy that the biological media is saline solution including various ions, such as Na^{+} and Cl^{-}. Under the electrostatic field, these positive and negative ions have completely different movement and gather near the corresponding electrodes, which induce non-uniform distribution of ions. In addition, the heterogeneous structure of ions, such as their different size, also leads to different ions' behaviors. These behaviors induced by the electrostatic field will affect the dynamical behavior of ice interface and strengthen the asymmetric growth of ice [4]. Thus, the effects of the ions under the electrostatic field are also indicated via modifying the surface tension. The effects of ions coupling with the electrostatic field on the dynamical behavior of ice interface should be incorporated to a general model. The present investigation focuses on the effects of the electrostatic field on the ice interface but is not to completely characterize the effects of ions, which will be addressed in further work.

4. The interaction between ice growths with cell

The interaction between the cell and ice during freezing is very important to investigate the cell cryoinjury during freezing. However, it is still a rather difficult task to track the evolution of the solidification front and characterize the interaction between the cell membrane and ice in detail. Recently, the progress [35, 36] gave some promising research directions to solve this issue.

Mao et al. [35] developed a sharp interface method to simulate the response of a biological cell during freezing. The cell is modeled as an aqueous salt solution surrounded by a semi-permeable membrane. The concentration and temperature fields both inside and outside a single cell are computed taking into account heat transfer, mass diffusion, membrane transport, and evolution of the solidification front. The external ice front is computed for both stable and unstable growth modes. It is shown that for the particular geometry chosen in this study, the instabilities on the front and the diffusional transport have only modest effects on the cell response. For the cooling conditions, solute and cell property parameters used, the low Peclet regime applies. The computational results are therefore validated against the conventional membrane-limited transport (Mazur) model. Good agreement of the simulation results with the Mazur model are obtained for a wide range of cooling rates and membrane permeabilities. A spatially non-isothermal situation is also considered and shown to yield significant differences in the cell response in comparison to the isothermal case.

Another important process is from Chang et al. [36]. They developed a modified level set method for modeling the interaction of biological cells with ice front during a directional solidification. The simulation result has shown that different types of cell interaction with the ice interface can lead to significant differences in the final volume of individual cells. The cell is easily exposed to high concentrations of electrolytes in the interdendritic regions due to cell entrapment. Cell pushing can be used to control the dehydration of the cell more easily, which avoids exposing the cell to high concentrations that develop in the interdendritic spaces. Cell engulfment protects the cell from high concentration regions and leaves the cell vulnerable to intracellular ice formation. However, it could be used to control the dehydration of the cell and provide limited interaction with high electrolyte concentrations through combination of cell pushing and cell engulfment. The numerical results also indicate that the interactions between cells are very important for cryopreservation conditions, where suspensions have higher volume fractions of cells. Level set method could deal with the interaction of clusters of cells with ice by assuming that the clusters act like a single particle with a radius that represents the overall size of the cell cluster. Increasing the radius of the representative particle would increase the probability of engulfment and pushing by the interface. Although the intermolecular forces between cells and the fluid flow around various arrays of single cells and cell clusters have not been included in the current model, these phenomena could be easily included in the simulations in order to study their effects on the partitioning of cells by the ice interface.

5. Summary

The present chapter has presented an overview on the theoretical modeling of the microscale phase change of biological system subject to freezing with special focus on the ice crystal formation and growth. In addition, the heat and solute transport, and interaction between the cell and ice during freezing have also been discussed. Although considerable investigation focus on ice crystal formation and its growth, there still exists some outstanding questions [37-40] including ice structure and ice phase diagram. Here we discuss some important issues and challenges about ice formation in biological system in brief. Firstly, the effects of cell membrane on ice nucleation and growth inside a cell with restricted space during freezing have not been investigated deeply, which is in fact very important to clarify the puzzle about the ice propagation between cells. In addition, the improved understanding on the effects of cell–cell interactions suggested that the mechanism of ice propagation should be mediated to some extent, but not at all, by the gap junction, if uncontrolled factors are eliminated. Secondly, a detailed three dimensional model needs to be developed to characterize the complicated freezing process that coupling micro heat and mass transfer, and ice nucleation and growth, which involves the moving water-ice interface and deformed cell membrane. Thirdly, the advanced experimental measures need to be designed to determine the parameter in the physical model and confirm the model validity. All the efforts will warrant a bright future for better understanding and practice of cryobiology.

Acknowledgements

This work is supported by the NSFC under Grant No. 51006114 & 81071255, the Specialized Research Fund for the Doctoral Program of Higher Education, and Research Fund from Tsinghua University under Grant 523003001.

Author details

Zhi Zhu He[1] and Jing Liu[1,2]

*Address all correspondence to: jliu@mail.ipc.ac.cn

1 Beijing Key Laboratory of Cryo-Biomedical Engineering, and Key Laboratory of Cryogenics, Technical Institute of Physics and Chemistry, Chinese Academy of Sciences, Beijing, China

2 Department of Biomedical Engineering, School of Medicine, Tsinghua University, Beijing, China

References

[1] Bischof JC: Quantitative Measurement and Prediction of Biophysical Response During Freezing in Tissues. *Annual Review of Biomedical Engineering* 2000, 2:257-288.

[2] Mazur P, Leibo SP, Chu EH: A Two-factor Hypothesis of Freezing Injury. Evidence from Chinese Hamster Tissue-culture Cells. *Exp Cell Res* 1972, 71:345-355.

[3] Mazur P, Seki S, Pinn IL, Kleinhans FW, Edashige K: Extra- and Intracellular Ice Formation in Mouse Oocytes. *Cryobiology* 2005, 51:29-53.

[4] Li FF, Liu J: Characterization of Micro-/Nano-Scale Ice Crystal Formation in Cryo-Biomedical Engineering: A Review. *Journal of Computational and Theoretical Nanoscience* 2010, 7:85-96.

[5] Toner M, Cravalho EG, Karel M: Thermodynamics and Kinetics of Intracellular Ice Formation during Freezing of Biological Cells. *Journal of Applied Physics* 1990, 67:1582-1593.

[6] Karlsson JOM, Cravalho EG, Toner M: A Model of Diffusion-Limited Ice Growth inside Biological Cells during Freezing. *Journal of Applied Physics* 1994, 75:4442-4445.

[7] Mazur P: Kinetics of Water Loss from Cells at Subzero Temperatures and Likelihood of Intracellular Freezing. *Journal of General Physiology* 1963, 47:347-369.

[8] Zhao G, Luo DW, Gao DY: Universal Model for Intracellular Ice Formation and Its Growth. *AICHE Journal* 2006, 52:2596-2606.

[9] Yang G, Zhang AL, Xu LX, He XM: Modeling the Cell-type Dependence of Diffusion-limited Intracellular Ice Nucleation and Growth During both Vitrification and Slow Freezing. *Journal of Applied Physics* 2009, 105:114701-114711.

[10] Yan JF, Liu J: Characterization of the Nanocryosurgical Freezing Process through Modifying Mazur's Model. *Journal of Applied Physics* 2008, 103: 084311- 084321.

[11] Karim OA, Haymet ADJ: The Ice Water Interface - a Molecular-Dynamics Simulation Study. *Journal of Chemical Physics* 1988, 89:6889-6896.

[12] Karim OA, Haymet ADJ: The Ice Water Interface. *Chemical Physics Letters* 1987, 138:531-534.

[13] Matsumoto M, Saito S, Ohmine I: Molecular Dynamics Simulation of the Ice Nucleation and Growth Process Leading to Water Freezing. *Nature* 2002, 416:409-413.

[14] Handel R, Davidchack RL, Anwar J, Brukhno A: Direct Calculation of Solid-liquid Interfacial Free Energy for Molecular Systems: TIP4P Ice-water Interface. *Physical Review Letters* 2008, 100:036104-036107.

[15] Smith EJ, Bryk T, Haymet ADJ: Free Energy of Solvation of Simple Ions: Molecular-dynamics Study of Solvation of Cl- and Na+ in the Ice/water Interface. *Journal of Chemical Physics* 2005, 123:03470601-03470616.

[16] Vrbka L, Jungwirth P: Brine Rejection from Freezing Salt Solutions: A Molecular Dynamics Study. *Physical Review Letters* 2005, 95:148501-148504.

[17] Carignano MA, Shepson PB, Szleifer I: Ions at the Ice/vapor Interface. *Chemical Physics Letters* 2007, 436:99-103.

[18] Nada H, Furukawa Y: Growth Inhibition at the Ice Prismatic Plane Induced by a Spruce Budworm Antifreeze Protein: A Molecular Dynamics Simulation Study. *Physical Chemistry Chemical Physics* 2011, 13:19936-19942.

[19] Wathen B, Kuiper M, Walker V, Jia ZC: A New Model for Simulating 3-D Crystal Growth and Its Application to the Study of Antifreeze Proteins. *Journal of the American Chemical Society* 2003, 125:729-737.

[20] Yang C, Sharp KA: The Mechanism of the Type III Antifreeze Protein Action: A Computational Study. *Biophysical Chemistry* 2004, 109:137-148.

[21] Nada H, Furukawa Y: Growth Inhibition Mechanism of An Ice-water Interface by A Mutant of Winter Flounder Antifreeze Protein: A Molecular Dynamics Study. *Journal of Physical Chemistry B* 2008, 112:7111-7119.

[22] Chen C, Li WZ, Song YC, Yang J: Molecular Dynamics Simulation Studies of Cryoprotective Agent Solutions: The Relation Between Melting Temperature and the Ratio of Hydrogen Bonding Acceptor to Donor Number. *Molecular Physics* 2009, 107:673-684.

[23] Chen C, Li WZ, Song YC, Yang J: A Molecular Dynamics Study of Cryoprotective Agent - Water-sodium Chloride Ternary Solutions. *Journal of Molecular Structure-Theochem* 2009, 916:37-46.

[24] Tao LR: Microscopic Experimental Study of the Freezing Process in Aqueous Solution. Ph.D. Thesis, University of Shanghai for Science and Technology, 2000.

[25] Sun W, Xu XB, Zhang H, Xu CX: Effects of Dipole Polarization of Water Molecules on Ice Formation under An Electrostatic Field. *Cryobiology* 2008, 56:93-99.

[26] Petersen A, Schneider H, Rau G, Glasmacher B: A New Approach for Freezing of Aqueous Solutions under Active Control of the Nucleation Temperature. *Cryobiology* 2006, 53:248-257.

[27] Svishchev IM, Kusalik PG: Crystallization of Liquid Water in a Molecular-Dynamics Simulation. *Physical Review Letters* 1994, 73:975-978.

[28] Braslavsky I, Lipson SG: Electrofreezing Effect and Nucleation of Ice Crystals in Free Growth Experiments. *Applied Physics Letters* 1998, 72:264-266.

[29] Boettinger WJ, Warren JA, Beckermann C, Karma A: Phase-field Simulation of Solidification. *Annual Review of Materials Research* 2002, 32:163-194.

[30] Li FF, Liu J, Yue K: Numerical Simulation on Ice Crystal Formulation in Cellular Level Based on Phase Field Theory. *Chinese Journal of Low Temperature Physics* 2008, 30:84-92.

[31] Xu Y, Mcdonough JM, Tagavi KA, Gao DY: Two-Dimensional Phase-Field Model Applied to Freezing into Supercooled Melt. *Cell Preservation Technology* 2004, 2:113-124.

[32] He ZZ, Liu J: Characterizing Ice Crystal Growth Behavior Under Electric Field Using Phase Field Method. *ASME Journal of Biomechanical Engineering* 2009, 131: 0745021-0745023.

[33] Wang SL, Sekerka RF, Wheeler AA, Murray BT, Coriell SR, Braun RJ, Mcfadden GB: Thermodynamically-Consistent Phase-Field Models for Solidification. *Physica D* 1993, 69:189-200.

[34] Karma A, Rappel WJ: Phase-field Model of Dendritic Sidebranching with Thermal Noise. *Physical Review E* 1999, 60:3614-3625.

[35] Mao L, Udaykumar HS, Karlsson JOM: Simulation of Micro-scale Interaction between Ice and Biological Cells. *International Journal of Heat and Mass Transfer* 2003, 46:5123-5136.

[36] Chang A, Dantzig JA, Darr BT, Hubel A: Modeling the Interaction of Biological Cells with a Solidifying Interface. *Journal of Computational Physics* 2007, 226:1808-1829.

[37] Bartels-Rausch T, Bergeron V, Cartwright JHE, Escribano R, Finney JL, Grothe H, Gutierrez PJ, Haapala J, Kuhs WF, Pettersson JBC, et al: Ice Structures, Patterns, and Processes: A View Across the Icefields. *Reviews of Modern Physics* 2012, 84:885-944.

[38] Libbrecht KG: The Physics of Snow Crystals. *Reports on Progress in Physics* 2005, 68:855-895.

[39] Blackford JR: Sintering and Microstructure of Ice: A Review. *Journal of Physics D-Applied Physics* 2007, 40:355-385.

[40] Petzold G, Aguilera JM: Ice Morphology: Fundamentals and Technological Applications in Foods. *Food Biophysics* 2009, 4:378-396.

Section 2

Inorganic Systems

Chapter 9

Crystal Growth by Electrodeposition with Supercritical Carbon Dioxide Emulsion

Masato Sone, Tso-Fu Mark Chang and
Hiroki Uchiyama

Additional information is available at the end of the chapter

1. Introduction

1.1. Electroplating with supercritical carbon dioxide emulsion

1.1.1. Introduction

1.1.1.1. Supercritical Carbon Dioxide

A supercritical fluid (SCF) is any substance at a temperature and pressure above its critical point, as shown in Fig. 1, where distinct liquid and gas phases do not exist [1]. It can effuse through solids like a gas, and dissolve materials like a liquid. In addition, close to the critical point, small changes in pressure or temperature result in large changes in density, allowing many properties of a SCF to be fine-tuned between a gas and a liquid. SCFs are suitable as a substitute for organic solvents in a range of industrial and laboratory processes.

CO_2 is non-polar, combining with the low surface tension property when it is in supercritical state; it is often used in extraction of organics in the food industries [2]. The extremely low surface tension also makes supercritical CO_2 (sc-CO_2) an ideal medium in drying of nano-porous structures [3]. CO_2 is non-toxic, and the critical point is relatively low when comparing with the other solvents, therefore, sc-CO_2 is an important commercial and industrial solvent. Critical temperature pressure of sc-CO_2 are 304.5K and 7.39 MPa, respectively. Comparison for the critical conditions of some commonly used solvents is shown in Table 1.

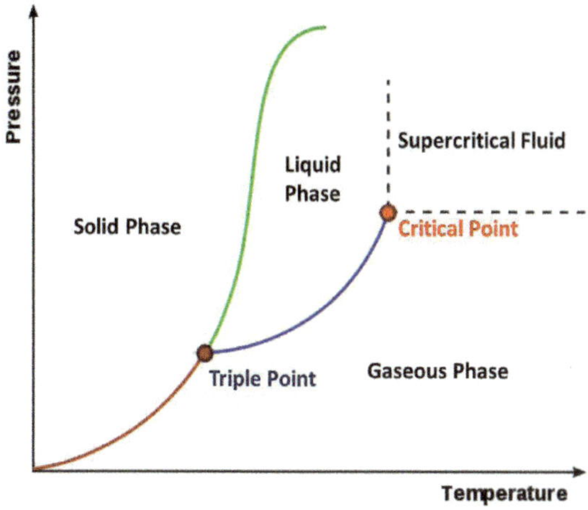

Figure 1. Phase diagram of a single substance.

Fluid	Critical Temperature (K)	Critical Pressure (MPa)
CO_2	304.1	7.39
NH_3	405.5	11.35
H_2O	647.3	22.12
n-Pentane	469.7	3.37
Toluene	591.8	4.10

Table 1. Critical conditions of commonly used solvents

1.1.1.2. Supercritical Carbon Dioxide Emulsion

Electrodeposition is a key technology for fabricating micro components used in micro-electro-mechanical systems (MEMS) [4,5]. Application of sc-CO_2 in electrodeposition process is believed to solve the problems encounter in miniaturization of the devices [6], such as reducing usage of organic solvents in the cleaning process, drying of the nano-structures after the electrodeposition process, and minimize problems caused by evolution of H_2. Evolution of H_2 is an inevitable size reaction when performing electrodeposition reaction with an aqueous electrolyte. H_2 gas bubbles adsorbed on the surface of cathode is one of the major causes for defects found in the electrodeposited materials [7]. CO_2 is non-polar, solubility of H_2 is high in CO_2 [8]. Therefore, desorption of H_2 gas bubbles from the surface of cathode could significantly enhanced in sc-CO_2.

However, electrical conductivity and metal salts solubility are both very low in sc-CO_2 [1], which are the basic requirements in electrochemistry. The limitations could be overcome by addition of a surfactant to form an emulsion composed of an aqueous electrolyte, sc-CO_2, and the surfactant [9,10]. Two types of the emulsion could be formed depending on the type and concentration of surfactant used and concentration of CO_2 in the system. One is H_2O in CO_2, where the continuous phase (CP) is CO_2 and the dispersed phase (DP) is H_2O. The other one is CO_2 in H_2O, where the CP is H_2O and the DP is CO_2 [11,12]. Structure of the DP in CO_2 in H_2O emulsion is similar to micelles in mixture of oil and water. CO_2 in H_2O emulsion is usually used for application in electrochemical reaction because the higher solubility of metal salts and electrical conductivity CO_2 in H_2O emulsion when compared with H_2O in CO_2 emulsion [11,12].

1.1.1.3. Electrodeposition with Supercritical Carbon Dioxide Emulsion (EP-SCE)

Surface smoothening, grain refinement, and hardness enhancement are the effects of applying sc-CO_2 emulsion (SCE) in electrodeposition of Ni [13-15]. Surface of the Ni films fabricated by electrodeposition with SCE (EP-SCE) is defect-free and pinhole free as shown in Fig. 2(a), and defects are found on the Ni films when only the aqueous electrolyte is used at ambient pressure as shown in Fig. 2(b) [13]. Ni films is electrodeposited with the aqueous electrolyte only at an elevated pressure, 42 MPa, to confirm the surface smoothening effect is not caused by the high pressure, and many defects are found on the Ni film as shown in Fig. 2(c). In order to further confirm the smooth surface is caused by SCE, emulsion made of n-hexane is studied [14]. Properties of n-hexane are considered to be close to sc-CO_2, such as electrical conductivity. Surface conditions of the Ni films fabricated by EP-SCE are much better than the Ni films fabricated by electrodeposition with n-Hexane. The results indicate that only emulsion made of sc-CO_2 is effective in increasing smoothness of the Ni films electrodeposited.

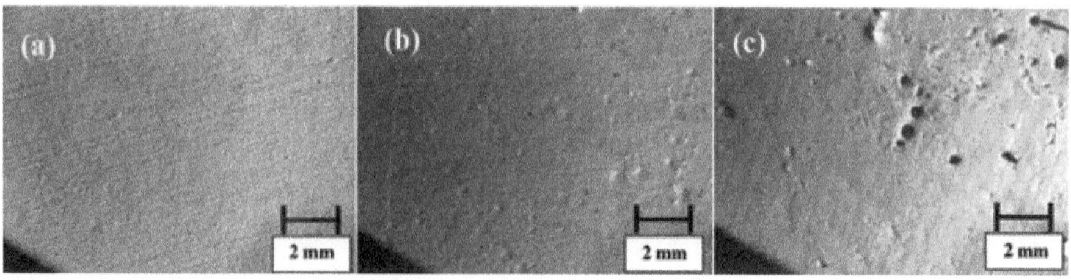

Figure 2. Ni films electrodeposited from (a)SCE, (b) the aqueous electrolyte only at atmospheric pressure and (c) the aqueous electrolyte only at 42 MPa.

For improvement in mechanical properties, grain refinement is believed to be the main cause as shown in Fig. 3 [15-17]. Ni films fabricated by EP-SCE are reported to have grain size in nano-scale. Because of the nano-grains, wear properties of the Ni films could be improved significantly [18]. Chung and Tsai proposed the grain refinement and improvement in mechanical strength are also caused by C impurity in the Ni film from decomposition of CO_2 in the electrolyte; evidence of the C impurity is detected from X-ray photoelectron spectra [19].

Many studies have been reported on application of SCE in electrodeposition. However, additive such as brightener is often used in the electrolyte in these studies, and properties of the materials electrodeposited could be influenced by the additives [20,21]. Therefore, we studied and proposed a mechanism called periodic-plating-characteristic (PPC) to be the main cause for the effects observed in the metal films fabricated by EP-SCE. Physical properties of SCE are expected to affect the PPC and properties of the metal films electrodeposited, and physical properties of SCE could be controlled by varying experimental pressure, volume fraction of CO_2 and surfactant in the system. In this study, physical properties of SCE are adjusted to study the influence on PPC and properties of the Ni films electrodeposited.

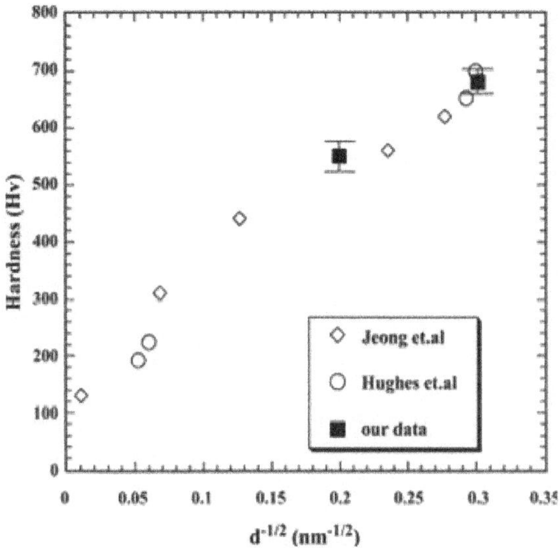

Figure 3. Effect of grain size on hardness.

1.1.2. Experimental section

1.1.2.1. Materials

CO_2 with a minimum purity of 99.9 % was used. Composition of the additive-free Watts bath was $NiSO_4 \bullet 6H_2O$ (300 g/l), $NiCl_2 \bullet 6H_2O$ (50 g/l), and H_3BO_3 (50 g/l). pH of the additive-free Watts bath was 3.31. A non-ionic surfactant, polyoxyethylene lauryl ether ($C_{12}H_{25}(OCH_2CH_2)_{15}OH$) was used to form the emulsion. Volume fraction of CO_2 with respect to the total volume of the reaction chamber and volume fraction of surfactant with respect to the volume of the aqueous electrolyte were varied from 10 to 50 vol% and 0 to 2.0 vol%, respectively. Cu plates with width and length at 1.0X2.0 cm^2 were used as the working substrate, and Ni plates with width and length at 1.0X2.0 cm^2 were used as the counter electrode.

1.1.2.2. Experimental apparatus

The high-pressure experimental apparatus is shown in Fig. 4. Temperature variation of each run was confirmed to be less than 1.0 K. Maximum working temperature and maximum pressure were 424 K and 50 MPa, respectively. The reaction chamber was a stainless steel 316 vessel (PEEK coating on the inner wall) with a volume of 50 ml, kept in a temperature controlled air bath. There were holes at chamber cap for inflow and outflow of CO_2 and wiring. Through the holes, platinum wires inserted in PEEK tube were used to position the substrates and connected to a programmable power supply. A magnetic agitator with a cross-shaped magnetic-stirrer-bar was placed in the reaction chamber for mixing.

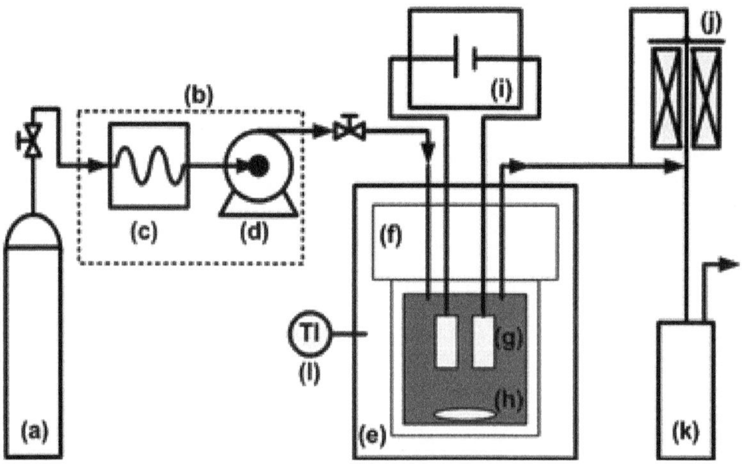

Figure 4. a) CO_2 gas tank, (b) liquidization unit, (c) liquidization pump, (d) high-pressure pump, (e) thermal bath, (f) reaction cell, with PEEK coating on the inner wall, (g) substrates, (h) stirrer, (i) programmable power supply, (j) back pressure regulator, (k) trap, (l) thermometer.

1.1.2.3. Electrodeposition

Cu plates were treated with 10 wt% degreasing solution and 10 wt% HCl solution for 1 min and 10 sec, respectively, prior to the reaction. Distance between the two substrates was 2.5 cm. Samples were electrodeposited at a constant temperature of 323 K with pressure varying from 9 to 18 MPa and current density from 0.01 to 0.20 A/cm^2. The deposition time was 30 min for all the samples. According to Faraday's Law, ca. 6 μm of Ni film would be electrodeposited if 100 % efficiency is achieved.

1.1.3. Effects of pressure

Physical properties of sc-CO_2, such as density, surface tension and viscosity, could be directly adjusted by pressure, which is also expected to affect physical properties of SCE. Pressure could also influence homogeneity of the emulsion, size of the DP, and population or concentration of the DP in the emulsion.

Circular marks/defects with diameter ranged from several to several tens of micrometers were found on surface of the Ni films electrodeposited at 9.0 MPa as shown in Fig. 5 (a) [22]. Size in diameter and total number of the circular marks decreased with increase in pressure. Overall surface uniformity of the Ni films improved significantly when pressure was rose to 18.0 MPa as shown in Fig. 5(c). Surface of the Ni films fabricated by EP-SCE was composed of nano-scaled particles as shown in Fig. 5, which was very different from electric field oriented conical-shape morphology of the Ni films electrodeposited without SCE at atmospheric pressure [23]. Average roughness (R_a)of the Ni films decreased from 19.22 to 12.10 nm as pressure was increased from 9.0 to 18.0 MPa, shown in Fig. 6. The relatively large standard deviation of R_a for the Ni films electrodeposited at 9.0 MPa was caused by the circular marks found on the surface. Standard deviation of R_a decreased significantly as surface uniformity of the Ni films improved with increase in pressure.

Chemical reaction in a reaction medium containing SCE is highly dependent on homogeneity of SCE [1]. High homogeneity of SCE is a prerequisite for fabrication of smooth film when applying SCE in electrodeposition. Homogeneity of SCE can be referred as stability, average size, and size distribution of the DP in the emulsion. SCE is a dynamically emulsified system; size of the DP could be continuously fluctuating in the emulsion. Stability of the DP is considered to be high when size fluctuation of the DP and tendency for phase separation to occur are both low. Quantitatively, the stability is high when interfacial tension between sc-CO_2 and the aqueous solution (γ) is low [24]. Growth in size of the DP is more likely to occur when γ is high, and continue growth in size of the DP would lead to phase separation. Creaming velocity (u_s) is another parameter that could be used to quantify stability of the DP [25]. Creaming is more likely to occur when u_s is high, and occurrence of creaming will lead to phase separation. u_s could be calculated using properties of the emulsion. Equation of u_s is shown in the following:

$$u_s = \frac{2r^2 \Delta\rho g}{9\mu_C} \tag{1}$$

where r is radius of the DP, $\Delta\rho$ is density difference between the DP and the CP, g is gravity, and μ_c is viscosity of the CP.

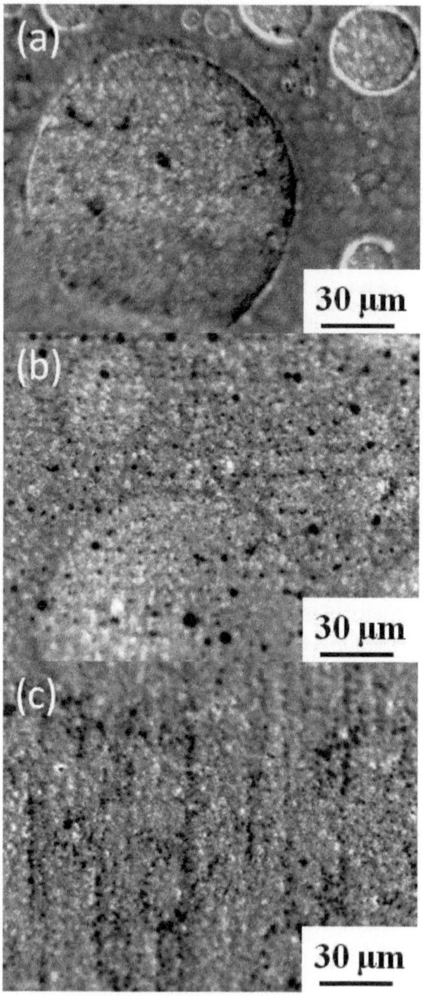

Figure 5. Ni films electroplated with SCE and pressure at (a) 9.0 MPa, (b) 12.0 MPa, and (c) 18.0 MPa.

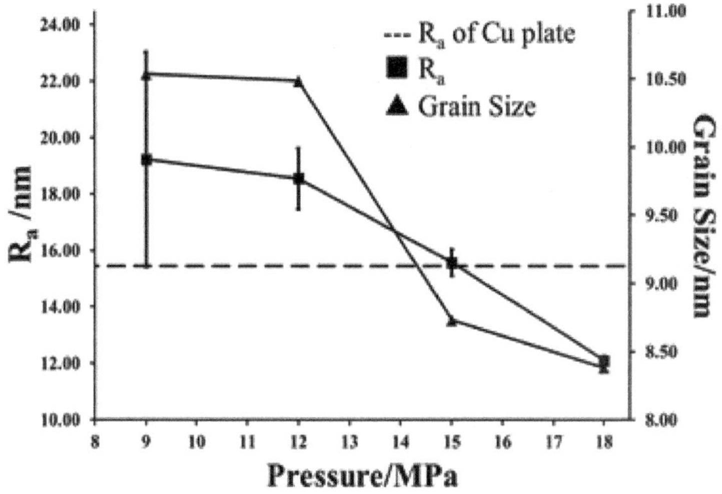

Figure 6. R_a and grain size of Ni films fabricated by EP-SCE at different pressure.

Both γ and u_s are highly related to density of CO_2 in SCE [24,25], and it has been reported that decrease in γ is observed when physical properties of the DP (mostly composed of CO_2) is adjusted to close to physical properties of the CP. For SCE, density and viscosity of the DP are both much lower than those of the CP, and both density and viscosity of the DP can be increased by increasing pressure. In addition, increase in density of the DP can also reduce $\Delta\rho$ in equation (1), which leads to decrease in u_s. Therefore, both γ and u_s are lowered and the stability is improved with increase in pressure. At 320K, density of sc-CO_2 increases from 320 to 770 kg/m^3 and viscosity from 24 to 65 μPa·s when pressure is increased from 9.0 to 18.0 MPa [1,26].

Major peak in XRD patterns of Ni films electrodeposited with SCE was (111) peak, which was caused by texture of the Cu plate, XRD patterns were shown in Fig. 7. Increase in pressure did not have significant influence on position of the (111) peak and relative intensity and position of both (200) and (220) peaks. Grain size calculated from Scherrer equation showed that grain refinement was observed with increase in pressure. Grain size decreased from 10.53 to 8.38 nm when pressure was increased from 9.0 to 18.0 MPa, shown in Fig. 6. Grain refinement is believed to be caused by the PPC when applying SCE in electrodeposition reaction, where adsorption and desorption of the DP from surface of the working electrode would cause a reaction-on and –off phenomenon, respectively.

More uniform size distribution of the DP can lead to more uniform on-time and off-time of PPC. Lee *et al*. and Dhanuka *et al*. both reported that average size of the DP reduced and the size distribution became more uniform with increase in pressure through dynamic light scattering measurement and direct SEM observation [12,24]. In addition, high pressure also favors monomer salvation over aggregates, which prevents aggregation of the DP.

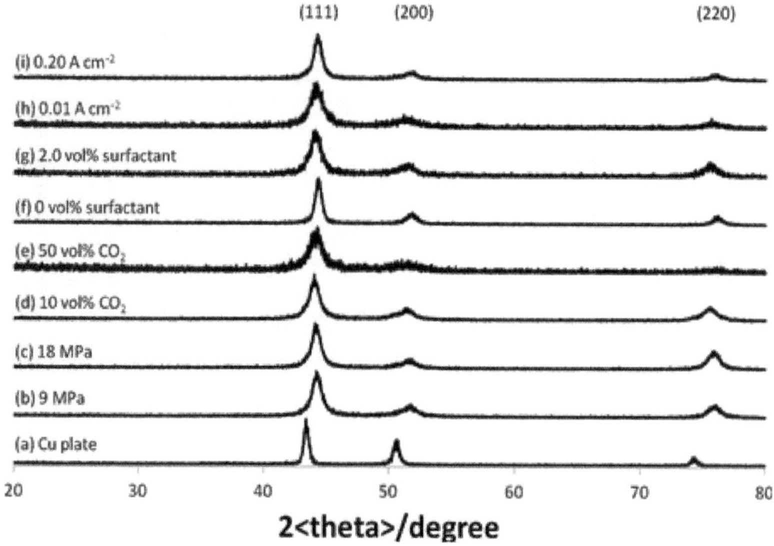

Figure 7. XRD patterns of the Cu plate and Ni films fabricated by EP-SCE with different conditions.

Decrease in size of the DP is expected to cause decrease in on-time of PPC. According to Einstein-Stokes equation, diffusion constant (*D*) of the DP is increased with a decrease in size of the DP. Einstein-Stokes equation is shown in the following:

$$D = \frac{k_B T}{6\pi\eta r} \tag{2}$$

where k_B is Boltzmann's constant, T is the absolute temperature, η is viscosity of the reaction medium, and r is the radius of the DP. Movement of the DP in SCE is faster with increase in *D*, and frequency of a particular region to have contact with the DP would be increased by faster mobility of the DP. This would be like decreasing on-time in pulse plating. Reduction in surface roughness and grain size of Ni film electrodeposited has been reported when on-time in pulse plating is decreased [27-29].

Desorption of H_2 gas bubbles from surface of cathode is also promoted with increase in pressure, because solubility of H_2 in sc-CO_2 is increased with increase in pressure. Combining the improved homogeneity, the PPC, the promoted H_2 gas bubbles desorption, R_a and grain size of the Ni film could be significantly reduced with increase in pressure for EP-SCE.

1.1.4. Effects of carbon dioxide volume fraction

Increasing volume fraction of CO_2 could increase total amount or concentration of the DP in SCE if enough surfactant is provided to maintain homogeneity of the DP. Desorption of H_2

gas bubbles from the surface of cathode is promoted when concentration of the DP is increased. Better desorption of H_2 gas bubbles then leads to improvement in uniformity of the surface and decrease in R_a of the Ni films electrodeposited. R_a decreased from 30.18 to 16.79 nm with CO_2 volume fraction increased from 10.0 to 50.0 vol%, shown in Fig. 8. Decrease in R_a is also contributed by decrease in on-time of the PPC, because frequency of adsorption and desorption between the DP and the working electrode is increased with increase in concentration of the DP. This decrease in on-time also leads to decrease in grain size of the Ni films electrodeposited. Grain size was found to reduce from 9.41 to 8.02 nm when CO_2 volume fraction was increased from 10.0 to 50.0 vol%, shown in Fig. 8.

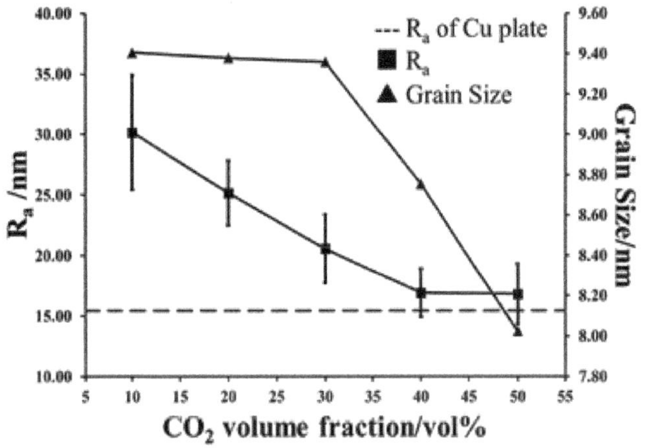

Figure 8. R_a and grain size of Ni films fabricated by EP-SCE with different CO_2 volume fraction.

Defects were found on the surface for Ni films electrodeposited with CO_2 volume fraction higher than 50.0 vol%. The defects were caused by adsorption of the DP on the surface of cathode when homogeneity of the emulsion is too low. Increase in size and size distribution of the DP occur when CO_2 volume fraction is increased while temperature, pressure, and volume fraction of the surfactant remain fixed in the system [12]. Growth in size of the DP is an indication of poor homogeneity, and continues increasing volume fraction of CO_2 in the system would eventually lead to phase separation. Some portion at the upper part of the Cu plate was found to be not electrodeposited with Ni when volume fraction of CO_2 was increased to 60.0 vol%. When 60.0 vol% of CO_2 was used, ca. 13% of the surface was found to be not covered by electrodeposited Ni, the value increased to 26% and 50% when CO_2 volume fraction was increased to 70.0 and 80.0 vol%, respectively.

1.1.5. Effects of the surfactant volume fraction

Usage of the surfactant in the system allows formation of the DP. Uniformity of the surface improved significantly when 0.1 vol% of the surfactant was used when comparing with the

Ni film electrodeposited without addition of the surfactant. R_a of the Ni films decreased dramatically from 89.67 to 18.93 nm as volume fraction of the surfactant increased from 0 to 0.1 vol% as shown in Fig. 9, and R_a was 16.90 and 17.04 nm when 1.0 and 2.0 vol% of the surfactant, respectively, was used. Increasing volume fraction of the surfactant is expected to cause reduction in size of the DP, because γ is reduced with increase in volume fraction of the surfactant [24], therefore, on-time of the PPC is decreased and causes grain size to decrease. Grain size decreased from 15.23 to 8.81 nm for surfactant volume fraction from 0 to 2.0 vol%., shown in Fig. 9. Volume fraction of CO_2 was fixed for the samples prepared with various volume fraction of the surfactant. Both R_a and grain size of the Ni films did not decrease much from 1.0 to 2.0 vol% of the surfactant used. This indicates concentration of the surfactant is close to saturation at ca. 1.0 vol%.

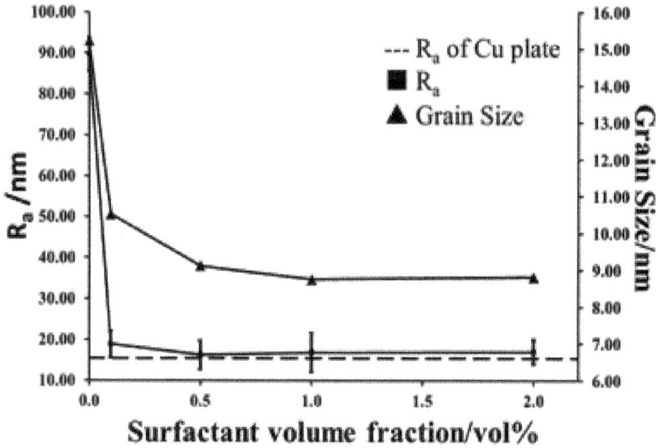

Figure 9. R_a and grain size of Ni films fabricated by EP-SCE with different surfactant volume fraction.

1.2. Periodic-plating-characteristic

Direct observation of PPC is extremely difficult because of the electrolyte is actually composed of heterogeneous emulsion and the high pressure. Indirect evidence of PPC was reported in Rahman et al.'s work [30] as shown in Fig. 10. The porous structures were expected to be caused by adsorption of the dispersed phase (DP), since electrodeposition reaction would be restrained in the region on the working electrode where it is covered by the DP. However, direct observation of PPC in EP-SCE is still required, but direct observation is extremely difficult because of the high pressure environment and heterogeneous emulsified electrolyte. In addition, size of the DP or the micelle is usually in the micro- or even nanoscale range [31,32]. This increases the difficulty for observation of the periodically reaction-on and –off phenomenon directly.

The periodically adsorption and desorption of the DP is expected to cause a difference in the actual surface area available for the electrochemical reaction. Therefore, a fluctuation in the potential response would be expected if a constant current is applied to the system if the difference in the actual surface area available is significant enough. Based on these assumptions, an electroanalytical method, chronopotentiometry (CP), and a modified working electrode are used to directly observed the PPC.

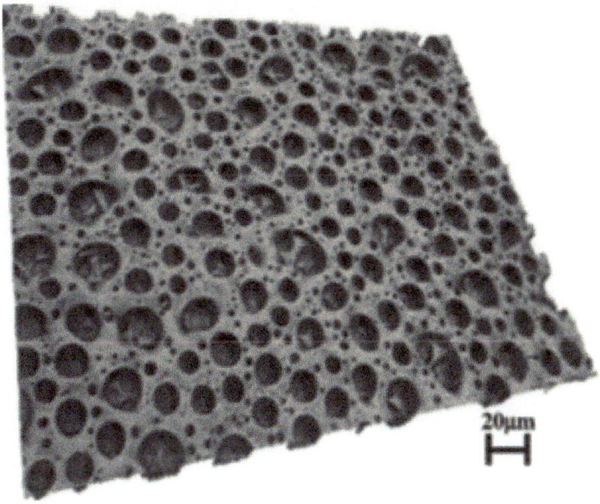

Figure 10. Porous Ni film fabricated by Ni ESCE.

In order to magnify the fluctuation in the potential response obtained from CP, small contact area between the working electrode and the electrolyte is used as shown in Fig. 11. Area of the working electrode having contact with the electrolyte would be only the tip of the Ø0.5 mm Cu wire, and the area is 0.196 mm^2.

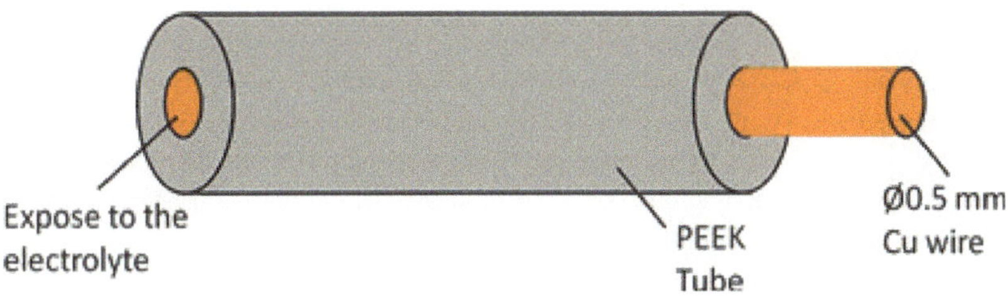

Figure 11. Working electrode used for direct observation of PPC.

Fluctuation in potential response was not observed for conventional electrodeposition as shown in Fig. 12(a). When EP-SCE was used, fluctuation in the potential response was observed as shown in Fig. 12(b). The fluctuation indicates the change on the surface condition of the working electrode, and this is most likely to be caused by adsorption and desorption of the DP. The surfactant used for form SCE was added to the Ni electrolyte without using the sc-CO_2 to confirm the cause of the fluctuation, and the no fluctuation was observed as shown in Fig. 12(c). This result showed the fluctuation was not caused solely by the surfactant. Hence, we could confirm that the fluctuation observed in Fig. 12(b) is a direct observation of PPC.

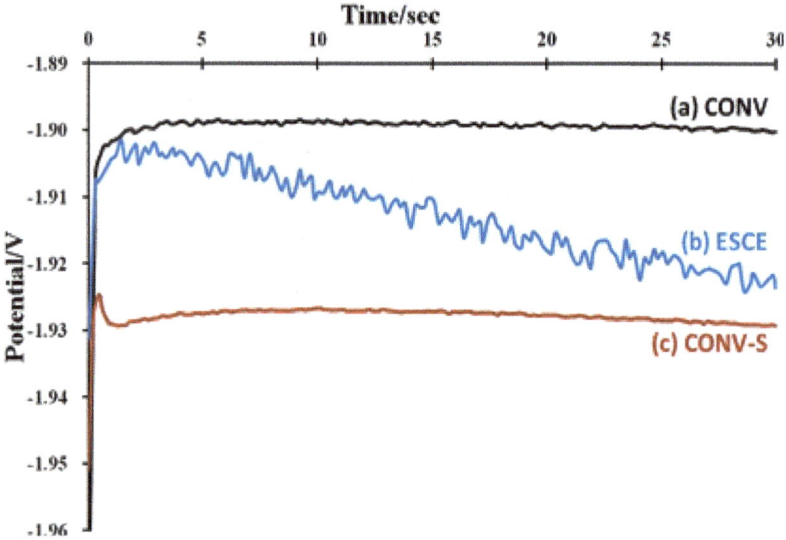

Figure 12. Working electrode used for direct observation of PPC.(a) Conventional electrodeposition, (b) ESCE, and (c) conventional electrodeposition with addition of the surfactant. Current density applied was 0.020 A/cm^2.

1.3. Application of EP-SCE

In fabrication of micro-structures used for Micro-Electro-Mechanical systems, electrodeposition with a template made up of photoresist patterns on top of a conductive substrate is often used as shown in Fig. 13. The photoresist patterns are used to confine dimensions of the structures electrodeposited on the conductive substrate. However, transport of H_2 gas bubbles away from the reaction site is less efficient at the bottom/holes of the photoresist patterns, and H_2 gas bubbles remained at the bottom of the photoresist patterns would cause formation of defects in the micro-structures fabricated. There are several methods to eliminate the problems caused by evolution of H_2. The most often applied method is reducing the evolution rate of H_2 using a lower current density, but growth rate of the micro-structures is also decreased with low current density. Removal of the H_2 gas bubbles from the reaction site is known to be improved by application of SCE. In this way, high current density and high growth rate of the micro-structures could be assured by application of EP-SCE.

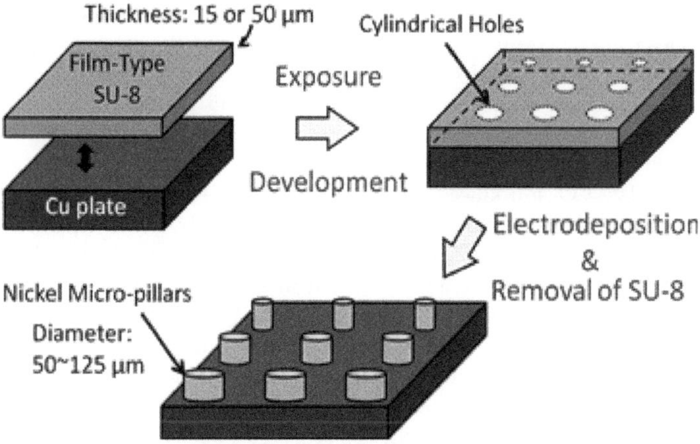

Figure 13. Procedures for fabrication micro-structures.

Ni micro-structures fabrication by conventional electrodeposition with a current density are defective, shown in Fig. 14(a) and (b). These defects were caused by H_2 gas bubbles remained inside the photoresist patterns. On the other hand, when EP-SCE was applied, defect-free Ni micro-structures could be obtained even when a high current density was used, shown in Fig. 14(c), and the average growth rate of the Ni micro-structures was about 5.1 μm/min.

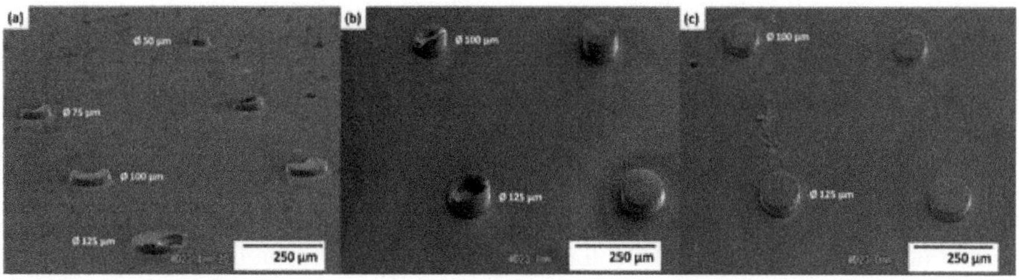

Figure 14. Ni micro-structures fabricated by (a) conventional electrodepositon at 0.100 A/cm_2 for 3 min, (b) conventional electrodepositon and (c) ESCE at 0.100 A/cm2 for 5 min.

1.4. Conclusion

Physical properties of SCE are found to be affected by pressure, volume fraction of CO_2 and the surfactant. Improvement in homogeneity of SCE is achieved by increasing pressure. Therefore, size and size distribution of the DP are reduced as shown in Fig. 15(a) and (c). Since volume fraction of CO_2 in the system is fixed, therefore, reduction in the size would lead to increase in concentration of the DP, and this is expected to cause decrease in on-time of PPC and decrease in R_a and grain size. Increase in volume fraction of CO_2 leads to increase in concentration of the DP, and the size and size distribution would remain small if enough surfactant is present in the system, shown in Fig. 15(b) to (c). Growth in size of the DP would occur if insufficient amount of the surfactant is present to stabilize CO_2 introduced into the system, and continue growth in size of the DP would lead to phase separation as shown in Fig. 15(d). Phase separation could be prevented by increasing pressure or volume fraction of the surfactant, which lowers γ and u_s. Improvement in stability and reduction in size and size distribution of the DP are achieved with increase in volume fraction of the surfactant as shown in Fig. 15(e) and (c). This is why both R_a and grain size is lowered with increase in volume fraction of the surfactant used. We also studied and proposed a mechanism called periodic-plating-characteristic (PPC) to be the main cause for the effects observed in the metal films fabricated by EP-SCE. Moreover, we directly observed PPC by an electroanalytical method, chronopotentiometry (CP), and a modified working electrode. On these experimental results, we applied EP-SCE on fabrication of microstructure and succeeded.

2. Crystal growth by electroless Ni-P plating using supercritical carbon dioxide emulsion

2.1. Electroless plating in a supercritical CO_2 emulsion (ELP-SCE)

2.1.1. Introduction

Electroless plating (ELP) has a low processing temperature, a high metal-ion transportation density, the ability to deposit on electrically nonconductive materials, a more uniform thickness for products of any shape, and a simple deposition mechanism.[33,34] A wet process such as ELP can serve as a superior alternative for a three-dimensional (3D) integration technology. The disadvantages of ELP are the high viscosity of the solution and anomalous growths in the plating film caused by the pretreatment condition. These disadvantages have interfered with the formation of microstructures for electronic devices and MEMS.

The key requirement for forming a uniform, conformal thin film over a complex 3D micro/nanostructure will be to improve the transport properties. To this end, we propose a technique based on the criteria mentioned above. Specifically, we proposed an ELP method using dense carbon dioxide (CO_2) beyond the critical point as a solvent. The changeable density of dense CO_2 enables excellent control of the intermolecular interactions, and the high density and diffusivity of the material assure that the plating films can form over nanoscale areas with outstanding reliability. Supercritical carbon dioxide (sc–CO_2), however, has been found to be unsuitable as a medium for plating reactions. Metal salts are generally soluble in water, but water and CO_2 tend to mix poorly. The problem can be solved by emulsifying sc–CO_2 and a plating solution, then adding a nonionic surfactant.[13] The ELP technique we propose in this study uses a dense CO_2 beyond the critical point. The ELP reaction in Electroless Plating in a Supercritical CO_2 Emulsion (ELP-SCE) [35] takes place in an emulsion containing dense CO_2. We discuss the surface morphology of the film plated by ELP-SCE, which turns out to have various advantages over the surface morphology of film plated by conventional ELP.

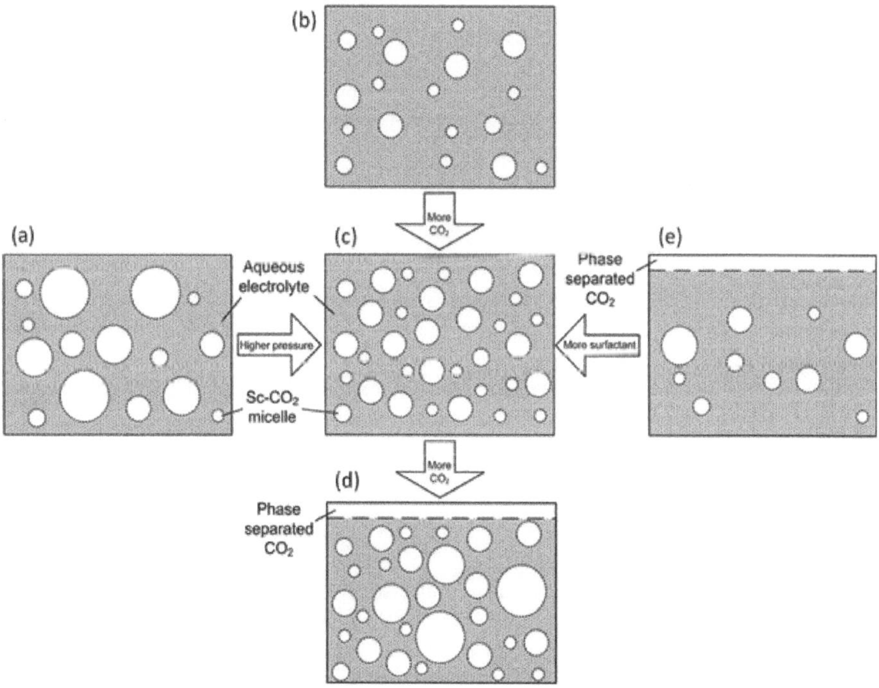

Figure 15. Illustration of conditions of the DP in SCE: (a) at relatively low pressure, (b) at relatively low CO_2 volume fraction, (c) at relatively high pressure, medium CO_2 volume fraction, and high surfactant volume fraction, (d) at relatively high CO_2 volume fraction, and (e) at relatively low surfactant volume fraction. .

	Reducing Agent						
Metal	$H_2PO_2^-$	N_2H_4	CH_2O	BH_4	RBH_3	Me ions	Others
Ni	Ni-P	Ni		Ni-B	Ni-B		
Co	Co-P	Co	Co	Co-B	Co-B		
Fe				Fe-B			
Cu	Cu	Cu	Cu	Cu	Cu	Cu	
Ag		Ag	Ag	Ag	Ag	Ag	Ag
Au		Au	Au	Au	Au		Au
Pd	Pd-P	Pd	Pd	Pd-B	Pd-B		
Rh		Rh					Rh
Ru				Ru			
Pt		Pt		Pt			Pt
Sn						Sn	
Pb			Pb				

Table 2. Coating Obtained by Electroless Plating

2.1.2. Electroless plating reaction and experimental method

The ELP methods currently known can be used to deposit 12 different metals (Table 2). [34] Decomposition products (phosphorous and boron) in the reducing agent precipitate as the metal deposits, leaving films of the respective alloys. Two or more metals can be deposited at once without much difficulty. ELP methods are used for the deposition of more than 50 alloys of different qualitative compositions, mostly based on nickel, cobalt, and copper. For this study we selected an electroless nickel-phosphorus (Ni-P) plating process. The most widespread type of Ni-P plating uses hypophosphite as the reducing agent.[33,34][4] Three major properties of electroless Ni-P, namely, its hardness, wear resistance, and corrosion resistance, have led to its technical application in many industries, from electronics, automaking, aerospace, and machinery to oil and gas production, power generation, printing, and textiles. The process is generally applied with a stable, acidic Ni-P plating solution, as contact with CO_2 acidifies water due to the formation and dissociation of carbonic acid.

2.1.2.1. Materials

CO_2 with a minimum purity of 99.99% was purchased from Nippon Tansan Co., Ltd. The experiments were performed with a nonionic surfactant polyoxyethylene lauryl ether ($C_{12}H_{25}(OCH_2CH_2)_{15}OH$) supplied by Toshin Yuka Kogyo. The electroless Ni–P plating solution had a chemical composition of nickel chloride (9%), sodium hypophosphite (12%), complexing agent (12%), and ion-exchanged water (67%) (Okuno Chemical Industries Co., Ltd.). The substrate was a 99.99% pure film of copper measuring 10×20 mm (Mitsubishi Shindoh Co., Ltd). The substrate was washed with acetone and rinsed in deionized water before each reaction. The grease was removed from the sample by successive dipping in a 10 wt % solution of NaOH and a 10 wt % solution HCl followed by rinsing in deionized water. The sample was added to an activator solution consisting of hydrogen chloride (18%), palladium chloride (0.04%), and ion-exchanged water (81.96%) (Okuno Chemical Industries Co., Ltd.)

at 303 K, then rinsed in deionized water. This pretreatment was applied to all copper substrates regardless of the condition of the ELP.

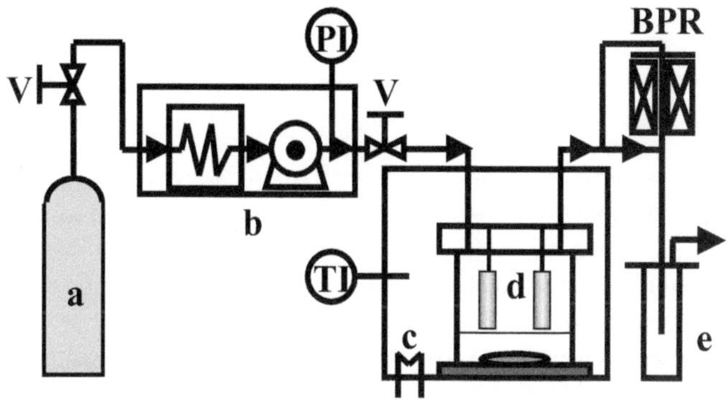

Figure 16. Experimental apparatus used for batch reaction in our electroless plating experiments beyond the critical point of CO_2; (a) CO_2 cylinder; (b) cooler and high pressure pump; (c) temperature controlled air bath; (d) reactor with magnetic stirrer; and (e) trap; BPR: back-pressure regulator; PI: pressure indicator; TI: temperature indicator; V: valves.

2.1.2.2. Experimental apparatus and procedure

Fig. 16 shows the high-pressure experimental apparatus (Japan Spectra Company) used for the ELP.[35] The temperature variation of each run was confirmed to be less than 1.0 K. The maximum working temperature and the maximum pressure were 424 K and 50 MPa, respectively. The reactor was a stainless steel 316 vessel with an internal volume of 50 mL, kept in a temperature-controlled air bath. A magnetic agitator with a cross-magnetic stirrer bar was placed within the reactor and the activated substrate was attached to the reactor with stainless wires. A plating reaction within a reactor starts only upon contact between the substrate and plating solution. As such, the substrate position in a reactor affects both the reproducibility and surface morphology of the ELP film. Yan et al. demonstrated the dispersion behaviors of a ternary system composed of dense CO_2, an electroplating (EP) solution, and a surfactant.[14] Their experiment was performed in a high-pressure view cell with an internal volume of 45 mL. In the absence of stirring, two separated phases, namely, a transparent upper CO_2 phase and a clear green lower phase (the nickel EP solution) were observed at 323 K and 10 MPa. The ternary system with the CO_2 volume fraction of 0.2 was stirred at 400 rpm. The CO_2 dispersed into the plating solution with stirring, and the light scattering from the small CO_2 drops in the solution increased the turbidity of the system. In our experiment, before the ELP reaction, the electroless Ni–P plating solution (30 mL) and surfactant (surfactant concentration: 1.0 wt % to the ELP solution) were placed in the reactor at atmospheric pressure. Next, liquid CO_2 was pumped into the cell by a high performance liquid chromatography (HPLC) pump until a predetermined pressure was reached. The ELP reaction was performed at a temperature of 353 K and a pressure of 15 MPa in a constantly agitating ternary system stirred at a speed of 500 rpm. The reaction commenced from the start of agitation. In the Results and Discussion of this chapter we describe the properties of the plated film fabricated by ELP-SCE in comparison with those of film fabricated by conventional ELP. The conventional method was performed under the following experimental conditions: 353 K, atmospheric pressure, a plating solution with a chemical composition identical that used in ELP-SCE, constant stirring at a speed of 50 rpm.

2.1.2.3. *Analysis*

An optical microscope (Digital Microscope VHX-500, Keyence. Co., Ltd.) and scanning electron microscope (FE-SEM, S-4500, Hitachi High-Technologies Co., Ltd.) were used to study the surfaces of the plated Ni–P films. A surface texture measuring instrument (Surfcom 480A, Tokyo Seimitsu Co., Ltd.) with a diamond-tipped detector (2 μm tip radius) was used to measure the average surface roughness (*Ra*) to a minimum height resolution of 1 nm (height measurement range: 80 μm). The average *Ra* was calculated from measurements at five points or more. The film thickness was measured directly from a cross-sectional scanning electron microscopy (SEM) image of a plated Ni–P film fabricated by a focused ion beam system (FB-2100, Hitachi High-Technologies Co., Ltd.). The phosphorus composition of the fabricated film was measured by an FE-SEM (S-4300SE, Hitachi High-Technologies Co. Ltd.) equipped for energy-dispersive X-ray spectroscopy (EDX). An accelerating voltage of 20 kV with a collecting time of more than 300 s was applied. X-ray diffraction (XRD) analysis (2θ-ω scans) was performed at room temperature (RT, 298 K) using a PANalytical X'pert Pro Galaxy system equipped with an X'celerator module. The X-ray source was CuKα, and the tube voltage and current were 45 kV and 40 mA, respectively.

2.1.3. *Electroless Ni-P plating in an emulsion of supercritical CO_2*

Fig. 17 shows optical microscope images of a copper substrate and Ni-P films plated over the substrate by ELP-SCE at 15 MPa and 6 MPa. Polishing trace was observable on the copper substrate before pretreatment. The film plated at 15 MPa was uniformly bright and covered both the front and back of the substrate (Fig. 17 (b)). The uniform brightness and coverage were attributable to the exact position of the substrate and the uniformity of the emulsion. The formation of emulsion was unstable in the film plated at 6 MPa, beyond the critical point, so the film was thin and even unformed in portions (Fig. 17 (c)). The surface roughness (*Ra*) of the plating films was 0.030 μm at 15 MPa and 0.059 μm at 6 MPa.

If CO_2 exceeds a critical point, the density will rise rapidly. While the CO_2 density at 323 K and 6 MPa is only 0.1 kg L^{-1}, it reaches 0.7 kg L^{-1} at the higher pressure of 15 MPa.[8] Moreover, Sone et al. reported that in a ternary system of water, surfactant, and CO_2, the CO_2 formation in the water emulsion by the surfactant is affected not only by the temperature and pressure, but also agitation.[4] They also found, in a similar experiment, that no emulsion was formed below the critical point. With ELP-SCE, a stable emulsion forms when the CO_2 density approaches the liquid phase, and a uniform plating film can be deposited.

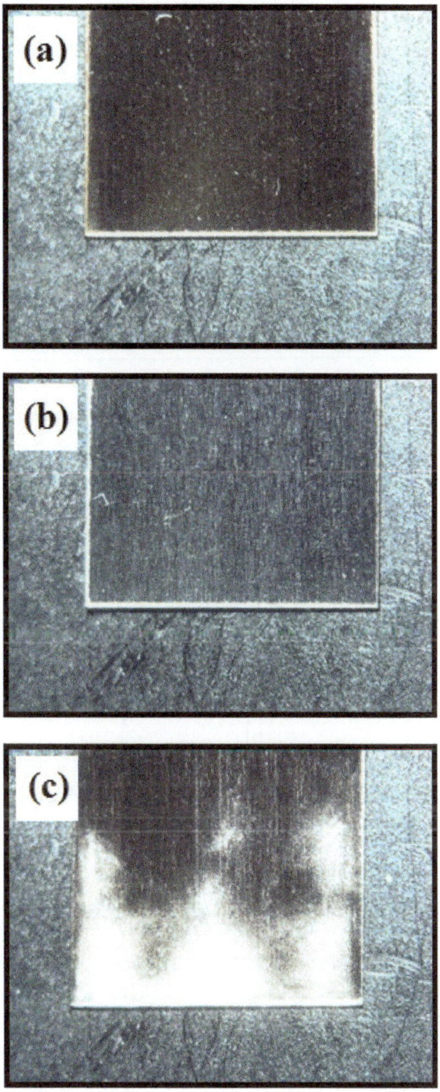

Figure 17. Optical microscopy images of (a) pure Cu substrate and Ni–P films plated from ELP-SCE at 353 K for 60 min (b) at 15 MPa and (c) at 6 MPa.

According to an as-deposited coating surface composition analysis by EDX, the Ni-P film formed by ELP-SCE was composed of 20 wt % phosphorus. When the phosphorus content increases, the microstructure of an electroless Ni-P deposit changes from a mixture of amorphous and nanocrystalline phases to a fully amorphous phase.[36] The structure of our Ni-P film was confirmed to be amorphous by XRD. Amorphous profiles with a wide angular range of 40-45° (2θ) appear nearby a 2θ position corresponding to the Ni {111} plane. The copper substrate underneath was responsible for the Cu diffraction peaks in the profiles of the Ni-P film formed by ELP-SCE. The peaks appear because the coated-deposit was too thin (1.0 μm less) to totally absorb the penetration of the X-ray beam (Fig. 18).

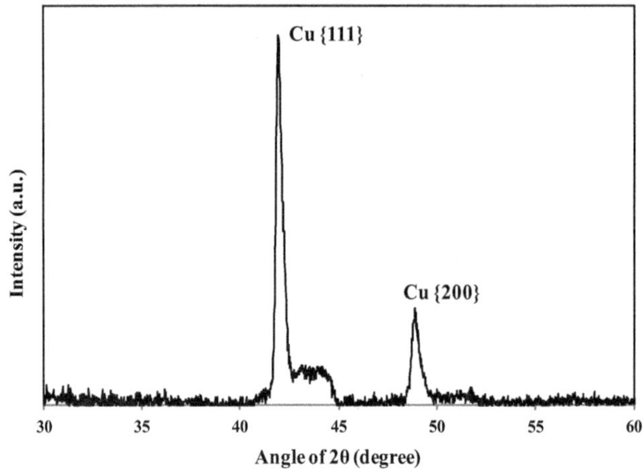

Figure 18. XRD spectrum of Ni-P film plated from ELP-SCE at 353K and 15 MPa for 180 min.

2.1.4. Surface morphology of Ni-P film by conventional ELP and ELP-SCE

We used an optical microscope to observe the surface features of a Ni–P film formed by conventional ELP, another Ni–P film formed by ELP-SCE, and a pure copper substrate (Fig. 19). The only clear difference in defects between the two Ni–P films was a polishing trace on the copper substrate. As is widely known, thin films fabricated by ELP have smaller pinholes and cracks than those fabricated by EP.

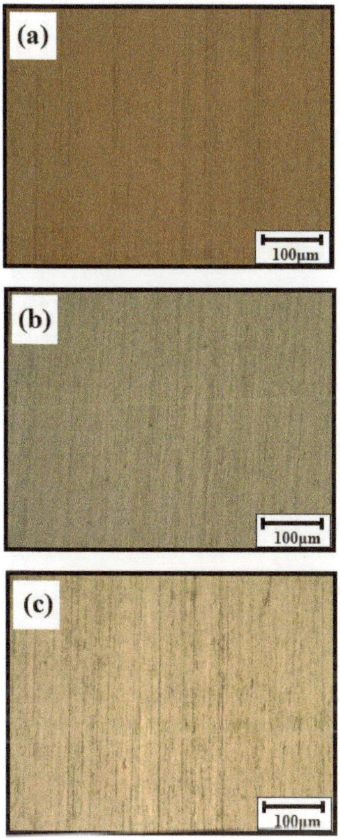

Figure 19. Optical microscopy images of Ni–P films plated from (a) ELP-SCE at 353 K and 15 MPa for 180 min, (b) conventional ELP at 353 K and atmospheric pressure for 2 min, and (c) pure copper substrate.

SEM observations reveal clear differences between the surface morphologies of the Ni–P films and the substrate activated by catalytic Pd, as shown in Fig. 20. The thickness of the Ni–P film fabricated by conventional ELP was 0.3 μm, while that fabricated by ELP-SCE was 0.8 μm. Nodules were observed on the Ni–P film fabricated by conventional ELP in the early stage of the reaction, as shown in Fig. 20 (b). This nodule formation was the result of concentrated nickel reactions over a localized area on Pd nucleus on the surface of the substrate as shown in Fig. 20(c). Earlier reports also have confirmed that the nickel nucleus appears and grows on the Pd nucleus on the surface of the substrate.[37-39] These nodules become a serious problem when they form in films fabricated by ELP on fine electronic devices and MEMS. Meanwhile, the thin film fabricated by ELP-SCE was free from nodules and pinholes, but its thickness was still more than double that of the thin film fabricated by conventional ELP. The only particles observed were extremely fine, with diameters of several tens of nm or less (Fig. 20 (a)).

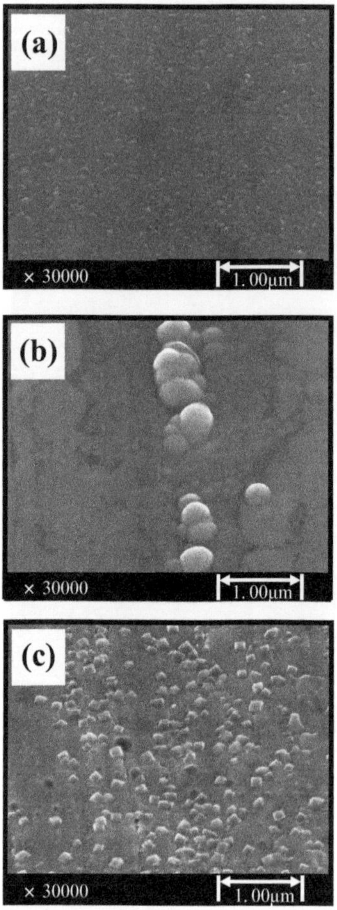

Figure 20. SEM images of Ni–P films plated from (a) ELP-SCE at 353 K and 15 MPa for 180 min, (b) conventional ELP at 353 K and atmospheric pressure for 2 min, and (c) substrate activated by catalytic Pd.

Fig. 21 shows how variations in the reaction time influence the surface features of the Ni–P films formed by conventional ELP and by ELP-SCE. The plating films formed by conventional plating for 5.0 and 30 min had thicknesses of 0.8 and 4.9 μm, respectively. The plating films formed by ELP-SCE for 360 and 540 min had thicknesses of 0.9 and 1.0 μm, respectively. The SEM images reveal nodules on all of the surfaces of the Ni–P films fabricated by conventional ELP (Fig. 20 (b), 21 (c), and 21 (d)). The images also show that the nodules increased, both in size (from several hundred nm to over several μm) and in number, as the reaction time increased.

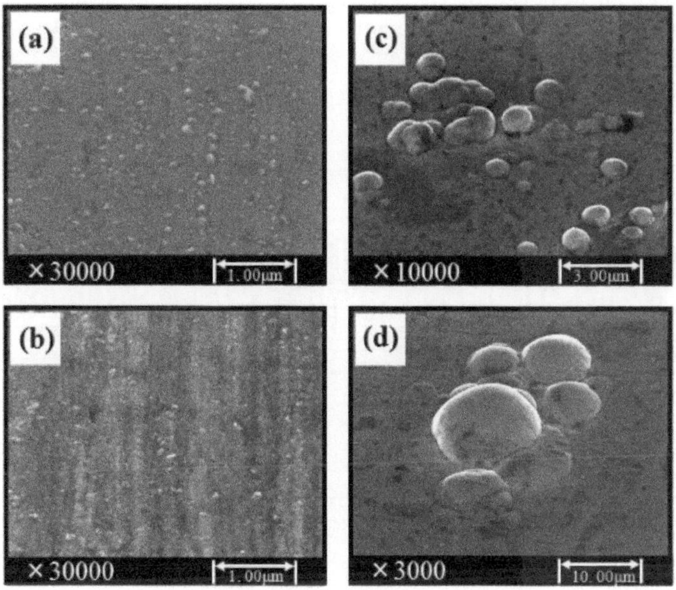

Figure 21. SEM images of Ni–P films plated from ELP-SCE at 353 K and 15 MPa for reaction times of (a) 360 min and (b) 540 min, and conventional ELP at 353 K and atmospheric pressure for (c) 5 min and (d) 30 min.

Meanwhile, the Ni–P film fabricated by ELP-SCE was free of nodules (Fig. 20 (a)) and had an extremely uniform surface (Figs. 21 (a) and (b)). These results differed considerably from the changes in the surface features of conventional ELP films brought about by adjustments in the reaction time and the processing methods for the substrate activation by Pd. [37-39] We also found that our ELP technique could fabricate superb, highly uniform plated films even when the substrate pretreatment, reaction temperature, and chemical composition of the electroless Ni–P plating solution were all identical to those use d in conventional ELP. On this basis, we surmise that our ELP technique may be effective in suppressing the growth of nodules.

Fig. 22 shows the roughness curves on the surfaces of the Ni–P films formed by ELP-SCE, by conventional ELP, and by surface activation of the substrate. The evaluation length of the surface roughness measurement was 1.250 mm. The activated substrate had an *Ra* of 0.040 μm. The conventional ELP had an *Ra* of 0.048 μm and a rougher surface than the activated substrate. Previous reports have shown how activation processing changes the surface morphologies and deposition behaviors of electroless Ni–P films. [37-39] Meanwhile, ELP-SCE formed a film with improved smoothness (*Ra* of 0.030 μm) under the same activation processing conditions used for conventional ELP. We also found that the film thickness conferred a strong influence on the surface roughness. The thin film formed by ELP-SCE was very smooth, though it was still more than twice as thick as the film fabricated by conventional ELP. These results demonstrate that ELP-SCE suppressed the deposition reaction of the locally concentrated nickel. We can also see, in Fig. 20 and Fig. 21, that no nodules were formed.

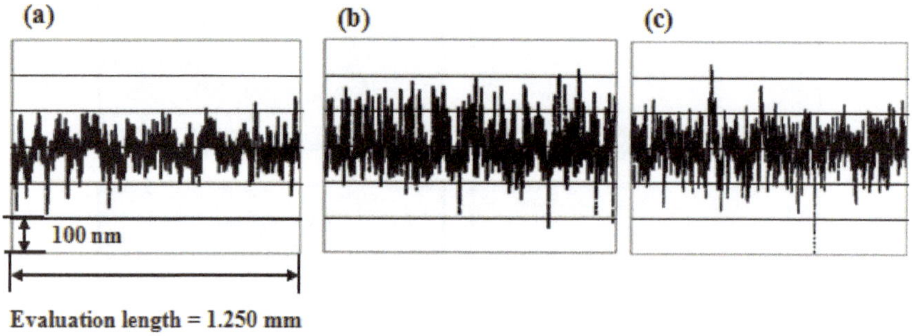

Figure 22. Roughness curves of the surfaces of Ni–P films plated by (a) ELP-SCE at 353 K and 15 MPa for 180 min, (b) conventional ELP at 353 K and atmospheric pressure for 2 min, and (c) substrate activated by catalytic Pd.

The conventional ELP as performed at 353 K and atmospheric pressure. ELP-SCE was performed at 353 K and 15 MPa. Fig. 23 shows the relationship between the surface roughness and reaction time. The surface of the Ni–P film fabricated by the conventional ELP roughened as the reaction time increased. The roughness of the Ni–P film formed by ELP-SCE, meanwhile, showed no dependence on the reaction time. Nodules appeared on the surfaces of the Ni–P films fabricated by conventional ELP at all reaction times, and the nodules grew as the reaction times increased. No nodules were observed on the surfaces of any of the Ni–P films fabricated by ELP-SCE, even at the maximum reaction times. The aforementioned results confirm that the excellent smoothness of ELP-SCE film mitigated the influence of the substrate pretreatment and was independent of the reaction time. Fig. 20, meanwhile, shows a suppression of the growth of the nodules generated by conventional ELP.

ELP-SCE produced a thin film with high smoothness and outstanding uniformity. The Ni–P film fabricated by conventional ELP with a reaction time of 5 min had a thickness of 0.8 μm, or about the same thickness as ELP-SCe film fabricated with a reaction time of 180 min. The *Ra* of the ELP-SCE film was 0.03 μm, while that of the conventional ELP film was 0.06 μm.

ELP films are generally smoother than EP films and have fewer defects.[40] Even with ELP, however, defects such as microscopic nodules, pits, and pinholes are difficult to suppress. [41-43] Although suppression of a through-hole like a pinhole need a thick film, more nodules form in a thicker film. Nodules also easily form when the underlayer has projecting parts, foreign objects, and nuclear growth sites. Conventional suppression of nodule method prepares the smoothness and cleanness of an underlayer, while nonlinear diffusion adds reactive species that interfere with film growth over the projecting parts of a plating film (see Fig. 24).[44] Further, the pulse electroplating controls the thickness of a diffusion layer and is available to suppress nodule growth. [45]

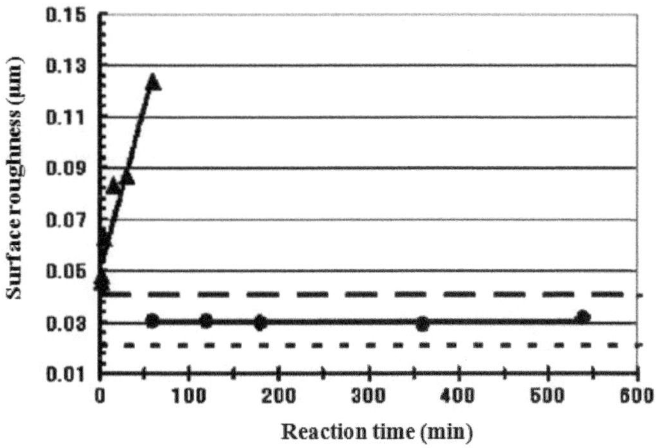

Figure 23. Relationship between Ra and reaction time. Ra of the Ni–P film made by ELP-SCE and by conventional ELP are plotted as circles and triangles, respectively. The dotted line shows Ra of the activated substrate and the fine dotted line shows the surface roughness of the pure Cu substrate.

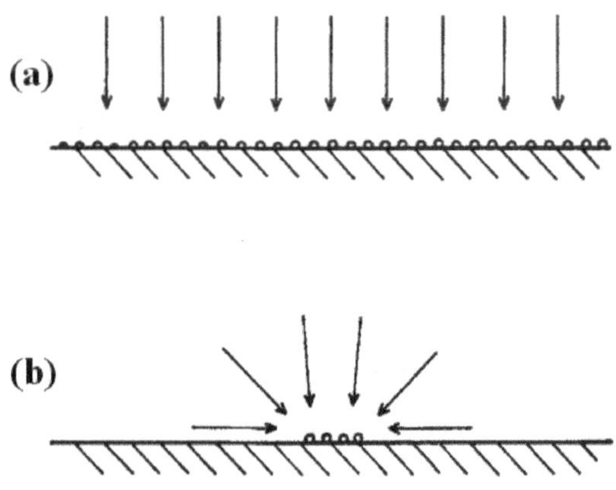

Figure 24. a) Linear O_2 diffusion to a large activated area. (b) Linear and nonlinear O_2 diffusion to a small pattern of nuclei.[15]

ELP-SCE formed very smooth thin films whose thicknesses suppressed both nodules and pinholes without exceeding even 1 μm. The pH, reaction temperature, pretreatment, stirring speed, additive, and reactive species concentration in the plating solution all influence the deposition behavior of the ELP film. Yet in our current work we used the same plating solution, reaction temperature, and pretreatment for both ELP-SCE and conventional ELP.

The Ni–P film plated by ELP-SCE was free of pits and pinholes because the hydrogen bubbles produced by the electrolysis of the water were dissolved in the dispersed sc-CO_2 phase of the emulsion.[13] Moreover, ELP-SCE plates the film under high pressure. High-pressure plating failed to deliver good results because hydrogen bubbles were less buoyant in the high-pressure system than at atmospheric pressure. Hence, the larger bubbles prevent the metal from covering the substrate. This, a characteristic effect of plating techniques that use sc-CO_2 emulsion, suppresses the formation of pits or pinholes via the mechanism shown in Fig. 25.

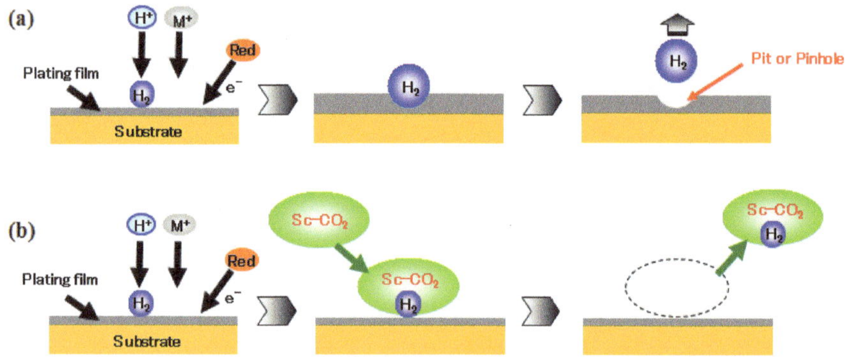

Figure 25. a) Pinhole formation in conventional electroless plating. (b) Suppressed pinhole formation in ELP-SCE. M: metal, Red: reducing agent, and sc-CO_2: supercritical carbon dioxide.

The best feature of ELP-SCE is its ability to suppress nodules and other abnormal growths formed by the plating reaction. The electroless metal deposition occurs by repeated 3D nucleation at catalytic sites on the substrate.[46] In 3D growth of the deposited Ni under a low-nucleation-density condition, the deposited Ni grows and the surface roughness increases. Further, an activation processing technique with Pd catalyst can be used to influence the growth of the Ni–P film.[38] In the Ni–P films fabricated by conventional ELP in our current experiments, the activation processing roughened the surface and nodule growth was confirmed. Nodule suppression is only attainable when the abovementioned factors exert their effects at the reaction site of the plating. That is, the growth suppression factor of the plating reaction and the state of the diffusion layer that conveys the film onto the substrate must both be influenced. ELP-SCE differs from the conventional method in three ways: the stirring speed, the addition of the surfactant for emulsion formation, and the decrease of the pH by the CO_2 dissolution in the plating solution. Henceforth, we will also need to consider how the collision phenomenon influences the plating film of the CO_2 phase. We will need to collect more evidence to formulate a detailed mechanism for our process. Even so, our experiments have demonstrated that ELP-SCE produces more outstanding results than conventional ELP, forming thin films with high smoothness and superb uniformity.

2.1.5. Conclusion

This chapter has proposed ELP-SCE, a hybrid technique combining ELP and supercritical fluid technology. The ELP reactions are carried out in an emulsion of sc-CO_2 and an ELP solution with surfactant. ELP-SCE formed a uniform Ni–P film free from the pinholes that typically form from the hydrogen bubbles produced by the electrolysis of water, and free from the nodules that form from the nuclear growth in the ELP reaction.

2.2. Direct observation of nodule growth on electroless Ni-P deposition in supercritical CO_2 emulsion

2.2.1. Introduction

The plated film obtained by ELP-SCE was extremely uniform, smooth and free from pinholes and nodules. The film growth speed of ELP-SCE was slower than the conventional ELP. It is reported that an effect of pulse electroplating-like mechanism by adsorption and desorption of the supercritical CO_2 (sc-CO_2) phase from the plated film, called as "Periodic-Plating-Characteristic (PPC)" is a cause in the smoothing mechanism of the electroplating using a supercritical carbon dioxide emulsion (EP-SCE).[22] PPC might not be the only cause of film smoothing and nodule suppression in ELP-SCE since the film formation mechanisms of electroplating and ELP are different, though the nodule formation of ELP-SCE can be affected by fast cycle of adsorption and desorption of dispersed CO_2 phases to a minute convex part, and for the growth to be suppressed.

In addition, PPC effect itself cannot completely explain the effect of higher P content in Ni-P film by ELP-SCE, and the film growth speed is slower than conventional ELP. As one of the factors of the phenomenon of ELP-SCE, CO_2 dissolves in the plating solution and causes decrease of the pH. When the pH of the plating solution decreases, it is expected to cause P content to increase in the plated film and decrease in the film growth speed.[33] Thus, in previous study we discussed that the increase of the proton concentration in the plating solution caused the effect of suppression of nodule growth.[47] However, it is necessary to clarify not only by an indirect method of observing the surface of the plated film via SEM or AFM, but also a direct method of observing growth of one nodule, in order to discuss the suppression mechanism of nodule growth of ELP-SCE clearly. Moreover, the slow film growth speed of ELP-SCE cannot be explained by only making a low pH plating solution. When a plated film fabricated by ELP in the pH=4.0 solution alone with agitation at the same speed used in ELP-SCE, the formation of the plated film was insufficient. For ELP-SCE, the sc-CO_2 phase distributed in the plating solution causes viscosity of plating solution to be low, and, as a result, plating under a high-speed agitation was enabled. This means thickness of diffusion layer close to surface of substrate and transport property of reactive materials at surface of substrate for ELP-SCE is different from conventional method.

In this chapter, we discuss what influences the characteristic reaction field of ELP-SCE to the plated film growth. In addition, a novel direct observation of nodule is proposed to clarify the nodule suppression mechanism of ELP-SCE. [48] At first, the Ni-P plated film is plated on Cu substrate by the conventional ELP method, and then re-plating by ELP-SCE or conventional ELP method is conducted on the Ni-P plated film in which nodules were formed by conventional ELP. Moreover, morphology of the nodules at a selected position in the Ni-P plated film is compared before and after the re-plating.

Chemicals	
Nickel chloride	9.0%
Sodium hyphosphite	12.0%
Complexing agent	12.0%
Ion-exchange water	67.0%
pH	5.3
Bath temperature	353K

Table 3. Bath composition and operating conditions of electroless Ni - P films

2.2.2. Base Ni-P (BNP) film preparation for direct observation of nodule

The substrate was a film of 99.99% pure copper measuring 10×20 mm (Mitsubishi Shindoh Co., Ltd.). The plated film was made by using the ELP solution shown in 2.1.2 at temperature 353K and reaction time 5 minutes. Process procedures from step 1 to step 8 (Pretreatment-A) was shown in Table 3. The activation agent, the degreasing agent and the ELP solution were purchased from the Okuno Chemical Co., Ltd. The plating solution was kept in a glass beaker in a temperature-controlled water bath agitated with a magnetic agitator and a cross-magnetic stirrer bar. Agitation speed was 50 rpm. The substrate was inserted to the beaker by stainless steel wires. A lot of nodules were formed on the surface for the plated film made with this condition, and the phosphorus content was 14wt% (Fig.26). The plated film is amorphous in as deposited condition and strongly support the observations made from X-ray diffraction (XRD) measurements (Fig.27). When plating was performed again, influence from the substrates is a little because it is an amorphous plated film. Thereafter, the Ni-P film made under this process condition is called base Ni-P film (BNP film).

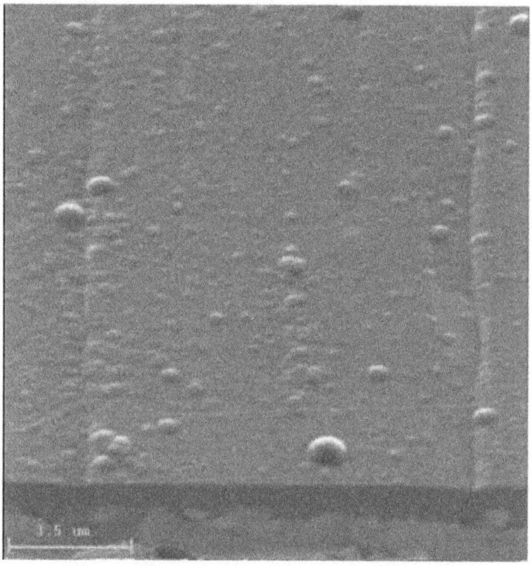

Figure 26. SIM image of Ni-P film by conventional ELP at 353 K and atmospheric pressure for 5min (with a film thickness of 0.8 μm).

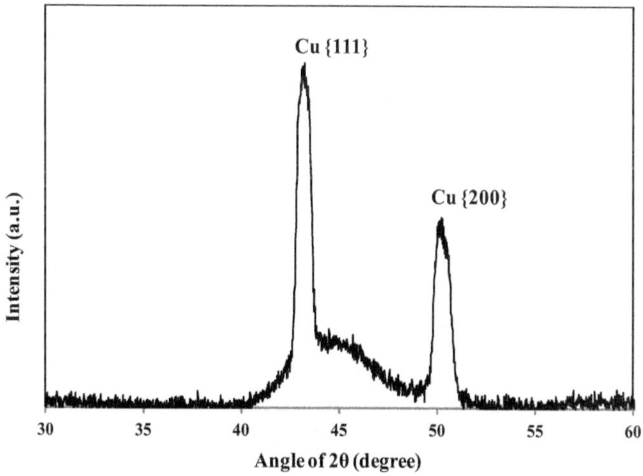

Figure 27. XRD spectrum of Ni-P film plated from conventional ELP at 353K and atmospheric pressure for 5 min.

A rectangular shape shown in Fig.28. was fabricated by focused ion beam system (FIB) (accelerating voltage 40kV) on the surface of BNP film. The fabricating area was made by four rectangles of 50×20 μm to make a square observation area of 50×50 μm. The fabrication program was set so that the gallium (Ga) ion beam was not irradiated to the observation area to prevent influence from irradiated Ga ion of FIB to the plated Ni-P film growth at the observation area. The observation area was measured by atomic force microscopy (AFM) before re-plating, and nodules that become observation candidates were decided on the observation area.

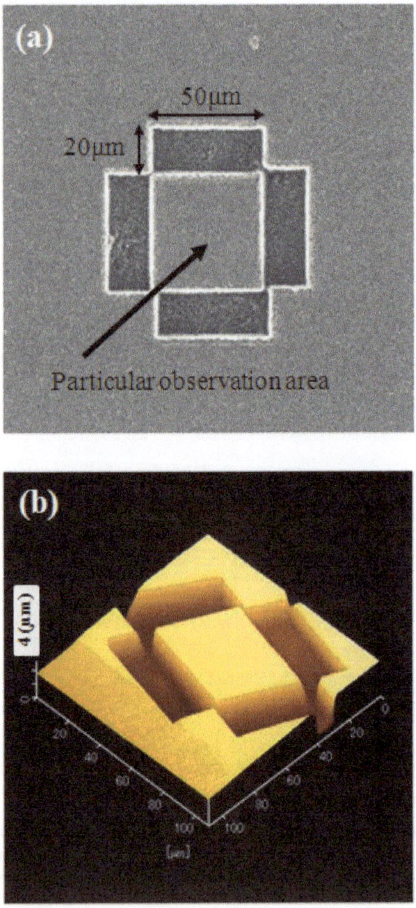

Figure 28. a) SIM and (b) AFM image of rectangular shaped Ni-P film plated by conventional ELP at 353 K and atmospheric pressure for 5min.

The activation treatment using Pd has a big influence on the surface morphology of plated film. This is undesirable to observe the morphological change of the fine nodules[39]. In this study, a pretreatment for the re-plating on the rectangular shaped BNP film was processed in order from step 1 to 6 of Table 4 (Pretreatment-B). In the result of the surface texture measuring instrument, Ra of the plated film was 0.029 μm before and after the pretreatment-B, and the AFM measurement result did not have a substantial change in the surface morphology either (Fig. 29).

Step		Solution composition	Temperature(K)	Duration(min)
1	Cleaning-1	Aceton with ultrasonic cleaning	298	15
2	Cleaning-2	Deionized water	298	1
3	Degreasing	Sodium metasilicate (91.0%)	323	1
		Sodium alkylbenzene sulphonate (5.1%)		
		Surfactant (3.9%)		
4	Rinsing	Deionized water	298	1
5	Acid solution treatment	HCl (10.0%)	298	1
		Ion-exchanged water (90.0%)		
6	Rinsing	Deionized water	298	1
7	Activation	Hydrogen chloride (18.0%)	303	1
		Palladium chloride (0.04%)		
		Ion-exchanged water (81.96%)		
8	Rinsing	Deionized water	298	1

Table 4. Composition of solutions and producers of catalyzing process.

2.2.3. *ELP-SCE on pretreated BNP film and on pretreated copper substrate*

The rectangular shaped BNP film after pretreatment-B was plated by the conventional ELP and ELP-SCE again respectively. The conventional ELP was the same as the fabrication condition of BNP film. Details on the apparatus and plating method of ELP-SCE can be found in 2.1.2. The observation area after re-plating was measured by AFM, and morphology change of the nodule was analyzed. For re-plating, not only rectangular shaped BNP film after pretreatment-B but also the copper substrate after pretreatment-A was performed regardless of each plating method.

2.2.4. *Material characterization*

Focused ion beam system (FIB, FB-2100, Hitachi High-technologies Co., Ltd.) has scanning ion microscope (SIM). The liquid-metal ion sources of this instrument used Ga ion sources. SIM was used to observe the surfaces of the plated Ni-P films. A cross section of the plated Ni-P film was fabricated by FIB and the thickness of the plated film could be measured directly from the SIM image on the screen. Moreover, FIB was used for fabricating of the area for carrying out direct observation of the nodule growth. The phosphorous composition of the base film was measured by an FESEM (S-4300SE, Hitachi High-technologies Co., Ltd.) equipped for energy-dispersive X-ray spectroscopy (EDX). An accelerating voltage of 20 kV with a collecting time of more than 300 s was applied. The surface morphology of plated film was examined using an atomic force microscopy (AFM, SPA-400, Seiko Instruments., Inc.) with a calibrated 20 μm xy-scan and 10 μm z-scan range PZT-scanner. A surface texture measuring instrument (Surfcom 480A, Tokyo Seimitsu Co., Ltd.) with a diamond-tipped detector (2 μm tip radius) was used to measure the average surface roughness (Ra) to a minimum resolution of 1 nm for height (height measurement range: 80 μm). The average Ra was calculated from measurements at three points. 2θ-ω X-ray diffraction (XRD) analysis was performed at room temperature (RT, 298 K) using a PANalytical X'pert Pro Galaxy system equipped with an X'celerator module. The X-ray source was CuKα, and the tube voltage and the current are 45 kV and 40mA, respectively.

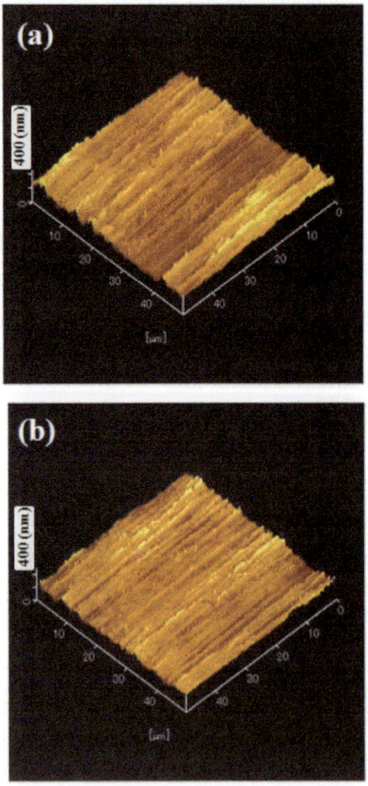

Figure 29. AFM image of (a) Ni-P film by conventional ELP at 353 K and atmospheric pressure for 5min, and (b) After pretreatment-B.

2.2.5. Direct observation of nodule growth in conventional method

Copper substrate after pretreatment-A performed with the re-plating procedure had 0.8 μm of the plated film thickness, and it showed a lot of nodules as that of the BNP film as shown in Fig. 30 (1-a). This means that decomposition of a plating bath did not occur, when re-plating is performed. Film thickness increased from 0.8 μm to 1.7 μm as a result of re-plating on BNP film after pretreatment-B by conventional ELP, and the number of nodules on the surface increased from the initial state as shown in Fig. 30 (1-b). It was possible to re-plating on BNP film with Pretreatment-B. Also, surface observation and film thickness of the plating films by the ELP-SCE and the conventional method using the plating solution with adjusted pH were shown in Figs. 30 (2-a), (2-b), (3-a), and (3-b). The stability of the plating solutions in each method were also confirmed by the observations of the plated surfaces. The plated film thickness is 0.6 μm, and the film has grown up on the copper substrate after Pretreatment-A set up simultaneously (Fig. 30 (2-a)). The film became a smooth film, although there were ditches resulted from the polishing ditches of the copper substrate. The change from initial film thickness of 0.8 μm could not be observed by SIM observation as a result of re-plating on BNP film after Pretreatment-B by ELP-SCE (Fig. 30 (2-b)). The phenomenon in Fig. 30 (2-b) is not peculiar and it also happens when the Ni-P film is obtained after Pretreatment-B without Pd activation and re-plated with low pH bath. A similar phenomenon was confirmed for the re-plating with conventional electroless plating adjusted to pH=4.0 by adding HCl (10wt%). The film was formed as for the copper substrate after Pretreatment-B (Fig. 30 (3-a)) but the film growth was difficult to be confirmed by the SIM observation from Fig. 30 (3-b).

The AFM observation of the BNP film fabricated by FIB before and after re-plating with conventional ELP was shown in Fig. 31. Before re-plating, the film was measured with AFM while raising the magnification from Fig. 31 (1-a) to (1-c), and nodules at a specific position were decided by three places. After re-plating, the change in the morphology of the specific nodule was observed with AFM from Fig. 31 (2-a) to (2-c). The height of nodule before re-plating was 40 nm or less, and the width was about 500 nm. The measurement of the width of nodule was conducted along the direction of the dotted line arrow along the ditch in Fig. 31 (1-d) and (2-d). The shape of nodule was not changed as the film thickness increased to twice that of the initial thickness. The prior growth of nodules can be considered to come from spherical diffusion layer surrounding the neighboring nodule cores on the surface. The localized c oncentration by spherical diffusion occurs at the convex part when the thickness of the diffusion layer is the same or thicker than that of the convex part on the surface, and the thickness of the diffusion layer in neighborhood on the plated film greatly influences the surface-roughness of the plated film.[49,50] The thickness of typical Nernst diffusion layer was reported to be about 0.2 mm, and thickness could be about 0.02 mm when agitation is added.[49] In this experimental condition, it was considered that the Nernst diffusion layer was larger than enough to the size of these nodules in resulting the spherical diffusion[51], and led to a surface morphology like Fig. 31 (2-d). Moreover, this evaluating method is an effective direct observation method to study the growth mechanism and the surface morphology of the plated film including nodule at a specific position.

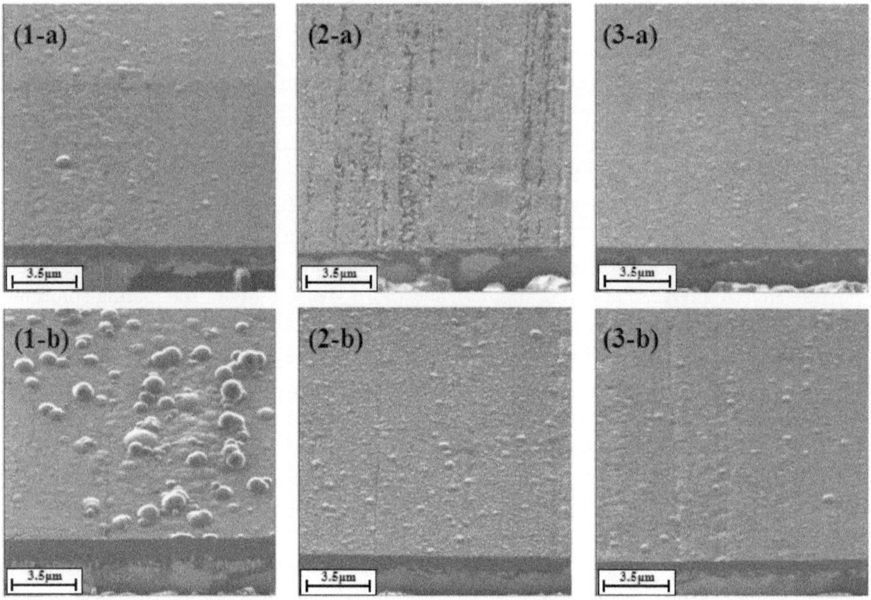

Figure 30. SIM image of Re-Ni-P films plated on the Cu substrate by (1-a) conventional ELP at 353 K and atmospheric pressure for 5 min, (2-a) ELP-SCE at 353 K and 15MPa for 180 min and (3-a) conventional ELP at 353 K and atmospheric pressure for 20 min and pH of plating solution is 4.0. (b) means SIM image that plates the BNP film being processed for pretreatment-A put at the same time as each (a) samples.

The effect of PPC proposed with EP-SCE and the mechanical agitation are the causes for decrease in the thickness of the diffusion layer of ELP-SCE. PPC for the sc-CO_2 phase to repeat adsorption and desorption on the surface of the plated film could appear as similar as that of the pulse electro deposition (PED).[52] Actually, Rahman have succeeded by using the effect of PPC in the sc-CO_2 emulsion and the perfluorocarbon surfactant in forming the porous film.[30] Size of the pores are roughly several μm or less depends on the size of the sc-CO_2 phase, which can be controlled by pressure, the amount of CO_2, and the amount of the surfactant. Moreover, the influence of the mechanical agitation is an important factor in the diffusion layer control. In the PED, the mechanical agitation and the duty cycle are optimized, and the thickness of the diffusion layer is controlled.[53] In ELP-SCE, the plating reaction field is made by making sc-CO_2 and the plating solution in an emulsion to lowered the viscosity. This gives the possibility to exert a difference in influence from the mechanical agitation like the conventional on the growth of the plated film.

2.2.6. Suppression of nodule growth in ELP-SCE

An extraordinary effect was confirmed for the morphological change of nodule (Fig. 32). This is the first report on of direct observation of the effect of nodule growth suppression by

ELP-SCE. Decrease in size of nodules and initiation of nucleation on the surface at random were observed though the plated film as the film grew up. The ditch parts were on the substrate before re-plating, which were filled after re-plating as shown in Fig.32 Random generation of the refined nucleus and leveling effect are similar to the phenomenon that happens because of PPC.[22] The effect of the leveling of ELP-SCE could be caused by the fact that the thickness of the diffusion layer is as thin as the size of these nodules. On the other hand, we discussed the influence of the proton for the growth suppression of nodule in the previous chapter. The re-plating experiment was done by conventional ELP that used the plating bath with pH adjusted to 4.0 by adding HCl (10 wt%) to confirm the influence of diffusion layer thickness of ELP-SCE and the proton. New nodules were observed to be hardly formed after re-plating as shown in Fig. 33. The plated film was observed to be from 183 nm to 254 nm as a result of measuring the width of ditch before and after re-plating, because of this re-plating. The growth of the convex part was suppressed under a plating condition even when it was thought that there was a thicker diffusion layer where spherical diffusion happened. However, all peculiar phenomena of ELP-SCE were not able to be shown. That is, the thickness change in the diffusion layer in addition to the influence of the proton greatly influences in ELP-SCE.

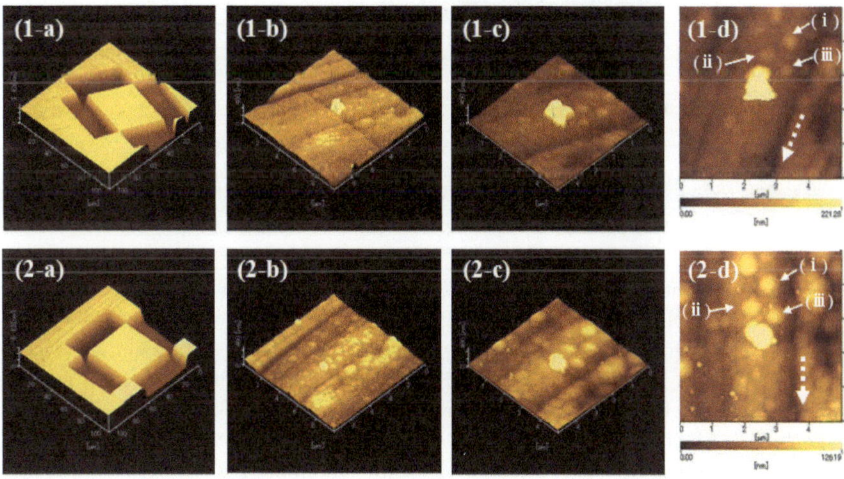

Figure 31. AFM images (1-a, b, c, d) of rectangular shaped Ni-P film plated by conventional ELP at 353 K and atmospheric pressure for 5 min, and (2-a, b, c, d) of re-plated Ni-P film plated by conventional ELP at 353 K and atmospheric pressure for 5 min on the (1-a) film. The magnification rises from (a) to (c). (d) is 2D image of (c). The dotted arrow indicates the direction where the width of nodules was measured.

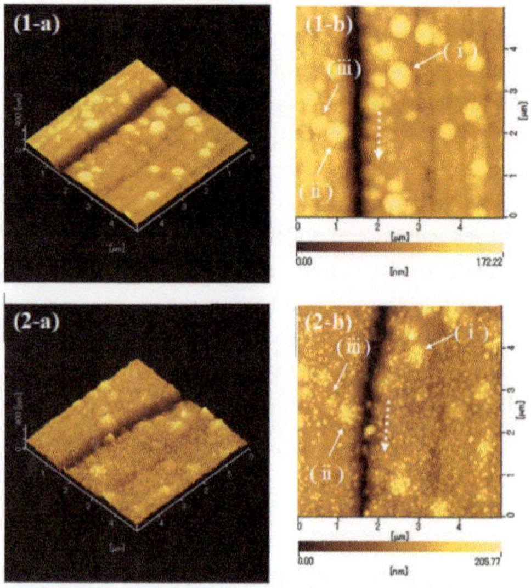

Figure 32. AFM images (1-a) of rectangular shaped Ni-P plating film plated by conventional ELP at 353 K and atmospheric pressure for 5min, and (2-a) of re-plated Ni-P film plated by ELP-SCE at 353 K and 15 MPa for 180 min on the (1-a) film. (number-b) is 2D image of (number-a). The dotted arrow indicates the direction where the width of nodules was measured.

Two influences mentioned above on the plating reaction field might be the formation factors of the refined nuclei observed. In the PED, the nucleation is promoted because the plating reaction can be performed at high overpotential, and it becomes easy for a minute crystallization to occur. On the other hand, the ELP that was not applied by the external power is the base reaction in ELP-SCE, and the electron necessary for the nucleation is only supplied by the reducing agent oxidized on the plated film. Also, crystallization overpotential that is higher than nuclear growth are necessary for the nucleation in the electrocrystallization.[54] Therefore, it will be necessary to supply a large amount of reducing agents to a reaction surface in ELP-SCE. The small film growth speed of ELP-SCE might have been caused by the reaction filed where the nucleation occur more frequently than nuclear growth. It causes high P content with plated film by ELP-SCE, moreover, because not only the reducing agent but also other reactive species such as the metal ions and protons are supplied easily voluminously so far because of PPC and the mechanical agitation.

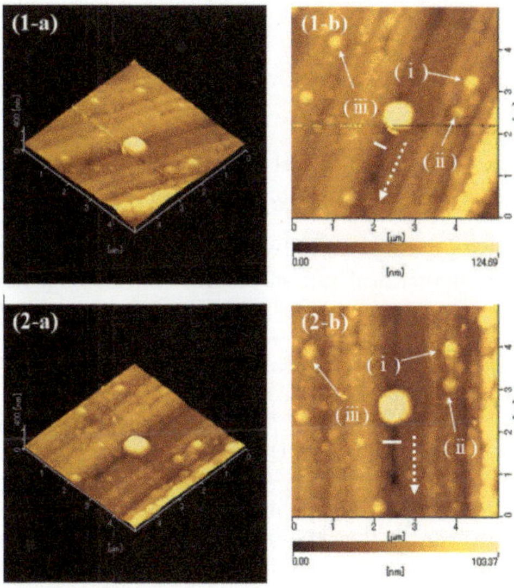

Figure 33. AFM images (1-a) of rectangular shaped Ni-P plating film plated by conventional ELP at 353 K and atmospheric pressure for 5min, and (2-a) of re-plated Ni-P film plated by conventional ELP at 353 K and atmospheric pressure for 20 min and pH 4.0 of plating solution on the (1-a) film. (number-b) is 2D image of (number-a). The dotted arrow indicates the direction where the width of nodules was measured

2.2.7. *Conclusion*

We examined a direct observation of selected nodule growth in ELP reaction using AFM on a square sample substrate of 50×50 μm fabricated by FIB. The Ni-P plated film is plated by the conventional method and ELP-SCE again on the Ni-P plated film with nodule formed. Changes in fine nodules and other areas at a specific position in the surface morphology were compared before and after the re-plating. In ELP-SCE, the dominant growth of nodules was suppressed and the nucleation occurred on the other surface of the nodules, although the convex part of nodules grew dominantly in conventional ELP using the electrolyte only.

Author details

Masato Sone*, Tso-Fu Mark Chang and Hiroki Uchiyama

*Address all correspondence to: msone@pi.titech.ac.jp

Precision & Intelligence Laboratory, Tokyo Institute of Technology, Yokohama, Japan

References

[1] Clifforo T. Fundamentals of supercritical fluids. United Kingdom: Oxford University Press; 1999.

[2] McHugh M, Krukonia V. Supercritical fluid extraction: principles and practice. United States of America: Butterworth Publishers; 1986.

[3] Shimoyama Y, Ogata Y, Ishibashi R, Iwai Y. Drying processes for preparation of titania aerogel using supercritical carbon dioxide. Chem. Eng. Res. Des. 2010; 88(10): 1427-1431.

[4] Gad-el-Hak M. The MEMS Handbook. Boca Raton, Fla.: CRC, Taylor & Francis; 2006.

[5] Schultze JW, Bressel A. Principles of electrochemical micro- and nano-system technologies. Electrochima Acta 2010; 47(1): 3-21.

[6] Ke J, Su W, Howdle SM, George MW, Cook D, Perdjon-Abel M, Bartlett PN, Zhang W, Cheng F, Levason W, Reid G, Hyde J, Wilson J, Smith DC, Mallik K, Sazio P. Electrodeposition of metals from supercritical fluids. P. Natl. Acad. Sci. 2009; 106(35): 14768-14772.

[7] Tsai WL, Hsu PC, Hwu Y, Chen CH, Chang LW, Je JH, Lin MH, Groso A, Margaritondo G. Electrochemistry: Building on bubbles in metal electrodeposition. Nature (2002); 417: 139.

[8] Howdle SM, Bagratshvili VN. The effects of fluid density on the rotational Raman-spectrum of hydrogen dissolved in supercritical carbon-dioxide. Chem. Phys. Lett. (1993); 214: 215-219.

[9] Darr JA, Poliakoff M. New directions in inorganic and metal-organic coordination chemistry in supercritical fluids. Chem. Rev. (1999), 99: 495-541.

[10] McClain JB, Betts DE, Canelas DA, Samulski ET, DeSimone JM, Londono JD, Cochran HD, Wignall GD, Chillura-Martino D, Triolo R. Design of nonionic surfactants for supercritical carbon dioxide. Science (1996); 274(5295): 2049-2052.

[11] Johnston KP, da Rocha SRP. Colloids in supercritical fluids over the last 20 years and future directions. J. Supercrit. Fluids (2009); 47(3): 523-530.

[12] Lee CT Jr, Ryoo W, Smith PG Jr, Arellano J, Mitchell DR, Lagow RJ, Webber SE, Johnston KP. Carbon dioxide-in-water microemulsions. J. Am. Chem. Soc. (2003);125(10): 3181-3189.

[13] Yoshida H, Sone M, Wakabayashi H, Yan H, Abe K, Tao XT, Mizushima A, Ichihara S, Miyata S. New electroplating method of nickel in emulsion of supercritical carbon dioxide and electrolyte solution to enhance uniformity and hardness of plated film. Thin Solid Film (2004); 446(2): 194-199.

[14] Yan H, Sone M, Sato N, Ichihara S, Miyata S. The effects of dense carbon dioxide on nickel plating using emulsion of carbon dioxide in electroplating solution. Surf. Coat. Tech. (2004); 182: 329-334.

[15] Wakabayashi H, Sone M, Sato N, Takada Y, Yan H, Abe K, Ichihara S, Miyata S. Nano-grain structure of nickel films prepared by emulsion plating with dense carbon dioxide. Surf. Coat. Tech. (2005); 190(2-3): 200-205.

[16] Jeong DH, Gonzalez F, Palumbo G, Aust KT, Erb U. The effect of grain size on the wear properties of electrodeposited nanocrystalline nickel coatings. Scr. Mater. (2001); 44(3): 493-499.

[17] Hugh GD, Smith SD, Pande CS, Johnson HR, Armstrong RW. Hall-petch strengthening for the microhardness of twelve nanometer grain diameter electrodeposited nickel. Scr. Metall. (1986); 20(1): 93-97.

[18] Rahman MdZ, Sone M, Eguchi M, Ikeda K, Miyata S, Yamamoto T. Wear properties of nickel coating film plated from emulsion with dense carbon dioxide. Surf. Coat. Tech. (2006); 201(3-4): 606–611.

[19] Chung ST, Tsai WT. Nanocrystalline Ni-C electrodeposits prepared in electrolytes containing supercritical carbon dioxide. J. Electrochem. Soc. (2009); 156(11): D457-D461.

[20] Rashidi AM, Amadeh A. The effect of saccharin addition and bath temperature on the grain size of nanocrystalline nickel coatings. Surf. Coat. Tech. (2009); 204(3): 353-358.

[21] Bakonyi I, Tóth-kádár E, Pogány L, Cziráki Á, Gerőcs I, Varga-Josepovits K, Arnold B, Wetzigdeh K. Preparation and characterization of d.c.-plated nanocrystalline nickel electrodeposits. Surf. Coat. Tech. (1996); 78(1-3): 124-136.

[22] Chang TFM, Sone M. Function and mechanism of supercritical carbon dioxide emulsified electrolyte in nickel electroplating reaction. Surf. Coat. Tech. (2011); 205(13-15): 3890-3899.

[23] Chang TFM, Sone M, Shibata A, Ishiyama C, Higo Y. Bright nickel film deposited by supercritical carbon dioxide emulsion using additive-free Watts bath. Electrochim. Acta (2010); 55(22): 6469-6475.

[24] Dhanuka VV, Dickson JL, Ryoo W, Johnston KP. High internal phase CO_2-in-water emulsions stabilized with a branched nonionic hydrocarbon surfactant. J. Colloid Interf. Sci. (2006); 298(1): 406-418.

[25] da Rocha SRP, Psathas PA, Klein E, Johnston KP. Concentrated CO_2-in-water emulsions with nonionic polymeric surfactants. J. Colloid Interf. Sci. (2001); 239(1): 241-253.

[26] Vesovic V, Wakeham WA. The Transport Properties of Carbon Dioxide. J. Phys. Chem. Ref. Data (1990); 19(3): 763-808.

[27] Yuan X, Wang Y, Sun D, Yu H. Influence of pulse parameters on the microstructure and microhardness of nickel electrodeposits. Surf. Coat. Tech. (2008); 202(9): 1895-1903.

[28] Choo RTC, Toguri JM, El-Sherik AM, Erb U. Mass transfer and electrocrystallization analyses of nanocrystalline nickel production by pulse plating. J. Appl. Electrochem. (1995); 25(4): 384-403.

[29] El-Sherik AM, Erb U, Page J. Microstructural evolution in pulse plated nickel electrodeposits. Surf. Coat. Tech. (1997); 88(1-3): 70-78.

[30] Rahman MdM, Sone M, Uchiyama H, Sakurai M, Miyata S, Nagai T, Higo Y, Yamamoto T. Novel porous film by electroplating with an emulsion of supercritical CO_2. Surf. Coat. Tech. (2007); 201(16-17): 7513-7518.

[31] Lee CT Jr, Ryoo W, Smith PG Jr, Arellano J, Mitchell DR, Lagow RJ, Webber SE, Johnston KP.Carbon dioxide-in-water microemulsions. J. Am. Chem. Soc. (2003);125(10): 3181-3189.

[32] Adkins SS, Chen X, Chan I, Torino E, Nguyen QP, Sanders AW, Johnston KP. Morphology and stability of CO_2-in-water foams with nonionic hydrocarbon surfactants. Langmuir (2010); 26(8): 5335-5348.

[33] Mallory G O, Hajdu J B, *Electroless Plating: Fundamentals and Applications*, American Electroplaters and Surface Finishers Society, Orlando, FL; (1990).

[34] Tracton A A, in *Coatings Technology Handbook*, 3nd ed., Vakelis A, Editor, p. 27-1, CRC Press Taylor & Francis Group, New York;(2006).

[35] Uchiyama H, Sone M, Ishiyama C, Endo T, Hatsuzawa T, Higo Y, Uniform Ni-P Film using an Electroless Plating Method with an Emulsion of Supercritical Carbon Dioxid.e J.Electrochem.Soc, (2007); 154; E91-E94.

[36] Imai Y, What is Supercritical Fluid ?, *Hyomen Gijutsu*, 2005; 56, 74-78.

[37] Watanabe H, Honma H, brication of Nickel Microbump on Aluminum using Electroless Nickel Plating *J. Electrochem. Soc.*, (1997); 144; 471.

[38] Chang S Y, Hsu C J, Fang R H, Lin S J, Electrochemical Deposition of Nanoscaled Palladium Catalysts for 65 nm Copper Metallization *J. Electrochem. Soc.*, (2003); 150; C603.

[39] Tsunoda T, Nagashima H, Nishinakayama H, Watanabe H, Honma H, Effect of Direct Electroless Nickel Plating Process on Surface Morphology of Nickel Plating, *Hyomen Gijutsu*, (2005);56; 463-467.

[40] Paunovic M, Schlesinger M, *Fundamentals of Electrochemical Deposition*, 2nd ed., John Wiley & Sons, Hoboken, NJ; (2006).

[41] Nakamaru Y, Joya T, Tashiro K, Honma H, Reduction of Pit and Nodule Defects on Thick Electroless Nickel Plating Film, *Hyomen Gijutsu*, (2009); 60; 661-667.

[42] Tashiro K, Yamamoto S, Hashimoto Y, Kawashima S, Honma H, Initial deposition morphologies of electroless nickel-phosphorus plating on a nonconductor and a conductor, *Hyomen Gijutsu,* (2002); 53; 459-465.

[43] Cheong W J, Luan B L, Shoesmith D W, The effects of stabilizers on the bath stability of electroless Ni deposition and the deposit, *Appl. Sur. Sci.,* (2004); 229; 282-300.

[44] Jacobs J W M, Rikken J M G, Oxygen-Diffusion-Size Effect in Electroless Metal Deposition, *J. Electrochem. Soc.,* (1988); 135; 2822-2827.

[45] Chandrasekar M S, Pushpavanam M, Pulse and pulse reverse plating—Conceptual, advantages and applications, *Electrochim. Acta.,* (2008); 53; 3313-3322.

[46] Hsu H H, Yeh J W, Lin S J, Repeated 3D Nucleation in Electroless Cu Deposition and the Grain Boundary Structure Involved, *J. Electrochem. Soc.,* (2003); 150, C813-C815.

[47] Uchiyama H, Shibata A, Sone M, Higo Y, Effects of CO_2 on Ni-P Electroless Plating in an Emulsion of Supercritical CO_2, *J.Electrochem.Soc,* (2010);157; D550-D552.

[48] Uchiyama H, Endo T, Sone M, Direct Observation of Nodule Growth on Electroless Ni-P Deposition in Supercritical CO_2 Emulsion, *J.Electrochem.Soc,* (2012); 159; D114-D118.

[49] Paunovic M, Schlesinger M, *Fundamentals of Electrochemical Deposition,* 2nd ed., John Wiley & Sons, Hoboken, NJ; (2006).

[50] Despic A R, Popov K I, *J. Appl. Electrochem.,* (1971); 1; 275.

[51] Chou Y H, Sung Y, Liu Y M, Pu N W, Ger M D, Defect-Free Copper Filling Using Nonisothermal Electroless Deposition with Fluorocarbon Surfactant, *J. Electrochem. Soc.,* (2008); 155; D791.

[52] Lam T L, Koike A, Ohno I, Haruyama S, *J. Met. Finish. Soc. Jpn.,* (1983); 34; 428.

[53] Kaishev R, Mutaftschiev B, An International Journal of the Science of Corrosion and Protection of Metals, *Electrochim. Acta.,* (1965); 10; 643-716

Chapter 10

Chemical Vapor Transport Reactions–Methods, Materials, Modeling

Peer Schmidt, Michael Binnewies,
Robert Glaum and Marcus Schmidt

Additional information is available at the end of the chapter

1. Introduction

1.1. Chemical vapor transport reactions

A variety of processes of crystal growth proceeds via the gas phase. A short comparative overview on gas phase transports is given here. However, in the main we deal with the concept of Chemical Vapor Transport Reactions [1, 2]. The term "*Chemical Vapor Transport*" (CVT) summarizes heterogeneous reactions which show one shared feature: a condensed phase, typically a solid, has an insufficient pressure for its own volatilization. But the pertinent phase can be volatilized in the presence of a gaseous reactant, the *transport agent*, and deposits elsewhere, usually in the form of crystals. The deposition will take place if there are different external conditions for the chemical equilibrium at the position of crystallization than at the position of volatilization. Usually, different temperatures are applied for volatilization and crystallization, Figure 1.

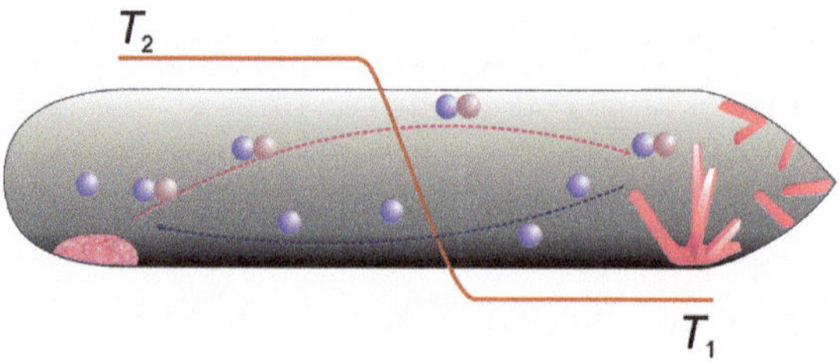

Figure 1. Scheme of CVT experiments for crystallization of solids in a temperature gradient.

Chemical vapor transport reactions address the formation process of pure and crystalline solids. Especially, the growth of single-crystalline material is of particular value because, among other things, it allows the determination of the crystal structure by diffraction methods. Beyond the aspect of basic research, chemical vapor transport reactions have also gained practical significance: they form the basis of the operating mode of halogen lamps. Furthermore, an industrial process is based on a chemical transport reaction, the Mond-Langer-Process for the production of ultrapure nickel [3]. Chemical vapor transports likewise occur in nature forming minerals without human influence, in particular at places of high temperatures. Bunsen was the first who observed and described it [4]. He noticed that the formation of crystalline Fe_2O_3 is associated with the presence of volcanic gases which contain gaseous hydrogen chloride. Van Arkel and de Boer were the first scientists who carried out specific transport reactions in the laboratory from 1925 onwards [5]. They were motivated by the huge interest in finding a process to fabricate pure metals like titanium at that time [6]. Van Arkel and de Boer used the so called *glowing wire method*. In the process, the contaminated metal *M* (e.g. a metal of the 4th group) transforms into a gaseous metal iodide (MI_n) in the presence of iodine as the transport agent. The iodide is formed at the metal surface in an exothermic reaction and vaporizes completely, thus reaching a glowing wire which was heated up to high temperatures. On the glowing wires surface, the back reaction (that is the endothermic reaction) is favored by Le Chatelier's principle. That way the decomposition of the metal iodide via the metals deposition proceeds and the metal is deposited on the hot wire.

A systematic research and description of chemical transport reactions was carried out by Schäfer in the 1950s and 1960s [1]. It became apparent that pure and crystalline species of various solids could be made with the help of chemical transport reactions: metals, metalloid, and intermetallic phases as well as halides, chalcogen halides, chalcogens, pnictides and many others. The current knowledge comprises of thousands of different examples for chemical vapor transport reactions. The results of different periods of investigations are recorded in some review articles [7-13]. Besides, the monographs [1, 2] and an extensive book chapter [14] give an overview on principles and applications of chemical transport reactions referred to the pertinent period of knowledge. To date the chemical vapor transport method developed to be an important and versatile preparative method of solid state chemistry.

Schäfer's endeavour also showed that chemical transport reactions follow thermodynamic regularities [15]; kinetic effects are rarely observed which makes a general description easier. Subsequently, the thermodynamic approaches for detailed description of chemical vapor transports became more sophisticated [16-22]. As a result, complex models for the description of vapor transports of phase mixtures, phases with variable composition, and transports with deposition sequences were established - the "Extended transport model" [18-21] and the "Co-operative transport model" [22]. Thus, the understanding of chemical vapor transport reactions is well developed; predictions on alternative transport agents, optimal reaction conditions and the amount of transported substance are possible and fairly easy accessible via computer programs [23, 24]. Indeed, the proper handling of these programs requires a profound knowledge on thermodynamic data (enthalpy, entropy, heat capacity) of all condenses and gaseous substances that are involved.

The following section shall provide an extensive overview on both principles and mechanisms of chemical vapor transport reactions and on characteristic examples of crystal growth of different substance classes by CVT. A simple thermodynamics basis is given in order to set you in ability to estimate the conditions of vapor transport experiments by own calculations; more complex calculations methods are presented for advanced investigations. Not at least, a short introduction for performing different CVT experiments (ampoule technique, oven setup, determination of transport rates, investigation of transport sequences,...) is given.

2. Vapor transport methods and principles

2.1. Overview on vapor transport methods

A vast number of reactions involving gas phases hardly differ from each other: If a condensed substance encounters a temperature gradient, it moves from the place of dissolution via the gas phase to the place of deposition, from **source** to **sink**. However, we do not "see" how the substance is led to the gas phase and deposited at another place. The mechanisms of gas phase transports can be deduced from experimental determination of the gas phase composition and/or from thermodynamic considerations of the pertinent heterogeneous equilibria between the solid and the gas phase [25].

Sublimation: Sublimations occurs without decomposition of the initial solid by forming only one dominating gas species. Substances showing sublimation are often solids constituting of molecular units, which are "bonded" by only weak interactions. The much stronger (covalent) bond in the molecular unit persist even under external energy stress, and the molecule can sublime undecomposed. Iodine, I_2 is a concise example of sublimation.

$$I_2(s) \rightleftarrows I_2(g) \quad (1)$$

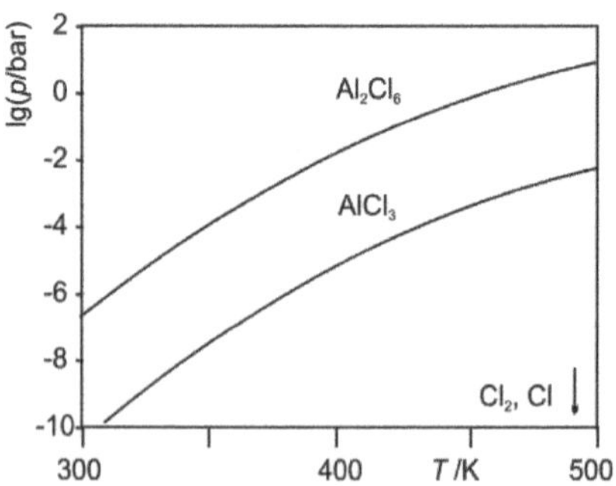

Figure 2. Composition of the gas phase of sublimation of $AlCl_3$.

The other possible gas species in the system, I(g) is of less importance due to the significantly lower partial pressures at the temperature of sublimation.

Saline solids can sublime, too. A well-known example is aluminum(III) chloride which is present in the gas phase in large proportion in form of dimeric molecule Al_2Cl_6, Figure 2. The additional systems gas species, such as $AlCl_3$(g), Cl_2(g), and Cl(g) show significantly lower partial pressures at the given temperature and thus not take part in the evaporation process.

$$2\ AlCl_3(s) \rightleftarrows Al_2Cl_6(g) \tag{2}$$

In a generalized form, the sublimation of a compound AB_x is described by equilibrium (3).

$$AB_x(s) \rightleftarrows AB_x(g) \tag{3}$$

Decomposition sublimation: If there is no stable molecular unit which is evaporable, a solid can decompose into various gaseous products while heating. Changing the equilibrium conditions, the initial solid can be recovered out of such a gas phase. This is called decomposition sublimation.

The gas phase transport of bimuth(III) selenide - Bi_2Se_3, an important constituent for thermoelectric materials - shows the characteristic of that. It decomposes into stoichiometric amounts of BiSe(g) and Se_2(g) in the vapor phase, the molecule of the initial composition Bi_2Se_3(g) does not occur in evaporation process, Figure 3. During cooling, the gas phase condenses completely and solid solid bimuth(III) selenide is formed (4).

$$Bi_2Se_3(s) \rightleftarrows 2BiSe(g) + 1/2\ Se_2(g) \tag{4}$$

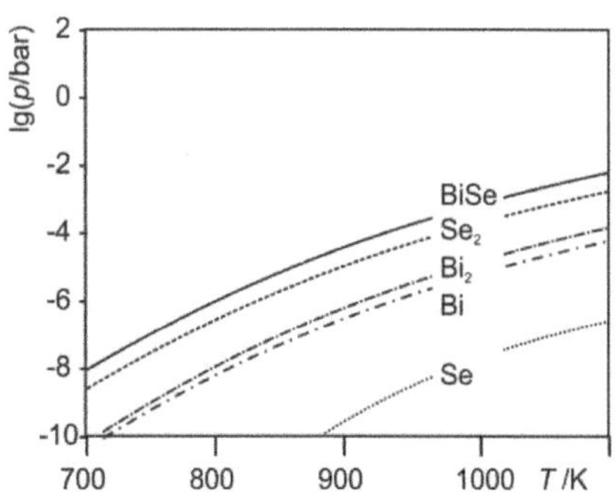

Figure 3. Composition of the gas phase of decomposition sublimation of Bi_2Se_3.

As a generalization, the decomposition sublimation of a compound AB_x can be described by the equilibria (5) or (6), respectively.

$$AB_x \rightleftarrows A(g) + xB(g) \tag{5}$$

$$AB_x \rightleftarrows AB_y(g) + (x-y)B(g) \tag{6}$$

The gas phase transport of Bi_2Se_3 by decomposition sublimation gives an example for congruent dissolution and condensation. Thus, a solid of always the same, constant composition is deposited. Nevertheless, a decomposition sublimation can be incongruent, too. Often, the product of an incongruent decomposition sublimation has the same composition as the initial solid. A simple example of this is copper(II) chloride. If heated at a running pump to several hundred degrees, the steam that is built in subsequent equilibria (7) and (8) contains the molecules CuCl, Cu_3Cl_3, Cu_4Cl_4 and Cl_2.

$$CuCl_2(s) \rightleftarrows CuCl(s) + 1/2\ Cl_2(g) \tag{7}$$

$$CuCl(s) \rightleftarrows CuCl(g) \tag{8}$$

The mechanism of decomposition sublimation is even more complex in the case of Bi_6Cl_7 [25, 26]. The initial solid is decomposed into a second solid – here elemental bismuth – and the dominating gas species $BiCl_3$ in equilibrium (9). Thereby, the compositions of the gas phase and thus the "solubility" of all components is unequal the initial composition of the solid, Figure 4.

$$3/7\ Bi_6Cl_7(s) \rightleftarrows 11/7\ Bi(s) + BiCl_3(g) \tag{9}$$

$$2\ Bi_6Cl_7(s) \rightleftarrows 11\ BiCl(g) + BiCl_3(g) \tag{10}$$

Nevertheless, the gas phase transport of Bi_6Cl_7 is realized by a second, subordinated equilibrium (10). In this case, the composition of the deposited solid strongly depends on the experimental conditions: Congruent deposition of Bi_6Cl_7 only occurs with low temperature gradients between source and sink. At higher temperature gradients pure $BiCl_3$ is deposited.

As has been shown, the mechanism of gas phase transport will get more complicated if the decomposition leads to a further condensed solid and a reactive gas phase. Subsequent, auto transport processes can result.

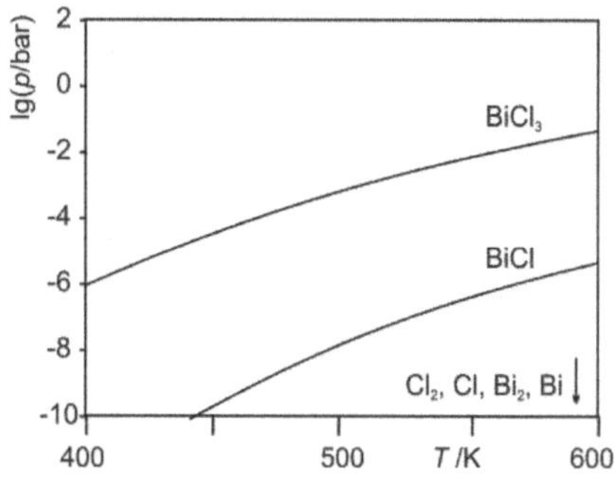

Figure 4. Composition of the gas phase of decomposition sublimation of Bi_6Cl_7.

Auto transport: The auto transport resembles the appearance of sublimation or decomposition sublimation. Nevertheless, the transported solid does not generate an effective partial pressure on its own. Rather, a transport agent is formed by (incongruent) thermal decomposition of the solid and the substance is transferred into the gas phase at a higher temperature without the addition of an external transport agent. The crystallization of $MoBr_3$ serves as an example of auto transport processes. The compound decomposes incongruently under formation of solid $MoBr_2$(s) releasing a gas phase of the dominant species $MoBr_4$ and Br_2 in equilibria (11) and (12), see Figure 5. The heterogeneous equilibria of $MoBr_4$ and Br_2 with the initial solid $MoBr_3$ can lead at higher temperature to the vapor transport equilibrium (13).

$$2MoBr_3(s) \rightleftarrows MoBr_2(s) + MoBr_4(g) \tag{11}$$

$$MoBr_3(s) \rightleftarrows MoBr_2(s) + 1/2\ Br_2(g) \tag{12}$$

$$MoBr_3(s) + 1/2\ Br_2(g) \rightleftarrows MoBr_4(g) \tag{13}$$

Based on this example, one can formulate the course of the auto transport in general terms [25]: A compound AB_x(s) does not generate a transport effective partial pressure of the gas species AB_x(g) or A(g)+B(g) (AB(g)+B(g)) on its own. The auto transport is based on two coexisting solid phases AB_x(s) and AB_{x-n}(s) as well as a gas phase which is generated through a decomposition reaction (14). The formed gaseous species B then reacts under the formation of only gaseous product AB_{x+n} in the sense of CVT reactions (15).

$$AB_x(s) \rightleftarrows AB_{x-n}(s) + nB(g) \tag{14}$$

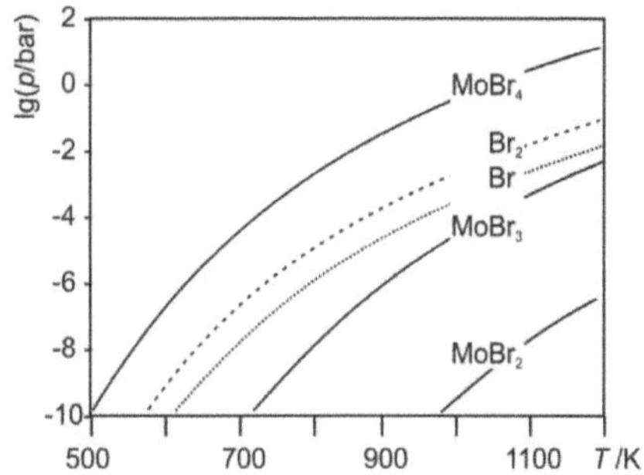

Figure 5. Composition of the gas phase for the auto transport of $MoBr_3$.

$$AB_x(s) + nB(g) \rightleftarrows AB_{x+n}(g) \tag{15}$$

Auto transports are generally endothermic reactions like sublimation and decomposition sublimation (deposition in the direction form source to sink: hot to cold). The transport equilibrium can only be effective if two conditions are met: First, the partial pressure of *B* must be sufficiently high and second, the transporting phase AB_x(s) must remain in equilibrium, thus AB_x(s) must not be decomposed completely. *B* stands for a gaseous decomposition product in the sense of equation (14). Thus, *B* can be an atom (e.g. an bromine atom), a homonuclear molecule (Cl_2, Br_2, O_2,...) or a heteronuclear molecule.

There may be a smooth transition of the described phenomena of sublimation or decomposition sublimation to the mechanism of auto transport. The dissolution of $CrCl_3$ in the gas phase represents such a complex behavior. One can find congruent sublimation, the formation of gaseous chromium(III) chloride, and of an incongruent decomposition at the same time. In a consecutive reaction chlorine can react with the primary solid $CrCl_3$, thus becoming the transport agent. The transport effective gaseous molecule is $CrCl_4$, Figure 6. In case of the chlorides $MoCl_3$ or VCl_3, the gas molecules MCl_3 are too unstable or unknown and the movement in the temperature gradient takes place only through auto transport according to the equilibria (16) and (17) (M = Cr, Mo, V).

$$MCl_3(s) \rightleftarrows MCl_2(s) + 1/2\ Cl_2(g) \tag{16}$$

$$MCl_3(s) + 1/2\ Cl_2(g) \rightleftarrows MCl_4(g) \tag{17}$$

The principle of auto transport is to apply also for other substance classes, such as oxides, chalcogenides and above all chalcogenide halides. As an example of the auto transport of an oxide, the crystallization of IrO_2 is presented. At temperatures of about 1050 °C the phase

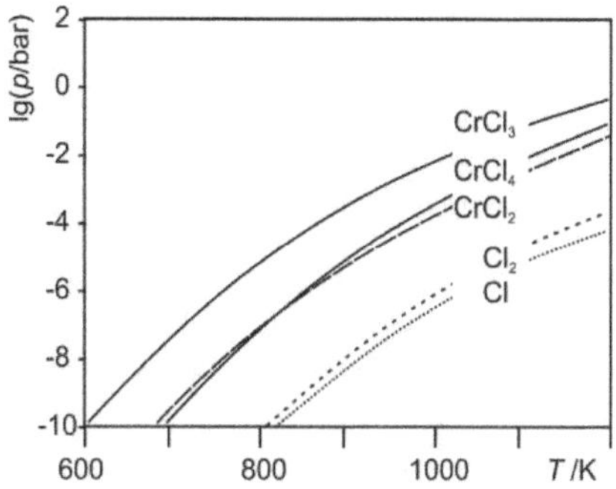

Figure 6. Composition of the gas phase for vapor transport processes of $CrCl_3$.

decomposes into the metal and molecular oxygen (18). In a subsequent heterogeneous equilibrium (19), oxygen reacts with the primary solid to form the transport effective gas species IrO_3, Figure 7. The back reaction takes place at lower temperature and IrO_2 is deposited.

$$IrO_2(s) \rightleftarrows Ir(s) + O_2(g) \tag{18}$$

$$2IrO_2(s) + O_2(g) \rightleftarrows 2IrO_3(g) \tag{19}$$

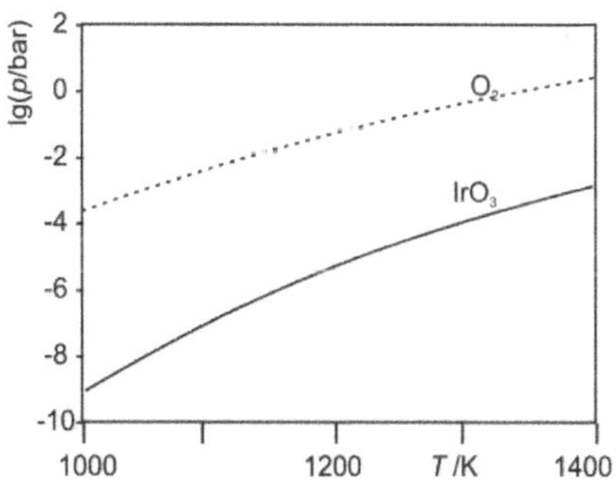

Figure 7. Composition of the gas phase for the auto transport of IrO_2.

Generally, all the auto transports are feasible as "regular" chemical vapor transport reactions. In these cases, the transport is possible as well through the addition of the transport agent without the preceding decomposition reaction. An important difference in both experiments can be observed: As the auto transport is based on a decomposition reaction, crystals of a different (metal rich) composition can be deposited. Thus the auto transport of IrO_2 leads to the deposition of IrO_{2-x} (depleted by oxygen). Otherwise, the regular vapor transport in an open systems oxygen stream ($p(O_2) = 1$ bar) leads to the formation of pure IrO_2 [1]. The transport behavior of RuO_2 is to discuss in the same way: by auto transport $RuO_{1.998}$ is obtained while $RuO_{2.0}$ crystallizes in vapor transport experiments with excess of oxygen [5]. Even if the change in composition is very small, the physical properties can differ significantly. Thus the specific electrical resistance of RuO_2 is of more than one order of magnitude higher than of RuO_{2-x} [27].

If at least one of the components of AB_x does not form a gas species with sufficient vapor pressure which is suitable for a substance transport (sublimation, decomposition sublimation, auto transport), the addition of an external transport agent will be necessary.

Chemical Vapor transport reaction: A chemical vapor transport reaction is characterized by the fact that another substance, the *transport agent*, is required for the dissolution of a solid in the gas phase. This characteristic is to illustrate by the example of vapor transport of zinc oxide by addition of chlorine. ZnO decomposes forming its own vapor phase with only small extent. Thus, the equilibrium pressure at a temperature of 1000 K is below 10^{-10} bar, Figure 8. By adding chlorine, the transport effective gas species $ZnCl_2$ and O_2 are formed in heterogeneous equilibrium (20). Here, the substances that appear in the vapor are different to those in the solid. Thereby transport effective vapor pressures ($p(i) > 10^{-4}$ bar) of the gas species can be observed already below 600 K, Figure 9.

$$ZnO(s) + Cl_2(g) \rightleftarrows ZnCl_2(g) + 1/2O_2(g) \quad (20)$$

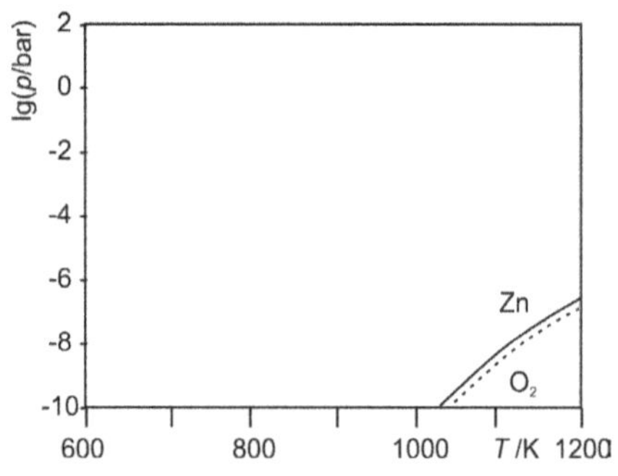

Figure 8. Composition of the gas phase for the thermal decomposition of ZnO.

The chemical vapor transport of ZnO is also possible by addition of hydrogen chloride. Likewise for the transport with chlorine $ZnCl_2$ is formed as the transport effective species for the transfer of zinc from source to sink, Figure 10. Otherwise the used transport agent HCl can react with oxygen, too. Thus the oxygen transferring species H_2O is formed in equilibrium (21).

$$ZnO(s) + 2HCl(g) \rightleftarrows ZnCl_2(g) + H_2O(g) \quad (21)$$

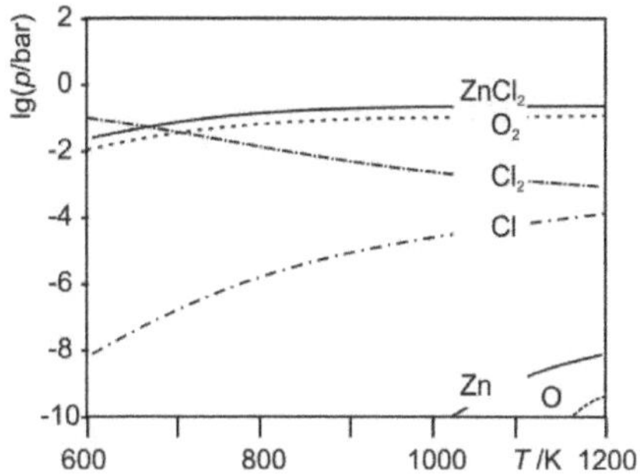

Figure 9. Composition of the gas phase for the CVT of ZnO with chlorine.

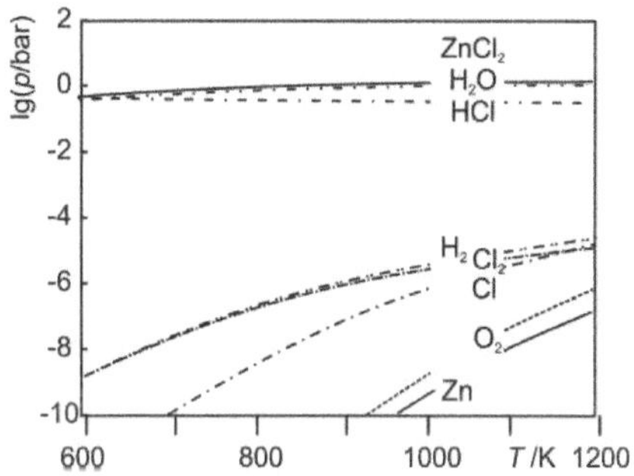

Figure 10. Gas phase composition for the CVT of ZnO with hydrogen chloride.

Here, the general principle of transport reactions can be seen clearly: The source material is transformed reversibly into gaseous products by the use of the transport agent. The transfer of the solid can be realized in different ways by formation of both heteronuclear species (like $ZnCl_2$ and H_2O) and atomar or homonuclear species (O_2).

$$AB_x(s) + nC(g) \rightleftarrows AC_n(g) + x/yB_y(g) \tag{22}$$

$$AB_x(s) + nCD(g) \rightleftarrows AC_n(g) + x/yB_yD(g) \tag{23}$$

2.2. Principles and thermodynamic considerations on CVT

In principle, two working methods are applied for the practical realization in the laboratory: the transport in open or closed systems. An open system is applied with an on both sides opened tube made of glass or ceramic material. Inside, a continuous flow of the transport agent is led over the source material; the solid, which is kept at a certain temperature, deposits at a different place with another temperature under the release of the transport agent. In a closed system, typically a sealed ampoule, the transport agent remains in the system and constantly re-enters the reaction. Thus, in a closed system, a much smaller amount of the transport agent is needed. In some cases only few milligrams of the transport agent are sufficient to cause a transport effect. In the laboratory one predominantly works with closed systems. An easy closed system is a sealed glass tube. Such a *transport ampoule* has a typical length of 100 to 200 mm and a diameter of 10 to 20 mm. It includes about one gram of the solid which is to be transported, and as much transport agent as is needed to raise the pressure in the ampoule to one bar or less during the reaction.

It is of prime interest for preparative working chemists whether a certain solid can be prepared with the aid of chemical vapor transport reactions, which transport agents are suitable and under which conditions a transport can be expected. At this point, we want to appoint some general qualitative considerations.

- **The suitable transport agent for the investigated system**

The vapor transport reaction has to realize, that all formed products are gaseous under the reaction conditions. Thus a suitable transport agent is to select, which can transfer all components of the initial solid into the gas phase.

- **The basic precondition for successful CVT**

The equilibrium position of the transport reaction must not be extreme, so that dissolution into the gas phase and re-condensation of the solid are possible under slightly changed experimental conditions. In cases of an extreme equilibrium no dissolution occurs (evaporation reaction unfavored) or the formation of gaseous products is not reversible (back reaction under re-condensation unfavored). In both cases no vapor transport is observed.

- **The suitable temperature**

The temperature at which the numerical value of the equilibrium constant K_p equals 1 ($\Delta_r G^0_T = 0$) is referred to as optimal transport temperature T_{opt} ($T_{opt} = \Delta_r H^0_T / \Delta_r S^0_T$).

- **The transport direction**

The transport is caused in almost every case by different temperatures and therefore changed equilibrium position in source and sink. It is common to characterize the volatilization (source) and the deposition temperature (sink) with T_1 and T_2, respectively, T_1 representing the lower temperature. The *transport direction* results from the sign of the reaction enthalpy of the transport reaction based on Le Chatelier's principle. Therefore, exothermic transport reactions always transport to the zone of higher temperature - from T_1 to T_2 ($T_1 \rightarrow T_2$), endothermic reactions transfer the solid to the cooler zone - from T_2 to T_1 ($T_2 \rightarrow T_1$).

- **The rate of mass transport**

A chemical vapor transport reaction can be divided into three steps: the *forward reaction* at the source material; the *gas motion;* and the *back reaction* leading to the formation of the solid in the crystallization zone. In most cases, the slowest and therefore the rate-determining step is the gas motion. At a pressure of about 1 bar, the gas motion mainly takes place through diffusion; thus, the diffusion laws determine the velocity of the whole process. As we are observing the diffusion of gases, it is practicable to introduce the partial pressure gradient dp/ds. Thus, the transported amount of substance per time is proportional to the partial pressure gradient.

To intensify the theoretical understanding of chemical vapor transport reactions in a comprehensible way the representative experiment of the transport of tungsten(IV) oxide is illustrated. With the help of the clear example of the transport of WO_2, the mentioned general considerations can be tackled:

The suitable transport agent for the investigated system. The solid WO_2 does not have an own measurable vapor pressure which would be suitable to transfer the compound to the gas phase in the sense of a sublimation. The phase rather decomposes at 1000 K with an oxygen partial pressure of 10^{-20} bar towards metallic tungsten. Here, the addition of a transport agent is necessary. As a general rule, the halogens chlorine, bromine and iodine (24) or halogen compounds (25), such as the hydrogen halides HX (X = Cl, Br, I) are suitable as transport agents. For the transport of WO_2 the gas species WO_2X_2 can be effective for vapor transport. Thus all the components of the system – tungsten, oxygen, and chlorine – are present in the gas phase. Finally, the transport equations can reflect the formation of only gaseous species.

$$WO_2(s) + X_2(g) \rightleftarrows WO_2X_2(g) \tag{24}$$

$$WO_2(s) + 2\,HX(g) \rightleftarrows WO_2X_2(g) + H_2(g) \tag{25}$$

Also, adding mercury halides, which are solid at room temperature, is potentially suitable to transport both components of the solid phase – tungsten as well as oxygen – into the gas phase (26).

$$WO_2(s) + HgX_2(g) \rightleftarrows WO_2X_2(g) + Hg(g) \tag{26}$$

At temperatures above 300 °C the mercury halides evaporate completely. Afterwards the gas species WO_2X_2 is formed in addition to mercury. Here, only gaseous species (WO_2X_2(g)+Hg(g)) are formed, too.

The basic precondition for successful CVT. Basic precondition for chemical vapor transport reactions is a balanced equilibrium position: For reactions which are described by *one* independent reaction equation, transports can be expected for equilibrium constants K_p in the range from 10^{-4} up to 10^4 respectively Gibbs energies Δ_rG^0 of approx. –100 to +100 kJ mol^{-1}. The partial pressure gradient Δp as a driving force for the material transport between dissolution and deposition site is achieved by a temperature gradient.

A highly exergonic reaction $\Delta_rG^0 < -100$ kJ mol^{-1} ($K_p > 10^4$) shows a high dissolution of the solid into the gas phase. That sounds great. But: The back reaction under deposition of the solid phase is not possible in thermodynamic terms. That means that on the source side the transporting compound is almost completely transferred into the gas phase without depositing on the sink side. During a highly endergonic reaction $\Delta_rG^0 > 100$ kJ mol^{-1} ($K_p < 10^{-4}$) the solid is hardly transferred into the gas phase, thus a transport cannot take place. With the help of thermodynamic data of the substances involved in the reaction, the values of the Gibbs energy, respectively the equilibrium constants of possible transport reactions can be calculated.

$$WO_2(s) + Cl_2(g) \rightleftarrows WO_2Cl_2(g) \tag{27}$$

$\Delta_rH^0_{1000} = -86.5\ kJ \cdot mol^{-1},\ \Delta_rS^0_{1000} = 73.2\ J \cdot mol^{-1} \cdot K^{-1}$

$\Delta_rG^0_{1000} = -159.7\ kJ \cdot mol^{-1},\ K_{p,\,1000} \approx 10^8$

$$WO_2(s) + Br_2(g) \rightleftarrows WO_2Br_2(g) \tag{28}$$

$\Delta_rH^0_{1000} = 13.0\ kJ \cdot mol^{-1},\ \Delta_rS^0_{1000} = 74.7\ J \cdot mol^{-1} \cdot K^{-1}$

$\Delta_rG^0_{1000} = -61.7\ kJ \cdot mol^{-1},\ K_{p,\,1000} \approx 10^3$

$$WO_2(s) + I_2(g) \rightleftarrows WO_2I_2(g) \tag{29}$$

$\Delta_rH^0_{1000} = 112.4\ kJ \cdot mol^{-1},\ \Delta_rS^0_{1000} = 84.6\ J \cdot mol^{-1} \cdot K^{-1}$

$\Delta_rG^0_{1000} = 27.8\ kJ \cdot mol^{-1},\ K_{p,\,1000} \approx 10^{-2}$

$$WO_2(s) + 2\ HCl(g) \rightleftarrows WO_2Cl_2(g) + H_2(g) \tag{30}$$

$\Delta_rH^0_{1000} = 31.0\ kJ \cdot mol^{-1},\ \Delta_rS^0_{1000} = 283.4\ J \cdot mol^{-1} \cdot K^{-1}$

$\Delta_rG^0_{1000} = -252.4\ kJ \cdot mol^{-1},\ K_{p,\,1000} \approx 10^{13}$

$$WO_2(s) + 2\ HBr(g) \rightleftarrows WO_2Br_2(g) + H_2(g) \tag{31}$$

$\Delta_r H^0_{1000} = 105.7\ kJ \cdot mol^{-1},\ \Delta_r S^0_{1000} = 296.1\ J \cdot mol^{-1} \cdot K^{-1}$

$\Delta_r G^0_{1000} = -190.4\ kJ \cdot mol^{-1} K_{p,1000} \approx 10^{10}$

$$WO_2(s) + 2\ HI(g) \rightleftarrows WO_2I_2(g) + H_2(g) \tag{32}$$

$\Delta_r H^0_{1000} = 174.6\ kJ \cdot mol^{-1},\ \Delta_r S^0_{1000} = 314.3\ J \cdot mol^{-1} \cdot K^{-1}$

$\Delta_r G^0_{1000} = -139.7\ kJ \cdot mol^{-1},\ K_{p,1000} \approx 10^7$

The calculations' results give a realistic outlook on the prospective results of transport experiments: Using halogens the transport with iodine seems to be promising (29). In the case of bromine, the transport seems at least possible (28) whereas chlorine causes an extreme equilibrium position under the formation of WO_2Cl_2(g) – a transport should not be possible (27). With the hydrogen halides equilibria are far on the side of the reaction products (30 – 32). This is due to clearly higher gain of entropy during the reaction. Although one can observe gradations in the equilibrium position for transports with HI and HBr compared to HCl, transports are principally not expected.

The transport of WO_2 with mercury halides seems possible for all three transport agents HgX_2 (X = Cl, Br, I): The equilibrium constants are within the limits of $10^{-4} < K_p < 10^4$. Thus, these systems are ideal to foster the understanding of a systematic approach and particularly to extend the understanding of chemical vapor transports. In the following, we shall focus on the transport of WO_2 with mercury halides. Using mercury bromide as a transport agent, the equilibrium position is least extreme (34) – in this case, the best transport results can be expected, see Figure 11. Using $HgCl_2$(g) for the transport, the equilibrium position is shifted to the right (33); that means, the solid WO_2 is transferred well into the gas phase. However, the deposition of the solid on the sink side is only possible to a limited degree. Even when the temperature is decreased, the equilibrium position is still on the product side. The equilibrium constant for the transport with HgI_2(g) indicates that the solid is hardly dissolved – the equilibrium position is shifted to the left (35). Thus, these are adverse conditions for a transport.

$$WO_2(s) + HgCl_2(g) \rightleftarrows WO_2Cl_2(g) + Hg(g) \tag{33}$$

$\Delta_r H^0_{1000} = 115.5\ kJ \cdot mol^{-1},\ \Delta_r S^0_{1000} = 171.9\ J \cdot mol^{-1} \cdot K^{-1}$

$\Delta_r G^0_{1000} - 56.4\ kJ \cdot mol^{-1},\ K_{p,1000} \approx 10^3 bar$

$$WO_2(s) + HgBr_2(g) \rightleftarrows WO_2Br_2(g) + Hg(g) \tag{34}$$

$\Delta_r H^0_{1000} = 190.5 kJ \cdot mol^{-1}$, $\Delta_r S^0_{1000} = 170.5\ J \cdot mol^{-1} \cdot K^{-1}$

$\Delta_r G^0_{1000} = 20.0\ kJ \cdot mol^{-1}$, $K_{p,\,1000} \approx 10^{-1} bar$

$$WO_2(s) + HgI_2(g) \rightleftarrows WO_2I_2(g) + Hg(g) \tag{35}$$

$\Delta_r H^0_{1000} = 249.6\ kJ \cdot mol^{-1}$, $\Delta_r S^0_{1000} = 179.2\ J \cdot mol^{-1} \cdot K^{-1}$

$\Delta_r G^0_{1000} = 70.4\ kJ \cdot mol^{-1}$, $K_{p,\,1000} \approx 10^{-4} bar$

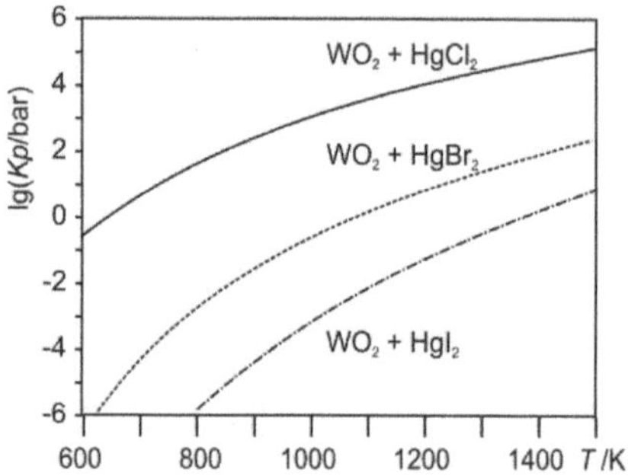

Figure 11. Equilibrium constants of transport reactions of WO_2(s) with HgX_2(g), X = Cl, Br, I.

The suitable temperature. The optimum average temperature [$\overline{T} = (T_2+T_1)/2$] for chemical vapor transports results from the requirement of $\Delta_r G^0 \approx 0$. If the thermodynamic data of the reaction are known, which can easily be obtained from the values of the involved species according to Hess's law, the optimum average temperature can be calculated from quotient of the reaction enthalpy and entropy (38). Vant`t Hoff`s equation (37) establishes the link between the equilibrium constant K and the thermodynamic data of the reaction enthalpy and entropy. The better the data, the more realistic are the results. With the help of standard data given for 298 K, the first estimation of the optimum transport temperature can be made. The results of this calculation are not to be met to the exact degree. One rather finds a range of ± 100 K which is suitable for the transport.

$$\Delta_r G_T^0 = - \Delta_r H_T^0 - T \cdot \Delta_r S_T^0 \tag{36}$$

$$\ln K = - \frac{\Delta_r H_T^0}{R \cdot T} + \frac{\Delta_r S_T^0}{R} \tag{37}$$

For the precondition of balanced equilibrium position at $K = 1$ ($\Delta_r G^0{}_T = 0$) results:

$$T_{opt} = \frac{\Delta_r H_T^0}{\Delta_r S_T^0} \tag{38}$$

Through differences in the temperatures of the source and sink side, the equilibrium position is brought towards the gaseous products when dissolving and shifted towards the solid when deposing. Calculations of the equilibrium constants were first made for an average temperature of 1000 K. If the temperatures vary, one will get the typical courses of the curve (see Figure 11). If the temperature is decreased, the equilibrium position in the transport system with $HgCl_2$ becomes less extreme. In contrast, the equilibrium position for the transport with HgI_2 becomes more favorable when the temperature is increased above 1000 K. The optimum, average temperature resulting from the quotient of the reaction enthalpy and entropy for the transport with $HgCl_2$ is at about 700 K respectively 400 °C; with $HgBr_2$ at about 1100 K respectively 800 °C and with HgI_2 1400 K (1100 °C, respectively). In this case, the calculation of the temperature on the basis of the standard values at 298 K as well as of the derived values for 1000 K lead to the same results; which means that an estimation is possible with simple calculations, (39 – 41).

$$WO_2(s) + HgCl_2(g) \rightleftharpoons WO_2Cl_2(g) + Hg(g) \tag{39}$$

$\Delta_r H^0{}_{298} = 122.5\ kJ \cdot mol^{-1},\ \Delta_r S^0{}_{298} = 183.6\ J \cdot mol^{-1} \cdot K^{-1}$

$T_{opt} = 125500\ J \cdot mol^{-1} / 183.6\ J \cdot mol^{-1} \cdot K^{-1}$

$T_{opt} \approx 700\ K\ respectively\ 400\ ^\circ C$

$$WO_2(s) + HgBr_2(g) \rightleftharpoons WO_2Br_2(g) + Hg(g) \tag{40}$$

$\Delta_r H^0{}_{298} = 1881\ kJ \cdot mol^{-1},\ \Delta_r S^0{}_{298} = 170.0\ J \cdot mol^{-1} \cdot K^{-1}$

$T_{opt} \approx 1100\ K\ respectively\ 800\ ^\circ C$

$$WO_2(s) + HgI_2(g) \rightleftharpoons WO_2I_2(g) + Hg(g) \tag{41}$$

$\Delta_r H^0{}_{298} = 238.4\ kJ \cdot mol^{-1},\ \Delta_r S^0{}_{298} = 165.4\ J \cdot mol^{-1} \cdot K^{-1}$

$T_{opt} \approx 1400\ K\ respectively\ 1100\ ^\circ C$

The transport direction. If a transport operation can be described in good approximation by only *one* reaction, the direction of the transport results from the heat balance of the heterogeneous equilibrium according to the van 't Hoff equation respectively to the Clausius–Clapeyron relation (42):

$$\frac{\mathrm{d}\ln K_p}{\mathrm{d}\frac{1}{T}} = -\frac{\Delta_r H_T^0}{R} \tag{42}$$

In a reaction with negative reaction enthalpy (exothermic dissolving reaction), the equilibrium constant K_p increases with decreasing temperatures – thus dissolution takes place at low, deposition at high temperatures. To put it in other words: The transport is directed to the hotter zone ($T_1 \rightarrow T_2$).

$$\Delta_r H^0{}_T < 0;\ d\ lnK_p \sim d1/T \tag{43}$$

In a reaction with positive reaction enthalpy (endothermic dissolving reaction), K_p increases with increasing temperatures – so dissolution takes place at higher, deposition at lower temperatures. Now, the transport proceeds to the cooler zone ($T_2 \rightarrow T_1$).

$$\Delta_r H^0{}_T > 0;\ d\ lnK_p \sim dT \tag{44}$$

The transport direction results only from the reaction enthalpy which is why the conclusion of all three investigated transport systems of WO_2 is clear: The reaction enthalpy is positive in each case – a transport to the cooler zone results. The total amount of the reaction enthalpy does not affect the decision if a transport is carried out. If the reaction enthalpy is close to zero one has to check the accuracy of the used data as they can contain errors of 10 to 20 kJ mol^{-1}.

The rate of mass transport. The substance transport via gas motion between dissolution and deposition site takes place by diffusion or convection. If the ampoule lies horizontally and if the total pressure is between 10^{-3} bar and 3 bar, the substance transport is affected by diffusion [1, 28]. In most cases, the diffusion is the velocity determining step, as it is much slower than the heterogeneous reaction of the solid with the transport agent. Using pressures above 3 bar, convection becomes dominating [28].

The term "*transport rate*" expresses the amount of deposited substance per time in the sink. For transports which run by only *one* reaction, one can describe the transport rate by Schäfer's transport equation (45) under the precondition that the chemical transport is solely determined by diffusion[1)] [1, 15]. High values of Δp result in a high transport rate. A large cross section effects the transport rate positively as does a short transport distance. According to the transport equation a high average temperature is formally advantageous for the transport rate; however, the influence of the temperature on the equilibrium constant, and thus Δp is more essential.

$$\dot{n}(A) = \frac{n(A)}{t'} = \frac{i}{j} \cdot \frac{\Delta p}{\Sigma p} \cdot \frac{\bar{T}^{0.75} \cdot q}{s} \cdot 0.6 \cdot 10^{-4} \left(mol \cdot l^{-1} \right) \quad (45)$$

$\dot{n}(A)$: Transport rate /mol h^{-1}

i,j,k stoichiometric coefficients in transport equation: *i* A(s)+*k* B(g) = *j* C(g)+…

Δp: partial pressure difference of the *transport effective* species C /bar

Σp: total pressure /bar

$\bar{T}$ average temperature along the diffusion path /K (practically $\bar{T}$ results as an average of T_1 and T_2)

q: cross-section of the diffusion path /cm^2

s: length of the diffusion path /cm

t' duration of the transport experiment /h

In most cases instead of the diffusion factor 0.6 10^{-4} a value of 1.8 10^{-4} is given which found entrance to the literature [1]. According to recent findings the factor 0.6 10^{-4} results in a smaller numerical value of the diffusion coefficient and corrects a mathematical error.

The calculation of transport rates for WO_2 by Schornstein and Gruehn [29, 30] at first show a clear dominance of transports with $HgBr_2$ in the average temperature range: The expected transport rates are ten times higher than for transports with $HgCl_2$ and HgI_2. Due to the balanced position of the equilibrium, high differences of partial pressures occur between the source and the sink. This way, the driving force for diffusion of the gas particles is high and thus for the substance transport as well. For the transport with $HgCl_2$ the transport rate decreases with increasing temperatures. As we have already seen, the equilibrium position, which is far to the right side, is responsible for it. Only if the temperature decreases, the equilibrium position can move to the left. The resulting, higher differences of partial pressures between dissolution and deposition side cause increasing transport rates at low temperatures. Using mercury iodide as transport agent, the equilibrium position is on the side of the source material at low temperatures. By increasing the temperature the equilibrium position is shifted to the side of the reaction products, the transport rate increases, Figure 12.

Corresponding to the simple estimation of the transport behavior of WO_2 with mercury halides, one gets the best results with the addition of $HgBr_2$. The chemical vapor transport of mercury bromide is possible in a wide temperature range. Transport rates above 30 mg h^{-1} are achievable, Figure 13. Temperatures of the source side of about 800 °C and of the sink side of 720 °C prove optimum. This result confirms the estimations of the optimum transport temperature. Due to the shift of the equilibrium position, the transport rate decreases at both, rising temperatures (880 → 800 °C respectively 960 → 880 °C; and falling temperatures (720 → 640 °C) [2]. Transports with $HgCl_2$ and HgI_2 clearly show smaller transport rates. Experiments with mercury iodides must be realized with higher temperatures according to the estimation. Temperatures up to 1000 °C are practicable; above, the silica glass ampoule will be heavily damaged by re-crystallization. Using an average transport temperature of 940 °C, transport

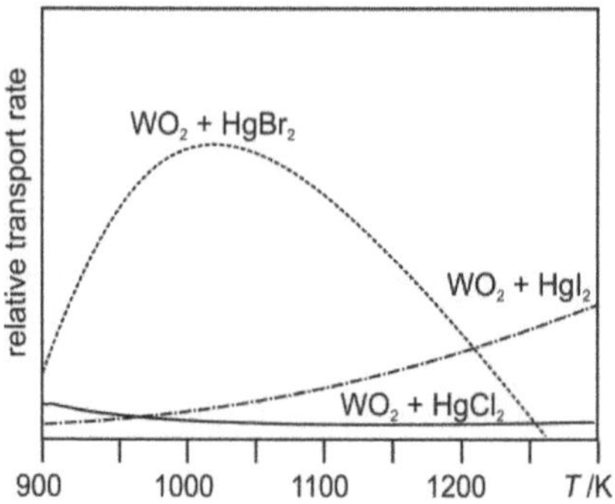

Figure 12. Progression of theoretical transport rates during the transport of WO_2 with HgX_2 (X = Cl, Br, I) according to [29, 30].

rates of up to 15 mg h^{-1} can be achieved, Figure 13. The transport rate decreases drastically with falling temperatures. With an average temperature of 640 °C the rate is even lower than 1 mg h^{-1}. Transport experiments with $HgCl_2$ show worst results as far as the transport rate is concerned: According to the calculation, lower temperatures are principally more favorable, however, in the range from 500 to 700 °C the transport rates are only in the range of 1 mg h^{-1}. The transport almost grinds to a halt at higher temperatures.

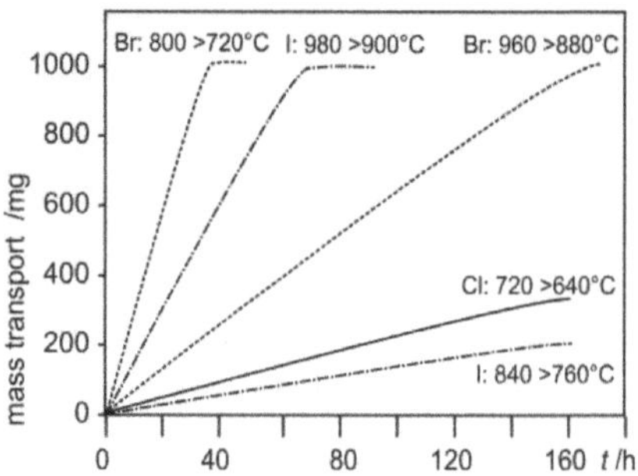

Figure 13. Experimental transport rates during the transport of WO_2 with HgX_2 (X = Cl, Br, I) according to [2].

One can come to a completely different evaluation if the quality of the crystals instead of the transport rate is given prominence. Relatively high transport rates cause uncontrolled nucleation and crystal growth. As a consequence, one gets highly epitaxial and rose-shaped crystal agglomerations for transports with $HgBr_2$. Frequently *one* compact solid of these epitaxial crystallites of WO_2 is found in the sink. Using average temperatures of approx. 800 °C in the transport system with HgI_2, one gets isolated, rod-shaped crystals of up to 1 mm edge length, Figure 14. The preparation of mono-crystals for crystal structure analysis is possible from these approaches, even though not every crystal is suitable. The low transport rate of 1 to 2 mg h^{-1} makes an undisturbed nucleation and crystal growth possible in this case. In the process, smaller, highly defected crystallites are dissolved in favor of other individuals.

Figure 14. Typical crystal morphology of single crystallites formed during the transport of WO_2 with HgI_2.

Finally, in selecting the transport agent, the temperature, and the temperature gradient, respectively, one should consider the aim of the transport. A high transport rate is undoubtedly advantageous for the synthesis of a compound or the purification of it. If crystals are to be grown, keep in mind the crystal quality and therefore rather choose a smaller transport rate.

Conclusion. Chemical vapor transport reactions are predictable. Simple estimations concerning the feasibility and the course of transport reactions are already possible with a basic understanding of the method and its thermodynamic backgrounds. It is worth the effort in every case – one avoids unnecessary experiments using "trial and error" procedure. Conclusions by analogy between similar transport systems are helpful as a first guide. However, the factual favorable parameter for a transport system should be estimated carefully in advance

3. Chemical vapor transport reactions – applications

3.1. Transport agents and gas species

It is of prime interest for preparative working chemists whether a certain solid can be prepared by chemical transport reactions, which transport agents are suitable and under which conditions a transport can be expected. If one wants to use transport reactions only in a preparative way – without the purpose of understanding the course of the reaction in detail – often it is sufficing to check on an empiric basis which solid can be transported by using what kind of transport agent. A further, quantitative description of the transport reaction requires knowledge of the thermodynamic data of the condensed phases and gaseous molecules that are involved. In this section, we will provide a short overview of the different kinds of gaseous inorganic molecule that can occur during chemical vapor transport reactions. Under the precondition of formation of only gaseous species, transport agents and transport effective species share the property of high volatility under experimental conditions. Thus, especially halogens and halogen compounds are qualified. Some elements, hydrogen compounds, and oxygen compounds are suitable as transportable species, too.

Halogen compounds. Due to its volatility, metal halides and non-metal halides play a central role for chemical transport reactions. Thus, halogens and many halogen compounds are effective and often used transport agents.

Halogens and halogen compounds as transport agents. The elemental halogens chlorine, bromine and iodine are used frequently as transport agent. Fluorine, in contrast, is not suited due to extreme equilibrium position of formation of fluorides in most cases. Besides, there are material problems because fluorine reacts with the ampoule materials if silica glass is used. Under the pertinent transport conditions, chlorine, bromine, and iodine react with solids of different substance classes, e.g. with metal, intermetallic compounds, semi-metals, metal oxides, sulfides, selenides, tellurides, nitrides, phosphides, arsenides, antimonides, silicides, germanides, some metal halogens and more. In the process, gaseous metal halides respectively semi-metal halides and the respective non-metal are formed as a rule. In some cases, non-metal halides are formed. Thus, during a reaction of metal phosphides with a halogen or a halogen compound, not only metal halides but also phosphoric halides can occur. Due to the fact that halides show oxidizing characteristics, metal halides or oxide halides are often formed as transport effective species in which the metal has a higher oxidation level than in the solid as shown for the transport of chromium(III) chloride with chlorine (46) or tungsten(IV) oxide with iodine (47).

$$CrCl_3(s) + 1/2\ Cl_2(g) \rightleftarrows CrCl_4(g) \tag{46}$$

$$WO_2(s) + I_2(g) \rightleftarrows WO_2I_2(g) \tag{47}$$

Hydrogen halides are versatile transport agents. The oxidation levels of the metal in the solid and in the transport effective gas species are generally equal because hydrogen halides do not have an oxidizing effect. Hydrogen halides are often used during the transport of oxides. Here, the gaseous metal halide and water vapor are formed. Halogen compounds, such as $TeCl_4$, PCl_5, $NbCl_5$ or $TaCl_5$ are also useful transport agents, especially for metal oxides. Reactions of the mentioned chlorides lead on the one hand to the formation of gaseous metal chloride or metal oxide chloride, on the other hand oxygen is fixed in form of volatile oxides (TeO_2, P_4O_6, P_4O_{10}) or oxide chlorides ($TeOCl_2$, $POCl_3$, $NbOCl_3$, $TaOCl_3$). Oppermann was able to show that tellurium(IV) chloride is a particular versatile transport agent [31]. According to the basic works of Schäfer, gaseous metal respectively semi-metal halides are formed as transport effective species during the reaction of different solids with halogens or halogen compounds [1].

$$Co(s) + I_2(g) \rightleftarrows CoI_2(g) \tag{48}$$

$$Si(s) + SiI_4(g) \rightleftharpoons 2\ SiI_2(g) \tag{49}$$

$$TiO_2(s) + 4\ HCl(g) \rightleftharpoons TiCl_4(g) + 2\ H_2O(g) \tag{50}$$

$$CrCl_3(s) + 1/2\ Cl_2(g) \rightleftharpoons CrCl_4(g) \tag{51}$$

$$2\ InP(s) + InI_3(g) \rightleftharpoons 3\ InI(g) + 1/2\ P_4(g) \tag{52}$$

The vapor of metal halides can consist of monomeric, dimeric and/or oligomeric molecules. With *M* = Al, Ga, In, Fe, Sc, Y, *Ln* (*Ln* = lanthanoides), one can observe particular large amounts of dimers M_2X_6. Trimers occur with copper(I) halides and silver halides. Metal halides of different components *A* and *B* can react in the gaseous state under the formation of *gas complexes* [32]. Such gas complexes are also formed during heterogeneous reactions between solid and gaseous halides. For example, gaseous aluminum(III) chloride reacts with a number of heavy volatile, solid metal chlorides under the formation of gas complexes. This way, their volatility is massively increased, often by orders of magnitude. These reactions can be used in vapor transport experiments [33]. This way, cobalt chloride can be transported with aluminum(III) chloride (53) far below the boiling temperature (e.g. 400 → 350 °C) [34]. To date, numerous of these examples are known [34 - 37]. Thermodynamic data were determined for a considerable number of gas complexes [38] and empirical rules which can help estimating the thermodynamic data of gas complexes are laid down [36, 38].

$$CoCl_2(s) + Al_2Cl_6(g) \rightleftharpoons CoAl_2Cl_8(g) \tag{53}$$

Oxide halides. Gaseous oxide halides are known of few metals only, in particular of transition elements [39]. These molecules appear in particular in case of metals with high oxidation numbers. VOX_3, $NbOX_3$, $TaOX_3$ (X = F…I), CrO_2X_2 (X = F…Br), MoO_2X_2, WO_2X_2 (X = F…I), $MoOX_4$, WOX_4 (X = F…Br), $ReOCl_4$, OsO_2Cl_2, $OsOCl_4$, RuOCl, ReO_3X (X = F…I). Gaseous oxide halides can play an important part as transport effective species during the transport of oxides [40].

$$SiO_2(s) + 2\ TaCl_5(g) \rightleftharpoons SiCl_4(g) + 2\ TaOCl_3(g) \tag{54}$$

$$MoO_3(s) + I_2(g) \rightleftharpoons MoO_2I_2(g) + 1/2\ O_2(g) \tag{55}$$

Some gaseous oxide halides are known of main group metals. Elements of group 13 form oxide halides, such as AlOCl, at very high temperatures around 2000 °C. Phosphorus forms several oxide chlorides and -bromides that are stable at high temperatures: POX_3, PO_2X, and POX (X = Cl, Br) [42]. For arsenic and antimony AsOCl and SbOCl are stable at high temperatures, however, oxide halides with the metal in the oxidation stage V are not known [42]. $TeOCl_2$ is the most important oxide halide of the main group elements. This gas species plays an important role during the transport of numerous oxides with tellurium(IV) chloride [31].

Elements in gaseous state. The importance of elemental halogens for chemical vapor transport reactions was already mentioned. Other elements also often occur in gaseous state during transport reactions, especially gaseous non-metals. Gaseous species of elements of groups 15 and 16 are formed during the transport of pnictides or chalcogenides with halogens (56) or halogen compounds in many cases. In contrast, metal gas species play a role in only a few reactions. In particular, this behavior occurs, if the transport agent reacts with the non-metal instead of the metal atoms and the metal is sufficient volatile (57). The formation of an unsaturated metal vapor during transport reactions can only be expected if the boiling temperature of the metal is below approx. 1200 °C. This only applies to the following metals: Na (881 °C), K (763 °C), Rb (697 °C), Cs (657 °C), Mg (1093 °C), Zn (906 °C), Cd (766 °C), Hg (356 °C), Yb (1194 °C) and Te (989 °C).

$$ZrAs_2(s) + 2\,I_2(g) \rightleftarrows ZrI_4(g) + 1/2\,As_4(g) \tag{56}$$

$$ZnO(s) + H_2(g) \rightleftarrows Zn(g) + H_2O(g) \tag{57}$$

In some cases, gaseous elements can work as transport agents. Hence, oxygen can cause the transport of some platinum metals [40]. Sulfur can transport a series of transition metal sulfides [43]. Here, gaseous polysulfides, such as TaS_3, are assumed transport effective species. There are similar observations for the chemical transport of some selenides. Sulfur is an effective transport agent for tellurium as well [44]. Compounds in which tellurium atoms were integrated in the different ring-shaped sulfur molecules were detected as transport effective species. Phosphorus can transport gallium phosphide, GaP, and indium phosphide, InP, probably via GaP_5 respectively InP_5 as transport effective species [45]. With the help of arsenic, the transport of gallium arsenide, GaAs, and indium arsenide, InAs, succeeded in a similar way [46].

Metal vapors predominantly consist of the atoms. The fraction of bi- or polyatomic molecules in the saturated vapor is between 10^{-5} and 10 % [47]. In contrast, the vapors of non-metals, apart from noble gases, consist of very stable polyatomic molecules which appear in the gas phase in great amounts, atoms appear only subordinated: N_2, P_4, P_2, As_4, As_2, Sb_4, Sb_2, O_2, S_2, S_3 … S_7, S_8, Se_2, Se_3 … Se_7, Se_8, Te_2, Cl_2, Br_2, and I_2. The ratio of different molecular species in the vapors of non-metals depends on the temperature and the pressure. Higher temperature and lower pressures abet the formation of small molecules respectively atoms.

Hydrogen compounds. Hydrogen compounds of non-metals play an important role for chemical vapor transport reactions. Hydrogen halides, which are often used as transport agents, are particularly important; for example during the transport of metal oxides. The example (58) shows that water vapor becomes transport effective species. However, water can also function as transport agent as for the transport of molybdenum(VI) oxide (59) and the one of germanium (60). Additionally, water can lead to the formation of transport effective gaseous hydroxides (61):

$$MgO(s)+2\ HCl(g) \rightleftarrows MgCl_2(g)+H_2O(g) \tag{58}$$

$$MoO_3(s)+H_2O(g) \rightleftarrows H_2MoO_4(g) \tag{59}$$

$$Ge(s)+H_2O(g) \rightleftarrows GeO(g)+H_2(g) \tag{60}$$

$$Li_2O(s)+H_2O(g) \rightleftarrows 2LiOH(g) \tag{61}$$

The transport reaction of tungsten with water and iodine is an important one in daily life. This reaction provides the basis of the operating mode of halogen lamps.

$$W(s)+2\ H_2O(g)+3\ I_2(g) \rightleftarrows WO_2I_2(g)+4\ HI(g) \tag{62}$$

Traces of water, often from the wall of the silica glass tubes, which were used during the transport, can be important for transport effects [48]. Hydrogen sulfide and hydrogen selenide also appear during the transport of sulfides respectively selenides with hydrogen halides. Hydrogen telluride is too unstable to develop under transport conditions. Ammonium chloride is particularly important. It decomposes to ammonia and hydrogen chloride during sublimation. Thus, it is a hydrogen chloride source which is easy to handle and easy to dose. Ammonia decomposes to the elements at higher temperature and thus creates a reducing atmosphere which effects the equilibria involved in the transport in different ways.

Oxygen compounds. The large amount of metal oxides decomposes completely or partly while heating to high temperatures. Gingerich provides a compilation of gaseous metal oxides and their stability [47]. Some metal oxides vaporize congruently: CrO_3, MoO_3, WO_3, Re_2O_7, IrO_3, RuO_3, RuO_4, OsO_4, GeO, SnO, and PbO. In the vapors of CrO_3, MoO_3 and WO_3 trimeric molecules appear as M_3O_9. SnO and PbO form dimers and trimers. Nevertheless, gaseous metal oxides play a subordinated role for chemical vapor transport reactions [40, 49]:

$$OsO_2(s) + OsO_4(g) \rightleftarrows 2\ OsO_3(g) \tag{63}$$

$$Pt(s) + O_2(g) \rightleftarrows PtO_2(g) \tag{64}$$

The role of non-metals is more important. Carbon monoxide can function as transport agent in different ways [50]:

$$SnO_2(s) + CO(g) \rightleftarrows SnO(g) + CO_2(g) \tag{65}$$

$$Ni(s) + 4\ CO(g) \rightleftarrows Ni(CO)_4(g) \tag{66}$$

Non-metal oxides occur as *transport effective species* for the transport of oxides. Thus, sulfuric vapor can cause in presence of iodine the transport of tin(IV) oxide [51].

$$SnO_2(s) + I_2(g) + 1/2\ S_2(g) \rightleftarrows SnI_2(g) + SO_2(g) \tag{67}$$

During the chemical transport of metal oxides, tellurium(IV) chloride plays a particular role, it reacts under the formation of a metal chloride or a metal oxide chloride and binds oxygen in form of TeO_2(g) or $TeOCl_2$(g) at the same time [31].

$$TiO_2(s) + TeCl_4(g) \rightleftarrows TiCl_4(g) + TeO_2(g) \tag{68}$$

$$MoO_3(s) + TeCl_4(g) \rightleftarrows MoO_2Cl_2(g) + TeOCl_2(g) \tag{69}$$

Further gaseous oxides that are important for the transport of oxide compounds are amongst others: B_2O_3, SiO, P_4O_6, P_4O_{10}, As_4O_6, Sb_4O_6, SeO_2.

Other substance groups. During some transport reactions, gaseous sulfides, selenides, tellurides or sulfide halides are of particular importance. Thus, boron can be volatilized in presence of sulfur or selenium; In this context, the molecules BS_2 [52] and BSe_2 [53] were detected. Aluminum also forms gaseous sulfides and selenides at high temperatures: Al_2S, AlS, Al_2S_2, Al_2Se, AlSe, Al_2Se_2 [54]. Gaseous sulfides, selenides and tellurides of group 14 are also known: SiS, SiSe, SiTe, GeS, GeSe, GeTe, SnS, SnSe, SnTe, PbS, PbSe, PbTe [54]. These molecules form dimers to a minor degree. Disulfides respectively diselenides of these elements are less stable. The gaseous monosulfide, PS, of sulfur is described [55].

Gaseous sulfide halides or selenide halides are known of only a few elements. For example, PSX_3 and PSX (*X* = F, Cl, Br) [42] for phosphorus are described. Niobium forms the gaseous sulfide halides respectively selenide halides $NbSCl_3$, $NbSBr_3$, $NbSeBr_3$ [56], for tantalum $TaSCl_3$ and $TaSeBr_3$ are known [56]; the transport effectiveness of MoSBr was mentioned [57]. Tungsten forms two volatile sulfide chlorides, WS_2Cl_2 [39] and $WSCl_4$ [58]. The thermodynamic data of WS_2X_2 (*X* = Cl, Br, I) were determined [59]. The transport effectiveness of the hydrated oxide halides, $Bi(OH)_2X$ (*X* = Cl, Br, I), was reported [60]. Finally, $Pt(CO)_2Cl_2$ is a gaseous compound that becomes transport effective during the transport of platinum with carbon monoxide and chlorine [61].

3.2. Chemical vapor transport of elements and compounds

To date the chemical vapor transport of almost all substance classes has been described. This chapter will show characteristic examples of different transport reactions. A comprehensive overview with more than 2000 references of CVT is not intended here. For more details and references of CVT of elements and compounds see [2]. The chemical vapor transport of elements has been studied and described in detail using metals and some semi-metals as examples; the transport of intermetallic phases principally follows the one of metals. The oxides are the largest group among all compounds which were crystallized by chemical transport reactions with more than 600 examples. The transport of chalcogenides clearly differs from the ones of the oxides. This is due to the higher thermodynamic stability of the metal oxides, compared to the sulfides, selenides, and tellurides. Finally, the chemical vapor transport provides a very good access to phosphides and arsenides, too.

3.2.1. *Chemical vapor transport of elements*

The chemical transport of elements has been studied and described in detail using metals and some semi-metals as examples. In the case of the typical non-metals phosphorus and sulfur, there is no need to increase their volatility in the sense of a chemical vapor transport reaction due to their high vapor pressures. This way, those metals and semi-metals which feature high vapor pressures can also easily be transferred into the gas phase through distillation or sublimation. The following elements belong to this group: Alkali and alkaline earth metals, zinc, cadmium, mercury, europium, ytterbium, arsenic, antimony, selenium and tellurium. Some metals' melting temperature is that low that they can be obtained in liquid form at the most. This, for example, applies for gallium, tin and lead. Thus, chemical vapor transports are relevant for high melting elements with low vapor pressures. These elements can be deposited from the gas phase in closed reaction vessels (ampoules), fluid systems, special reactors (hot-wire process according to (Van Arkel und De Boer), or through CVD-processes [62]. All of these processes are based on the same thermodynamic basic principles. This way, more than 40 elements can be crystallized with chemical transport reactions, more than 25 with iodine as transport agent [63, 64].

In addition to iodine as most important transport agent for elements, compounds such as aluminum(III) chloride, gallium(III) chloride and iron(III) chloride as well as aluminum(III) iodide and indium(III) iodide are described as transport effective additive [65]. These can act halogenating and thus form gaseous halides with the transporting elements. Additionally, they stabilize them by forming gaseous complexes. Halogens such as fluorine, chlorine and bromine as well as the hydrogen halides, water, the chalcogens oxygen, sulfur, selenium, tellurium as well as carbon monoxide are other transport agents which can be used in individual cases. Although carbon monoxide can be used for the transport of nickel only, the industrial purifying process according to Mond and Langer found its way into chemistry textbooks, making carbon monoxide particularly prominent as a transport agent [66]. More details and references of CVT of the elements are presented in [2].

Iodine as transport agent. The exothermic transport of an element with iodine from T_1 to T_2 is the most frequently described transport reaction involving metals. This reaction type were studied and described extensively for titanium, zirconium, hafnium, and thorium (process of van Arkel and de Boer) [67 -70]. Further elements which can be transported this way are yttrium, vanadium, niobium, tantalum, chromium, iron, cobalt, nickel, copper, boron, silicon, germanium and tin as well as uranium and protactinium [63, 64, 71 - 75]. This kind of transport to the hotter zone shall be illustrated with the example of the transport of zirconium with iodine. The temperatures of the dissolving side T_1 can vary between 200 and 650 °C. The most suitable temperature is between 350 and 400 °C. The temperatures of the decomposing side T_2 can be between 1100 and 2000 °C; whereby temperatures around 1400 °C are usually applied. Often, a wire heated by current flow is the place of decomposition. The transport behavior is described by the equilibria (70 – 72), Figure 15.

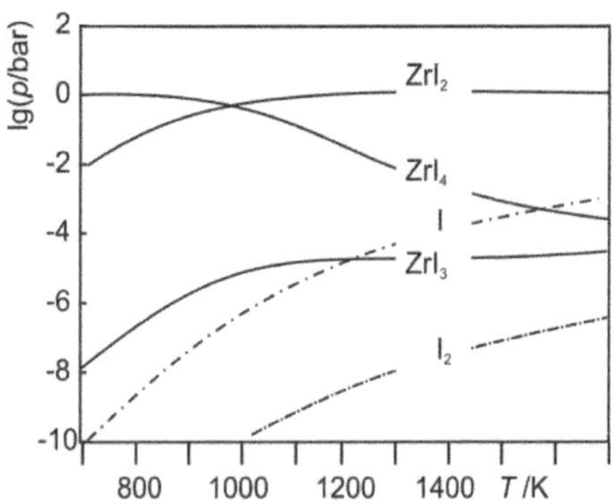

Figure 15. Partial pressures of species in the transport system Zr/I.

$$Zr(s) + 2\,I_2(g) \rightleftarrows ZrI_4(g) \tag{70}$$

$$Zr(s) + I_2(g) \rightleftarrows ZrI_2(g) \tag{71}$$

$$I_2(g) \rightleftarrows 2\,I(g) \tag{72}$$

According to Van Arkel, iron can be transported exothermically with iodine from 800 to 1000 °C. At first, the following transport equation (73) comes into consideration. This reaction, however, is endothermic. According to Le Chatelier's principle, transport from T_2 to T_1 is expected. Because of the strict validity of this principle, one has to assume that the transport obviously cannot be described, or at least not alone, by the reaction formulated above. A detailed investigation showed that other reactions take place as well. Accordingly, iron(II) iodide forms monomer and dimer molecules, FeI_2 and Fe_2I_4, in the vapor. The reaction of iron and iodine with the formation of gaseous Fe_2I_4 molecules (74) can be described. This reaction equation has the character of a transport equation, too. As the reaction (74) is exothermic, one expects transport from T_1 to T_2.

$$Fe(s) + I_2(g) \rightleftarrows FeI_2(g) \tag{73}$$

$\Delta_r H^0_{298} = 24\ kJ \cdot mol^{-1}$

$$2\,Fe(s) + 2\,I_2(g) \rightleftarrows Fe_2I_4(g) \tag{74}$$

$\Delta_r H^0_{298} = -116\ kJ \cdot mol^{-1}$

Using the example of the transport of germanium with iodine, Oppermann and colleagues investigated the proportion of diffusion and convection of the gas movement at different total pressures. In comparative experiments, the transport behavior was determined at normal gravity on earth and under microgravity in space [28]. At microgravity conditions, the convection is negligibly small; substance transport takes place by diffusion only. The experiments indicated that in the gravitational field of earth the gas movement above 3 bar occurs not only by diffusion, but increasingly by convection.

The knowledge gained for exothermic transports with iodine also applies for the other halogens. However, their meaning as transport agents for elements is of low importance due to their unsuitable equilibrium situation. Because the stability of halides increases from iodides to fluorides, their decompositions temperatures increase as well in that direction. Higher decomposition temperatures become necessary which are more difficult to put into practice in experiments.

Conproportionation reactions. Besides the formation and decomposition of the halides, also conproportionation reactions (dissolution) respectively disproportionation (deposition) can be used for the chemical transport of elements. During such reactions, at least two halides of different composition appear in the gas phase (75).

$$Si(s) + SiX_4(g) \rightleftarrows 2\ SiX_2(g) \quad (75)$$

$(X = F,\ Cl,\ Br,\ I,\ 1100 \rightarrow 900\,°C)$

The increase of entropy is the driving force of the endothermic formation of the low halide. The transport always takes place from T_2 to T_1. Elements which can be transported by these reactions are amongst others beryllium, zinc, cadmium, boron, aluminum, gallium, silicon, germanium, tin, antimony and bismuth [8]. The according halides can directly be used as transport agent. Instead the halogen is added in many cases. In this process, the halides are formed by a primary reaction. The given examples can be generalized as follows:

$$M(s) + MX_2(g) \rightleftarrows 2MX(g),\ (M = Be,\ Cd,\ Zn) \quad (76)$$

$$2M(s) + MX_3(g) \rightleftarrows 3MX(g),\ (M = B,\ Al,\ Ga,\ In,\ Sb,\ Bi) \quad (77)$$

$$M(s) + MX_4(g) \rightleftarrows 2MX_2(g),\ (M = Si,\ Ge) \quad (78)$$

$$M(s) + 4MCl_5(g) \rightleftarrows 5MCl_4(g),\ (M = Nb,\ Ta) \quad (79)$$

The principle of conproportionation can also be used for transport reactions with chalcogenides [76] as shown by the following examples:

$$4\ Al(s) + Al_2Q_3(s) \rightleftarrows 3\ Al_2Q(g),\ (Q = S,\ Se) \quad (80)$$

$$Si(s) + SiQ_2(g) \rightleftarrows 2\ SiQ(g),\ (Q = S,\ Se,\ Te) \quad (81)$$

$$Ge(s) + GeQ_2(g) \rightleftarrows 2\ GeQ(g),\ (Q = S,\ Se,\ Te) \quad (82)$$

Reversal of the transport direction. If several reactions are necessary in order to describe the transport of an element respectively a solid, endothermic and exothermic reactions can be relevant. Which of these reactions is the dominant and thus the direction determining one is dependent on the total pressure and the temperature. Thermodynamic discussion shows that the transport direction can be reversible if the transport conditions are varied. There is a change of the transport direction during the deposition of titanium when iodine is added. At lower temperature (< 1000 °C), the endothermic equilibrium (83) predominates while the transport at higher temperatures is determined by the exothermic equilibrium (84).

$$Ti(s) + TiI_4(g) \rightleftarrows 2\ TiI_2(g) \tag{83}$$

$\Delta_r H^0_{298} = 237.9\ kJ \cdot mol^{-1}$

$$Ti(s) + 4\ I(g) \rightleftarrows TiI_4(g) \tag{84}$$

$\Delta_r H^0_{298} = -704.3\ kJ \cdot mol^{-1}$

Formation of gas complexes. Besides pure halogenating equilibria, halogenating equilibria in combination with complex formation equilibria are of importance for the chemical transport of elements [65]. In the process, the formation of gas complexes leads to an increase of the solubility of the respective element in the gas phase. AlX_3, GaX_3, InX_3 and FeX_3 (X = Cl, Br, I) are used as complexing agents whereby the chlorides are used most often. In the gas phase, the mentioned halides can be present as dimeric molecules to a considerable extent. Amongst others, silver, gold, cobalt, chromium, copper, nickel osmium palladium, platinum, rhodium and ruthenium can be transported via complex formation equilibria. In many cases, in particular at temperatures below 500 °C, the transport effective equilibria can be generally described by the following equations. The transport equation is the sum of the equilibria (85) and (86). The formation of gas complexes according to (87) is always endothermic.

$$M^a(s) + a/2X_2(g) \rightleftarrows MX_a(g) \tag{85}$$

$$MX_a(g) + M'_2X_6(g) \rightleftarrows MM'_2X_{6+a}(g) \tag{86}$$

$$M^a(s) + a/2X_2(g) + M'_2X_6(g) \rightleftarrows MM'_2X_{6+a}(g) \tag{87}$$

(*M* = Co, Cu, Ni, Pd, Pt, *M'* = Al, Ga, In, Fe, X = Cl, Br, I)

Transport with addition of hydrogen halides and water. As far as the transport of metals is concerned, the hydrogen halides are of minor importance. Only chromium, iron, cobalt, nickel and copper can be endothermically transported with hydrogen chloride. Iron can also be transported with hydrogen bromide (1020 → 900 °C) [77]. The transport equation (88) exemplarily describes the processes.

$$Ni(s) + 2\ HCl(g) \rightleftarrows NiCl_2(g) + H_2(g) \tag{88}$$

Some elements, such as molybdenum, tungsten, rhenium, gallium, germanium, tin and antimony can be transported with water via the gas phase. The transport is based on the formation of volatile oxides respectively acids. Besides the volatile acids H_2MoO_4 respectively H_2WO_4, one has also to consider gaseous oxides as transport effective species for the transport of molybdenum and tungsten in the given temperature range. Molybdenum and tungsten can be crystallized by adding iodine *and* water via exothermic chemical transport reactions (Mo: 1050 → 1150 °C, W: 800 → 1000 °C) [78, 79].

Oxygen as transport agent. Oxygen can function as transport agent for a series of noble metals – ruthenium, rhodium, iridium, platinum and silver [80 - 82]. In doing so, the transport always takes place at relatively high temperatures in strong endothermic reactions under the formation of volatile oxides. Thus, platinum is transported from 1500 °C to T_1, silver from 1400 °C to T_1, iridium from 1325 to 1125 °C. In particular, the chemical transport of iridium takes place at low oxygen partial pressure and high transport temperatures. Under these conditions, the formation of the solid iridium (IV) oxide can be suppressed.

$$Pt(s) + O_2(g) \rightleftarrows PtO_2(g) \tag{89}$$

$$Ag(s) + 1/2\ O_2(g) \rightleftarrows AgO(g) \tag{90}$$

$$Ir(s) + 3/2\ O_2(g) \rightleftarrows IrO_3(g) \tag{91}$$

3.2.2. *Chemical vapor transport of intermetallic phases*

If one refers to intermetallic phases, solids are meant which are built up by two or more metal atoms. Sometimes there is a differentiation between alloy and intermetallic compounds. In the literature, however, these terms are not used uniformly. In order to avoid misunderstanding, we solely use the term intermetallic phase. It includes metallic solids that are composed stoichiometrically as well as those with phase ranges respectively solid solutions. Solids, which are formed from metals and the semi-metals boron, silicon, germanium and antimony, can be dealt with as well due to their behavior during chemical transports.

Figure 16. Crystal of $FeSi_{0.25}Ge_{0.75}$ grown by chemical vapor transport.

The chemical transport of intermetallic phases principally follows the one of metals. Nowadays, a variety of examples of transports of intermetallic phases is known [2, 83]. In intermetallic phases *all* components of the solid have to be transferred to the gas phase under formation of volatile gas species under the same conditions. Intermetallic phases can be obtained in particular if the elements are also obtained with the same transport agent. Examples can be found in the systems molybdenum/tungsten, cobalt/nickel, and copper/silver. Exceptions to this general rule can be found if the amount of the free enthalpy of formation of the intermetallic phase is especially high, e.g. in the systems chromium/germanium, cobalt/germanium, iron/germanium, nickel/tin or copper/tin. Chemical transport reactions are not only an alternative method for synthesis and crystal growth of intermetallic phases with high melting temperatures. They are preferable in particular for the just mentioned processes:

–One or more components of the intermetallic phase have a high vapor pressure at melting temperature.

–The intermetallic phase decomposes, e.g. peritectically before the melting temperature is reached.

–The intermetallic phase shows one or more phase changes before the melting temperature is reached.

Plenty intermetallic systems show the characteristic of the appearance of numerous solid phases with similar stabilities. Thus, often incongruent vapor transports with different composition of source and sink solid can be observed. The directed deposition of the solid with defined composition can be influenced by the composition of the source solid, the kind of the transport agents and its concentration, and the temperatures of the source and sink side as well as the resulting temperature gradient [2]. Thus, it is possible to obtain low temperature modifications of polymorphic phases in form of single-crystal. Their preparation only rarely succeeds with other methods. FeGe is an example of this in the cubic modification (575 → 535°C, [85, 86]). Likewise, the crystallization of Fe_3Ge is possible by vapor transport [850…900 → 950…1000°C; [85]) despite of the peritectoid behavior of this phase, see Figure 17.

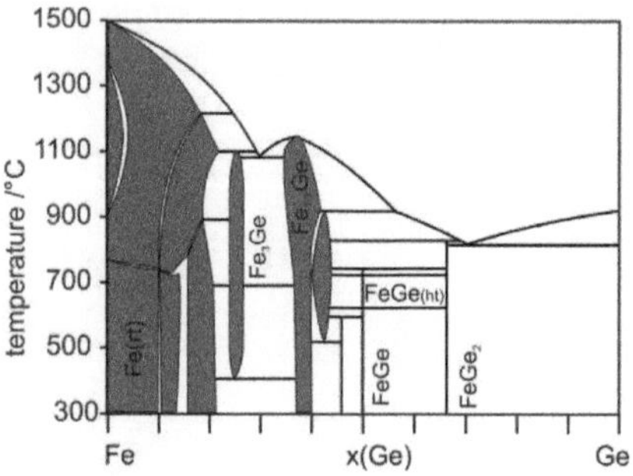

Figure 17. Phase diagram of the system Fe/Ge, according to [84, 85].

Iodine as transport agent. Due to the fact that the transport behavior of the intermetallic phases often follows that of the elements, iodine is the most commonly used transport. Apart from iodine, combinations of iodine and aluminum(III), gallium(III), or indium(III) iodide are used as transport agents as well. Other transport agents and transport-effective additives, respectively, are the halogens chlorine and bromine as well as hydrogen chloride, copper(II) chloride, manganese(II) chloride, the mercury(II) halides, tellurium(IV) chloride, and iron(II) bromide in individual cases.

As an example, the system iron-silicon can be presented. All binary phases, Fe_2Si, Fe_5Si_3, FeSi, and $FeSi_2$, can be crystallized by CVT reactions with iodine [87]. On the iron-rich side to FeSi, the exothermic transport takes places ($T_1 \rightarrow T_2$) with deposition temperatures between 700 and 1030 °C. The transport behavior parallels that of the elemental iron. If the transport efficiency of the individual gas species is considered for the reaction of FeSi with iodine the transport equation (92) can be derived.

$$FeSi(s) + 7/2\ I_2(g) \rightleftharpoons FeI_3(g) + SiI_4(g) \tag{92}$$

Halogen as transport additive – halides as transport agent. Most often, the added iodine is not the transport agent but the silicon(IV) iodide that was formed from it. Thus this transport is similar to that of silicon with SiI_4. The following transport equation (93) can be formulated for the *endothermic* transport of the silicon-rich phase $FeSi_2$ [87]:

$$FeSi_2(s) + 2\ SiI_4(g) \rightleftharpoons 3\ SiI_2(g) + FeI_2(g) \tag{93}$$

The CVT in the Cr-Si system by adding the halogens chlorine, bromine, and iodine is well examined and thermodynamically understood [89, 90]. Cr_3Si, Cr_5Si_3, CrSi, and $CrSi_2$ can be deposited by adding chlorine from 1100 to 900 °C. At the same temperatures, Cr_3Si, Cr_5Si_3, CrSi, and $CrSi_2$ can be deposited with bromine. The transport with iodine, on the other hand, takes place exothermically from 900 to 1100 °C. In this process, Cr_3Si, Cr_5Si_3, CrSi, and $CrSi_2$ can be deposited. In all three cases, transport mechanisms are clearly different. If one considers the transport efficiency of the individual gas species, the following transport equations are derived:

$$Cr_5Si_3(s) + 19\ SiCl_4(g) \rightleftarrows 5\ CrCl_2(g) + 22\ SiCl_3(g) \tag{94}$$

$$CrSi_2(s) + 8\ SiCl_4(g) \rightleftarrows CrCl_2(g) + 10\ SiCl_3(g) \tag{95}$$

$$CrSi_2(s) + 3\ SiBr_4(g) \rightleftarrows CrBr_2(g) + 5\ SiBr_2(g) \tag{96}$$

$$CrSi_2(s) + 10\ I(g) \rightleftarrows CrI_2(g) + 2\ SiI_4(g) \tag{97}$$

In the first three cases, the transport agent is not the added chlorine or bromine, respectively, but the silicon(IV) chloride or bromide, respectively, which was formed in a simultaneous reaction, Figure 18. In contrast to this, iodine functions directly as the transport agent when added.

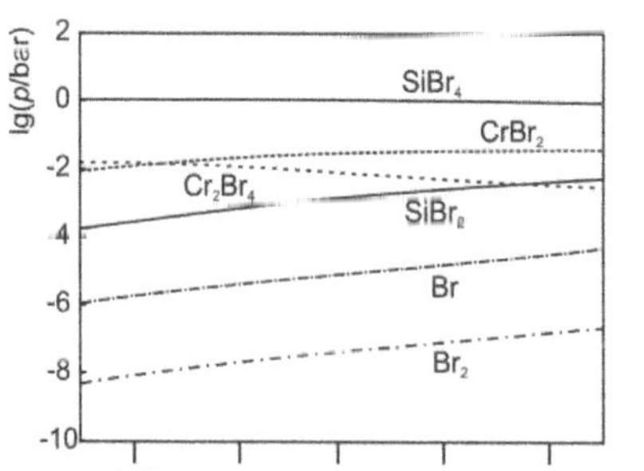

Figure 18. Composition of the gas phase for the transport of CrSi using bromine, according to [89].

Intermetallic phases with wide phase range. Intermetallic systems often show the formation of solid solutions or at least vast regions of solubility of the components. For these characteristic phase relations, the crystallization of phases with defined composition is demanding. As a special example, molybdenum and tungsten are two metals with very high melting points. They are isotypic and completely mixable in the solid and liquid state. Here the formation of specified compositions of mixed crystals molybdenum-tungsten from the melt requires great experimental effort due to the exsolution within the solidus-liquidus-region. With the help of

CVT reactions, this succeeds far below the liquidus curve. As both metals can be transported with the same transport agent under the same conditions, molybdenum-tungsten mixed-crystals can be deposited by CVT from 1000 to 900 °C when mercury(II) bromide is added [91]. The transport equation (98) describes the process:

$$Mo_{1-x}W_x(s) + 2\,HgBr_2(g) \rightleftharpoons (1-x)\,MoBr_4(g) + xWBr_4(g) + 2\,Hg(g) \tag{98}$$

In the following binary systems, mixed-crystals are transportable in an analogous way: cobalt-nickel [91, 92], iron-nickel, silver-copper, gold-copper, copper-nickel, gold-nickel [93], and copper-gallium [94].

3.2.3. Chemical vapor transport of halides

The majority of metal halides are sufficiently stable to evaporate undecomposed. Thus, most of them can be volatilized by distillation or sublimation; the deposition occurs at lower temperatures. Some metal halides decompose at higher temperatures either to the elements or to a metal-rich halide and the according halogen. In this manner platinum(II) chloride decomposes notably above 500 °C forming solid platinum and gaseous chlorine. Otherwise, copper(II) chloride decomposes above 300 °C under the formation of copper(I) chloride and chlorine. The tendency of decomposing generally increases from the fluorides to the iodides. Some metal halides disproportionate while heating: molybdenum(III) chloride essentially dissociates above 600 °C under the formation of solid molybdenum(II) chloride and gaseous molybdenum(IV) chloride.

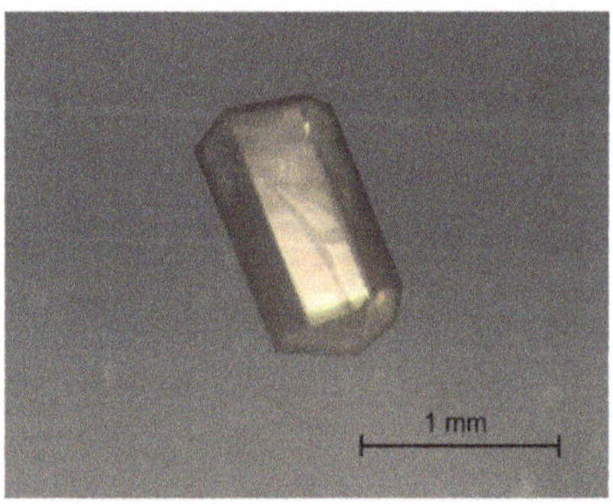

Figure 19. Crystal of $CeCl_3$ grown by chemical vapor transport

Beside the sublimation processes, metal halides can be obtained by CVT reactions, too. Four different types of solid-gas reactions are of relevance. Additionally, further reactions of different kinds are known, which can be used for the transport of metal halides. However, their application is limited so far. An overview on the CVT of halides is provided by Oppermann [95]; for more current references of CVT of the halides see [2].

Formation of halogen-rich halides. If halides of a concerned metal exist in different oxidation states, decreasing boiling temperatures for increasing oxidation numbers (halogen-rich halides) can be observed. This behavior is effected by the covalence of the metal-halogen-compound, which increases with higher oxidation number. Thus the formation of higher volatile halide species is possible by reaction of a metal-rich halide with a halogen. This kind of transport reaction is often observed with halides of the transition metals (99).

$$CrCl_3(s) + 1/2\ Cl_2(g) \rightleftarrows CrCl_4(g) \tag{99}$$

However, the tendency of decomposition of halides with higher oxidation numbers increases at the same time. For this reason, the vapor transport by halogenation is restricted in temperature. Or, in another way, a high partial pressure of the halogen is needed to form a sufficiently high pressure of the transport effective metal halide species. Gaseous ruthenium(IV) bromide is formed during the transport of ruthenium(III) bromide with bromine (100); [96]. However, a high bromine pressure of 15 bar is required to cause a sufficient transport effect.

$$RuBr_3(s) + 1/2\ Br_2(g) \rightleftarrows RuBr_4(g) \tag{100}$$

Generally, the halogen is used as a transport agent, which is also contained in the solid. Sometimes, however, another halogen is used as the transport agent (101), [97]. Here, during the crystallization a small amount of transport agent bromine condenses and a solid of the composition $VCl_{2.97}Br_{0.03}$ forms in the sink.

$$VCl_3(s) + 1/2\ Br_2(g) \rightleftarrows VCl_3Br(g) \tag{101}$$

Conproportionation reactions. Transition metals can appear in different binary halides, which are similarly stable under vapor transport conditions. This particularly applies for metals of group 5 and 6. This behavior can be used in order to transport a solid metal halide in which the metal has a low oxidation number with a gaseous metal halide in which the metal has an oxidation number that is higher by two units or more. The transport of niobium(III) chloride with niobium(V) chloride as transport agent is given as an example [1]. Gaseous niobium(IV) chloride is formed, which disproportionates in the sink of transport to form solid niobium(III) chloride and gaseous niobium(V) chloride. Frequently, the according halogen is used as transport additive instead of the transport agent that was formulated in the transport equation. Thereby, the actual effective transport agent forms in a preliminary reaction of the transport additive with the solid. Additional examinations and/or thermodynamic model calculations are necessary in order to decide whether the added halogen or a higher halide that is formed by the halogen is the actual transport agent.

$$NbCl_3(s) + NbCl_5(g) \rightleftarrows 2NbCl_4(g) \tag{102}$$

Formation of gas complexes. The term *gas complex* refers to a gaseous metal-halogen compound in which several metal atoms are bonded with each other by halogen bridges. Gas complexes with several identical metal atoms, such as Al_2Cl_6, are also called *homeo complexes.* Those with different metal atoms, such as $NaAlCl_4$, are labeled *hetero complexes* [98 - 100]. The CVT of metal halides under formation of gas complexes is realized for solid metal halides with a high boiling temperature under addition of a high volatile halide, particularly often an aluminum halide. The aluminum halides have low boiling temperatures and form stable gas complexes with a variety of metal halides. Gallium(III) halides, indium(III) halides, and iron(III) halides are used as transport agents as well.

Metal monohalides, like the alkali metal halides *MX* form gas complexes of the composition $MAlX_4$ by adding the aluminum halides AlX_3. These complexes are characterized by an extremely high stability. However, the solid and liquid ternary halides of these compositions are to stable to reverse the transport equilibrium under crystallization of the alkali metal halides. That is because in the sink not the alkali metal halides but a different ternary phase is always deposited. Accordingly, this also applies when gallium(III) halides, indium(III) halides, and iron(III) halides are used as transport agents.

Metal dihalidesMX_2 form gas complexes of the composition MAl_2X_8 and $MAlX_5$ when the trihalides of aluminum, gallium, and iron are used as a transport agent. As an example, the transport of manganese(II) chloride with aluminum-chloride, gallium-chloride, and indium-chloride is discussed in detail [102]. Additionally, the formation of larger gas complexes of the composition MAl_3Cl_{11} and MAl_4Cl_{14} has been reported [101, 102]. At moderate temperatures of about 400 °C the dimer Al_2Cl_6 acts as a transport agent and the transport of metal dihalides takes place via $MnAl_2Cl_8$ as transport-effective species (103).

$$MnCl_2(s) + Al_2Cl_6(g) \rightleftarrows MnAl_2Cl_8(g) \tag{103}$$

The enthalpy of formation of the complex does not entirely compensate the sublimation enthalpy of metal halide so that the transport reaction (103) is always endothermic. Under different transport conditions (> 500 °C) the transport direction can change in presence of the transport agent aluminum(III) chloride as monomeric $AlCl_3$(g). Additionally, the formation of complexes of the composition $MnAlCl_5$ becomes more important. This transport takes place to the hotter zone in an exothermic reaction (104).

$$MnCl_2(s) + AlCl_3(g) \rightleftarrows MnAlCl_5(g) \tag{104}$$

Metal trihalidesMX_3 can form gas complexes of the composition MM'_2X_9, MM'_3X_{12}, and MM'_4X_{15} by adding the trihalides $M'X_3$ of aluminum, gallium, and iron [32, 103, 104]. By the formation of these gas complexes even the transport of non- volatile trihalides is possible in endothermic reactions at 500 to 400 °C [105, 106]. This way, chromium(III) chloride [105] and the trihalides of the lanthanoid metals [107] can be obtained in crystalline form with aluminum(III) chloride as transport agent.

$$LnCl_3(s) + 2M'Cl_3(g) \rightleftarrows LnM'_2X_9(g) \quad (105)$$

$$LnCl_3(s) + 2M'Cl_3(g) \rightleftarrows LnM'_3X_{12}(g) \quad (106)$$

$$LnCl_3(s) + 2M'Cl_3(g) \rightleftarrows LnM'_4X_{15}(g) \quad (107)$$

The formation of gas complexes plays an important role for the separation of the halides of the lanthanoids. In a gas stream of aluminum(III) chloride the individual lanthanoid halides form gas complexes of different stability. These complexes decompose under the formation of the halides LnX_3 at different places. This way, the halides of the lanthanoids can be separated in a "fractionalized chemical vapor transport" [109-118]. During this process the solid oxides can be used (108).

$$Ln_2O_3(s) + 3\ C(s) + 3\ Cl_2(g) \rightleftarrows 2LnCl_3(s) + 3\ CO(g) \quad (108)$$

Metal tetrahalides UCl_4 and $ThCl_4$ realize a vapor transport under addition of aluminum(III) chloride [106] probably by endothermic formation of UAl_2Cl_{10} and, respectively, $ThAl_2Cl_{10}$.

Metal pentahalides are not very common. Their vapor pressure is relatively high so that they can be sublimed without any problems and are of minor interest for transport reactions.

Halogen transfer reactions and formation of interhalogen compounds. In contrast to the semi- and non-metals, the fluorides of the metals have essentially higher boiling temperatures compared to the chlorides, bromides, and iodides: The boiling temperatures of aluminum fluoride is 1275 °C, those of the other halides 181 °C ($AlCl_3$), 254 °C ($AlBr_3$), and 374 °C (AlI_3). Thus, the variety of metal fluorides cannot be crystallized by sublimation. In a few cases, transport with silicon(IV) chloride succeeded [119, 120]. The transport reaction works due to the fact that silicon(IV) fluoride as well as silicon(IV) chloride are highly volatile compounds.

$$4\ AlF_3(s) + 3\ SiCl_4(g) \rightleftarrows 4\ AlCl_3(g) + 3\ SiF_4(g) \quad (109)$$

Principally, the CVT of fluorides with halogens as a transport agent is not possible via equilibria, such as (110), due to their unfavorable position. The release of fluorine, which occurs during the reaction, is thermodynamically unfavorable. Nevertheless, magnesium fluoride can be crystallized with iodine as transport agent [121]. Thermodynamic model calculations with data for the gaseous iodine fluorides IF_n ($n = 1, 3, 5, 7$) reflect the observed transport effect. These calculations suggest a significant participation of IF_5 in the transport process according to (111).

$$MgF_2(s) + X_2(g) \leftarrow MgX_2(g) + F_2(g) \tag{110}$$

(X = Cl, Br, I)

$$5MgF_2(s) + 6\,I_2(g) \rightleftarrows 5\,MgI_2(g) + 2\,IF_5(g) \tag{111}$$

3.2.4. *Chemical vapor transport of oxides*

Oxides represents the most reported substance group with more than 600 examples of chemical vapor transports. For more details and particularized references of CVT of the oxides see [2]. Simple binary oxides, such as zinc(II) oxide and iron(III) oxide, have been crystallized as well as oxides with complex anions, such as phosphates or sulfates, and oxides with several cations, such as $ZnFe_2O_4$ or $Co_{1-x}Ni_xO$. Metal oxides are thermodynamically very stable compounds. However, only a few of them evaporate undecomposed; among them are CrO_3, MoO_3, WO_3, Re_2O_7, GeO, SnO, PbO, and TeO_2. Most of the metal oxides decompose by evaporation of oxygen. Besides, the respective metal or a metal-rich oxide is formed. The latter can be present as condensed phase or as a gas. The following three examples show the different thermal behavior of metal oxides:

$$2\,ZnO(s) \rightleftarrows 2\,Zn(g) + O_2(g) \tag{112}$$

$$2\,SiO_2(s) \rightleftarrows 2\,SiO(g) + O_2(g) \tag{113}$$

$$6\,Fe_2O_3(s) \rightleftarrows 4\,Fe_3O_4(s) + O_2(g) \tag{114}$$

Figure 20. Crystal of ZnO grown by chemical vapor transport.

The thereby released oxygen partial pressure is called the **decomposition pressure**. It is the leading determinant of the transport behavior and the composition of the crystallized oxide. The transport of Fe_2O_3 can be served as an example: If the oxygen partial pressure in the system is higher than the decomposition pressure of Fe_2O_3, this compound is stable as a solid; if it is lower, the reduced solid, Fe_3O_4 will be formed. If the oxygen partial pressure is identical with the decomposition pressure at a certain temperature, both solid phases will co-exist. If the logarithm of a co-existence decomposition pressure is plotted against the reciprocal temperature for different oxides of a metal, the **phase barogram** of the system will be obtained. By means of the phase relations presented in the phase barograms, the choice of suitable parameters for phase pure transports respecting the temperature and the temperature gradient becomes possible [2].

A variety of transport agents has been investigated for oxides, but chlorinating equilibria proved most suitable. Apart from chlorine and hydrogen chloride, tellurium(IV) chloride is an important transport agent. Tellurium(IV) chloride is used especially when the oxygen partial pressure in the system varies, and the setting of the oxygen partial pressure is of essential importance for the transport behavior. Some other chlorinating additives include phosphorus(V) chloride, niobium(V) chloride, selenium(IV) chloride, and tetrachloromethane as well as mixtures of sulfur/chlorine, vanadium(III) chloride/chlorine, and chromium(III) chloride/chlorine. Due to unfavorable equilibrium positions, brominating and iodinating equilibria are of minor importance for the CVT of oxides. Here, transport agents or transport effective additives, respectively, are: bromine and iodine, hydrogen bromide and hydrogen iodide, phosphorus(V) bromide, niobium(V) bromide and -iodide as well as sulfur+iodine. Iodine as a transport agent and iodinating equilibria are of interest if chlorine is too oxidizing or if, as is the case with rare-earth metal oxides, stable solid oxide chlorides form. Some further transport agents or transport effective additives, respectively, are hydrogen, oxygen, water, carbon monoxide and in special cases, fluorine or hydrogen fluoride. In some cases, the solid oxides can form gaseous oxide halides: transport effective species, which contain both oxygen and halogen atoms.

Halogens as transport agents. The process of dissolution of an oxide in the gas phase by heterogeneous reaction with a halogen can be split into two partial reactions of decomposition of the oxide and the halogenation of the metal:

$$M_aO_{1/2a}(s) \rightleftarrows M_a(s) + 1/4aO_2(g) \tag{115}$$

(a = oxidation number of the metal)

$$M_a(s) + 1/2aX_2(g) \rightleftarrows MX_a(g) \; (X = Cl, Br, I). \tag{116}$$

Due to the higher stability of the chloride gas species compared to the bromide and the resulting equilibrium position, mostly chlorine is used as the transport agent for the CVT of oxides. In the process, sufficiently stable gas species are formed with adequately high partial pressures ($p > 10^{-5}$ bar) and sufficiently high partial pressure differences along the temperature gradient. The transports of Fe_2O_3 and $NiGa_2O_4$ with chlorine shall be used as examples.

$$Fe_2O_3(s) + 3\ Cl_2(g) \rightleftarrows 2\ FeCl_3(g) + 3/2\ O_2(g) \tag{117}$$

$$NiGa_2O_4(s) + 4\ Cl_2(g) \rightleftarrows NiCl_2(g) + 2\ GaCl_3(g) + 2\ O_2(g) \tag{118}$$

Halogens are also suited as transport agents for oxides when gaseous oxide halides are formed. This way, for example, the transport of molybdenum(VI) oxide with chlorine succeeds:

$$MoO_3(s) + Cl_2(g) \rightleftarrows MoO_2Cl_2(g) + 1/2\ O_2(g) \tag{119}$$

Instead of introducing pure halogens, decomposition of less stable halides, such as PtX_2 (X = Cl, Br, I), can be used to form halogens. If mercury halides are employed as transport agents, the equilibrium position of the transport reaction will shift compared to elemental halogen. By decomposing gaseous mercury halides to the elements (120), additional gas species are formed. There will be a change of the entropy balance shifting the equilibrium position to the side of the reaction products.

$$HgX_2(g) \rightleftarrows Hg(g) + X_2(g) \tag{120}$$

Transport with addition of hydrogen halides. Hydrogen chloride, and less frequently hydrogen bromide and hydrogen iodide, are often used and are effective transport agents for the CVT of oxides. In special cases, as for silicates, also hydrogen fluoride is used as a transport agent. During the transport of a binary oxide with a hydrogen halide, a gaseous metal halide is formed besides water (121). The transport of zinc oxide with hydrogen chloride is an example:

$$ZnO(s) + 2\ HCl(g) \rightleftarrows ZnCl_2(g) + H_2O(g) \tag{121}$$

The simple transport equation by forming the respective chloride and water only applies if no volatile acids, such as $H_2MoO_4(g)$, hydroxides, and oxide halides, respectively, are formed. Using hydrogen halides, often a more favorable equilibrium position can be achieved instead of halogens. As a feasible hydrogen halide source, the ammonium halides (NH_4X, X = Cl, Br, I) can be used. These solids are easy to handle and to dose. They decompose to ammonia and hydrogen halide at increased temperature. However, the formation of ammonia creates a reducing atmosphere. This can lead to a reduction of the gas species and/or the solid phase. In some cases, in which the transport of oxides with moisture-sensitive halides, such as aluminum(III) chloride or tellurium(IV) chloride, is described, hydrogen chloride can be expected as the transport agent. Traces of water, which can never be excluded completely, cause the formation of hydrogen halide.

Tellurium(IV) halides as transport agents. Tellurium(IV) chloride is a flexible transport agent, which can especially be used for oxides of the transition metals and compounds with complex anions [31, 122]. The simplified transport equation (122) can be assumed:

$$ZrO_2(s) + TeCl_4(g) \rightleftarrows ZrCl_4(g) + TeO_2(g) \tag{122}$$

In this simplification, however, the equilibria (123) to (128) in the system Te/O/Cl are not considered. Reichelt discussed the complex reaction behavior of tellurium(IV) chloride in detail [123].

$$TeCl_4(g) \rightleftarrows TeCl_2(g) + Cl_2(g) \tag{123}$$

$$TeCl_2(g) + ½\ O_2(g) \rightleftarrows TeOCl_2(g) \tag{124}$$

$$TeOCl_2(g) \rightleftarrows TeO(g) + Cl_2(g) \tag{125}$$

$$TeO_2(g) \rightleftarrows 1/2\ Te_2(g) + O_2(g) \tag{126}$$

$$Te_2(g) \rightleftarrows 2\ Te(g) \tag{127}$$

$$Cl_2(g) \rightleftarrows 2\ Cl(g) \tag{128}$$

Creating such a complex red-ox system, tellurium(IV) chloride is specially suited as a transport additive for oxide systems with a wide range of oxygen partial pressures between 10^{-25} and 1 bar. Thereby, at low oxygen partial pressures, the reduced gas species $TeCl_2$, Te_2, Te, Cl_2, and Cl dominate. The transport of Mn_3O_4 with tellurium(IV) chloride can be served as an example: The gas phase consists of the dominating gas species $MnCl_2$, Te_2, Mn_2Cl_4, Te, TeO, and $TeCl_2$ (with $p(i) > 10^{-4}$ bar in the temperature range of about 1000 °C; see Figures 3.4). The partial pressures of the other oxygen-containing gas species TeO_2 and $TeOCl_2$ are clearly below 10^{-5} bar [124]. At higher oxygen partial pressures, the amount of higher oxidized gas species

TeO_2 and $TeOCl_2$ becomes significantly higher, for example during the transport of manganese(III) oxide. Here, at1000 °C, the gas phase contains the species O_2, Cl_2, $TeOCl_2$, TeO_2, $MnCl_2$, Cl, $TeCl_2$, and TeO in the pressure range between 1 and 10^{-5} bar [124].

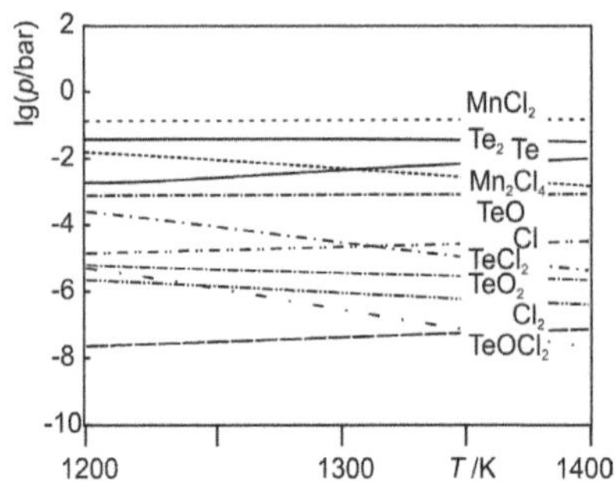

Figure 21. Composition of the gas phase for the transport of Mn_3O_4 using $TeCl_4$, according to [124].

In particular, tellurium(IV) chloride proves an ideal transport additive for those oxides that differ only slightly in their composition and stability and thus are thermodynamically stable only in narrow ranges of the oxygen partial pressure. Thus, the chemical vapor transport of the *Magnéli*phases of vanadium, V_nO_{2n-1} ($n = 2 \dots 8$), succeeded with tellurium(IV) chloride [123, 125, 126]; see Figure 22. Tellurium(IV) chloride is also suitable for the transport of oxide phases that show homogeneity ranges that are dependent on the oxygen partial pressure, such as "VO_2" [31, 127] and for oxides of transition metals that have similar stabilities, such as MnO and Mn_3O_4 [123, 128]. Similar redox systems form when tellurium(IV) bromide ($TeBr_4$) [129] and TeI_4 [130] are used as transport agents.

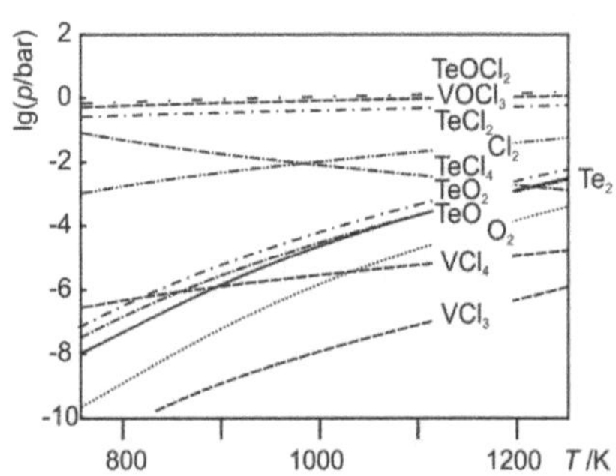

Figure 22. Composition of the gas phase for the transport of VO_2 using $TeCl_4$, according to [2, 125].

Reactions with combined transport additives. The combination of two transport additives is often used, for example the combination of sulfur, selenium or carbon in addition to the halogens. These gas mixtures form complex redox systems and can be treated in a similar way to tellurium(IV) chloride. The mechanism of the combination of sulfur+iodine is described exemplarily by equation (129). Here, sulfur transfers oxygen; iodine transfers gallium. The transport agent combinations carbon+chlorine and carbon+bromine are introduced in the form of CCl_4 (130) and CBr_4, respectively. The formation of gaseous SO_2 and CO, respectively, causes a more balanced equilibrium position compared to the reactions in which oxygen is formed. Due to the stability of the formed oxide gas species the reaction equilibrium position is shifted to the side of the gaseous reaction products.

$$2\ Ga_2O_3(s) + 3/2\ S_2(g) + 6\ I_2(g) \rightleftarrows 4\ GaI_3(g) + 3\ SO_2(g) \qquad (129)$$

$$Y_2O_3(s) + 3\ CCl_4(g) \rightleftarrows 2\ YCl_3(g) + 3\ CO(g) + 3\ Cl_2(g) \qquad (130)$$

In some cases, a transport agent combination consisting of a halide and a halogen is applied, for example for the transport of SiO_2 with $CrCl_4$+Cl_2. In this process, the surplus of halogen leads to formation of more volatile oxidized gas species (CrO_2Cl_2).

$$SiO_2(s) + CrCl_4(g) + Cl_2(g) \rightleftarrows SiCl_4(g) + CrO_2Cl_2(g) \qquad (131)$$

The usage of phosphorus(III) halides, PCl_3 and PBr_3, in addition to the respective halogens causes the formation of the pentahalides. The phosphorus(V) halides proved to be suitable transport agents as well as the analogues $NbCl_5$ and $TaCl_5$, as they have both a halogenating effect on the metal and a transport effective for oxygen (132, 133).

$$LuPO_4(s) + 3\ PCl_3(g) + 3\ Cl_2(g) \rightleftarrows LaCl_3(g) + 4\ POCl_3(g) \qquad (132)$$

$$Nb_2O_5(s) + 3\ NbCl_5(g) \rightleftarrows 5\ NbOCl_3(g) \qquad (133)$$

Aluminum(III) chloride is not suited for the transport of oxides because aluminum oxide is formed. Observed transport effects can most often be traced back the formation of hydrogen chloride.

3.2.5. Chemical vapor transport of oxides with complex anions

The chemical vapor transport of oxides with complex anions represents few examples of crystallization of

- sulfates, selenates, and tellurates
- phosphates, arsenates, and antimonates

- silicates
- borates

Figure 23. Crystals of $CuTe_2O_5$ grown by chemical vapor transport.

These complex oxides differ from other multinary oxides ("double oxides") by their high heats of reaction for the formation from binary oxides. In terms of their chemical structure, they are different by the low co-ordination number of the non-metal. For more details and references of CVT of compounds with complex anion see [2].

The crystallization of anhydrous *sulfates* is quite challenging. Most of the representatives of this compound class show a comparatively low thermal stability (decomposition to SO_3 and SO_2+O_2), only the sulfates of the alkali metals melt and sublime without decomposing. For CVT, chlorine or hydrogen chloride can be used as transport agents for sulfates [2]. The transport of $ZnSO_4$ (134) can be served as an example [131, 132]. In some cases, the vapor transport could be observed when I_2, NH_4Cl, $HgCl_2$, $PbCl_2$ (135), $PbBr_2$, or $SOCl_2$ were added.

$$ZnSO_4(s) + Cl_2(g) \rightleftarrows ZnCl_2(g) + SO_3(g) + 1/2\ O_2(g) \tag{134}$$

An oxidizing equilibrium gas phase is the requirement for the use of $PbCl_2$ as transport additive for some anhydrous sulfates, such as $NiSO_4$ or $CuSO_4$ [133]. In the process, chlorine is released in a pre-reaction (135); the formed chlorine functions as the actual transport agent for $NiSO_4$ (136).

$$2\ NiSO_4(s) + PbCl_2(l) \rightleftarrows PbSO_4(s) + 2\ NiO(s) + SO_2(g) + Cl_2(g) \tag{135}$$

$$NiSO_4(s) + Cl_2(g) \rightleftarrows NiCl_2(g) + SO_3(g) + 1/2\ O_2(g) \tag{136}$$

While the volatilization of aluminum (III) oxide with chlorine (as with other transport agents) in a temperature gradient is impossible due to the unfavorable equilibrium position of reaction, the crystallization of aluminum sulfate by CVT is successful using $SOCl_2$ as a transport agent [134]. The resulting transport reaction avoids the formation of free oxygen. Thus a favorable position of the heterogeneous transport equilibrium (137) is caused. The crystallization of $Cr_2(SO_4)_3$, $Ga_2(SO_4)_3$, and $In_2(SO_4)_3$ can be realized in the same way.

$$Al_2(SO_4)_3(s) + 3\ SOCl_2(g) \rightleftarrows 2\ AlCl_3(g) + 3\ SO_2(g) + 3\ SO_3(g) \quad (137)$$

The CVT of phosphates is a preparative method for crystallization of even thermally delicate phosphates, like $Re_2O_3(PO_4)_2$ [135] and CuP_4O_{11} [136]. Phosphates of transition metals with oxidation states that are not easily accessible in another ways (low numbers) can be synthesized in sealed silica ampoules and crystallized in "one-pot reactions" by CVT (e. g., $TiPO_4$, $V_2O(PO_4)$, $Cr_3(PO_4)_2$, and $Cr_2P_2O_7$). Apart from the elemental halogens Cl_2, Br_2, and I_2, halogen compounds (NH_4X and HgX_2; X = Cl, Br, I) as well as mixtures $P+X_2$ (X = Cl, Br, I) are used. In some cases, such as $Fe_3O_3PO_4$ or UP_2O_7 chlorinating compounds, such as VCl_4, $ZrCl_4$, $HfCl_4$, and $NbCl_5$ are suitable transport agents [137]. The best results, as far as transport rates and crystal growth of anhydrous phosphates are concerned, were achieved with chlorine or mixtures of phosphorus+iodine as transport agents [138].

$$Co_2P_2O_7(s) + 2\ Cl_2(g) \rightleftarrows 2\ CoCl_2(g) + 12\ P_4O_{10}(g) + O_2(g) \quad (138)$$

$$Ni_2P_4O_{12}(s)\ + 2/3\ P_4(g) + 4/3\ PI_3(g) \rightleftarrows 2\ NiI_2(g) + 2\ P_4O_6(g) \quad (139)$$

The transport of anhydrous phosphates with iodine and reducing additives does not take place via P_4O_{10}. Observations during the transport of $Cr_2P_2O_7$ [138] with iodine in the presence of a surplus of CrP are as remarkable in this context as the transport of $WOPO_4$ and WP_2O_7 adjacent to WP [139]. In all three cases, a *simultaneous transport* of phosphides and phosphates due to an endothermic reaction is found experimentally. Experiments with the transport balance show that phosphide and phosphate migrate from the source to the sink in a single stationary state if the two condensed phases are provided in a certain ratio with respect to their amounts of substance. This behavior indicates a *coupled vapor transport reaction* of the two phases.

$$Cr_2P_2O_7(s)\ + 8/3\ CrP(s)\ + 14/3\ I_2(g) \rightleftarrows 14/3\ CrI_2(g)\ + 7/6\ P_4O_6(g) \quad (140)$$

In contrast to anhydrous phosphates, metal *arsenates(V), antimonates(V), and vanadates(V)* show a clearly lower thermal stability. The compounds tend, more easily than phosphates, toward the formation of oxygen and gaseous As_4O_6, Sb_4O_6, and VO_2, respectively. The limited stability

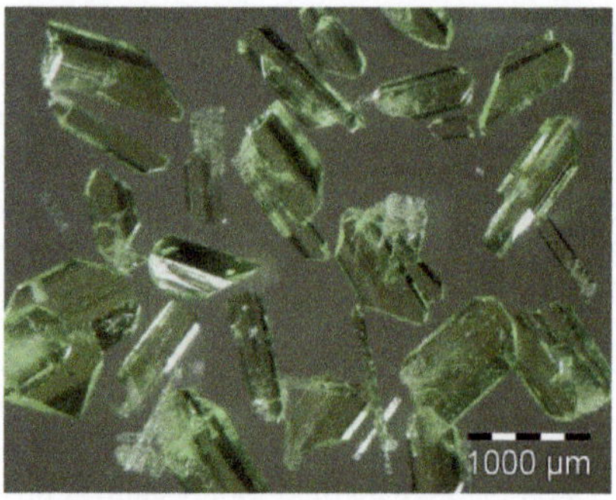

Figure 24. Crystals of $PrPO_4$ grown by chemical vapor transport.

of arsenates and antimonates, combined with the volatility of As_4O_6 and Sb_4O_6, seems to be favorable for CVT of these compounds. Hence, Weil [140] describes the successful CVT experiments aiming at the crystallization of different anhydrous arsenates with chlorine, a mixture of HCl +H_2 (addition of NH_4Cl), and $HgCl_2$ as the transport agent. For more details and references of CVT of the phosphates, arsenates, antimonates, and vanadates see [2].

While there are no indications on chemical vapor transport of *carbonates*, a number of reports on the CVT of *silicates* are given. Early on, the assumption was made that transport reactions with participation of the gas phase are involved in mineral-forming processes of silicates in nature [141]. Indeed, only the migration of europium(II) silicates (Eu_2SiO_4, $EuSi_2O_5$) at high temperatures with HCl as transport agent (141) and the crystallization of Be_2SiO_4 with SiF_4 (142) are based on transport reactions of the minerals [142].

$$Eu_2SiO_4(s) + 6\ HCl(g) \rightleftarrows 2\ EuCl_2(g) + SiCl_2(g) + 3\ H_2O(g) + 1/2\ O_2(g) \tag{141}$$

$$Be_2SiO_4(s) + 3\ SiF_4(g) \rightleftarrows 2\ BeF_2(g) + 4\ SiOF_2(g) \tag{142}$$

In contrast to reversible CVT reactions in the direct sense, the crystallization of silicates with participation of the gas phase can be traced back to partial transport reactions. The formation of *zircon* $ZrSiO_4$ from zirconium dioxide in silica ampoules when silicon(IV) fluoride is added, has been discussed by Schäfer [1]. The reactions (143) (over the solid $ZrO_2/ZrSiO_4$) and (144) (over the solid $SiO_2/ZrSiO_4$) describe the process completely. The partial equilibria allow sufficiently high pressures for SiF_4 and ZrF_4, so that the interdependent transport of silicon and zirconium becomes possible. Here, the migration takes place via the fluorides under isothermal (!) conditions in the gradient of the respective chemical potentials. The formation of other silicates, like *topaz* ($Al_2SiO_4F_2$) from AlF_3 and SiO_2 [143] takes place in a similar way.

$$2\ ZrO_2(s) + SiF_4(g) \rightleftarrows ZrF_4(g) + ZrSiO_4(s) \quad (143)$$

$$2\ SiO_2(s) + ZrF_4(g) \rightleftarrows SiF_4(g) + ZrSiO_4(s) \quad (144)$$

Boron(III) oxide forms numerous ternary and multinary oxido compounds. Nevertheless, there are hardly any indications on the chemical vapor transport of **borates**. Only the migration of $CrBO_3$, $FeBO_3$, BPO_4, and $Cr_2BP_3O_{12}$ in the temperature gradient is detected for sure [2, 144].

$$FeBO_3(s) + 3\ HCl(g) \rightleftarrows FeCl_3(g) + HBO_2(g) + H_2O(g) \quad (145)$$

Several other borates were obtained as a by-product during the synthesis of *boracites* $M_3B_7O_{13}X$ (M = metal atom with the oxidation number II, X = Cl, Br, I). Boracites can be crystallized well with the help of CVT reactions, in contrast to the halogen-free borates. [145]. Boracites are transported with water and the corresponding hydrogen halide. The constituent compounds MO, MX_2, and B_2O_3 are separately put in a *two-crucible technique* apparatus [146] or a *three-crucible technique* apparatus [145, 147, 148]. The transport takes place isothermally at about 900 °C along an *activity gradient* of the components. The metal dihalildes MX_2 as well as BX_3, $B_3O_3X_3$, and HBO_2 are considered as active transport species [149].

$$MX_2(s) \rightleftarrows MX_2(g) \quad (146)$$

$$MO(s) + 2\ HX(g) \rightleftarrows MX_2(g) + H_2O(g) \quad (147)$$

$$B_2O_3(s) + 6\ HX(g) \rightleftarrows 2\ BX_3(g) + 3\ H_2O(g) \quad (148)$$

$$3\ B_2O_3(s) + 6\ HX(g) \rightleftarrows 2\ (BOX)_3(g) + 3\ H_2O(g) \quad (149)$$

$$B_2O_3(s) + H_2O(g) \rightleftarrows 2\ HBO_2(g) \quad (150)$$

3.2.6. Chemical vapor transport of chalcogenides

Chemical vapor transports of metal sulfides, selenides, and tellurides have been examined in detail. The first investigations were made in the 1960s by Nitsche [150]. To date, the number of examples that are known from the literature [2] is only exceeded by those of the oxides.

Nonetheless, the CVT of chalcogenide compounds clearly differs from that of the oxides. Due to the lower thermodynamic stability of the metal sulfides, selenides, and tellurides compared to the oxides most often iodine or iodine compounds are used as transport agents. Thus more balanced equilibria of the transport reactions of sulfides, selenides, and tellurides can be achieved, in contrast to the one of the respective oxide (151).

$$ZnQ(s) + X_2(g) \rightleftarrows ZnX_2(g) + 1/2Q_2(g), \ (X = Cl, Br, I; Q = O, S, Se, Te) \tag{151}$$

$\Delta_r G^0_{1000} / kJ \cdot mol^{-1}$	Cl,	Br,	I
ZnO	−26	+22	+103
ZnS	−104	−56	**+25**
ZnSe	−123	−57	**+7**
ZnTe	−170	−122	**−41**

Figure 25. Crystal of TaS_2 grown by chemical vapor transport.

Transport of Sulfides. A large number of examples of binary and ternary sulfides as well as quaternary and even multinary sulfides, such as $FeSn_4Pb_3Sb_2S_{14}$ [151] are available by CVT reactions [2]. This is indeed noteworthy, because in these cases the transport agent is apparently able to transfer *all* cations that are present in the compound to the gas phase and deposit them at another temperature. Sulfides with a phase range, such as FeS_x, can be transported systematically as well [18 – 20, 152]. Mixed-crystals with substitution in the cationic sublattice, such as $Co_{1-x}Fe_xS$ [153]; in the anionic sublattice, such as $TiS_{2-x}Se_x$ [154 – 156]; or in the cationic *and* anionic sublattice, such as $Ge_xPb_{1-x}S_{1-y}Se_y$ [157] are accessible with defined compositions and in crystalline form. Here, sulfides and selenides behave in very similar ways. This is because of the similar ionic radii of the sulfide and selenide ions and the same electronegativity. Both properties cause similar chemical behavior and similar thermodynamic stabilities. Thus often the metal sulfides and selenides have the same structure types. Additionally, sulfides and selenides are often mixable completely in the solid state. Thus, the essential aspects that apply for the transport of sulfides, also apply for the selenides.

While heating, most of the metal sulfides decompose completely or partly to the elements. If the metal has a sufficiently high vapor pressure at the decomposition temperature, one can observe in some cases a *decomposition sublimation* (152).

$$MS(s) \rightleftarrows M(g) + 1/2\ S_2(g) \tag{152}$$

(M = Zn, Cd)

Only a few metal sulfides can be sublimed undecomposed. Examples are gallium(I) sulfide, germanium(II) sulfide, tin(II) sulfide, lead(II) sulfide:

$$Ga_2S(s) \rightleftarrows Ga_2S(g) \tag{153}$$

$$MS(s) \rightleftarrows MS(g) \tag{154}$$

(M = Ge, Sn, Pb)

Some sulfides decompose to a metal-rich solid and gaseous sulfur, for example pyrite, which forms FeS(s) and S_2(g) (155) at high temperatures. In some cases, the metal-rich sulfides, which were formed by thermal decomposition, can appear in the gas phase as well. These compounds show noticeable effects of the gas phase transport by decomposition sublimation (156).

$$FeS_2(s) \rightleftarrows FeS(s)\ + 1/2\ S_2(g) \tag{155}$$

$$MS_2(s) \rightleftarrows MS(g)\ + 1/2\ S_2(g) \tag{156}$$

(M = Si, Ge)

Transport of sulfides with iodine as transport agent. Mainly iodine is used as the transport agent for sulfides (as for selenides and tellurides). During the CVT of sulfides with iodine or iodine compounds, the corresponding metal iodide and sulfur are generally formed as transport effective species. In the temperature range that is often used for transport reactions (around 800 to 1000 °C), sulfur is mostly present as the S_2-molecule. At lower temperatures, the formation of larger sulfur molecules (S_2, S_3 … S_8) is additionally expected. The CVT of homogeneously composed crystals of sulfide-selenide solid solutions succeeds by similar vapor pressures of the respective selenium gas species (Se_2,… Se_8). In this respect, the system of cubic mixed-phases ZnS/SnSe has been examined in detail. Zinc sulfide and zinc selenide are completely mixable in the solid state and, moreover, can be transported under the same conditions. Consequently, the transport of $ZnS_{1-x}Se_x$ by adding iodine [158] results in large crystals without significant changes of the composition between source and sink [159], Figure 26. The concentration effects become clear by thermodynamic modeling of $ZnS_{1-x}Se_x$ mixed-phases and their transport behavior [159].

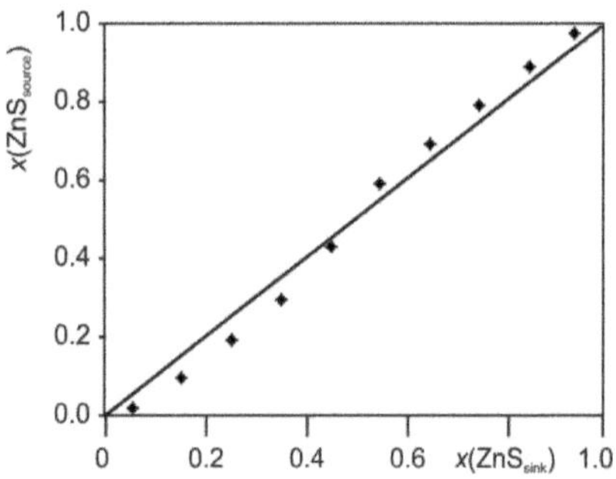

Figure 26. Relation between the composition of the solid in the source and sink during the transport of $ZnS_{1-x}Se_x$ mixed phases with iodine, according to [159].

The crystallization of iron(II) sulfide plays an important role for the understanding of vapor transports for compounds with a considerable homogeneity range. The transport of "FeS" with iodine was already reported in early times [1, 8, 160, 161]. Nevertheless, the transport does not always succeed under the given conditions, as it is dependent on the composition of the initial solid FeS_x, too [18]. When iodine is added, the gas phase over FeS_x basically contains FeI_2, Fe_2I_4, FeI_3, I, I_2, and S_2. Their partial pressures are dependent on the temperature and the composition of the solid, Figures 27, 28. Thus, at 1000 °C, sulfur is volatized in noteworthy scale only for solids of iron sulfide with $x > 0.05$ but the sulfur pressure is very low for stoichiometrically composed $FeS_{1.0}$. Consequently, due to the insufficient amount of sulfur as transport effective species, the transport of $FeS_{1.0}$ using iodine is not possible [18, 19].

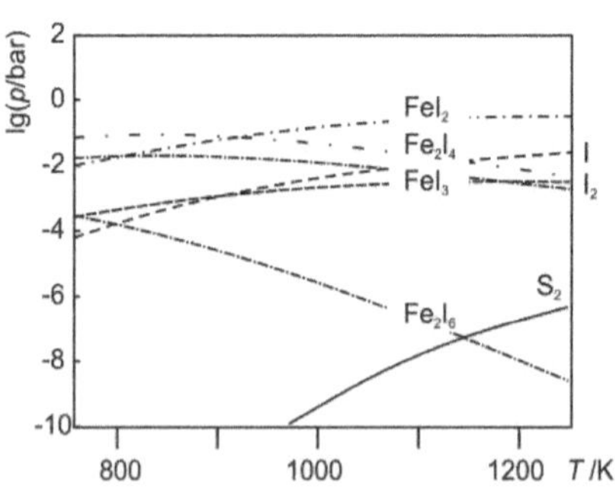

Figure 27. Composition of the gas phase for the transport of $FeS_{1.0}$ using iodine, according to [2, 18].

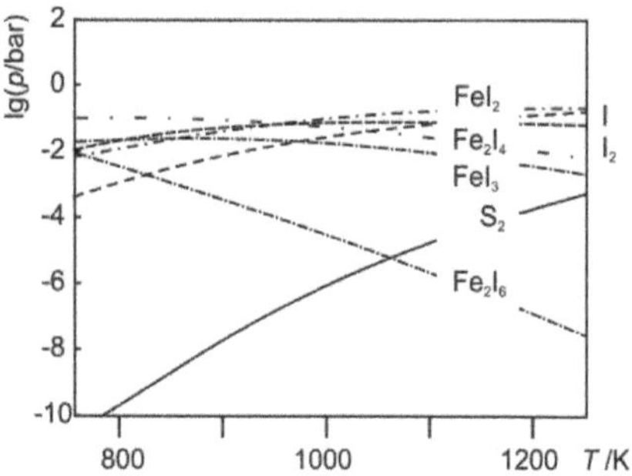

Figure 28. Composition of the gas phase for the transport of $FeS_{1.1}$ using iodine, according to [2, 18].

Transport of sulfides with addition of hydrogen halides. However, the transport of $FeS_{1.0}$ succeeds with hydrogen halides as transport agents (HCl, HBr, HI). In these cases, sulfur is not present in the gas phase in elemental form but as H_2S. Thus the solubility of sulfur increases by some orders of magnitude, and the transport succeeds with transport rates of some milligrams per hour [20]. The transport of hydrogen halide can be advantageous for sulfur-poor compounds when there is no transport effective solution due to the low partial pressure of sulfur.

$$FeS(s) + 2\ HCl(g) \rightleftarrows FeCl_2(g) + H_2S(g) \tag{157}$$

The use of hydrogen chloride was even successful to optimize the transport behavior of mixed-crystals $ZnS_{1-x}Se_x$ [162, 163]. Instead of pure HCl ammonium chloride, NH_4Cl, can be used as an additive. Then, hydrogen chloride is formed during heating of the transport ampoule.

$$ZnS_{1-x}Se_x + 2\ HCl(g) \rightleftarrows ZnCl_2(g) + (1-x)\ H_2S(g) + xH_2Se(g) \tag{158}$$

A series of studies report the CVT reactions of sulfides with halogenating additives $CrCl_3$, $AlCl_3$, $CdCl_2$, or $TeCl_4$. At least for transports with $AlCl_3$ and $TeCl_4$ the formation of hydrogen chloride (159) as an effective transport agent can be expected, too.

$$2\ AlCl_3(g) + 3\ H_2O(g) \rightleftarrows Al_2O_3(s) + 6\ HCl(g) \tag{159}$$

Transport of sulfides with hydrogen and other elements as transport agents. During a few transport reactions, the transport agent does not react with the metal atoms of the solid but with the sulfur atoms instead. In particular, hydrogen is one of these transport agents, which can be used successfully for cadmium and zinc compounds. Transport reactions in which hydrogen is used as transport agent are unusual. Here, the transport agent reacts with sulfur atoms of the solid under formation of hydrogen sulfide [164]. The vapor transports are made possible by the fact that zinc and cadmium, respectively, can be formed elementally in gaseous form during these reactions. The transport with hydrogen is also suited to grow larger single-crystals [165, 166], compared to iodine. Additionally, the contamination of the obtained crystals by transport agent is excluded.

$$CdS(s) + H_2(g) \rightleftarrows Cd(g) + H_2S(g) \tag{160}$$

Transport reactions in which the transport agent reacts solely with the non-metal of the solid are exceptions. Thus, for the transport of zinc sulfide with phosphorus gaseous PS is formed [167].

$$CdS(s) + 1/4\ P_4(g) \rightleftarrows Cd(g) + PS(g) \tag{161}$$

In some cases (SiS_2, TiS_2, TaS_2), CVT with sulfur as transport additive was successful. The transport effect was ascribed to the formation of gaseous polysulfides [168].

Transport of selenides and tellurides. To date, many examples of CVT of selenides and tellurides of the main group elements (groups 2, 13, 14, and 15) as well as almost all transition metal elements are known; some lanthanoids are included, too [2]. The first reports of on the preparation and purification of selenides and tellurides coincide with the methodological development of the CVT [150]. The alkali metal selenides and tellurides cannot be transported with halogens or halogen compounds due to their high stability. As the selenides and even more the tellurides are less stable than the analogous sulfides, the transport reactions are less endothermic. As a consequence higher partial pressures of the transport effective species and lower temperatures of volatilization, respectively, can be applied. Some selenides and tellurides sublime undecomposed. This applies for the compounds of groups 13 and 14, MQ (M = Ge, Sn, Pb, Q = Se, Te) and M_2Q (M = Ga, In, Tl), respectively. A vast number of compounds show noticeable effects of dissolution by decomposition reactions in the gas phase. In the process, high volatile chalcogen-poor chalcogenides as well as the gaseous chalcogen are formed (162 – 164).

$$M_2Q_3(s) \rightleftarrows M_2Q(g) + Q_2(g) \tag{162}$$

(M = Al, Ga, In)

$$GeSe_2(s) \rightleftarrows GeSe(g) + 1/2\ Se_2(g) \tag{163}$$

$$n/2\ M_2Q_3(s) \rightleftarrows M_nQ_n(g) + n/4Q_2(g) \tag{164}$$

(M = As, Sb, Bi)

The thermal decomposition of ZnSe and CdSe (and similarly ZnTe, CdTe, HgTe) to the elements is of importance, too. Applying the decomposition equlibria (165), the deposition of crystalline ZnSe and CdSe over the gas phase is possible at temperatures above 1000 °C ($Kp > 10^{-4}$ bar) without adding a transport agent. Thus the thermodynamic basis of physical vapor deposition (PVD) processes for deposition of layers of these two compounds is provided [169 – 171].

$$MSe(s) \rightleftarrows M(g) + 1/2\ Se_2(g) \tag{165}$$

(M = Zn, Cd)

More than three quarters of all known CVT reactions of selenides and tellurides take place with the addition of iodine. At temperatures above 600 °C, Se_2 dominates in the gas phase (166). Below this temperature, the higher condensed molecules Se_n (n = 3 … 8) have to be considered. The transport of tellurides dominantly runs by formation of Te_2 as effective species.

$$CdSe(s) + I_2(g) \rightleftarrows CdI_2(g) + 1/2\ Se_2(g) \tag{166}$$

Besides transports by using hydrogen or hydrogen halides have been reported.

$$MnSe(s) + 2\ HCl(g) \rightleftarrows MnCl_2(g) + H_2Se(g) \tag{167}$$

$$ZnSe(s) + H_2(g) \rightleftarrows Zn(g) + H_2Se(g) \tag{168}$$

As already mentioned, the use of hydrogen or hydrogen halides as transport agent is important for the transport of oxides and sulfides because the solubility of oxygen and sulfur, respectively, in the gas phase is supported by the formation of water and hydrogen sulfide, respectively. However, the stability of hydrogen compounds H_2Q (Q =O, S, Se, Te) constantly decreases, the participation of H_2Se and, in particular, of H_2Te in CVT reactions must be discussed critically. H_2O as well as H_2S is still stable above 1000 °C. H_2Se, however, decomposes already between 700 and 800 ° ($Kp_{,1000}(H_2Se) = 1$ bar). $H_2Te(g)$ is unstable in the entire temperature range ($Kp_{,T}(H_2Te) = 10^2$ bar), Figure 29. Consequently, the partial pressure of $Te_2(g)$ resulting from the equilibrium (169) is higher by orders of magnitude. Accordingly, the described transport of cadmium telluride in the presence of hydrogen [172, 173] should rather be seen as a decomposition sublimation. Transport reactions that take place with added halogen/hydrogen mixtures occur under formation of the respective metal halides and Te_2 (170), [172 – 174].

$$2/3\ H_2Q(g) \rightleftarrows 2/3\ H_2(g) + 1/3Q_2(g) \tag{169}$$

(Q = O, Se, Se, Te)

$$MTe(s) + 2\ HX(g) \rightleftarrows MX_2(g) + H_2(g) + 1/2\ Te_2(g) \tag{170}$$

(M = Cd, Pb, Zn)

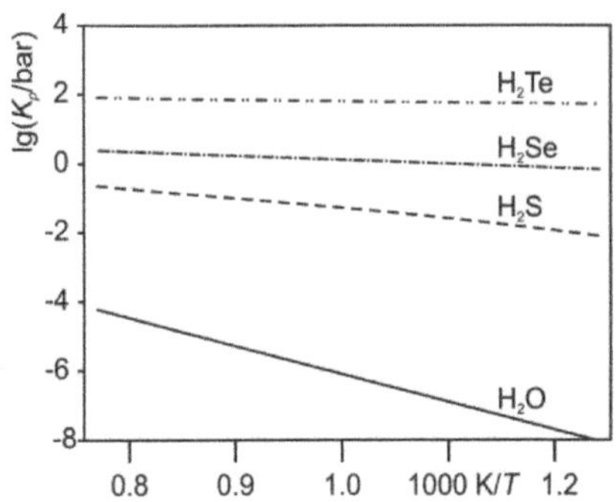

Figure 29. Equilibrium constants Kp for the decomposition of hydrogen chalcogenides H_2Q (Q = O, S, Se, Te) in equilibrium (169), according to [2].

3.2.7. Chemical vapor transport of pnictides

The character of chemical bonding of metal pnictides is very variable and ranges from the metallic, ionic, and covalent nitrides and phosphides through the rather covalent or metallic arsenides and antimonides to the typical metallic bismutides. Thus the transport behavior changes significantly. There is only one example of the CVT of a binary nitride, TiN [175]. The chemical vapor transport of phosphides and arsenides is documented by numerous examples [2] while there are only a few examples of the transport of antimonides and only one of a bismuth-containing intermetallic phase, NiBi [176].

Elemental halogens, in particular iodine, and halogen compounds are preferred as transport additives. While nitrogen, phosphorus, and arsenic have sufficiently high saturation pressures to be transport effective in elemental form, it is necessary to generate transport- effective compounds for the antimonides and bismuthides. This becomes possible by the increasing tendency of pnicogens to form halogen compounds. Concerning the transport of phosphides, in the gas phase mostly phosphorus(III) halides occur. For the transport of arsenides and

Figure 30. Crystal of $ZrAs_2$ grown by chemical vapor transport.

antimonides one has to expect, at rising temperatures, the formation of monohalides, too. This applies in particular for the heavy halogens.

Transport of phosphides with iodine as transport agent. In most cases, the transport of phosphides of the transition metals by adding iodine is possible. As a special feature, transports of phosphides require a comparatively high transport agent density of iodine of about 5 mg cm^{-3}. Depending on the thermodynamic stability of the phosphide and the volatile metal iodide, the vapor transport can occur in a temperature gradient via exothermic (e. g., VP (171), MnP, Cu_3P (172)) or endothermic (e. g., CrP (173), CoP (174), CuP_2) reactions [2]. The addition of phosphorus(III) iodide and hydrogen iodide as transport agent induces similar transport reactions (175, 176).

$$VP(s) + 7/2\, I_2(g) \rightleftarrows VI_4(g) + PI_3(g) \tag{171}$$

$$Cu_3P(s) + 3\, I(g) \rightleftarrows Cu_3I_3(g) + 1/4\, P_4(g) \tag{172}$$

$$CrP(s) + I_2(g) \rightleftarrows 1/2 Cr_2I_4(g) + 1/4\, P_4(g) \tag{173}$$

$$CoP(s) + 5/2\, I_2(g) \rightleftarrows CoI_2(g) + PI_3(g) \tag{174}$$

$$CuP_2(s) + 1/3\, PI_3(g) \rightleftarrows 1/3\, Cu_3I_3(g) + 7/12\, P_4(g) \tag{175}$$

$$MnP(s) + 2\, HI(g) \rightleftarrows 1/2\, Mn_2I_4(g) + 1/4\, P_4(g) + H_2(g) \tag{176}$$

Experimental results suggest that phosphides show the best results (high transport rates; large crystals) at a ratio of $n(M) : n(\mathrm{P})$ close to 1 : 1. Chemical vapor transports of metal-rich and phosphorus-rich phosphides can only be conducted with lower efficiency. This is due to the unbalanced chemical activities of the components in the respective binary compounds: If the activity of the metal component in a phosphide is high but that of phosphorus very low (metal-rich phosphide), the only reaction that will occur is that of the transport agent iodine with the metal under formation of the volatile metal iodide. In some cases, even its saturation pressure is exceeded so that condensed metal iodides appear as well. Phosphorus is kept and enriched in the solid; the simultaneous volatilization of both components is impossible. The reactions of $Cr_{12}P_7$, Fe_2P, and Co_2P with iodine can be served as examples of this behavior [138]. Otherwise, the formation of very stable metal iodides, as described above, can lead to the development of phosphorus-rich phosphides (incongruent volatilization of phosphides) even without high metal activity in a phosphide. Thus in experiments with sufficiently high initial amounts of iodine TiP_2 adjacent to TiP [177]; ZrP_2 adjacent to ZrP [177]; as well as CuP_2 adjacent to Cu_3P and CuI(l) [178] appeared (177, 178).

$$2MP(s) + 2\,I_2(g) \rightleftarrows MI_4(g) + MP_2(s) \tag{177}$$

(M = Ti, Zr)

$$2\,Cu_3P(s) + 5/2\,I_2(g) \rightleftarrows 5\,CuI(l) + CuP_2(s) \tag{178}$$

Transport of phosphides with mercury bromide as transport agent. If the phosphorus coexistences pressure is too low to be transport effective and additionally does the thermodynamic stability of the phosphorus iodides P_2I_4 and PI_3 not suffice to keep phosphorus in the gas phase, $HgBr_2$ can be applied as transport agent. Thus for metal-rich phosphides, Mo_3P, Mo_4P_3, and Fe_2P, the transfer of phosphorus through the gas phase takes place via the more stable phosphorus bromide (179-181) [138, 179].

$$Mo_3P(s) + 9/2\,HgBr_2(g) \rightleftarrows 3\,MoBr_2(g) + PBr_3(g) + 9/2\,Hg(g) \tag{179}$$

$$Mo_4P_3(s) + 17/2\,HgBr_2(g) \rightleftarrows 4\,MoBr_2(g) + 3\,PBr_3(g) + 17/2\,Hg(g) \tag{180}$$

$$Fe_2P(s) + 7/2\,HgBr_2(g) \rightleftarrows 2\,FeBr_2(g) + PBr_3(g) + 7/2\,Hg(g) \tag{181}$$

Transport of phosphides with phosphorus. In contrast to the above discussed transport reactions of phosphides with halogens or halogen compounds, the transport of InP and GaP succeeds by adding an excess of phosphorus. *Ab initio* calculation of the stability of different gas species in the system are indicating the formation of MP_5(g) (*M*: In, Ga) [180].

$$MP(s) + P_4(g) \rightleftarrows MP_5(g) \tag{182}$$

(M = In, Ga)

Transport of arsenides. The CVT of arsenides is referred for many examples [2]. Because of the technical applications of gallium arsenide, the arsenides of group 13 are experimentally examined in a comprehensive manner. Compared to the other pnictides, the transport of arsenides behaves similar to that of phosphides but markedly different to those of the antimonides and bismutides. This is due to the high saturation vapor pressure of phosphorus and arsenic at rather low temperatures: 1 bar at 277 °C and 602 °C, respectively. Hence phosphorus as well as arsenic can be transferred to the gas phase in considerable amounts at relatively low temperatures without exceeding the saturation vapor pressure and thus condensing again. The saturation vapor pressure of antimony, in contrast, reaches the value of 1 bar at 1585 °C. As far as the thermodynamic stability of the pnictides is concerned, phosphides and arsenides are similar as well. Consequently, the most important transport agent for the crystallization of the arsenides is iodine as well (183).

$$NdAs(s) + 3\,I(g) \rightleftarrows NdI_3(g) + 1/2\,As_2(g) \tag{183}$$

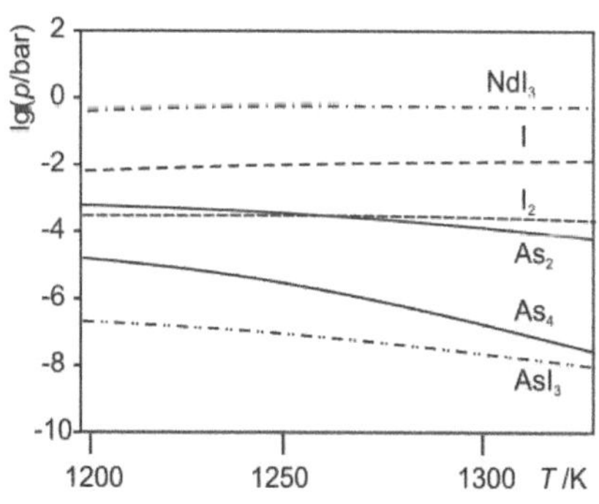

Figure 31. Composition of the gas phase for the transport of NdAs using iodine, according to [2].

Thermodynamic model calculations make clear, that the *transport additive* iodine (or other halogens), not necessarily acts as the effective *transport agent*. Often the arsenic trihalides or the metal or semi-metal halides, respectively, which are formed from the halogens and the solids, function as such. In the transport equilibria of FeAs (184) and GaAs (185), iodine is added but arsenic(III) iodide is the effective transport agent.

$$FeAs_2(s) + AsI_3(g) \rightleftarrows FeI_2(g) + 3/4\ As_4(g) + 1/2\ I_2(g) \tag{184}$$

$$GaAs(s) + 1/2\ GaI_3(g) \rightleftarrows 3/2\ GaI(g) + 1/4\ As_4(g) \tag{185}$$

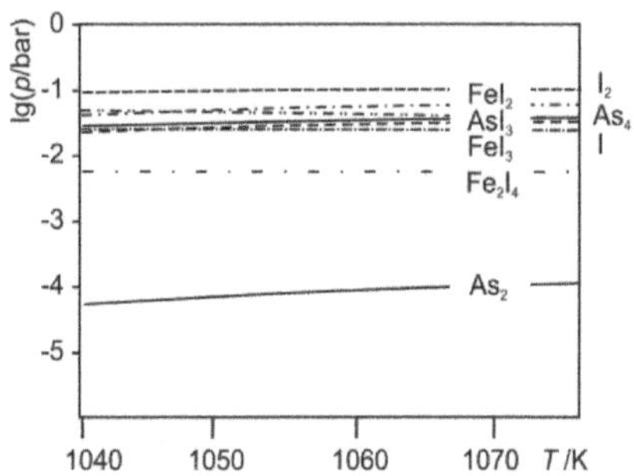

Figure 32. Composition of the gas phase for the transport of $FeAs_2$ using iodine, according to [2].

Arsenic is transferred into the gas phase mainly in elemental form due to the high saturation pressure and the comparatively low stability of gaseous arsenic iodides. Up to approximately 900 to 1000 °C the gas phase is mostly dominated by As_4, above that temperature by As_2. The species As_3 and As are of minor importance to the CVT.

The endothermic transport of silicon arsenide, SiAs can be described by the formation of SiI_4 as effective transport agent (186).

Otherwise, an exothermic transport can be described by HI as transport agent (187), which is formed by traces of water desorbed off the ampoule walls [181].

$$SiAs(s) + SiI_4(g) \rightleftarrows 2\ SiI_2(g) + 1/4\ As_4(g) \tag{186}$$

$$SiAs(s) + 4\ HI(g) \rightleftarrows SiI_4(g) + 1/4\ As_4(g) + 2H_2(g) \tag{187}$$

Additionally, hydrogen halides, hydrogen chloride in particular, are important for the transport of arsenides of group 13 (BAs, GaAs, and InAs). The transport of gallium arsenide with hydrogen chloride (188) and hydrogen bromide, respectively, is well investigated experimentally and by thermodynamic calculations. Here, the formation of AsH_3 has to be taken into account for complex description of the transport behavior. Additionally, GaAs, InAs, $Ga_{1-x}In_xAs$ and $InAs_{1-x}P_x$ can be transported with water. The transport occurs via the equilibrium (189):

$$GaAs(s) + HX(g) \rightleftharpoons GaX(g) + 1/4\ As_4(g) + 1/2\ H_2(g) \quad (188)$$

(X = Cl, Br)

$$2\ GaAs(s) + H_2O(g) \rightleftharpoons Ga_2O(g) + As_2(g) + H_2(g) \quad (189)$$

The transport reaction is always coupled with a redox equilibrium in which a gaseous suboxide [182], arsenic, and hydrogen are formed. Finally, GaAs can be transported with a mixture of water and hydrogen. The mentioned transport agents are used especially in open systems with flowing gases [183].

4. Advanced concepts and thermodynamic modeling of CVT

The course of chemical vapor transports can be understood by thermodynamic considerations (see chapter 2.2). Here various thermodynamic models will be explained in detail. It is state-of-the-art to use computer programs for modeling and quantitative description of transport reactions. Thus, optimum experimental conditions, the direction of a transport, and transport rates can be obtained for many transport systems, frequently even in a predictive way. For more complicated cases, however, a detailed treatment of the underlying thermodynamics will be required. Such a treatment is particularly necessary when a condensed phase with homogeneity range or multi-phasic equilibrium solids do occur in a transport experiment. In addition to the influence of thermodynamic data and phenomena, the transport behavior can be affected by kinetic effects. While the mass flow via the gas phase is generally assumed to be rate determining, some examples have been observed where the kinetics of one or more elementary reaction steps in the transport process exert a dominating influence.

In all cases, the simple looking as well as the more complicated ones, prior to an experiment the experimenter has to develop some idea of which condensed equilibrium phases and gaseous species are to be expected for the transport system under consideration. This knowledge is an essential prerequisite if modeling of transport experiments is to have an outcome close to reality. The most important characteristics for various transport processes are summarized by the following schematics.

Congruent vaporization of a condensed phase: The ratio of the elements in the condensed phase and the gaseous phase of the source are identical. Because of the congruent dissolution of all components into the gas phase always a congruent deposition at the sink occurs. Thus a stationary (steady state), not time-dependent transport behavior result.

The model of simple transport behavior: The vapor transport process can be fully described by a *single* heterogeneous equilibrium reaction. The assessment of the equilibrium state can be realized by calculation of Kp and subsequently of Δp (see chapter 2.2). Hence, the estimation of the transport direction succeeds using the sign of $\Delta_r H^0$ ($\Delta_r H^0 > 0$, $T_2 \rightarrow T_1$ or $\Delta_r H^0 < 0$, $T_1 \rightarrow T_2$).

The model of complex transport behavior The gas-phase composition is formed by *several* independent equilibria. The assessment of the equilibrium state requires the calculation of the gas phase solubility λ of regarding components, see section 4.1. Consequently, the change of solubility $\Delta\lambda$. describes the direction of the transport ($\Delta\lambda_{T2-T1} > 0$, $T_2 \rightarrow T_1$; $\Delta\lambda_{T2-T1} < 0$, $T_1 \rightarrow T_2$).

Incongruent dissolution of the source's condensed phase in the gas phase: During an incongruent dissolution, the molar ratio of the elements in the condensed phase and in the gas phase of the source are not identical. This behavior is always caused by simultaneous occurrence of *several* independent equilibria. In this case the calculation of the mass flow of the components *A* and *B*, *J*(*A.B*), between the equilibrium regions (volumes) is of decisive importance. Hence, the transport is to describe by the *flux relation.*

The extended transport model: This thermodynamic model represents the "quasi-stationary transport" behavior. Thereby, constant phase relations and equilibrium conditions are assumed. Actually, this assumption only applies for the first moment of the experiment. Nevertheless, the thermodynamic description by extended transport model fits very well, if time-independent behavior is experimentally observed. The determination of the composition of the sink's condensed phase succeeds by applying the condition for steady state with ε = constant (ε: relation of stationarity).

The co-operative transport model: If the composition of the deposited solid at the sink changes time-dependently it is called a "sequential transport". This non-stationary transport behavior can be described by the co-operative transport model. The determination of the composition of the sink's condensed phases and of the deposition sequence is realized by an iteration procedure.

4.1. Complex congruent transports

There are many examples, where chemical transport of a solid cannot be completely described by *just one* reaction, since a more complex gas phase is formed. For these cases several unique equilibrium reactions have to be considered. Their number r_u has to be derived by using equation (190). Here, *s* is the number of gas species, *k* the number of components (according to Gibbs's phase rule the number of elements).

$$r_u = s - k + 1 \tag{190}$$

The transport of iron with iodine corresponding to van Arkel [5, 6] might serve as an example for complex congruent transport behavior. The gas species FeI_2, Fe_2I_4, I_2, and I might occur. According to $r_u = 4 - 2+1 = 3$ the partial pressures of all gas species are determined by three unique equilibria (191 - 193).

$$Fe(s) + I_2(g) \rightleftarrows FeI_2(g) \tag{191}$$

$$2\ Fe(s) + 2\ I_2(g) \rightleftarrows Fe_2I_4(g) \tag{192}$$

$$I_2(g) \rightleftarrows 2\ I(g) \tag{193}$$

The first transport equilibrium (191) is endothermic ($\Delta_r H^0_{298} = 24\ kJ \cdot mol^{-1}$). According to Le Chatelier's principle, a transport from T_2 to T_1 is expected. Reaction (192) has the character of a transport equation, too. This reaction runs exothermic ($\Delta_r H^0_{298} = -116\ kJ \cdot mol^{-1}$). The situation becomes even more complicated because iodine is present partly in atomic form at high temperatures (193). Applying atomic iodine, further transport reactions under the formation of $FeI_2(g)$ and $Fe_2I_4(g)$ can be formulated. In all cases the molecules FeI_2 and Fe_2I_4 function as transport effective species. Thus, below 1000 °C, we deal with two opposing processes – the increasing formation of FeI_2 and the decreasing formation of Fe_2I_4, both because of rising temperature. The first process lets us expect transportation towards the cooler zone, the second one to the hotter zone. It is not predictable which process dominates. A new term – the **gas phase solubility** [15] – is helpful for answering this question.

The gas phase solubility. The term gas phase solubility λ refers back to the term solubility of a substance in a liquid. Solutions of solid substances are used for the purification of the dissolved substance through recrystallization. One uses the temperature dependency of the solubility, respectively the solubility equilibrium, and produces an in heat saturated solution. Through cooling, a recrystallization of the solid substance is achieved. A chemical vapour transport reaction works basically the same way. Here, one also uses the temperature dependency of the equilibrium position of the reaction in order to crystallize and to purify. In both cases, one deals with heterogeneous equilibria; in the first case between a solid and a liquid, in the second between a solid and a gas phase [15].

The example of the transport of iron shows the advantage of the term solubility in the description of complicated transport reactions. According to the transport equations (191) and (192), iron can be solved into the gas phase forming the species FeI_2 and Fe_2I_4. The solvent is the gas phase, i.e. all gaseous species together. The quantitative description of the solubility of iron in the gas phase considers that one molecule Fe_2I_4 includes two Fe-atoms, whereas the FeI_2 molecule only includes one. Hence the partial pressure of Fe_2I_4 is multiplied by the factor 2. If analogically same applies to the solvent gas phase, the solubility of iron in the gas phase can be described by equation (194):

$$\lambda(\mathrm{Fe}) = \frac{p(\mathrm{FeI_2}) + 2 \cdot p(Fe_2I_4)}{p(\mathrm{I}) + 2 \cdot p(\mathrm{I_2}) + 2 \cdot p(\mathrm{FeI_2}) + 4 \cdot p(Fe_2I_4)} \tag{194}$$

The temperature dependency of the solubility of iron in the gas phase takes both ferrous molecules FeI_2 and Fe_2I_4 into consideration. Figure 33. As the solubility of iron decreases with growing temperatures, less iron is dissolved at higher temperatures in the gas phase than at lower temperatures. Thus iron must be transported from lower to higher temperatures. This is in accordance with experimental observations of iron transport with iodine from 800 to 1000 °C.

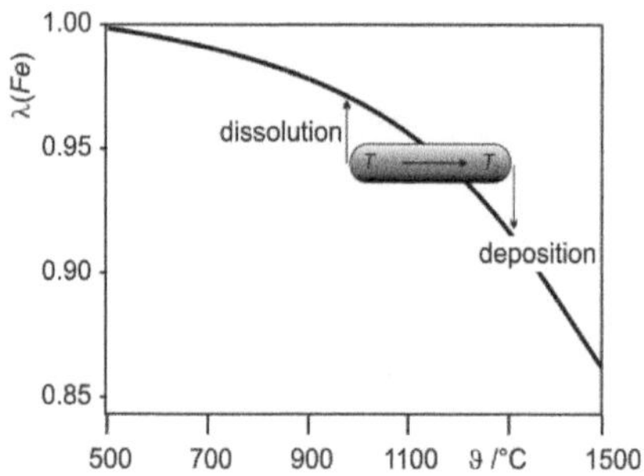

Figure 33. Temperature dependence of the solubility of iron and direction of the transport, according to [2].

Conclusion. With the aid of *solubility of a solid in the gas phase* several possible transport reactions including a variety of gas species can be considered for a complex transport system. The solubility λ can be described by the expression $\lambda = n^*(A)/n^*(L)$ or, using the relation between n and p given by the ideal gas law, by $\lambda = p^*(A)/p^*(L)$. L is meaning the solvent, which can be the transport agent or even an inert gas. The balance of A and L is expressed by the sum of all involved species (195). The numbers $\nu(A)$ and $\nu(L)$ denominate the stoichiometric coefficients of A and L in the gas species. The equation (195) for the solubility of a solid in a gas phase holds for systems of any order of complexity in closed as well as in open systems.

$p^*(A) = \Sigma(\nu(A) \cdot p(A))$

$$\lambda(A) = \Sigma(\nu(A) \cdot p(A)) / \Sigma(\nu(L) \cdot p(L)) \quad (195)$$

Given that equilibrium has been established, the transport direction depends on the difference $\Delta\lambda$ (196):

$$\Delta\lambda = \lambda(T_2) - \lambda(T_1) \quad (196)$$

$\Delta\lambda > 0$ transport direction $T_2 \rightarrow T_1$

$\Delta\lambda < 0$ transport direction $T_1 \rightarrow T_2$

4.2. Incongruent stationary transports (Extended transport model)

The thermodynamic description and modeling of transport systems get increasingly complicated if the transported compound shows a homogeneity range $AB_{x\pm\delta}$ or the transport occurs in a system with several coexisting condensed phases, e. g. AB_y and AB_z. Their vapor transports rather often occur under incongruent dissolution. This case is characterized by different molar ratios of the components (elements) in the source solid and the corresponding gas phase. Consequently, the ratio $(n(B)/n(A))$ of the solid $AB_{x,\ T(\mathrm{sink})}$ is not longer identical to the ratio of the balance pressures $p^*(A)/p^*(B)$ of the components A and B at the sink. Hence, the ratio of the components of the deposited phase (at the sink) does not need to be identical to that of the dissolved phase (source). This behavior is comparable to the peritectic melting of a solid and the compositional shift that accompanies the re-formation of a solid from this melt upon cooling. The composition of melt and solid are different.

The general task to describe transport reactions with phases of variable composition can be treated in a vivid way for the transport within the homogeneity range of $TiS_{2-\delta}$ [16]. The actual transport equilibrium (197) is attended by the decomposition reaction (198).

$$TiS_{2-\delta}(s) + 2\ I_2(g) \rightleftharpoons TiI_4(g) + (2\ -\delta)/2\ S_2(g) \tag{197}$$

$$TiS_2(s) \rightleftharpoons TiS_{2-\delta}(s) + \delta/2\ S_2(g) \tag{198}$$

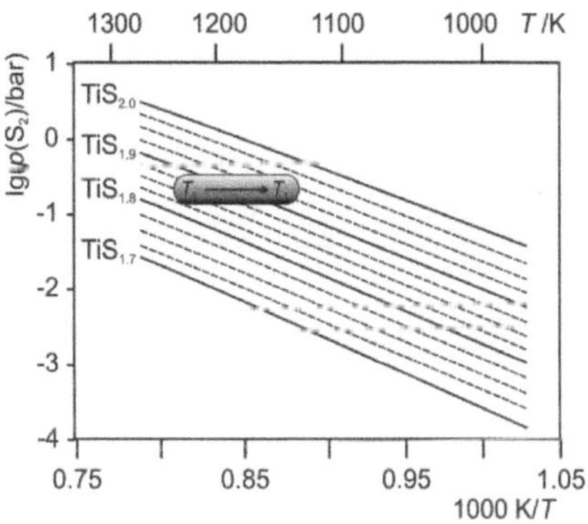

Figure 34. Phase barogram for the system Ti/S showing the co-existence pressures (according to 198) in the homogeneity range $TiS_{2-\delta}$. The phase relations in CVT experiments (950 to 850 °C) are visualized; graphic according to [16, 2].

For transport experiments in the temperature gradient 950 to 850 °C, independent on the starting composition $TiS_{2-\delta}$ of the source solid, at T_1 a sulfur-enriched phase will always be deposited. At the same time a sulfur-depleted phase forms at the source. Thus, the vapor transport starting from an initial composition $TiS_{1.889}$ yields crystals of $TiS_{1.933}$. Consequently, the solid at the source is depleted of sulfur, see Figure 34. The thermodynamic description of the observed phase relations is possible in a rather simple approach by independent calculation of the equilibrium conditions for source and sink. As both equilibrium regions are linked to each other via the gas phase, solids of corresponding compositions are obtained at T_2 and T_1 with the precondition $p(S_2)$ = constant (above $TiS_{2-\delta 1}$ at T_1 and $TiS_{2-\delta 2}$ at T_2; $\delta_2 < \delta_1$). Using the phase barogram ($\lg(p/p^0) = f(x, T)$) of the corresponding system the determination of the stoichiometric coefficients x succeeds along an isobar for given temperatures T_2 and T_1 (Figure 34).

A farther-reaching, general treatment of the phase relations encountered in transport systems with incongruent dissolution of a solid is based on the fact that the two equilibrium regions (source and sink) are indeed not independent to each other: In a system of two components A and B, the solid AB_x will be transferred by the transport agent X into the gas phase. According to Gibbs' phase rule the system with three components and two phases (solid+gas phase) possesses three degrees of freedom for its thermodynamic description: Δp, T_{source}, and $x(T_{source})$

$$AB_{x,source}(s) + X(g) \rightleftarrows AX(g) + x\,B(g) \tag{199}$$

$$F = C - P + 2 \rightarrow F = 3 - 2 + 2 = 3$$

From the considerations follows that the composition of $AB_{x,sink}$ at the sink temperature T_{sink} might be variable, but not independent of the equilibrium conditions valid for the dissolution (source) region (T_{source}, x_{source}, Δp). For a congruent chemical vapor transport, modeling of the transport effect is possible via *independent equilibrium calculations* for source and sink region followed by determination of the differences of partial or balance pressures. In contrast to this situation the equilibrium calculations for source and sink of an incongruent transport have to be linked to each other. Only in doing so, it becomes possible to determine the composition $AB_{x,\,sink}$ at T_{sink}. The relation between the two equilibrium regions at T_{source} and T_{sink} can be described by the mass flow via the gas phase from source to sink. Thereby, not the total substance amounts $n(A)$, $n(B)$ are considered but the resulting differences of the molar numbers in the gas phases of source and sink $n_{source} - n_{sink}$. As a consequence, the composition of $AB_{x,sink}$ is determined by the ratio of the molar flow for A and B, but not by the ratio of the balance pressures (200)

$$\left(\frac{n(B)}{n(A)}\right)_{T_{sink}} = \left(\frac{\text{flux}(B)}{\text{flux}(A)}\right)_{T_{source} \to T_{sink}} = \left(\frac{J(B)}{J(A)}\right) = x_{sink} \tag{200}$$

For a congruent transport equation (200) is valid, too. Obviously, a transfer with constant molar ratio of the components will occur between the equilibrium regions if the ratio of the balance pressures between source and sink is constant. The validity of the flux relation is assumed for *all* chemical transport reactions. The steady-state is given for an incongruent transport only as long as the equilibrium state at the source remains constant. For different values of x_{sink} and x_{source}, the composition of the source solid and the sources gas phase have to change during the course of the transport experiment. The compositional change of the source solid might proceed by discontinuous compositional change by formation of two co-existing phases or continuous compositional change within a homogeneity range.

According to Krabbes, Oppermann, and Wolf the steady-state of a transport system involving incongruent dissolution of a solid AB_x is determined by linking the balance of molar numbers for A in the sink $[n(A(s))_{sink}+n(A(g))_{sink}]$ to the molar number of A in the gas phase of the source [18 - 21]. The molar number of A in the sources solid does obviously not contribute to the flux (201).

$$n\big(A(g)\big)_{source} = n\big(A(s)\big)_{sink} + n\big(A(g)\big)_{sink} \tag{201}$$

The **stationarity relation** ε (202) is linking the fluxes $J(B)$ and $J(A)$ of the individual components of the system assuming that the net flux of the transport agent X will vanish; $J(X) = 0$ [18].

$$\left(\frac{p^*(\mathrm{B}) - x_{sink} \cdot p^*(\mathrm{A})}{p^*(\mathrm{X})}\right)_{T_{source}} = \left(\frac{p^*(\mathrm{B}) - x_{sink} \cdot p^*(\mathrm{A})}{p^*(\mathrm{X})}\right)_{T_{sink}} = \varepsilon \tag{202}$$

The statement of the stationarity relation becomes applicable for the description of a chemical vapor transport by equation (203).

$$\frac{\left(\frac{p^*(\mathrm{B})}{p^*(\mathrm{X})}\right)_{T_{source}} - \left(\frac{p^*(\mathrm{B})}{p^*(\mathrm{X})}\right)_{T_{sink}}}{\left(\frac{p^*(\mathrm{A})}{p^*(\mathrm{X})}\right)_{T_{source}} - \left(\frac{p^*(\mathrm{A})}{p^*(\mathrm{X})}\right)_{T_{sink}}} = \frac{\Delta\lambda(B)}{\Delta\lambda(A)} = x_{sink} \tag{203}$$

Consequently, the fluxes $J(A)$ and $J(B)$ are proportional to the differences of the corresponding balance pressures in source and sink, normalized by the balance pressures for the solvent. In the same way the ratio $J(B) : J(A)$ is equal to the ratio of differences of the components gas phase solubilities [18]. Based on the extended transport model, this approach to the theoretical treatment of chemical vapor transport reactions is realized in the software package TRAGMIN [23]. In addition to calculation of equilibrium partial pressures and condensed phases the extended transport model offers further information on experimental realization and theoretical understanding of transport reactions [2]:

- Calculation of the transport efficiency of gas species and deduction of the prevailing transport reaction(s)
- Calculation of the influence of experimental conditions on the deposition of solids with homogeneity range, see FeS_x [18-20].
- Calculation of the influence of experimental conditions on the deposition of multi-phasic solids, see V_nO_{2n-1} [125 - 127].

4.3. Non–stationary transports (Co–operative transport model)

Using rather large amounts of a solid as source material together with sufficiently short experiment duration will yield quasi-stationary transport behavior (composition almost independent on time). Thus, deposition of a single phase solid of constant composition will be possible. Non-stationary behavior occurs, if sequential migration of several different solids to the sink will be observed. The vapor transport of solids with homogeneity range, too, might be accompanied by a variation of the composition of the sink solid over time. Experimental evidence for *non-stationary* transport behavior can be obtained from series of transport experiments allowing for variable duration of the experiments. Much easier experimental access to non-stationary transport behavior is possible by using the so-called transport balance (see section 5). Despite charging a single-phase solid into a transport ampoule, a multi-phase equilibrium solid might form at the source region, due to the setting of chemical equilibrium at the beginning of the experiment [2, 139, 184 – 187]. Formation of *multi-phase* equilibrium solids at the source region of a transport ampoule can result from three reasons.

- Reaction between starting material and transport agent.
- Thermal decomposition of the starting material at the conditions of the transport experiment.
- Reaction between the starting material and the ampoule material (possibly involving the transport agent).

The observations made for the transport of copper(II) oxide by iodine [2, 185] can serve as an example for the complex phase relations and deposition sequences in chemical vapor transports. The transport behavior is characterized by partial thermal decomposition (204, 205) and the formation of condensed metal halides (206) occurring besides the actual transport reaction (207). Directed and reproducible syntheses depend not only on the appropriate molar ratios for the various components (copper, oxygen, iodine). The absolute amounts of starting materials and the ampoule volume are decisive too – since all components are solved at a substantial, however not equal, amount in the gas phase. The presence of multi-phase solids at the source at the beginning of the transport experiment leads to sequential migration of copper(II) oxide and copper(I) oxide, Figure 35.

$$2\ CuO(s) \rightleftarrows Cu_2O(s) + 1/2\ O_2(g) \tag{204}$$

$$Cu_2O(s) \rightleftarrows 2\ Cu(s) + 1/2\ O_2(g) \tag{205}$$

$$CuI(l) \rightleftarrows 1/3\ Cu_3I_3(g) \tag{206}$$

$$CuO(s) + \tfrac{1}{2}\ I_2(g) \rightleftarrows 1/3\ Cu_3I_3(g) + 1/2\ O_2(g) \tag{207}$$

Subsequent to initial equilibration the source solid consists of CuO and Cu_2O, the gas phase of O_2 (204, 205) and Cu_3I_3 (207). After transfer of the gas phase to the sink, cooling to the sink temperature leads to supersaturation of the gas phase, which eventually results in crystalli-

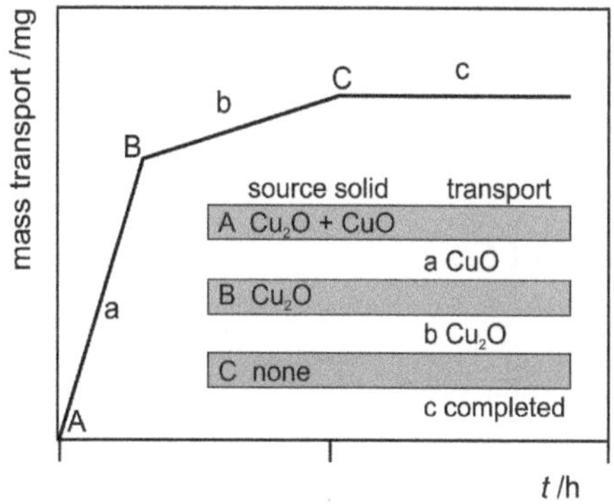

Figure 35. Non-stationary transport behavior of the CuO/I_2 system (1050 → 950 °C; 10 mg iodine), according to [185].

zation of the thermodynamically most stable phase, which is under the given conditions copper(II) oxide. Dissolution of copper(II) oxide at the source and its deposition at the sink result in the steady state (section a), which is characterized by constant ratio of fluxes from source to sink: $J(O)/J(Cu) = 1$ and $(1/2\ J(O_2))/(1/3\ (Cu_3I_3)) = 1$, respectively. After complete consumption of copper(II) oxide at the source, copper(I) oxide will be dissolved in a second steady state (section b). During the deposition of Cu_2O, the ratio of the fluxes are $J(O)/J(Cu) = 1/2$ and $(1/2\,J(O_2))/(1/3\,J(Cu_3I_3)) = 1/2$, since only O_2(g) and Cu_3I_3(g) are effective for the transport.

The model of co-operating equilibrium zones ("model of co-operative transport"). The calculation of the equilibrium solid(s) and gas phase in source and sink becomes possible applying the *model of co-operative transport* [22, 193]. It involves the minimization of the Gibbs energy according to Eriksson [188] for the two equilibrium regions of the transport ampoule. However, the main conceptual problem in modeling CVT experiments lies in the linking of the equilibrium calculations for the source and sink regions. In section 4.2 it has been described how the *extended transport model* can be applied to incongruent evaporation (and deposition) of solids in *quasi-stationary* transport experiments. In order to describe the complete (time dependent) transport behavior as a non *non-stationary* process, the model of co-operative transport uses an iterative calculation procedure [24].

For this purpose, the equilibrium condensed phase(s) of the source and sink obtained by a calculation cycle are kept at these regions. The source calculation of the subsequent cycle is performed without the molar amounts of the elements deposited at the sink in the preceding cycle. The stepwise ("cyclewise") transfer of the source solid(s) to the sink is simulated by repeated calculation cycles. The calculation is finished once no condensed phase is left at the source. Alternatively, the calculation is terminated when the source solid's composition remains stable from one cycle to the next – only the molar number of the solid is decreased. According to the stationarity criterion a steady state has been reached when the gas phase (in the source and sink) remains constant from one calculation cycle to the next. With respect to the mass transfer from the source to the sink, this means that within one calculation cycle the molar numbers of the source solid's components dissolved in the gas phase and deposited at the sink are equal. If more than one condensed phase is involved in this process, we find simultaneous transport. This procedure gets by without explicit balancing of the fluxes of the individual gas species, in contrast to the flux relation [18 - 21].

5. Experimental setup of CVT

Vapor transport experiments can be realized with different complexity. What kind of technique is used depends on the aim of the experiment. As shown in chapter 2.2, the setup as well as the specific parameters (substance amount of transport agent, temperature, temperature gradient) greatly influence the rate of mass transport. Accordingly, a high transport rate usually is chosen for the synthesis of a compound or the purification of it. If crystals are to be grown, the crystal quality is kept in mind and therefore rather smaller transport rates are aspired. In principle, two working methods can be applied for the practical realization in the laboratory: the transport in open or closed systems. In an open system a continuous flow of the transport agent is led over the source material; the solid, which is kept at a certain temperature, deposits at a different place with another temperature under the release of the transport agent. Transport reactions in an open system are often used for substance separation and purification. Due to the loss of the transport agent in the continuous gas flow only time-limited experiments in the range of some hours are realizable. Of course, high transport rates are intended for these experiments. In a closed system, typically a sealed ampoule, the transport agent remains in the system and consistently re-enters the reaction. Thus investigation periods of some days are attainable.

In most cases, transport reactions are executed in tubes or ampoules (diameter 10 to 20 mm) of a suitable glass. Today silica glass is frequently used, which is stable up to 1100 °C and quite inert to corrosive fillings. It is important to note that water is released during the heating of silica glass (water content up to 50 ppm). In order to avoid this, careful baking out of the ampoule in vacuum is recommended. Containers made from ceramic materials or glassy carbon can be integrated in a silica ampoule when highly corrosive materials have to be transported.

Vapor transport reactions take place in a temperature gradient. In order to set up the gradient in a controlled manner, tube furnaces with at least two independent heating zones are used, Figure 36. The transport furnace should be in a horizontal position in order to keep convection as part of the gas motion as small as possible. However, if the aim of the transport is the preparation of large amounts of substance by an endothermic transport, the furnace can be tilted so that the sink side is higher than the source side. This increases the transport rate. These experiments, however, cannot be described by the thermodynamic models that are based on gas motion by diffusion. The so-called *short-distance transport*, which was described by Krämer, uses specifically a high convective contribution to the gas motion [189]. The transport takes place in a vertical direction, over a distance of approximately 3 cm only. The ampoule cross section in such experiments is particularly large, 30 cm^2, Figure 37. This way, CVT also succeeds in systems that otherwise have a low transport rate due to an unfavorable equilibrium position.

This variation of the standard experimental set up is particularly effective for endothermic transport reactions. In cases of exothermic transport reactions, the convective part is omitted and the transport rate is exclusively determined by diffusion.

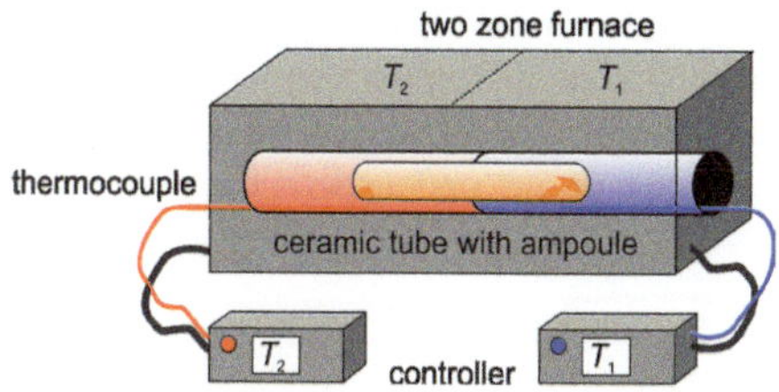

Figure 36. Experimental set up for chemical vapor transport in a conventional two-zone furnace, according to [2].

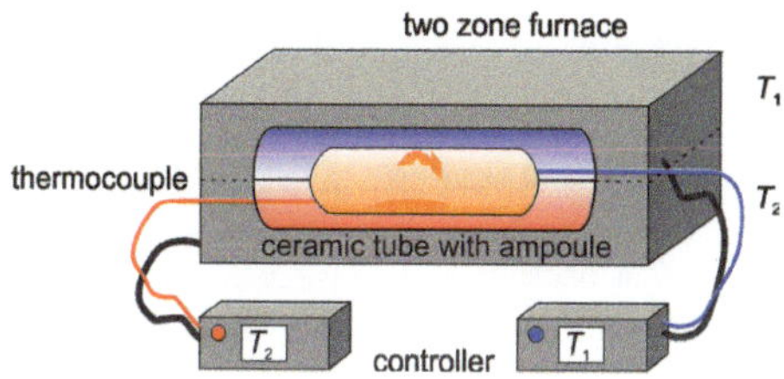

Figure 37. Experimental set up for chemical vapor transport in a short-distance two-zone furnace, according to [2, 189].

The experimental procedures for preparing transport ampoules can be different. Above all, they are dependent on the physical and chemical properties of the transport agent. First, the prepared ampoules are filled with approximately 0.5 up to 1 gram of the initial solid that is to be transported. For this purpose one uses a funnel long enough that the outlet is near the ampoule bottom. In the same way the transport agent can be added. Its amount is often selected so that the pressure (approximately expressed by the initial pressure of the transport agent) in the ampoule is 1 bar at the experiment temperature (calculated using the gas law). The transport ampoule and the vacuum line can be joined with a ground-glass joint. Alternatively, "quick-fit" joints have been established. Usually the contents of the ampoule must be cooled with liquid nitrogen before evacuation in order to avoid vaporization or sublimation of the respective transport agent. If iodine is used as transport agent, cooling is obligatory. If transport agents shall be used, which are already gaseous at room temperature (HCl, HBr, Cl_2, Br_2), more advanced techniques have to be applied for filling the ampoules [2]. These procedures can be avoided by using the ammonia halides as a source for the hydrogen halides and $PtCl_2$ or $CuCl_2$ for Cl_2. Finally, the reaction ampoule is evacuated and sealed under dynamic vacuum.

Safety advice: It is absolutely essential to avoid the condensation of liquid oxygen or moisture within the ampoule prior to sealing. If the ampoule is sealed in this state, a strong explosion will be the result after the removal of the cooling agent due to very high pressure in the ampoule.

The prepared transport ampoule is placed in the middle of the furnace reaching both temperature zones. Before the actual transport experiments, usually a back transport or transport in a reverse temperature gradient is applied. This way, the ampoule walls on the sink side are freed of small crystallization seeds. Finishing the experiment the ampoule is taken out carefully. In order to obtain crystals without being contaminated by the condensed gas phase, one has to make sure that the gas phase condenses on the source side.

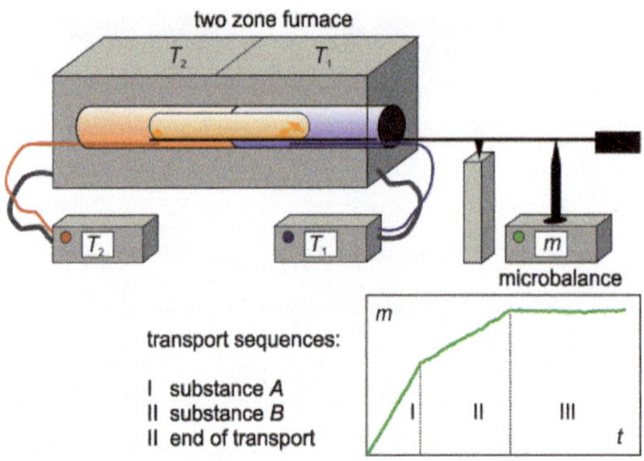

Figure 38. Determination of time dependent rates of mass transport using a transport balance, according to [2, 197].

For quite simple transport experiments, the determination of the transport rate is realized by weighing the crystals and calculation of an average rate within the total experimental time. More advanced, a transport balance can be applied, which is a measuring device for recording the time dependence of mass transports. In the process, the changes of the tracking force of the balance is recorded and graphically represented during the entire transport experiment. This way, the transport action can be followed online [133].

Author details

Peer Schmidt[1], Michael Binnewies[2], Robert Glaum[3] and Marcus Schmidt[4]

1 Lausitz University of Applied Sciences, Faculty of Science, Senftenberg, Germany

2 Leibniz-University Hannover, Inorganic Chemistry, Hannover, Germany

3 University Bonn, Inorganic Chemistry, Bonn, Germany

4 Max-Planck-Institute of Chemical Physics of Solids, Dresden, Germany

References

[1] H. Schäfer, Chemische Transportreaktionen. Weinheim: Verlag Chemie; 1962.

[2] M. Binnewies, R. Glaum, M. Schmidt, P. Schmidt, Chemical Vapor Transport Reactions. Berlin: De Gruyter; 2012.

[3] F. Holleman, N. Wiberg, Lehrbuch der Anorganischen Chemie. Berlin: de Gruyter; 102. Edt. 2007.

[4] R. Bunsen, J. Prakt. Chem. 1852; 56, 53.

[5] E. van Arkel, J. H. de Boer, Z. Anorg. Allg. Chem. 1925; 148, 345.

[6] E. van Arkel, Reine Metalle. Berlin: Springer; 1939.

[7] R. Nitsche, Fortschr. Miner. 1967; 442, 231.

[8] H. Schäfer, J. Cryst. Growth 1971; 9, 17.

[9] E. Kaldis, Principles of the vapour growth of single crystals. In: C. H. L. Goodman (Ed.), Crystal Growth, Theory and Techniques. Vol. 1. 1974, 49.

[10] J. Mercier, J. Cryst. Growth, 1982; 56, 235.

[11] M. Lenz, R. Gruehn, Chem. Rev. 1997; 97, 2967.

[12] M. Binnewies, Chemie in uns. Zeit 1998; 32, 15.

[13] R. Gruehn, R. Glaum, Angew. Chemie 2000; 112, 706; Angew. Chem. Int. Ed. 2000; 39, 692.

[14] K.-Th. Wilke, J. Bohm, Kristallzüchtung. Frankfurt: Harri Deutsch; 1988.

[15] H. Schäfer, Z. Anorg. Allg. Chem. 1973; 400, 242.

[16] M. Saeki, J. Crystal Growth 1976; 36, 77.

[17] M. W. Richardson, B. I. Noläng, J. Crystal Growth 1977; 42, 90.

[18] G. Krabbes, H. Oppermann, E. Wolf, Z. Anorg. Allg. Chem. 1975; 416, 65.

[19] G. Krabbes, H. Oppermann, E. Wolf, Z. Anorg. Allg. Chem. 1976; 421, 111

[20] G. Krabbes, H. Oppermann, E. Wolf,. Z. Anorg. Allg. Chem. 1976; 423, 212.

[21] G. Krabbes, H. Oppermann, E. Wolf, J. Crystal Growth 1983; 64, 353.

[22] R. Gruehn, H. J. Schweizer, Angew. Chem. 1983; 95, 80.

[23] G. Krabbes, W. Bieger, K.-H. Sommer, T. Söhnel, U. Steiner, Computerprogramm TRAGMIN, Version 5.0, IFW Dresden, TU Dresden, HTW Dresden, 2008. http://www2.htw-dresden.de/~webchem/steiner/Tragmin/tragmin.htm

[24] O. Trappe, R. Glaum, R. Gruehn, Das Computerprogramm CVTRANS, Universität Gießen, 1999. http//www.glaum.chemie.uni-bonn.de/forschung/cvtrans

[25] H. Oppermann, M. Schmidt, P. Schmidt, Z. Anorg. Allg. Chem. 2005; 631, 197.

[26] S. Hampel, M. Ruck, P. Schmidt, Z. Anorg. Allg. Chem. 2005; 631, 272.

[27] K. H. Berthel, R. Hedderich, B. Pietraß, In: Proc. 5th Internat. Symp. High Purity Materials in Science and Techn., Dresden, 1980; R. Hedderich, PhD thesis. Academy of Sciences of GDR, Berlin; 1980; D. Rhede, PhD thesis, Academy of Sciences of GDR, Berlin, 1982.

[28] H. Oppermann, A. S. Okhotin, Adv. Space Res. 1981; 1, 51; H. Oppermann, In: Freiberger Forschungshefte, VEB Deutscher Verlag für Grundstoffindustrie 1987; A 767, 97.

[29] H. Schornstein, R. Gruehn, Z. Anorg. Allg. Chem. 1988; 561, 103.

[30] H. Schornstein, R. Gruehn, Z. Anorg. Allg. Chem. 1989; 579, 173.

[31] H. Oppermann, M. Ritschel, Kristall Techn. 1975; 10, 485.

[32] H. Oppermann, P. Schmidt, Z. Anorg. Allg. Chem. 2005; 631, 1309.

[33] M. Binnewies, Z. Anorg. Allg. Chem. 1977; 435, 156.

[34] H. Schäfer, Z. Anorg. Allg. Chem. 1974; 403, 116.

[35] H. Schäfer, J. Cryst. Growth, 1975; 31, 31.

[36] H. Schäfer, Angew. Chem. 1976; 88, 775; Angew. Chem., Int. Ed. 1976; 15, 713.

[37] H. Schäfer, Adv. Inorg. Radiochem. 1983; 26, 201.

[38] M. Binnewies, H. Schäfer, Z. Anorg. Allg. Chem. 1974; 407, 327.

[39] L. H. Ngai, F. E. Stafford, In: L. Eyring (Ed.) Advances in High Temperature Chemistry. New York, Academic Press, 1971, 313.

[40] H. Schäfer, In: Nat. Bur. Standards, Spec. Publ. 364, Solid State Chem. Proceed.5th Materials Research Sympos., 1972, 413.

[41] S. Lieke und M. Binnewies, Z. Anorg. Allg. Chem. 1997; 623, 1705.; M. Binnewies, K. Jug, Eur. J. Inorg. Chem. 2000; 1127.

[42] M. Binnewies, H. Schnöckel, Chem. Rev. 1990; 90, 32.

[43] H. Schäfer, F. Wehmeier, M. Trenkel, J. Less-Common Met. 1968; 16, 290.

[44] M. Binnewies, Z. Anorg. Allg. Chem. 1976; 422, 43.

[45] R. Köppe, J. Steiner, H. Schnöckel, Z. Anorg. Allg. Chem. 2003; 62, 2168.

[46] R. Köppe, H. Schnöckel, Angew. Chem. 2004; 116, 2222; Angew. Chem., Int. Ed. 2004; 43, 2170.

[47] K. A. Gingerich, E. Kaldis, Eds., Current Topics in Material Science, North-Holland Publishing Company, 1980; 6, 345.

[48] H. Schäfer, Z. Anorg. Allg. Chem. 1986; 543, 217.

[49] H. Schäfer, A. Tebben, Z. Anorg. Allg. Chem. 1960; 304, 317.

[50] S. Tamaru, N. Ando, Z. Anorg. Allg. Chem. 1929; 184, 385; S. Tamaru, N. Ando, Z. Anorg. Allg. Chem. 1931; 195, 309.

[51] K. Matsumoto, S. Kaneko, K. Katagi, J. Cryst. Growth. 1977; 40, 291.

[52] J. M. Brom, W. Weltner, J. Mol. Spectrosc. 1973; 45, 82.

[53] M. Binnewies, Z. Anorg. Allg. Chem. 1990; 589, 115.

[54] K. C. Mills, Thermodynamic Data for Inorganic Sulfides, Selenides and Tellurides. London; Butterworths, 1974.

[55] K. Dressler, Helv. Phys. Acta. 1955; 28, 563.

[56] H. Schäfer, W. Beckmann, Z. Anorg. Allg. Chem. 1966; 347, 225.

[57] G. Krabbes, H. Oppermann, Z. Anorg. Allg. Chem. 1981; 481, 13.

[58] D. Britnell, G. W. A. Fowles, R. Mandycewsky, Chem. Commun. 1970; 608.

[59] E. Milke, R. Köppe, M. Binnewies, Z. Anorg. Allg. Chem. 2010; 636, 1313.

[60] H. Oppermann, M. Schmidt, H. Brückner, W. Schnelle, E. Gmelin, Z. Anorg. Allg. Chem. 2000; 626, 937.

[61] H. Schäfer, U. Wiese, J. Less-Common Met. 1971;, 24, 55.

[62] K. L. Choy, Progress in Materials Science 2003; 48, 57.

[63] R. F. Rolsten, Z. Anorg. Allg. Chem. 1960; 305, 25.

[64] R. F. Rolsten, Iodide Metals and Metal Iodides. New York, J. Wiley, 1961.

[65] H. Schäfer, M. Trenkel, Z. Anorg. Allg. Chem. 1975; 414, 137.

[66] L. Mond, C. Langer, F. Quincke, J. Chem. Soc. 1890; 749.

[67] E. van Arkel, Physica 1923; 3, 76.

[68] E. van Arkel, Metallwirtschaft 1934; 13, 405.

[69] J. H. de Boer, J. D. Fast, Z. Anorg. Allg. Chem. 1926; 153, 1.

[70] J. H. de Boer, J. D. Fast, Z. Anorg. Allg. Chem. 1930; 187, 193.

[71] R. A. J. Shelton, Trans. Faraday Soc. 1966; 62, 222.

[72] T. Hashino, T. Kawai, Trans. Faraday Soc. 1967; 63, 3088.

[73] J. Cueilleron, J. C. Viala, J. Less-Common Met. 1978; 58, 123.

[74] H. C. Spirlet, J. Phys. Colloq. 1979; 40, 87.

[75] D. M. Speros, R. M. Caldwell, W. E. Smyser, High Temp. Sci. 1972; 4, 99.

[76] H. Schäfer, M. Trenkel, Z. Anorg. Allg. Chem. 1979; 458, 234.

[77] M. Kotrbová, Z. Hauptman, Krist. Tech. 1967; 2, 505.

[78] J. H. Dettingmeijer, B. Meinders, L. M. Nijland, J. Less-Common Met. 1974; 35, 159.

[79] H. Schäfer, T. Grofe, M. Trenkel, J. Sol. State. Chem. 1973; 8, 14.

[80] H. Schäfer, A. Tebben, W. Gerhardt, Z. Anorg. Allg. Chem. 1963; 321, 41.

[81] H. Schäfer, W. Gerhardt, Z. Anorg. Allg. Chem. 1984; 512, 79.

[82] L. Hannevold, O. Nilsen, A. Kjekshus, H. Fjellvåg, J. Cryst. Growth 2005; 279, 206.

[83] R. Loitzl, C. Schüler, German Patent DE 2051404, 1972.

[84] M. Richardson, Acta Chem. Scand. 1967; 21, 2305.

[85] O. Bosholm, H. Oppermann, S. Däbritz, Z. Naturforsch. 2001; 56b, 329.

[86] H. Wilhelm, M. Schmidt, R. Cardoso-Gil, U. Burkhardt, M. Hanfland, U. Schwarz, L. Akselrud, Science Technol. Adv. Mater. 2007; 8, 416.

[87] O. Bosholm, H. Oppermann, S. Däbritz, Z. Naturforsch. 2000; 55b, 614.

[88] O. Bosholm, H. Oppermann, S. Däbritz, Z. Naturforsch. 2000; 55b, 1199.

[89] R. Krausze, M. Khristov, P. Peshev, G. Krabbes, Z. Anorg. Allg. Chem. 1989;579, 231.

[90] R. Krausze, M. Khristov, P. Peshev, G. Krabbes, Z. Anorg. Allg. Chem. 1990; 588, 123.

[91] R. Neddermann, S. Gerighausen, M. Binnewies, Z. Anorg. Allg. Chem. 1996;622, 21.

[92] H. Schäfer, J. Nowitzki, Z. Anorg. Allg. Chem. 1977; 435, 49.

[93] J. J. Nickl, J. D. Koukoussas, A. Mühlratzer, J. Less-Common Met. 1973; 32, 243.

[94] T. Plaggenborg, M. Binnewies, Z. Anorg. Allg. Chem. 2000; 626, 1478.

[95] H. Oppermann, Solid State Ionics 1990; 39, 17.

[96] K. Brodersen, H. K. Breitbach, G. Thiele, Z. Anorg. Allg. Chem. 1968; 357, 162.

[97] R. E. Mc Carley, J. W. Roddy, K. O. Berry, Inorg. Chem. 1964; 3, 50.

[98] H. Schäfer, Angew. Chem. 1976, 88; 775, Angew. Chem. Int. Ed. 1976; 15, 713.

[99] H. Schäfer, Adv. Inorg. Nucl. Chem. 1983; 26, 201.

[100] H. Schäfer, J. Cryst. Growth 1975; 31, 31.

[101] H. Schäfer, Z. Anorg. Allg. Chem. 1980; 469, 123.

[102] R. Krausze, H. Oppermann, Z. Anorg. Allg. Chem. 1987; 550, 123.

[103] H. A. Oye, D. M. Gruen, J. Amer. Chem. Soc. 1969; 91, 2229.

[104] D. M. Gruen, H. A. Oye, Inorg. Nucl. Chem. Lett. 1967; 3, 453.

[105] K. Lascelles, H. Schäfer, Z. Anorg. Allg. Chem. 1971; 382, 249.

[106] H. Schäfer, M. Binnewies, W. Domke, J. Karbinski, Z. Anorg. Allg. Chem. 1974; 403, 116.

[107] H. Gunsilius, W. Urland, R. Kremer, Z. Anorg. Allg. Chem. 1987; 550, 35.

[108] K. Shinozaki, Y. Hirashima, G. Adachi, Trans. Nonferrous Met. Soc. China 1990; 16, 42.

[109] G.-Y. Adachi, K. Shinozaki, Y. Hirashima, K.-I. Machida, J. Less-Common Met. 1991; 169,

[110] K. Murase, K. Shinozaki, K.-I. Machida, G.-Y. Adachi, Bull. Chem. Soc. Jpn. 1992; 65, 2724.

[111] K. Murase, K. Shinozaki, Y. Hirashima, K.-I. Machida, G.-Y. Adachi, J. Alloys Compd. 1993; 198, 31.

[112] J. Yu, Y. Yu, Z. Wang, Jinshu Xuebao (Acta Metallurgica Sinica) 1997; 3, 391.

[113] T. Ozaki, K.-I. Machida, G.-Y. Adachi, Rare Earths 1998; 32, 154.

[114] T. Ozaki, J. Jiang, K. Murase, K.-I. Machida, G.-Y. Adachi, J. Alloys Compd. 1998; 265, 125.

[115] T. Ozaki, K.-I. Machida, G.-Y. Adachi, Metall. Mater. Trans. 1999; 30 B, 45.

[116] Z.-C. Wang, Y.-H. Sun, L. Guo, J. Alloys Compd. 1999; 287, 109.

[117] Y. H. Wang, L. S. Wang, J. Chinese Rare Earth Soc. 2000; 18, 74.

[118] D. Yang, J. Yu, J. Jiang, Z. Wang, J. Mater. and Metall. 2003; 2, 113.

[119] Bonnamy, J. C. Launay, M. Pouchard, Rev. Chim. Min. 1978; 15, 178.

[120] W. Redlich, T. Petzel, Rev. Chim. Min. 1983; 20, 54.

[121] L.-P. Zenser, Dissertation, University of Gießen, 1999.

[122] J. Mercier, G. Fourcaudot, Y. Monteil, C. Bec, R. Hillel, J. Cryst. Growth 1982; 59, 599.

[123] W. Reichelt, PhD thesis. Academy of Sciences of GDR, Berlin; 1977.

[124] Rossberg, H. Oppermann, Z. Anorg. Allg. Chem. 1988; 556, 109.

[125] H. Oppermann, W. Reichelt, G. Krabbes, E. Wolf, Krist. Tech. 1977; 12, 717.

[126] H. Oppermann, W. Reichelt, G. Krabbes, E. Wolf, Krist. Tech. 1977; 12, 919.

[127] H. Oppermann, W. Reichelt, E. Wolf, Z. Anorg. Allg. Chem. 1977; 432, 26.

[128] H. Schäfer, M. Binnewies, H. Rabeneck, C. Brendel, M. Trenkel, Z. Anorg. Allg. Chem. 1977; 435, 5.

[129] H. Oppermann, V. A. Titov, G. Kunze, G. A. Kokovin, E. Wolf, Z. Anorg. Allg. Chem. 1978; 439, 13.

[130] H. Oppermann, G. Kunze, E. Wolf, G. A. Kokovin, I. M. Sitschova, G. E. Osigova, Z. Anorg. Allg. Chem. 1980; 461, 165.

[131] M. Spieß, R. Gruehn, Naturwissenschaften 1978; 65, 594.

[132] L. Bald, R. Gruehn, Z. Anorg. Allg. Chem. 1984; 509, 23.

[133] V. Plies, T. Kohlmann, R. Gruehn, Z. Anorg. Allg. Chem. 1989; 568, 62.

[134] T. Dahmen, R. Gruehn, Z. Anorg. Allg. Chem. 1995; 621, 417.

[135] M. S. Islam, R. Glaum, Z. Anorg. Allg. Chem. 2009; 635, 1008.

[136] R. Glaum, M. Weil, D. Özalp, Z. Anorg. Allg. Chem. 1996; 622, 1839.

[137] Kostencki, PhD thesis, University of Gießen; 1997.

[138] R. Glaum, Neue Untersuchungen an wasserfreien Phosphaten der Übergangsmetalle, Habilitation, University of Gießen; URL: http://www.geb.uni-giessen.de.

[139] Litterscheid, Diploma thesis, University of Bonn; 2003.

[140] M. Weil, Z. Naturforsch. 2000; 55b, 699; M. Weil, Acta Crystallogr. 2001; E57, i22; M. Weil, Acta Crystallogr. 2001; E57, i28; M. Weil, C. Lengauer, E. Fueglein, E. J. Baran, Cryst. Growth Des. 2004; 4, 1229; M. Weil, Z. Anorg. Allg. Chem. 2004; 630, 213; M. Weil, Acta Crystallogr. 2004; E60, i139; M. Weil, U. Kolitsch, Z. Kristallogr. 2005; 22 Suppl., 183.

[141] M. A. Daubrée, Ann. Mines 1841; 20, 65. Cited in: W. Schrön, Eur. J. Mineral. 1989; 1, 739. M. A. Daubrée, Compt. Rend. 1849; 29, 227. Cited in: W. Schrön, Eur. J. Mineral. 1989; 1, 739.

[142] Kaldis, R. Verreault, J. Less-Common Met. 1970; 20, 177; E. Kaldis, J. Cryst. Growth, 1971; 9, 281.

[143] R. Schober, E. Thilo, Chem. Ber. 1940; 73, 1219.

[144] R. Diehl, A. Räuber, F. Friedrich, J. Cryst. Growth 1975; 29, 225.

[145] H. Schmid, J. Phys. Chem. Solids 1965; 27, 973.

[146] K. Nassau, J. W. Shiever, J. Cryst. Growth 1972; 16, 59.

[147] G. Castellanos-Guzman, M. Trujillo-Torrez, M. Czank, Mater. Science Engin. 2005; B 120, 59.

[148] J. Campa-Molina, S. Ulloa-Godınez, A. Barrera, L. Bucio, J. Mata, J. Phys.: Condens. Matter 2006; 18, 4827

[149] T. Takahashi, O. Yamada, J. Cryst. Growth 1976; 33, 361.

[150] R. Nitsche, J. Phys. Chem. Solids 1960, 17, 163. R. Nitsche, H. U. Bölsterli, M. Lichtensteiger, J. Phys. Chem. Solids 1961; 21, 199.

[151] Menzel, R. Kaden, D. Spemann, K. Bente, T. Butz, Nucl. Instr. Meth. Phys. Res. B 2006; 249, 478.

[152] Wolf, H. Oppermann, G. Krabbes, W. Reichelt, Current Topics in Materials Science 1978; 1, 697.

[153] Krabbes, H. Oppermann, Z. Anorg. Allg. Chem. 1979; 450, 27.

[154] R. Nitsche, J. Phys. Chem. Solids, Suppl. 1967; 1, 215.

[155] P. B. Rimmington, A. Balchin, B. K. Tanner, J. Cryst. Growth 1972; 21, 171.

[156] U. Hotje, R. Wartchow, M. Binnewies, Z. Anorg. Allg. Chem. 2005; 631, 403.

[157] K. Kimoto, K. Masumoto, Y. Noda, K. Kumazawa, T. Kiyosawa, N. Koguchi, Jpn. J. Appl. Phys., Part 1, 1993, 32 (Suppl. 32–3, Proceedings of the 9th International Conference of Ternary and Multinary Compounds, 1993; 187.

[158] E. Catano, Z. K. Kun, J. Cryst. Growth 1976; 33, 324.

[159] U. Hotje, C. Rose, M. Binnewies, Z. Anorg. Allg. Chem. 2005; 631, 2501.

[160] B. Van den Berg, J. E. van Delden, J. Boumann, Phys. Status Solidi 1969; 36, K 89.

[161] P. Gibart, A. Begouen-Demeaux, Compt. Rend. Acad. Sci. Paris C 1969; 268, 111.

[162] Hartmann, R. Mach, N. Testova, J. Cryst. Growth 1987; 84, 199.

[163] Matsumoto, Shizuoka Daigaku Denshi Kogaku Kenkyusho Kenkyu Hokoku 1988; 23, 127.

[164] Ito, Y. Matsuura, Proc. Electrochem. Soc. 1979; 79, 281.

[165] G. Attolini, C. Paorici, L. Zanotti. J. Cryst. Growth 1982; 56, 254.

[166] G. Attolini, C. Paorici, Mater. Chem. Phys. 1983; 9, 65.

[167] S. Locmelis, U. Hotje, M. Binnewies, Z. Anorg. Allg. Chem. 2005; 631, 672.

[168] H. Schäfer, Z. Anorg. Allg. Chem. 1982; 486, 33.

[169] W. Kleber, I. Mietz, U. Elsasser, Kristall Technik 1967; 2, 327.

[170] A.G. Sigai, H. Wiedemeier, J. Electrochem. Soc. 1972; 119, 910.

[171] Y. G. Sha, C. H. Su, W. Palosz, M P. Volz, D. C. Gillies, F. R. Szofran, S. L. Lehoczky, H. C. Liu, R. F. Brebrick, J. Cryst. Growth 1995; 146, 42.

[172] Paorici, Proc. Int. Symp. Cadmium Telluride, Mater. Gamma-Ray Detectors 1972; VII-1–VII-4.

[173] Y. G. Akhromenko, G. A. Il'chuk, I. E. Lopatinskii, S. P. Pavlishin, Izv. Akad. Nauk SSSR, Neorg. Mater. 1981; 17, 2016.

[174] V. V. Tokmakov, I. I. Krotov, A. V. Vanyukov, Izv. Akad. Nauk SSSR, Neorg. Mater. 1979; 15, 1546.

[175] Munster, G. Rinck, W. Ruppert, Z. Phys. Chem. 1956; 9, 228.

[176] Ruck, Z. Anorg. Allg. Chem. 1999;, 625, 2050.

[177] J. Martin, R. Gruehn, Z. Kristallogr. 1988; 182, 180.

[178] Özalp, PhD thesis, University of Gießen; 1992.

[179] Lenz, PhD thesis, University of Gießen; 1995

[180] R. Köppe, H. Schnöckel, Z. Anorg. Allg. Chem. 2003; 629, 2168.

[181] P. Bolte, R. Gruehn, Z. Anorg. Allg. Chem. 1994; 620, 2077.

[182] C. N. Cochran, L. M. Foster, J. Electrochem. Soc., 1962; 109, 144.

[183] C. J. Frosch, J. Electrochem. Soc. 1964; 111, 180.

[184] K. Nocker, R. Gruehn, Z. Anorg. Allg. Chem. 1993; 619, 699; K. Nocker, PhD thesis, University of Gießen; 1993.

[185] Trappe, Diploma thesis, University of Gießen; 1994

[186] J. Seiwert, R. Gruehn, Z. Anorg. Allg. Chem. 1984; 510, 93.

[187] H.-J. Schweizer, PhD thesis, University of Gießen; 1983.

[188] Eriksson, Acta Chem. Scand. 1971; 25, 2651.

[189] V. Krämer, R. Nitsche, M. Schumacher J. Cryst. Growth 1974; 24.25, 179.

Chapter 11

Growth and Development of Sapphire Crystal for LED Applications

Huili Tang, Hongjun Li and Jun Xu

Additional information is available at the end of the chapter

1. Introduction

LEDs (Light emitting diodes) are considered as the most promising green lighting sources in 21st Century for the advantages in high brightness, long lifetime (more than 50,000 hours), low energy consumption, short corresponding time, good shock resistance, non-toxic, recyclable, safety. LEDs have already been extensively used in outdoor displays, traffic lights, high-performance back light units in liquid-crystal displays, general lighting. Strategies Unlimited Company predicted that the compound annual growth rate (CAGR) of the LED market would increase to 30.6%, up to $20.2 billion in 2014. It is obvious that incandescent bulbs and fluorescent lamps will be replaced by LEDs, which could alleviate the increasingly serious global energy crisis. Therefore, the development of semiconductor lighting industry is of great significance. Many countries have already launched National Semiconductor Lighting Plan, investing heavily in researching and developing the LEDs industry. In 1998, Japan made a "Light for the 21st Century" plan with the budget of 6 billion yen. In July 2000, European Union implemented "Rainbow project bring color to LEDs" plan, setting up ECCR and promoting the application of white light LED through the EU BRITE/ EURAM-3 program. U.S. Department of Energy established "National research program on semiconductor lighting" plan. It is expected that in 2025, the use of solid state lighting will reduce half of the lighting electricity consumption and save $35 billion per year. In June 2003, the Chinese Ministry of Science and Technology launched an "National Semiconductor Lighting Project" in support of the "863" Project. In 2009, ministry of Science and Technology started "Ten thousand LED lights in ten cities" semiconductor lighting demonstration program. It is expected that in 2015, semiconductor lighting will occupy 30% of the domestic general lighting market.

The luminescent property of LED is being continuously improved under the research all over the world. In 2002, Lumileds Company made the LED with the luminous efficiency of 18-22 lm/W. In February 2009, Nichia Company fabricated a white LED with luminous efficiency of 249 lm/W under 20mA driving current. As a leading manufacturer in the field of LED, Cree Company has produced mass production products with the highest luminous efficiency of 161 lm/W in 2011. Later after that, in April 2012 they announced that the white power type LED with luminous efficiency up to 254 lm/W under 350mA driving current has been manufactured successfully, which again refreshed the industry record.

LED industry is divided into upstream, midstream and downstream chains. Upstream chain includes substrate material, epitaxial wafers and chip manufacturing; midstream chain includes packaging and devices; and various LED application products belong to downstream chain. As a cornerstone of the development of semiconductor lighting industry, to some extent, the development of substrate material determines the route of the development of semiconductor lighting technology. Therefore, substrate material is a critical core issue of current semiconductor lighting industry.

The main factors determining the appropriate substrate materials are matched lattice parameters and thermal expansion coefficients as well as good crystallinity, chemical, physical and mechanical properties. Many materials were investigated as substrates, such as sapphire (α-Al_2O_3), SiC, Si, GaAs, $MgAl_2O_4$, $ScAlMgO_4$, γ-$LiAlO_2$ and β-$LiGaO_2$ etc. Table 1 shows the related parameters of some substrate materials for GaN and ZnO epitaxy [1]. Among them, sapphire and SiC are the main commercial substrates. Due to SiC substrate is very expensive, sapphire is the most important semiconductor LED lighting industry substrate. According to a conservative estimation, the demand of epi-ready sapphire substrate in international market is 600,000 pieces per month. In order to reduce Metal-organic Chemical Vapor Deposition (MOCVD) epitaxial cost, the requirement of substrate wafer size is getting larger and larger, from 2″ to 4″, 6″ and 8″. Therefore, the growth and development of large size sapphire crystal have attracted increasing attention all over the world.

This chapter investigates the strengthening and toughening of sapphire crystal by ion doping. Besides, the chapter is devoted to review the raw material, seed crystal, growth direction and growth methods, which have influence on the quality of sapphire crystal. The latest progress in the main growth methods of sapphire substrate: Kyropoulos method, heat exchanger method, Czochralski method, edge-defined film-fed growth method and temperature gradient technique are systematical illustrated. Finally, the overall evaluation on the advantage and disadvantage of each method is briefly outlined.

2. Properties of sapphire crystal

2.1. Crystal structure

Sapphire crystal is a simple coordinated type oxide crystal, whose chemical constituent is Al_2O_3 and crystalline form is α-Al_2O_3. Sapphire, also named corumdum, belongs to trigonal

Substrate	Structure	Space group	Lattice constant (Å)	Thermal expansion ($\times 10^{-6}$ K^{-1})	Lattice mismatch	
					GaN	ZnO
w-GaN	wurtzite	$P6_3mc$	a=3.188 c=5.185	5.59 3.17	0%	-1.9%
ZnO	wurtzite	$P6_3mc$	a=3.250 c=5.206	2.9 4.75	1.9%	0%
α-Al_2O_3	rhombohedral	$R\bar{3}c$	a=4.757 c=12.983	7.5 8.5	-14%	18.4%
6H-SiC	6H (W)	$P6_3mc$	a=3.081 c=15.117	4.46 4.16	-3.3%	3.5%
Si	diamond	$Fd\bar{3}m$	a=5.430	3.59	20.4%	18.1%
GaAs	zincblende	$F\bar{4}3m$	a=5.6533	6.0	-20%	-19%
$MgAl_2O_4$	spinel	$Fd\bar{3}m$	a=8.083	7.45	-10.3%	-12.1%
$Mg_{0.4}Al_{2.4}O_4$	spinel	$Fd\bar{3}m$	a=7.984	5.62	-11.4%	-13.1%
$ScMgAlO_4$	tetragonal	$R\bar{3}m$	a=3.246 c=25.195	6.2 12.2	1.8%	0.09%
γ-$LiAlO_2$	tetragonal	$P4_12_12$	a=5.169 c=6.268	7.1 15	-1.7%	-3.5%
β-$LiGaO_2$	orthorhombic	$Pna2_1$	a=5.402 b=6.372 c=5.007	1.7 11.0 4.0	0.2%	2%

Table 1. Related parameters of the substrate materials for GaN and ZnO epitaxy.

system, $D_{3d}^6 - R\bar{3}C$ space group. It has symmetry elements as follows: mirror-turn axis of the sixth order (ternary inversion axis), three axes of the second order normal to it, three symmetry planes normal to the axes of the second order and intercrossing along the axis of the highest order and symmetry center [2.3]. The crystal lattice of sapphire is formed by Al^{3+} and O^{2-} ions. The crystal lattice takes the form of O^{2-} ions closest hexagonal packing and Al^{3+} cations locate in the octahedral hollows between the closely packed O^{2-} ions, filling two thirds of these hollows (see Figure 1). Polarity, semi-polarity and non polarity GaN films can be grown on different orientation sapphire substrates. The epitaxial relationship of $(11\bar{2}2)$ semipolar GaN film on $(10\bar{1}0)$ m-plane sapphire substrate is $[10\bar{1}0]_{GaN} \| [\bar{1}2\bar{1}0]_{sapphire}$, $[\bar{1}2\bar{1}\bar{1}]_{GaN} \| [0001]_{sapphire}$, and $(11\bar{2}0)$ nopolar GaN film on $(1\bar{1}02)$ r-plane sapphire substrate is $[1\bar{1}00]_{GaN} \| [11\bar{2}0]_{sapphire}$, $[0001]_{GaN} \| [\bar{1}101]_{sapphire}$ [5-7].

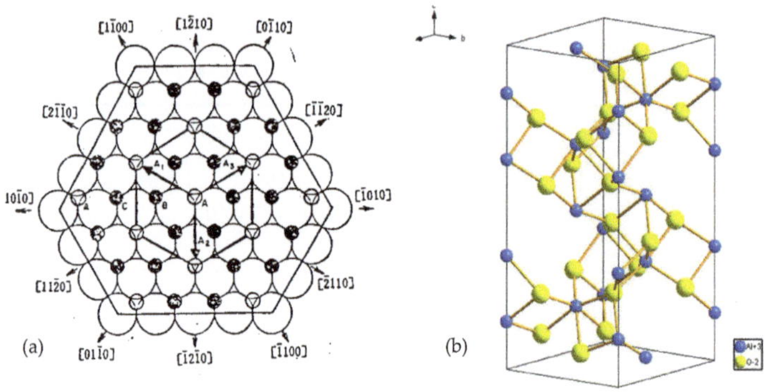

Figure 1. a) Schematic of the packing of O^{2-} ions in the sapphire cell,(b) Rhombodedral unite cell of the sapphire crystal.

2.2. Physical, thermal, optical and electrical properties

Sapphire is a high melting point oxide crystal (2050℃), which can be used at the highest temperature of 1900℃. Table 2 presents the main properties of the sapphire crystal. Sapphire has a high refractive index and a broad transmission band from 0.14 to 6.0 μm, spanning the UV, visible, and IR bands. Sapphire also has a high hardness (next to diamond) and surface smoothness, very good tensile strength, thermal conductivity, electric insulation, wear resistance, and thermal shock resistance [10-12]. The chemical properties of sapphire are very stable. Generally, sapphire is insoluble in water; insoluble in nitric acid (HNO_3), sulfuric acid (H_2SO_4), hydrochloric acid (HCL), hydrofluoric acid (HF) and phosphoric acid (H_3PO_4) up to 300°C; and insoluble in alkalis up to 800°C. The favorable combination of excellent optical and mechanical properties of sapphire, together with high chemical durability, makes it a desirable substrate material for LED applications.

Physical Properties	
Chemical Formula	Al_2O_3
Structure	hexagonal-rhombodedral
Molecular weight	101.96
Lattice Constants Å	a=4.765, c=13,000
Crystal density (g/cm^3)	3.98
Melt density (g/cm^3)	3.0
Hardness	9 Mohs 1800 knoop parallel to C-axis 2200 knoop perpendicular to C-axis

Young Modulus (GPa)	379 at 30° to C-axis 352 at 45° to C-axis 345 at 60° to C-axis 386 at 75° to C-axis
Shear Modulus (GPa)	145
Bulk Modulus (GPa)	240
Bending Modulus/ Modulus of Rupture (MPa)	350 to 690
Tensile strength	400 at 25°C 275 at 500°C 345 at 1000°C
Elastic Coefficient	C=496, C12=164, C13=115, C33=498, C44=148
Apparent Elastic Limit (MPa)	448 to 689
Flexural Strength (GPa)	2.5 - 4.0
Poisson ratio	0.25 - 0.30
Friction Coefficient	0.15 on steel 0.10 on sapphire
Abrasion resistance	8 times higher than steel
Thermal Properties	
Melting Point (°C)	2050
Specific Heat J/(kg ·K)	105 at 91 K 761 at 291 K
Thermal coefficient of linear expansion at 323 K (K^{-1})	66.66×10^{-6} parallel to optical axis 5×10^{-6} perpendicular to optical axis
Thermal conductivity (W/m °K) at 20°C	41.9
Thermal Expansion (20-1000°C)	Parallel to C-axis: 9.03×10^{-6} °C Perpendicular to C-axis: 8.31×10^{-6} °C 60° to C-axis: 8.4×10^{-6} °C
Optical Properties	
Transission Range	0.2 - 5.5 microns
Reflection loss	14% at 1 micron (2 surfaces)
Restrahlen Peak	13.5 micron
dN/dT	$+13\times10^{-6}$ °C
Refractive index	1.7122
T_t%	87.1

Electrical Properties	
Resistivity, Ohm•cm at 20 - 500°C	10^{11} - 10^{16}
Dielectric Constant	11.5 parallel to C axis 9.4 perpendicular to C axis
Dielectric strength (V/cm)	4×10^5
Loss Tangent	10^{-4}

Table 2. Main physical, thermal, optical and electrical properties of the sapphire crystal.

2.3. Improving mechanical properties of sapphire by ion doping

Because of excellent physical and chemical properties and outstanding spectrum transmission performance in wide bands range, sapphire crystal has been widely applied in various kinds of high-end window materials and LED substrates. But sapphire crystal is prone to brittle fracture at room temperature, causing the decline of mechanical properties and thermal shock resistance. So strengthening and toughening of sapphire at room temperature have important value on both scientific research and practical applications.

2.3.1. Strengthen and toughen sapphire by carbon doping

Figure 2 shows the carbon-doped sapphire (C:sapphire) single crystals grown by TGT method and the crystal was colored because of carbon doping. The C:sapphire crystal possessed remarkable absorption peaks at 206 nm and 256 nm (see Figure 3). The fracture strength and fracture toughness of C:sapphire and sapphire crystals along [11$\bar{2}$0] direction on (0001) plane are listed in Table 3. The fracture strength and fracture toughness of 1000 ppm C:sapphire crystal (No.2) were 752.0 MPa and 2.81 Mpa $m^{1/2}$, respectively. The fracture strength and fracture toughness of undoped sapphire crystals (No. 0) were 488.25 MPa and 1.99 Mpa $m^{1/2}$, respectively. It is demonstrated that the mechanical properties of sapphire crystal can be greatly improved by carbon doping.

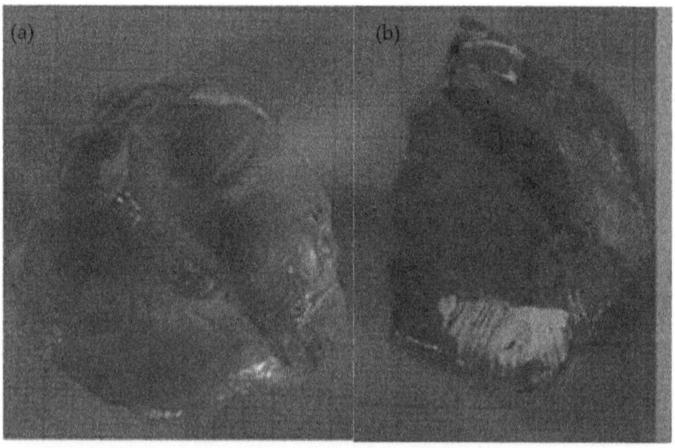

Figure 2. C:sapphire crystal grown by TGT method: (a) 2000 ppm C:sapphire, (b) 5000 ppm C:sapphire

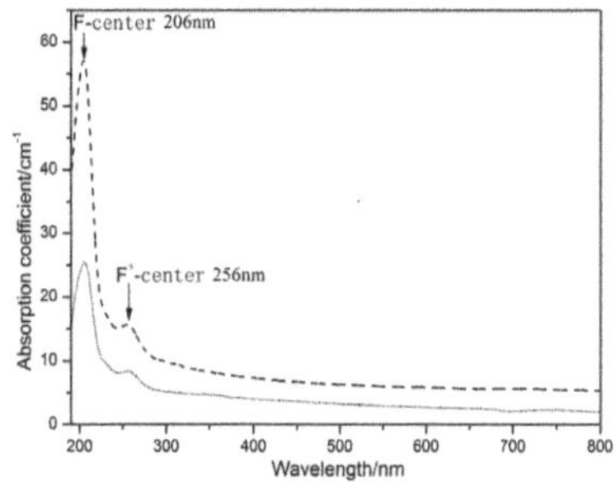

Figure 3. Absorption spectrum of C:sapphire.

Sample/Test	No.0	No.1	No.2
Fracture strength (MPa)	488.25	597.75	752.0
Fracture toughness(MPa·m$^{1/2}$)	1.99	2.41	2.81

Table 3. Mechanical properties of C:sapphire and sapphire crystals at room temperature

The doped carbon itself occurred disproportionating reaction at high temperature in the process of crystal growth, i.e. $3C \rightarrow 2C^{2+} + C^{4-}$. C^{4-} ions substituted negative oxygen ions (O^{2-}) and generated oxygen vacancy defects (Vo). Vo captured one or two electronic and formed F^+ or F color centers. Consequently, the absorption peaks at 206 nm and 256 nm of C:sapphire were strengthened significantly. And the C^{2+} whose radius is only 0.16 Å entered the lattice in the form of interstitial ion, which created blocking effect to the sapphires cracking and improved fracture strength and fracture toughness of sapphire crystal at room temperature.

2.3.2. Strengthen and toughen sapphire by titanium doping

Figure 4(a) shows titanium doped sapphire single crystal grown by Vertical Bridgman Method and the crystal was colored because of Ti^{3+} doping. After annealing at 1600°C for 24h in the air atmosphere, the crystal was colorless and the absorption peak of Ti^{3+} ion significantly weakens in the absorption spectrum. Meanwhile, we found that the mechanical properties of sapphire can be improved through titanium doping, as is shown in Figure 5. At room temperature, the fracture strength and fracture toughness of Ti^{3+}: sapphire along [11$\bar{2}$0] direction on (0001) plane were 560 MPa and 2.29 MPa m$^{1/2}$ respectively. And the fracture strength and fracture toughness of Ti^{4+}: sapphire single crystal (annealing from Ti^{3+}: sapphire under 1600°C for 24h) are 600 MPa and 2.35 MPa m$^{1/2}$ respectively. It has been found that Ti^{3+} and Ti^{4+} doping was beneficial to improve mechanical property of the sapphire crystal. Solid solution reaction of doped titanium ions occurred during crystal growth processes and the aluminum ions of the matrix were substituted by titanium ions. The solid solution makes the fracture strength and fracture toughness of doped sapphire improved. After annealing at 1600°C for 24h, doped Ti^{3+} ions were oxidized to Ti^{4+} ions. While Al^{3+} was substituted by Ti^{4+} in the crystal lattice, the defects such as aluminium ions vacancies appeared. These defect structures can improved fracture surface energy of Ti^{4+}:sapphire, so the strength and toughness were further improved.

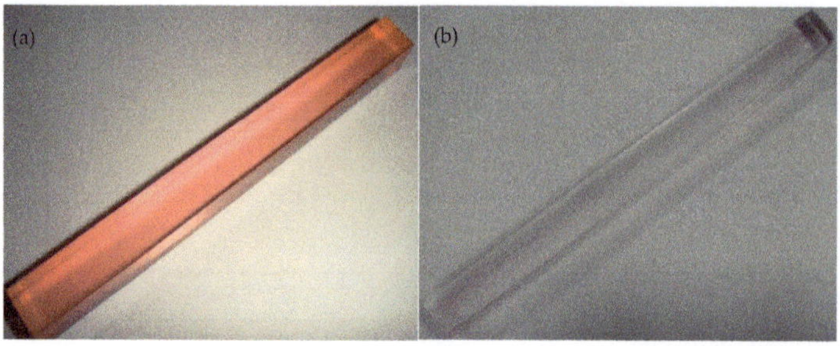

Figure 4. (a) Ti^{3+}:sapphire single crystal, (b) Ti^{4+}:sapphire (annealing 1600°C for 24h from Ti^{3+}:sapphire).

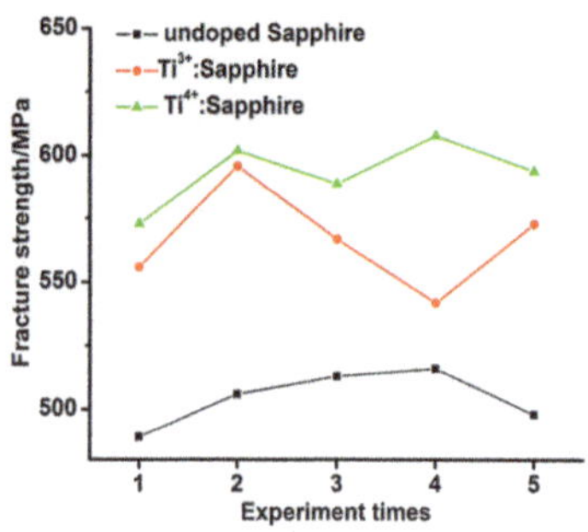

Figure 5. Fracture strength of Ti:sapphire and sapphire crystals.

3. Sapphire substrate crystal growth

3.1. Raw material

The raw materials (Al_2O_3) used for growing sapphire substrate crystal are divided into powder, sintered charge, cracked crystal and polycrystalline ingot, as is shown in Figure 6. In order to obtain LED grade sapphire crystal, Al_2O_3 raw material should be of high purity (≥99.996%) and high density. Table 4 presents the impurity concentrations of suitable Al_2O_3 analyzed by the inductively coupled plasmas optical emission spectroscopy (ICP-OES).

Cracked crystal stuff grown by Verneuil method without additives is commonly used. Due to undergoing a crystallization process, the purity and density of the cracked crystal raw material are much higher. Al_2O_3 polycrystalline ingot is prepared by the cold crucible induction skull melting technique (ISM) in recent years. The technical process is provided in Figure 7[13]. Al_2O_3 polycrystalline ingot could be made as a whole block according to the inside shape of the crucible used for crystal growth, which is convenient for charging.

The purities of sintered charge, cracked crystal and polycrystalline ingot are related to the purity of Al_2O_3 powder. Usually, the preparation methods of high purity Al_2O_3 powder include thermal decomposition of ammonium aluminum sulfate, thermal decomposition of ammonium aluminum carbonate hydroxide, aluminum isopropoxide hydrolysis method, aluminum choline hydrolysis method and high purity aluminum active hydrolysis method, etc[14,15]. The purity of the powder prepared by aluminum isopropoxide hydrolysis method is higher than that prepared by other methods. Impurities in raw material will reduce the transparency of crystal, make sapphire crystal pink or yellow, and increase dislocation defects that will cause LED luminous efficiency to decrease. Aluminum sulfate impurity from cracked crystal raw material decomposes again during crystal growth, which will make bubbles and insoluble remain in the sapphire. Excess Ca and Mg impurity ions will induce sapphire to crack, and excess K impurity ion will form scattering particles in the sapphire.

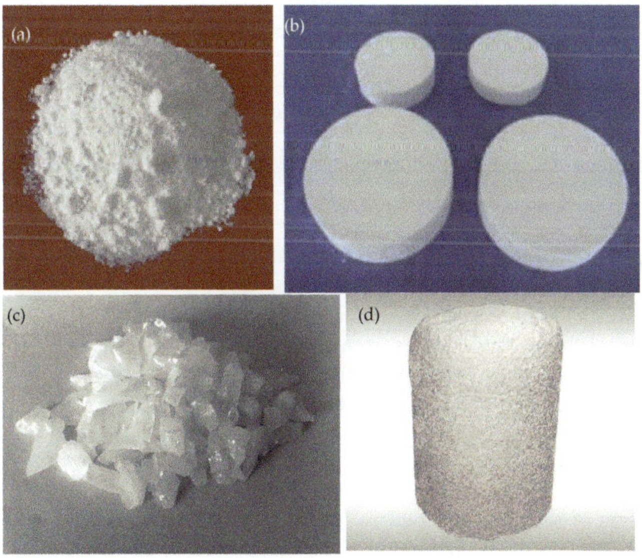

Figure 6. Al_2O_3 raw materials: (a) powder, (b) sintered charge, (c) cracked crystal, (d) polycrystalline ingot.

Figure 7. The process of Al_2O_3 polycrystalline ingot prepared by ISM.

Element	Na	Mg	Si	K	Ca	Ti	Cr
Concentration (μg/g)	<1	<1	<4	<1	<1	<1	<1
Element	Mn	Fe	Ni	Cu	Zn	Ga	Zr
Concentration (μg/g)	<1	<4	<1	<1	<1	<1	<1

Table 4. Impurity concentration analysis of suitable Al_2O_3 raw material.

3.2. Seed crystal

The selection of Al_2O_3 seed crystal is an important step before crystal growth. Sapphire crystal will inherit some defects in the seed crystal, and therefore seed crystal of high crystalline quality is absolutely necessary. The seed crystal should be colorless transparent, free from bubbles, inclusions, precipitations, twin crystal, grain boundary, microcrack and scattering particles.

3.3. Growth direction

The slip systems of sapphire reported so far are $(0001)1/3\langle 11\bar{2}0\rangle$ basal slip, $\{11\bar{2}0\}\langle 1\bar{1}00\rangle$ prism slip and $\{10\bar{1}1\}1/3\langle \bar{1}101\rangle$ pyramidal slip [16]. Slip is easy to occur along the direction of high atomic density. Among these systems, basal slip is the easiest slip system at high temperatures. When the obliquity between growth interface and (0001) plane is small, basal slip is prone to induce a large number of grain boundaries, even slip band and twin crystal.

On the contrary, it is not easy to slip and generate grain boundaries. Therefore, sapphire crystal is usually grown along a[11$\bar{2}$0], m[10$\bar{1}$0] or r[1$\bar{1}$02] direction to reduce sub-grain and twin boundaries defects.

3.4. Growth methods

3.4.1. Kyropoulos method

Kyropoulos (KY) method was put forward by Kyropoulos as early as 1926, and is used to prepare and study on large-size halogen group crystals, hydroxide crystals and carbonate crystals. Modified by Musatov in 1970's, Kyropoulos method is applied to grow sapphire single crystal. A schematic diagram of the KY furnace is shown in Figure 8. The crystal growth systems include vacuum system, heating system, cooling system, insulation system and control system. This method adopts the tungsten resistance-heated element, and the heat is transferred to the whole thermal field through radiation. Crystallization occurs when the seed contacts the melt at a temperature slightly lower than the melting temperature. The crystal gradually grows by water cooling from the seed holder and stably programmed reduction of the heat power to adjust the temperature distribution of the melt. The growth procedure is divided into seeding, shouldering, equal diameter growth, annealing and cooling. The crystals are usually boules, and the diameters could be 10~30 mm smaller than the crucible inner diameters. During the whole growth process, the crystal is in the crucible but no contact with the crucible wall, and located in the hot zone, thus the thermal stress in the crystal is small. At present, Kyropoulos method is one of the effective and mature methods to grow large diameter sapphire substrate crystals. In 2009, 200 Kg weight sapphire boule

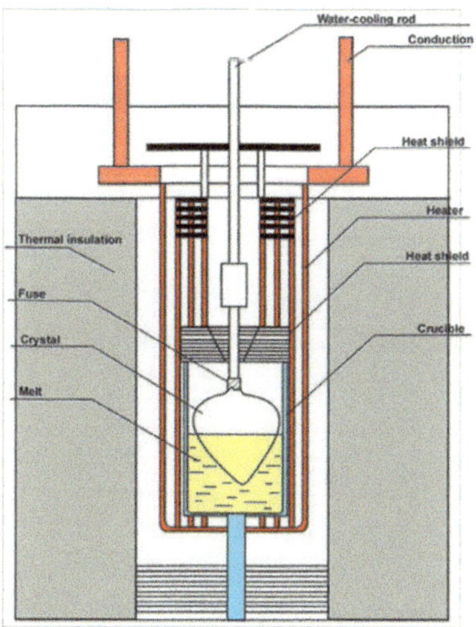

Figure 8. Schematic diagram of the resistance-heated KY furnace.

that is current world's largest sapphire crystal was successfully grown by Rubicon Technology Inc. using improved Kyropoulos method, proprietary ES2 crystal growth technology. Figure 9 shows the sapphire boule. Now, Rubicon has produced in mass 85 Kg grade sapphire boules, which is world leading level.

In China, the sapphire industry is developing rapidly. Many enterprises have built complete processes including sapphire growing, drilling, slicing and polishing which cover the entire industry chain. Under the leading of Prof. Xu, our team has developed advanced Kyropoulos method by optimizing furnace automatic closed-loop control system, thermal field system and growth technology. Melt convection, melt/crystal interface shape, temperature distribution and radiative heat exchange during the sapphire crystal growth process are analyzed systematically with CGSim software package of STR. Flow pattern in the melt at the melting temperature before modifications is displayed in Figure 10. The melt flow had a two-vortex structure: one is a larger vortex occupying the melt core, which is the normal melt convection; and the other is a smaller vortex of lower intensity located near the melt free surface, which is the abnormal melt convection. The melt flow direction of the smaller vortex is opposite to that of the main vortex. It is easy to cause overcooled melt to enrich in the area of the smaller vortex. The sapphire crystal will adhere to the crucible at the shouldering stage. The theoretical predictions are consistent with the experimental results. Isothermal changes at different growth stages and the small vortex gradually disappears at the cylindrical growth stage, which is demonstrated in Figure 11. After a series of modification in thermal field structure, the melt flow pattern of abnormal small vortex disappears completely, as is displayed in Figure 12. The thermal field is of reasonable temperature distribution, stable melt vortex, and has long service life and low cost. Our advanced Kyropoulos method has been successfully applied to the mass

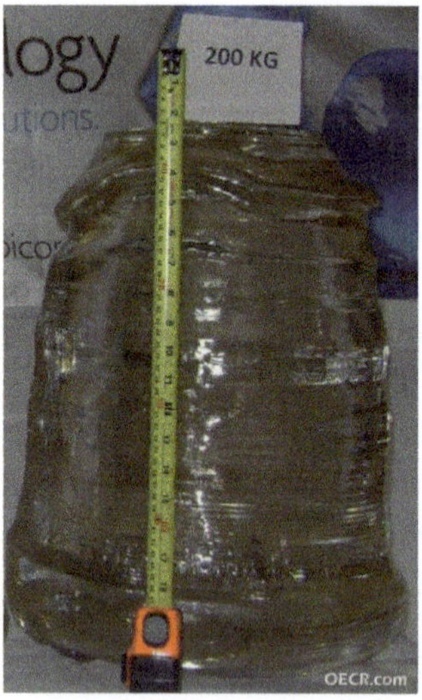

Figure 9. Sapphire boule grown by Rubicon Technology Inc.[17].

production of 65K g grade sapphire boules and pilot production of 85 Kg grade sapphire boules in several domestic companies. Figure 13 shows the 65 Kg and 85 Kg grades sapphire boubles. The KY furnace is designed and developed by our team and Shenyang Scientific Instrument Co., Ltd., Chinese Academy of Sciences (SKY), and manufactured by SKY. The full width at half maximum of X-ray double crystal diffraction of 4″ sapphire wafer is less than 10 arc sec, and the average density of dislocations is about $1.99{\times}10^2$ cm^{-2}.

Figure 10. Flow pattern in the melt before modifications at the melting temperature.

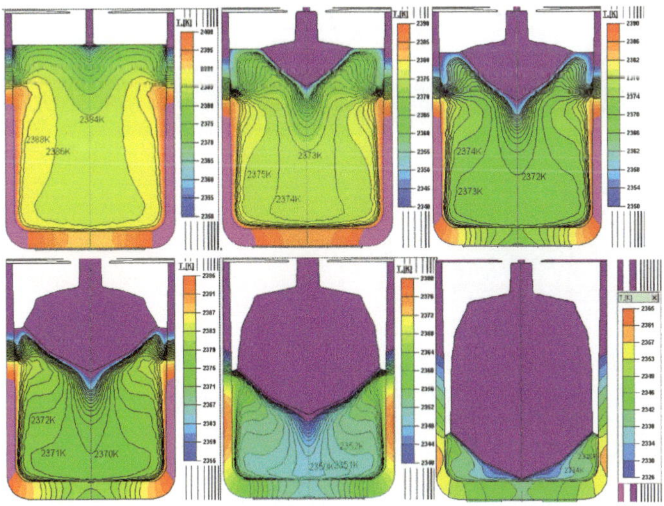

Figure 11. Isotherms at different growth stages.

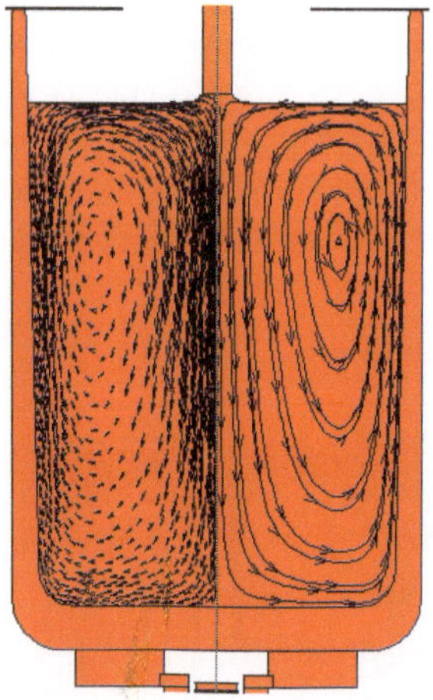

Figure 12. Flow pattern in the melt after modifications at the melting temperature.

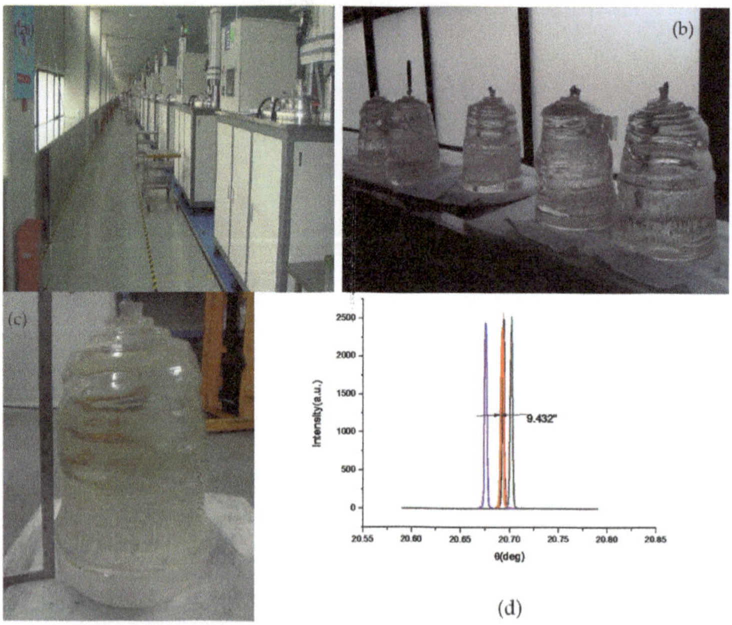

Figure 13. (a) KY furnaces manufactured by SKY, (b) 60 Kg, 85 Kg grades sapphire boules, (c) 83 Kg sapphire boule,(d) X-ray rocking curves of 4" sapphire wafer in the center and at the edge.

3.4.2. Heat exchanger method

The heat exchanger method (HEM) for growing large sapphire crystal was invented by Fred Schmid and Dennis Viechnicki at the Army Materials Research Lab in Watertown, Massachusetts in 1967[18]. The modern implementation of Schmid and Viechnicki's heat-exchanger method at Crystal Systems in Salem, Massachusetts is depicted in Figure14. A high temperature heat exchanger is introduced from the bottom of the furnace into the heat zone. As the heat exchange medium, helium gas circulates in the heat exchanger to take out the heat. The control of the crystal growth process is achieved by adjusting the helium flow. After partial melting of the seed, gas flow is increased to initiate crystallization of melt onto the seed. When the crystallization is complete, the furnace temperature and the helium flow are decreased, and the boule is slowly annealed in situ. During the whole growth process, the crystal is surrounded by the high-temperature melt. It is difficult to release crystalline latent heat, so the growth rate of the crystal is constrained.

The application and development of HEM is mainly in the United States. GT-Solar and ARC Energy are two leading companies in this field. GT-Solar has grown 115 Kg sapphire crystal along a-direction. Heat is supplied by the graphite resistance heat; so that the crystal interior forms oxygen vacancies and the sapphire crystal is pink. Besides, the molybdenum crucible is disposable, thus the cost is high. Figure 15 shows the 111.40 Kg sapphire grown by Guizhou Haotian Optoelectronics Technology Co., Ltd. using the GT-Solar furnace. As we all know, growing high-quality sapphire crystals along c-direction is a great challenge. ARC Energy has achieved a certain degree of technological breakthroughs. They have grown 37 Kg and 47 Kg grades sapphire crystals along c-direction. Figure 16 shows the Φ170 mm and Φ260 mm sapphire ingots grown by ARC Energy. Compared with GT-Solar furnace, heat is supplied by the tungsten resistance-heated element on ARC Energy furnace, thus the sapphire crystal is colorless transparent. Due to sub-grain boundary defect, it is difficult to grow large size and high-quality sapphire crystal along c direction.

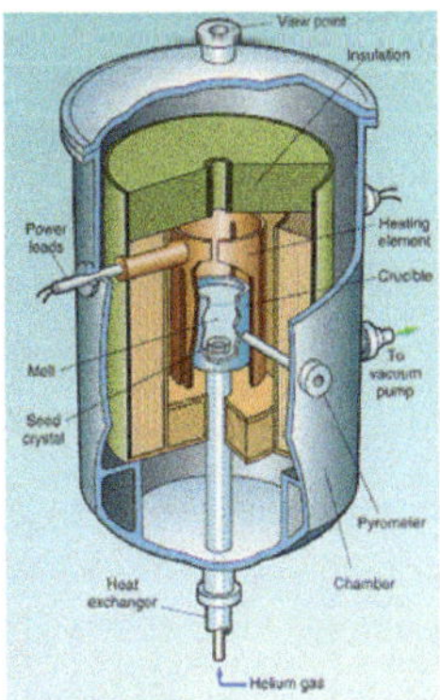

Figure 14. Schematic of a HEM furnace [19].

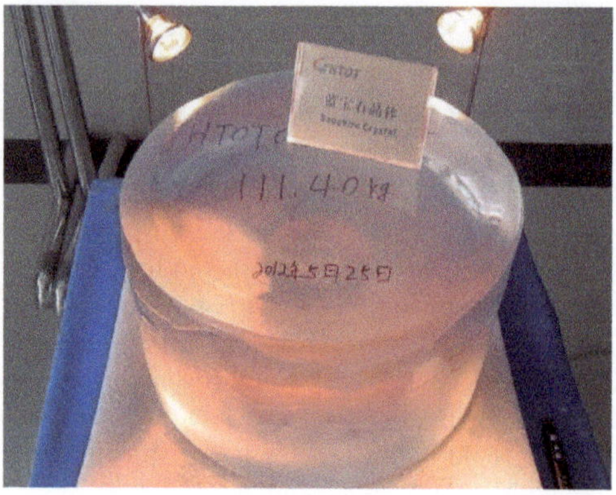

Figure 15. 111.40 Kg sapphire ingot.

Figure 16. Sapphire ingots (Φ170 mm & 260 mm) grown by ARC Energy.

3.4.3. Czochralski method

The Czochralski method is named after Jan Czochralski who introduced an early version in 1916, and published it as a method for studying the crystallization rate of metals [20]. Further modifications by Teal and Little brought the technique closer to the process known as the Czochralski or CZ method [21]. This technique was first applied to grow sapphire crystal by Poladino and Rotter in 1964, and the quality of the as-grown sapphire crystal was relatively high. Figure 17 presents the Czochralski furnace geometry. The crucible is heated by means of an rf generator operating in the range of 10-30 kHz. Considering the melting point of sapphire is 2050°C, Ir crucible whose highest working temperature is up to 2200°C is commonly used. The growth and control of the sapphire crystal is dependent upon the automatic diameter control system that weighs the crystal while it is growing. The change in crystal weight is used to generate a control signal that modifies the generator output power to the crucible, thereby controlling the diameter is achieved through small changes in the liquid temperature. The crystal is constantly rotated and pulled during the growth process, which is different from Kyropoulos method. The main advantages of CZ method are automatic diameter control, convenient to observe the growth status of crystal. Through technique improvement, UNC Co.Ltd. has grown 4″ and 6″ sapphire crystals along a or c direction by CZ method.

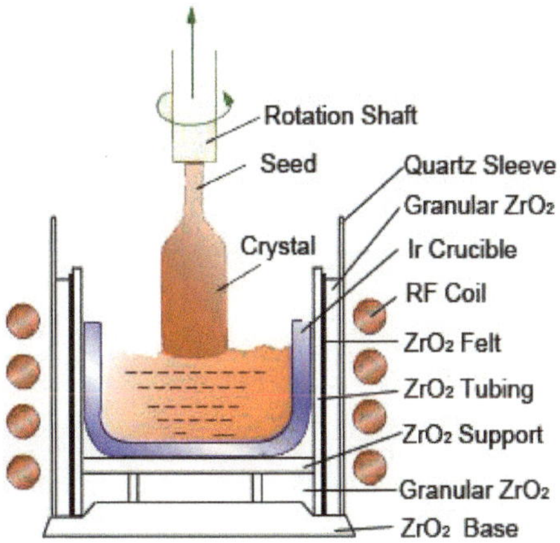

Figure 17. Schematic of a typical Czochralski furnace.

3.4.4. Edge-defined film-fed growth method

Edge-defined film-fed growth method was invented by Harold Labelle and Stepanov in 1960's [23]. This technique is used to grow special-shaped crystals, such as ribbon, tubal and nemaline crystals. Figure 18 shows the schematic diagram of the EFG furnace. The mold with capillary slit is put in the melt. The melt rises to the top of the mold due to the capillarity, forming a layer of thin film and spreading to the surrounding, at the same time crystallization starts from the seed. Melt rising height H is calculated by $H=2\gamma\cos\theta/(\varrho dg)$, where γ denotes the melt surface tension coefficient (N/cm), ϱ denotes the melt density (g/cm^3), d denotes the capillary diameter (cm), g denotes acceleration of gravity (N/g) and θ denotes solid-liquid wetting angle ($0°<\theta<90°$). The wetting angle of melt and mold should be less than 90° so as to form melt fluid film of a certain thickness. The advantages of EFG are that growth rate is fast, the cost is low, and multiple-piece crystals can be grown simultaneously. Figure 19 exhibits the substrate-grade sapphire ribbons grown by Namiki company whose technique is advanced in the world. The magnitude of dislocation density is about $10^{3\text{-}5}$cm^{-2} which is two magnitudes lower than that of grown by conventional EFG method. The main defects of sapphire crystal grown by EFG are bubble, grain boundary, dislocation and heat stress, resulting from overcooling temperature at the solid-liquid interface.

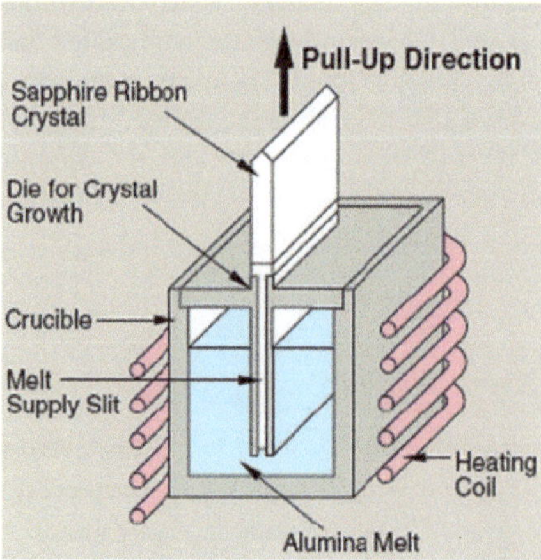

Figure 18. Schematic diagram of sapphire crystal grown by the edge-defined film-fed growth method[23].

Figure 19. (a) Sapphire ribbons grown by Namiki company, (b) Dislocation density of sapphire wafers grown by conventional EFG, (c) Dislocation density of sapphire wafers grown by Namiki.

3.4.5. Temperature gradient technique

The temperature gradient technique (TGT) is a typical static directional solidification technique, invented by Cui et al. from Shanghai Institute of Optics and Fine Mechanics(SIOM) in

1979[22]. A schematic diagram of the TGT furnace is shown in Figure 20. It consists of molybdenum crucible, graphite heating element and molybdenum heat shields. The cylindrical graphite heat element is designed as an electric circuit with proper linear resistance from the top to the middle by making holes in certain distribution. The cylinder is placed in the graphite water cooled electrodes. The temperature gradient of the upper part is built by the linear resistance of the heating element, and that of the lower depends on the extraction of heat by water flowing in the tubes through the electrodes. Besides, the temperature field near the seed is influenced by the heat conductivity of the water-cooled centre rod. The temperature gradient is in the opposite direction of gravity, and there are no moving parts in TGT. The growth process is accomplished by dropping temperature at designed rates with a high precision temperature program controller.

Figure 21 shows the sapphire crystals grown by TGT. The graphite volatilizes to a certain extent, which makes the atmosphere oxygen-lacked. In the reducing atmosphere, transition metal impurity ions exist in the low valence form (e.g.: Cr^{3+}, Ti^{3+}), so that sapphire crystals are pink or light green. In order to improve optical transmittance and uniformity, the sapphire crystals must undergo annealing treatment of decarbonisation and decolorisation. The annealing treanment has two steps. At first, the sapphire crystals are annealed in oxidizing atmosphere at 1700°C, and then annealed in strong reducing atmosphere at 1600-2200°C. After the special annealing process, the color can be eliminated and sapphire crystals revert to high optical property.

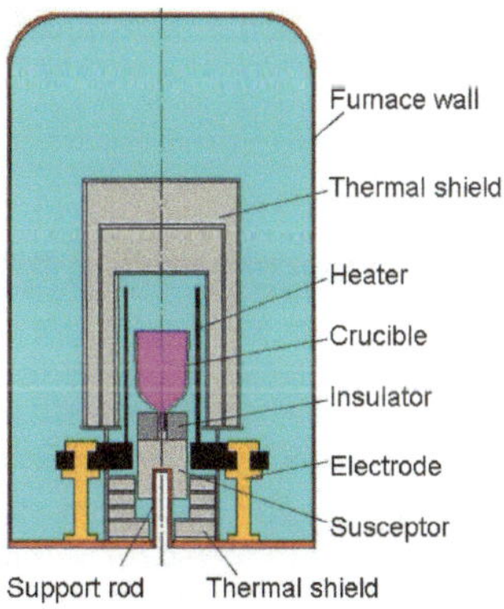

Figure 20. Schematic diagram of a TGT furnace.

3.5. Crystal defects

With the fast development of modern science and technology, the requirements of sapphire materials are getting higher and higher. A popular standard of sapphire substrate wafer for GaN epitaxy is as follows [24]: purity >99.999%, orientation accuracy <±0.5°, dislocation density $<1\times10^3$ cm^{-2}, carbon content $<5\times10^9$ cm^{-2}, number of particles bigger than

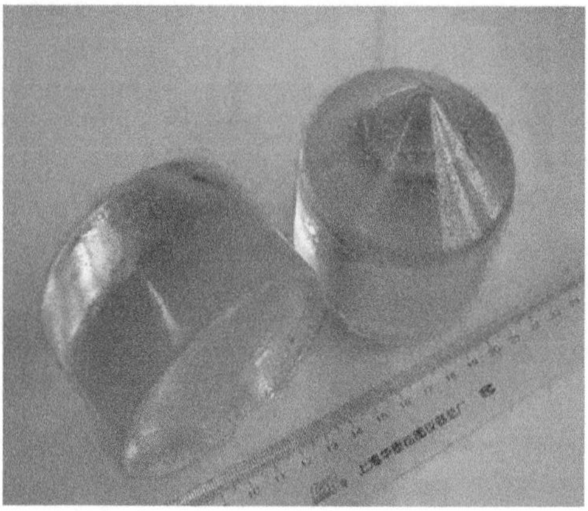

Figure 21. Sapphire crystals grown by TGT.

0.1μm <20/piece. Sapphire crystals grown from the melt usually contain various types of macro or micro defects, among which dislocations, sub-grain boundaries, inclusions, crack, bubbles are common. The distribution of each kind of defects in sapphire depends on each process of the production, from raw material preparation to crystal growth and subsequent heat treatment. Studying the origin and the distribution behavior of defects can deliver a better understanding on crystal growth process, which may provide solid information to production improvement.

3.5.1. Dislocations

Dislocation is one of the common microscopic defects in the sapphire crystals. Figure 22 shows the dislocation corrosion morphology of sapphire wafer grown by TGT [25]. In TGT, the crystal touches crucible, which can cause a stress field, especially during cooling process. Such stress field can cause plastic deformation of the crystal by activating the slip systems. And more dislocations are formed. The typical etch pits generated on (0001) plane are triangles, and on $(11\bar{2}0)$ plane are rhombuses. The dislocation corrosion morphology is determined by the point group and structure of the crystal. The role of the chemical etchant is to destroy the bonds between molecules or atoms inside the crystal, and the smaller bond strength is first to be destroyed, thus forming a particular shape of the corrosion spots. Figure 23 reveals that the density of etch pits on the different position of the sapphire crystals is not uniform. It is higher at the shoulder and lower at the equal-diameter position of the crystals.

The origins of the dislocations in the sapphire crystals are as follows: (1) dislocations inheriting from the seed, including the seed dislocations and the dislocations introduced during seed machining process and seeding process; (2) dislocation nucleation and multiplication produced when the thermal stress in the crystal exceeds the critical stress; (3) external thermal and mechanical fluctuation.

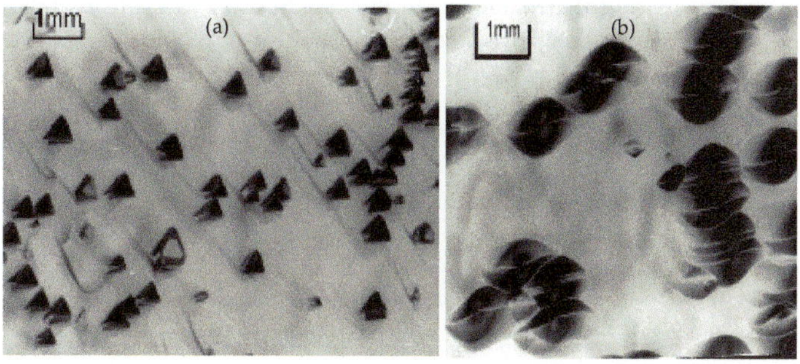

Figure 22. Dislocation corrosion morphology of sapphire wafers: (a) (0001) plane, (b) (11$\bar{2}$0) plane.

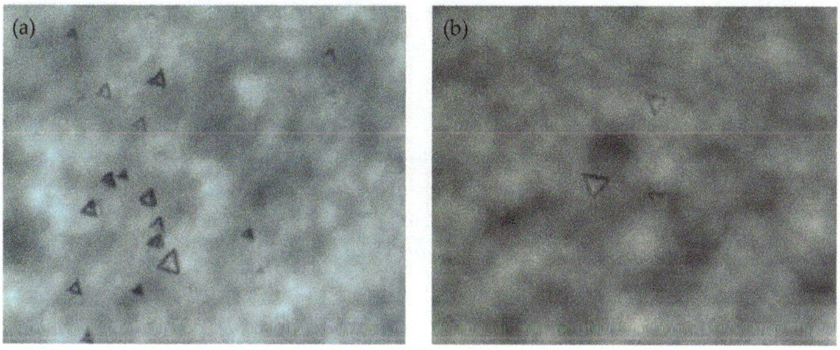

Figure 23. Dislocation etch pits on the (0001) plane of sapphire wafers grown by KY method:(a) shoulder position, (b) equal diameter position.

3.5.2. Sub-grain boundaries

Mosaic structure is the fine structural regions of the crystallographic misorientation smaller than 10°C. The boundaries between the two regions are known as sub-grain boundaries. Figure 24 shows the mosaic structure on (0001) plane of the sapphire blocks under the polariscope. The area of the mosaic structure has colored interference fringe, which is different in color from the main part of the crystal. Sub-grain boundaries could also be observed by chemical etching method, as is shown in Figure 25.

Dislocation sub-grain boundaries are usually formed on cooling or annealing procedures. During these periods, the dislocations acquire sufficient excitation energy to migrate. Then under the stress-field interaction, the dislocations on the different slip planes stop at equilibrium positions, and form the sub-grain boundaries. In addition, the lineage structure appearing in the melt crystallization process can also lead to the formation of sub-grain boundaries [26]. Therefore, the basic method to eliminate sub-grain boundaries is to reduce the dislocation density and thermal stress in the sapphire crystals. Meanwhile, the growth direction should be a$[11\bar{2}0]$, m$[10\bar{1}0]$ or r$[1\bar{1}02]$ direction, not c [0001] direction as possible.

Figure 24. Optical image of the mosaic structure on (0001) plane of the sapphire blocks.

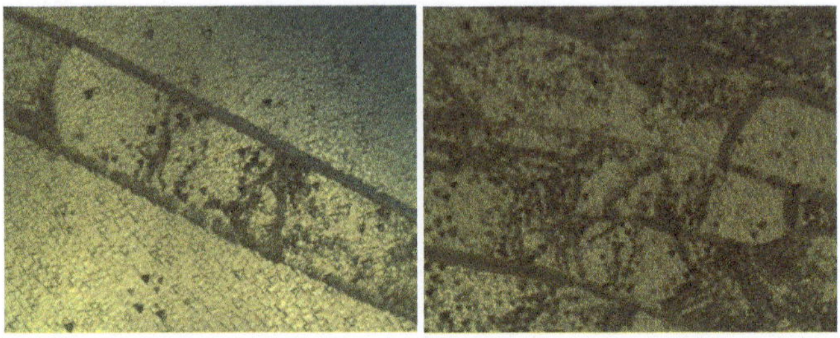

Figure 25. Optical image of sub-grain boundary obtained after the (0001) plane slice has been etched in KOH melt at 380℃ for 10 min.

3.5.3. Inclusions

Inclusion is a common macroscopic defect in the sapphire crystal grown by melt method. The impurity elements of Al_2O_3 raw material and crucible material, as well as tungsten, molybdenum fragments from the crucible, heating unit and insulations will form inclusions in the crystals. Due to the impurity discharging phenomenon, impurities in the raw materials discharge to the edge, and impurities in the crucible diffuse to the centre. These impurities precipitate when the concentration exceeds the saturated concentration in the melt, forming inclusions. Therefore, high purity of Al_2O_3 raw material, crucible material and insulation material are important to grow high quality sapphire crystal.

3.5.4. Cracking

Cracking is one of the main defects of large size sapphire crystal. It is divided into stress cracking and polycrystalline cracking, as is shown in Figure 26. The crystal with stress is also easy to crack along cleavage plane during drilling rod process. Figure 27 reveals the cracked sapphire rod. In addition, because the crystal and crucible material are different in the thermal expansion coefficients, the sapphire crystal is prone to crack on the annealing or cooling procedure if the crystal touches the crucible.

The crystal growth process is mainly achieved by controlling the heat transport system, thus the crystal itself is bound in a certain temperature gradient in the thermal field. When the crystal growth rate or cooling rate is too fast, too large temperature gradient is bound to lead to large thermal stress inside the crystal. Stress in the crystal causes the strain. The cracks initiate, expand and the crystal will crack when the strain exceeds the yield limit. Figure 28(a) presents the interference photographs of the (0001) plane sapphire crystals with stress. The colored strip is asymmetric, and the dark line of the extinction figure shows distortion.

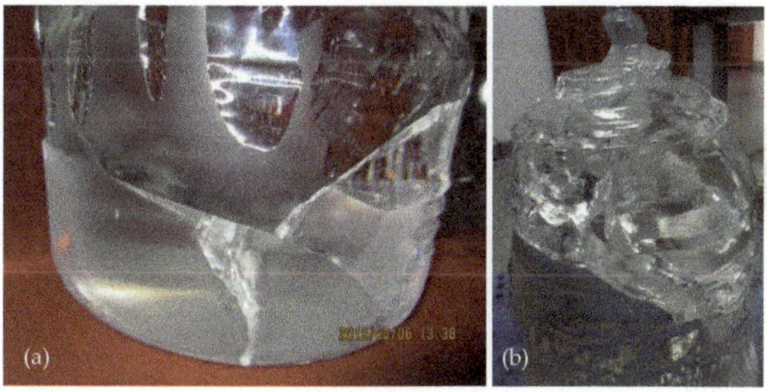

Figure 26. Cracking in large size sapphire crystals: (a) stress cracking, (b) polycrystalline cracking.

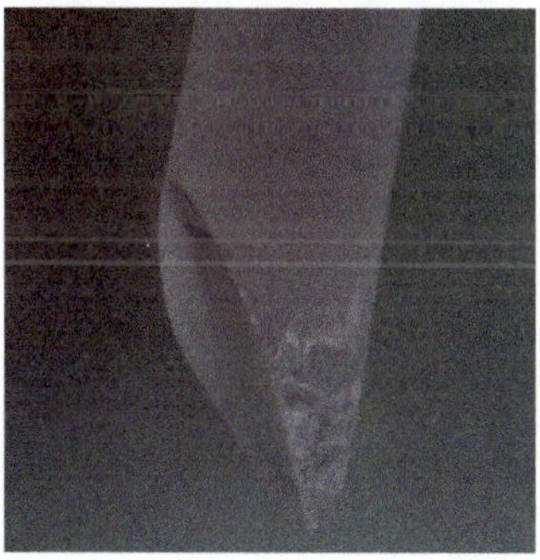

Figure 27. Cracking in the sapphire rod with stress.

3.5.5. Bubbles

Figure 29 presents the bubble distribution of the sapphire boule examined by He-Ne laser. The bubble content is higher at the shoulder and center of the crystal, and the quantity and size of the bubble decrease on the equal diameter position. When the crystallization rate is

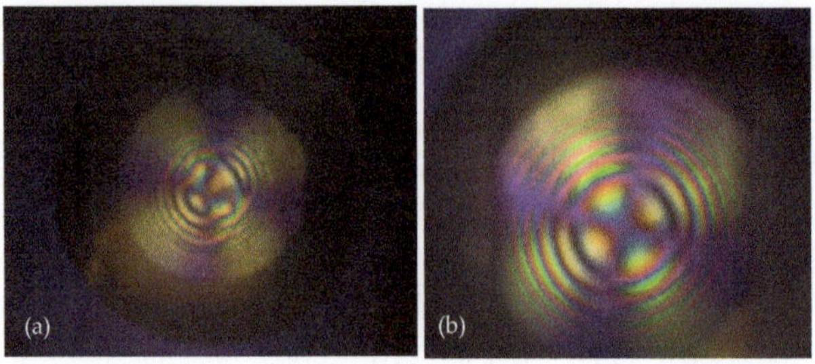

Figure 28. Interference photograph of (0001) plane sapphire rods under the polarizing microscopy:(a) sapphire bar with stress, (b) sapphire bar without stress.

too fast, the bubbles in the melt cannot be removed thoroughly, and bubble layer will form in the crystal. That the separation coefficients of the gas in the solid and melt are different is the reason for the formation of bubbles.

Mycatob et. al thought that micro-thermal decomposition of the alumina melt release O_2, and tungsten, molybdenum and oxygen react chemically. The chemical reaction equations are as follows [27]:

$$Mo(s) + \frac{3}{2}O_2(g) \rightarrow MoO_3(g)$$

$$Wo(s) + \frac{3}{2}O_2(g) \rightarrow WO_3(g)$$

When the crystalline rate is too fast or the fluctuation of crystal growth rate is too large, the gas can be easily captured and wrapped from the solid-liquid interface into the crystal.

Bunoiu et. al have put forward a model to explain the origin and distribution of bubbles in EFG sapphire. According to this model, the origin of the bubble formation is attributed to the reactions described as follows:

$$Al_2O_3(l) \rightarrow 2O(g) + Al_2O(g)$$

$$Mo(s) + Al_2O_3(l) \rightarrow MoO(g) + Al_2O(g) + O(g)$$

$$Mo(s) + O(g) \rightarrow MoO(g)$$

$$Mo(s) + Al_2O(g) \rightarrow MoO(g) + 2Al(g)$$

$$O(g) + C(s) \rightarrow CO(g)$$

$$Al_2O(g) + C(s) \rightarrow CO(g) + 2Al(g)$$

$$3MoO(g) + 5C(s) \rightarrow Mo_3C_2(g) + CO(g)$$

The main composition of the bubble is CO, which has been verified through spectroscopy. They also used numerical simulation to explain the distribution of bubbles and also developed a way to confine bubble distribution near the surface, so it can be easily removed through polishing [28].

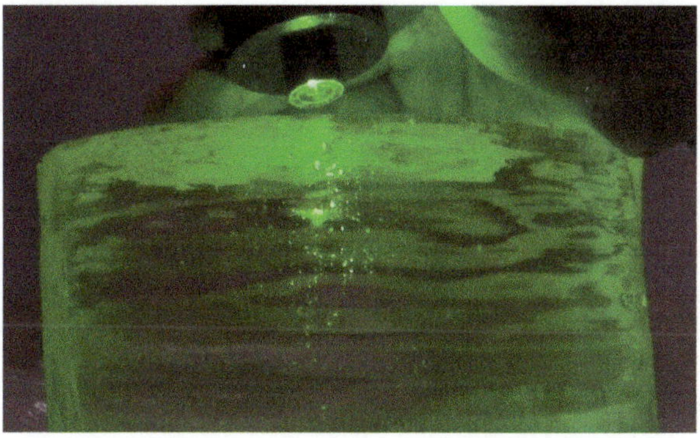

Figure 29. The bubble distribution of the sapphire boule grown by KY method.

4. Conclusions

The growth methods of sapphire nearly cover all of the crystal growth technology, and each method has its own uniqueness. However, Kyropoulos method, edge-defined film-fed growth method, heat exchanger method are the main methods used in mass production in the world, and Kyropoulos method and heat exchanger method are more universal. Table 5 lists the advantages and disadvantages of sapphire crystal growth methods. CZ method is mainly used to grow 3″ sapphire crystals, and HEM, TGT and KY methods are mainly used to grow larger than 4″ sapphire crystals. As the requirements of sapphire substrate constantly improve, the techniques to grow larger sapphire crystal are under research.

Method	Advantages	Disadvantages
KY	High quality and large-size crystal Fast growth rate	Low automation degree High technical requirement on operator
HEM	High automation degree Large size crystal	High cost
CZ	High automation degree	High cost Small size crystal
EFG	Fast growth rate Low cost	The orientation adjustment of large size crystal is difficult.

Table 5. The characteristics of sapphire substrate growth methods from the industrial point of view.

Author details

Huili Tang, Hongjun Li and Jun Xu

*Address all correspondence to: xujun@mail.shcnc.ac.cn

Shanghai Institute of Ceramics, Chinese Academy of Sciences, Shanghai, China

References

[1] Huili Tang. Study of growth and characterization on $ScAlMgO_4$, $Mg_{0.4}Al_{2.4}O_4$ novel substrate crystals. PhD thesis. Shanghai Institute of Ceramics, Chinese Academy of Sciences;2010.

[2] Elena R. Dobrovinskaya, Leonid A. Lytvynov, Valerian Pishchik, Sapphire Material, Manufacturing, Applications. 2009,P 55-56.

[3] Kecong Zhang, Lehui Zhang. Crystal Growth (version l).1981. P464-465.

[4] Conglu Wang. Sapphire single crystal (version l). 1982.P4-6.

[5] Ni X, Özgür Ü, Baski A. A., Morkoç H., Zhou Lin, David J. Smith and Tran C. A..Epitaxial Lateral Overgrowth of Semipolar GaN on (11−00) m-plane Sapphire by Metal-organic Chemical Vapor Deposition. Appl. Phys. Lett.90 2007; 182109.

[6] P. Vennégués and Z. Bougrioua. Epitaxial orientation of III-nitrides grown on R-plane sapphire by metal-organic-vapor-phase epitaxy. Appl. Phys. Lett. 2006; 89, 111915.

[7] B. A. Haskell, F. Wu, S. Matsuda, M. D. Craven, P. T. Fini, S. P. DenBaars, J. S. Speck and Shuji Nakamura. Structural and Morphological Characteristics of Planar (112¯0) a-plane Gallium Nitride Grown by Hydride Vapor Phase Epitaxy. Appl. Phys. Lett. 2003;83(8), 1554-1556.

[8] T. Paskova, R. Kroeger, S. Figurege, D. Hommel, V. Darakchieva, B. Monemar, E. Preble, A. Hanser, N. M. Williams, and M. Tutor. High-quality Bulk a-plane GaN Sliced from Boules in Comparison to Heteroepitaxially Grown Thick Films on r-plane apSphire. Appl. Phys. Lett. 2006;89, 051914.

[9] X. Nia, Y. Fu, Y.T. Moon, N. Biyikli, H. Morkoc. Optimization of (112¯0) a-plane GaN Growth by MOCVD on (112¯0) r-plane Sapphire. J. Cryst. Growth 2006; 290 166-170.

[10] Chun-Hung Chen, Jyh-ChenChen, Chung-WeiLu, Che-MingLiu. Numerical Simulation of Heat and Fluid Flows for Sapphire Single Crystal Growth by the Kyropoulos Method. J. Cryst. Growth 2011; 318 162-167.

[11] Hosseini S M, Aliabad H A R, Kompany A.. First-principles Study of the Optical Properties of Pure α-Al_2O_3 and La aluminates. Eur Phys J B 2005; 43(4): 439-444.

[12] http://www.mkt-intl.com/index.shtml.

[13] Jiao Wang. Synthesis of Polycrystalline α-Al2O3 Prepared by Cold Crucible Method and Study of Defects on Sapphire crystal by Kyropous. Master. Zhengzhou University;2012.

[14] Gaofeng Fu, Shiwen Bi, Xudong Sun, Yihong Yang. The Preparation Technique of Ultrafine Alumina Powders. Non-ferrous Mining and Metallurgy. 2000; 16(1), 39-41.

[15] Jianliang Liu, Jialin Sun, An Shi, Jin Hu, Xianyong Lu, Fuqian Zheng. Recent Development of the Methods to Make the High Purity Fine Alumina. Joumal of Kunming University of Science and Technology(Science and Technology). 2003; 28(3),22-24.

[16] Atsutomo Nakamura, Takahisa Yamamoto and Yuichi Ikuhara. Direct Observation of Basal Dislocation in Sapphire by HRTEM, Acta Materialia 2002;50 101-108.

[17] http://tech.oecr.com/.

[18] F. Schmid and D. Viechnicki. Growth of Sapphire Disks from the Melt by a Gradient Furnace Technique. J. Am. Ceram. Soc. 1970;53 528.

[19] D. C. Harris. A peek into the history of sapphire crystal growth,.Proc. SPIE 2003;5078 1-11.

[20] J. Czochralski. A new method for the measurement of crystallization rate of metals. Zeitschrift des Vereines Deutscher Ingenieure 1917;61 245-351.

[21] G. K. Teal and J. B. Little. Growth of germanium crystals. Phys. Rev. 1950;78 647.

[22] F. Cui, Y. Zhou, J. Qiao. Growth of high quality monocrystal sapphire by seed-induced Temperature Gradient Technique (STGT). J. Chin. Ceram. Soc. 1980;8 109-113.

[23] Bunoiu O M, Nicoara I, Santailler J L, et al. On the void distribution and size in shap ed sapphire crystals. Cryst Res Technol,2005;40(9) 852-859.

[24] Hai Zhou, Shaofeng Yao. Study of the system of quality inspection for sapphire wafer. Applied Science and Technology 2005;32(11):21-24.

[25] Guoqing zhou. Study of Defects and Growths on Sapphire and α-BaB_2O_4 Crystals, PhD thesis. Shanghai Institute of Optics and Fine Mechanics, Chinese Academy of Sciences;1997.

[26] HongJun Li, GugJuann Zhao, XiongHui Zeng, ZhenYing Qian, JuPing Guo, ShengMing Zhou, Jun Xu. Low-angle Boundary in High-temperature Scintillating Crystal Ce:YAP. Journal of Inorganic Materials 2004; 19(5):1186-1190.

[27] М.И. Мусатов. Причины образования пузырей в кристаллах корунда. Неорганические материалы. 1979; 15(10):1806-1810.

[28] Bunoiu, O.M., T. Duffar, and I. Nicoara. Gas bubbles in shaped sapphire. Progress in Crystal Growth and Characterization of Materials 2010; 56(3–4) 123-145.

Chapter 12

Inorganic Nanostructures Decorated Graphene

Hong Ngee Lim, Nay Ming Huang,
Chin Hua Chia and Ian Harrison

Additional information is available at the end of the chapter

1. Introduction

1.1. Why use graphene for the assembly of nanostructures?

Graphene, with zero energy gap between the highest occupied molecular orbit and the lowest unoccupied molecular orbit (HOMO-LUMO), offers a unique two-dimensional (2-D) environment for fast electron transport and has potential applications in electronic devices [1, 2]. Other consequences of the band structure is the opacity which is wavelength independent [3], and thermal conductivity [4]. The four edges of a graphene sheet provide significant number of centres for fast heterogeneous electron transfer [1], when compared to single-walled carbon nanotubes (SWCNTs) for which heterogeneous electron transfer occurs only at the two ends of the nanotube [5]. Consequently, graphene sheets may have wider applicability in electrochemistry [6]. While graphite is brittle, graphene's flexibility is beneficial for use in electromechanical devices [7] and energy storage devices [8]. In the energy storage devices, its weight is of extremely important and the specific area per unit weight is an important figure of merit. Graphene exhibits a theoretical surface area of 2630 $m^2\ g^{-1}$, which is ~ 260 times greater than graphite and twice that of CNTs [9]. Thus, graphene provides a way of enhancing the electrochemical catalytic activity of materials by greatly increasing the high surface area [10]. The intriguing electronic, optical, electrochemical, mechanical and thermal properties of graphene make it an indispensable material in various kinds of synthesis processes.

Various inorganic nanostructures have been prepared over the last two decades due to the special properties of nanostructured materials. Previously, 2-D ZnO nanoplates [11], 1-D PbS nanorods [12], 0-D semiconductor materials [13] and core-shell magnetic nanoparticles have been produced [14]. However, heavy aggregation of the nanostructures, resulted by van der Waals forces between the particles, may limit their special properties and cause structural instability, thus reduce their applicability. To prevent this clustering from occurring, nanocomposites, consisting of the nanoparticles embedded within a matrix compound, can be used.

These nanocomposites preserve the unique properties of the nanoparticles whilst often having the additional performance benefits of the matrix compound itself.

With large surface area and the unique properties given above, graphene is an attractive choice as the matrix for inorganic nanostructures [15]. Functionalization of graphene sheets with various nanostructures can further enhance the properties of graphene. Heterostructures consisting of nanostructures distributed on the surface of graphene could potentially display not only the unique properties of nanostructures [16] and those of graphene [2, 17, 18], but also additional novel functionalities and properties due to the interaction between them. Moreover, the growth of the nanostructures can take place easily when graphene is used as a matrix due to the planar structure of graphene, in comparison to the hollow tubal-shaped CNTs. The particle size and size distribution of nanostructures are small and narrow when graphene is used as a support for the growth of nanostructures [19-21].

2. Inorganic nanostructures decorated graphene

Graphene nanocomposites can be made from graphene oxide (GO) which is essentially a graphene sheet containing oxy-functional groups, such as epoxy, hydroxyl, carbonyl, and carboxylic, on the surface. The reactive functional groups provide a means of attaching the nanoparticles to the graphene sheet [22, 23]. The GO starting material can be simply and economically synthesized using chemical oxidation of graphite [24]. Graphene manufactured using this highly scalable synthesis route is also popularly known as reduced GO (rGO). Using GO as a support for metallic ions, many types of different nanocomposites have been made. The overview of the types of materials and synthesis methods pertaining to the graphene-based nanostructures are presented in Table 1 along with the potential application for the nanocomposites.

Inorganic Nanostructure	Morphology and Dimension	Synthesis Method	Potential Application	Reference
Metal				
Ag	16.9±3.5 nm	Rapid thermal treatment	Antibacterial agent, nanofluids for cooling technology, water treatments, surface-enhanced Raman scattering (SERS), transparent and conductive film, electrochemical immunosensor, and catalysis	[25]
Ag	~420 nm	Laser assisted photocatalytic reduction	Solar energy conversion	[26]
AgAu	50–200 nm	In situ chemical synthesis	Electrochemical immunosensing	[27]
Au	30-70 nm	Laser assisted photocatalytic reduction	Solar energy conversion	[26]

Inorganic Nanostructure	Morphology and Dimension	Synthesis Method	Potential Application	Reference
Au	30 nm	Cyclic voltammetry scanning	Detection of mercury	[28]
Ni	Single-layered	Electroless Ni-plating	Electrodes, sensors, hydrogen-storage, production of fuel cell	[29]
Pd	13±2 nm	Electrochemical codeposition	An anode catalyst for formic acid electrooxidation	[30]
Pd	5–7 nm	Laser irradiation	CO oxidation	[31]
Pt	5–7 nm	Laser irradiation	CO oxidation	[31]
PtRu	2 nm	Microwave	Electrocatalysts for methanol oxidation	[32]
Metal oxide				
Ag_2O	45 nm	In situ oxidation route	Supercapacitor	[33]
CoO	5–7 nm	Laser irradiation	CO oxidation	[31]
CuO	< 20 nm	In situ chemical synthesis	Glucose biosensor	[34]
CuO	30 nm	Hydrothermal	Anode for lithium-ion batteries	[21]
Fe_3O_4	20, 30 and 40 nm	In situ chemical synthesis	Sensors, supercapacitors, drug delivery systems, waste water treatment	[20]
Fe_3O_4	12.5 nm	Gas/liquid interface reaction	Anode for lithium-ion batteries	[35]
Mn_3O_4	10 nm	Ultrasonication	Supercapacitors	[36]
NiO	32 nm	Hydrothermal	Anode for lithium-ion batteries	[37]
PbO	image not shown	Electrochemical route	Detection of trace arsenic	[38]
SnO_2	10 nm	Microwave	Detection of mercury(II)	[19]
SnO_2	4-5 nm	One-step wet chemical method	Detection of cadmium(II), lead(II), copper(II), and mercury(II)	[39]
SnO_2	2–3 nm	Microwave	Supercapacitor	[32]
SnO_2	4–5 nm	Microwave autoclave	Anode for lithium-ion batteries	[40]
SnO_2	2–6 nm	Gas/liquid interface reaction	Anode for lithium-ion batteries	[41]
SnO_2	4–6 nm	In situ chemical synthesis	Anode for lithium-ion batteries	[42]
TiO_2	~20 nm	Hydrothermal	Detection of mercury	[43]
TiO_2	4–5 nm	Sonochemical	Photocatalyst	[44]
TiO_2 (P25)	~30 nm	Hydrothermal	Photocatalyst	[45]
TiO_2 (P25)	~30 nm	Hydrothermal	Photoelectrocatalytic degradation	[46]
ZnO	nanorods, an approximated diameter of ~90 nm and length of ~3 μm	Hydrothermal	Solar cells, gas sensors, transparent conductors, catalysis	[47]

Inorganic Nanostructure	Morphology and Dimension	Synthesis Method	Potential Application	Reference
ZrO_2	~42 nm	Electrochemical route	Enzymeless methyl parathion sensor	[48]
Metal oxides				
Bi_2WO_6	nanoparticles containing square nanoplates, 100–300 nm	Reflux	Photocatalyst	[49]
$La_2Ti_2O_7$	nanosheets, comprehensively integrated	Expansion and UV irradiation	Photocatalyst	[50]
$NiFe_2O_4$	6.5 nm	Hydrothermal	Anode for lithium-ion batteries	[51]
Pd-CoO	5–7 nm	Laser irradiation	CO oxidation	[31]
Others				
Ag/TiO_2	TiO_2 layer coexists with Ag	Dipping-lifting in sol-gel solution, reducing process and interface reaction	Photoelectrochemical conversion	[52]
CdS	7.5–20 nm	Solvothermal	Photocatalyst	[53]
Core-shell $Fe@Fe_2O_3@$ Si-S-O	mean of 22 nm	One-pot thermodecomposition	Chromium removal	[54]
Fe_2O_3-ZnO	Fe_2O_3 ~50 nm, ZnO <10 nm	Hydrothermal	Photocatalyst	[55]
FeS_2	spherical, ~50 nm	Hydrothermal	Solar energy conversion	[56]
SnSb	quasi spherical, 30–40 nm	Solvothermal	Anode for lithium-ion batteries	[57]

Table 1. A summary of inorganic nanostructures decorated graphene and their potential applications. (Note: If the inorganic nanostructures are nanoparticles, the shape is not mentioned under the column of Morphology and Dimension.)

3. Interaction of nanostructures with graphene

The widely accepted mechanism for the synthesis of inorganic nanostructures decorated graphene is the attraction of the positively-charged metal ions by the polarised bonds of the functional groups on the GO. The attachment of the metal ions to the surface and edges of the GO results in a redox reaction and the formation of nucleation sites, which eventually leads to the growth of nanostructures on the 2-D graphene sheets. An example of this process is the redox hybridization process which occurs between GO and Fe^{2+} to form Fe_3O_4/rGO nanocomposite and this is shown in Figure 1 [58]. The Fe^{2+} ions are first attached to the surface by the functional groups on the surface of the GO sheets (Figure 1a). The GO acts as an oxidizing agent (Equation 1), effectively increasing the oxidation state of the Fe ions from Fe^{2+} to Fe^{3+}. This is followed by the reaction of the Fe^{3+} ions, in an alkaline condition, into Fe_3O_4 nanoparticles (Equation 2) on the surface of the rGO. The complete stoichiometry is depicted by Equation 3. During the redox reaction, the polar oxygenated functional groups on the GO sheets serve as the anchoring sites for the Fe_3O_4 nanoparticles, consequently preventing serious agglomeration of the magnetic nanoparticles (Figure 1b).

$$2Fe^{2+} + GO \rightarrow 2Fe^{3+} + rGO \tag{1}$$

$$Fe^{2+} + 2Fe^{3+} + 8OH^{-} \rightarrow Fe_3O_4 + 4H_2O \tag{2}$$

$$3Fe^{2+} + GO\left(8OH^{-}\right) \rightarrow Fe_3O_4 + 4H_2O + rGO \tag{3}$$

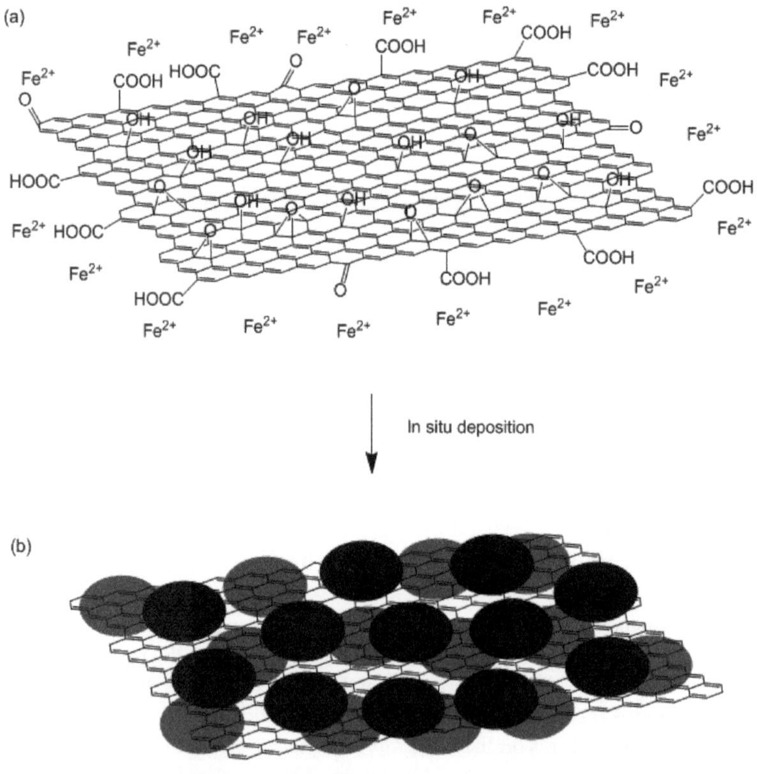

Figure 1. Schematic illustration of the formation of Fe_3O_4/rGO nanocomposite via a one-step in situ chemical deposition method [20].

In the case of metal/graphene, the nanocomposites are produced through simultaneous reduction of GO and metal ions [29, 31, 32]. In the absence of a reducing environment as in pure water but under the illumination of a radiating energy such as laser, the reduction mechanism of the metal ions (M^{2+}) involving the photogenerated electrons from GO are shown by Equations 4–7, where GO is partially reduced as portrayed by Equations 6 and 7 [26].

$$\mathrm{GO} + hv \rightarrow \mathrm{GO}\left(h^{+} + e^{-}\right) \tag{4}$$

$$4h^{+} + 2\mathrm{H_2O} \rightarrow \mathrm{O_2} + 4\mathrm{H^{+}} \tag{5}$$

$$\mathrm{M^{+}} + \mathrm{GO}\left(\mathrm{e^{-}}\right) \rightarrow \mathrm{GO} + \mathrm{M} \tag{6}$$

$$\mathrm{GO} + 4\mathrm{e^{-}} + 4\mathrm{H^{+}} \rightarrow \mathrm{rGO} + 2\mathrm{H_2O} \tag{7}$$

In the presence of a reducing agent such as sodium borohydride, $NaBH_4$ used for the synthesis of Ni/graphene, Ni ions and the COOH groups on the surface of the GO sheets were reduced to Ni metal and CH_2OH, respectively. The corresponding equation may proceed as represented by Equation 8 [29].

$$4Ni_2^{\,2+} + BH_4^{\,-} + 8OH^{-} \rightarrow 4Ni + BO_2^{\,-} + 6H_2O, -COOH \xrightarrow{[BH_4^{\,-}]} -CH_2OH \tag{8}$$

The incorporation of nanoparticles could be through chemisorption, physisorption, electrostatic interaction, van der Waals or covalent bonding with rGO [21, 22]. The attachment of nanostructures onto the surface of the graphene reduces the attractive interactions between the rGO sheets and, minimizes the aggregation and restacking of rGO during the reduction process [53]. Moreover, trace quantity of nanostructures on the basal planes of rGO allows uniform dispersion of the nanocomposite in polar solvents, which is otherwise impossible for rGO [20].

4. A new class of graphene-based inorganic nanostructures

The morphological structure of graphene nanocomposites is varied and depends on the synthesis route. Ag/graphene was synthesized using a thermal expansion method and yielded uniformly distributed Ag nanoparticles on graphene sheets [25] (Figure 2a). Whereas the hydrothermal approach gave an assemble pyrite structured FeS_2 nanospheres [56] (Figure 2b), CuO nanospheres [59] (Figure 2c), TiO_2 nanoparticles [43] (Figure 2d) and ZnO nanoflowers [60] (Figures 2e and 2f). In the last example, some of the ZnO nanoflowers were brighter in the SEM image than others and seemed to be enveloped by a thin film of graphene. Since the functional groups, such as hydroxyl and epoxy groups, are attached to both sides of GO sheets, the ZnO nanoflowers appear on both sides of the graphene support [61]. Another method based around microwave heating has been used to make SnO_2/graphene nanocomposite [19] (Figure 2g). At a higher magnification (Figure 2h), the SnO_2 nanoparticles were observed to uniformly adhere on the graphene sheets, with high density [19]. A simple, cost-effective, efficient, and green in situ deposition to synthesize Fe_3O_4 nanoparticles on graphene [20] (Figure 2i) has also been explored. The magnetic property of the Fe_3O_4/graphene nanocomposite allows the separation of composite from the solution by applying an external magnetic field (Figure 2j).

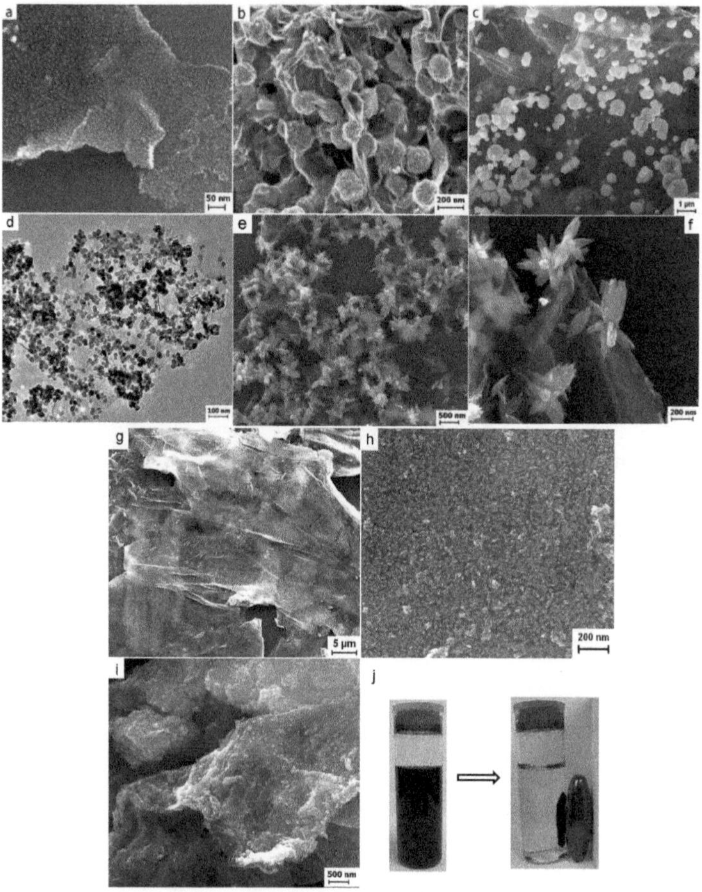

Figure 2. Electron micrographs of various inorganic nanostructures decorating graphene; (a) Ag, (b) FeS_2, (c) CuO, (d) TiO_2, (e) ZnO, (f) a higher magnification showing flower-like shape ZnO nanoarchitecture, (g) SnO_2, (h) a higher magnification revealing SnO_2 nanoparticles, and (i) Fe_3O_4. (j) Photo image of Fe_3O_4/graphene dispersed uniformly in water and attracted under the external magnetic field.

rGO has proven to be an effective matrix for the adhesion of nanostructures due to the rich content of oxide functional groups on the basal planes and edges of the 2-D material. Good evidence of rGO being an accomplished support is depicted through an FESEM image of Fe_3O_4/graphene, in which the nanocomposite was prepared at low concentration of Fe^{2+} ions. The micrograph of the sample shows no evidence for the formation of Fe_3O_4 nanoparticles (Figure 3a). This is in contrast to the sample prepared using a higher concentration of Fe^{2+} ions where nanoparticles are apparent (Figure 2i). To investigate this observation further, elemental mapping of C, O, and Fe using energy dispersive x-ray (EDX) analysis (Figure 3b) was undertaken (Figure 3c). The area of bright contrast correlates with the Fe signal map. This result, coupled with the XRD result [20], provides evidence for the presence of Fe_3O_4 on the surface of the graphene. The exact form of the Fe_3O_4 cannot be determined. It is possible that a layer of Fe_3O_4 has formed on the surface of the rGO or more likely that very small nanoparticles have formed.

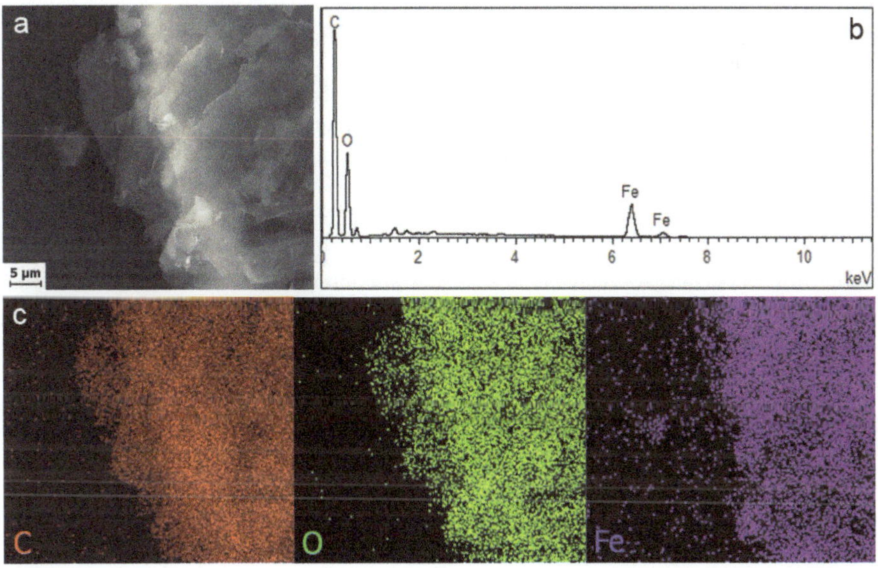

Figure 3. Fe_3O_4/graphene prepared at a low concentration of Fe^{2+} ions: (a) FESEM image, (b) EDX spectrum, and (c) elemental mapping [20].

5. Growth processes of inorganic nanostructures on graphene sheets

5.1. In situ chemical synthesis

In situ chemical synthesis is a robust route for the formation of graphene decorated with inorganic nanostructures. The attraction of the positively charged metal ions towards the negatively charged electron cloud surrounding the oxygen atoms of the GO sheets is the initial step in the nucleation of the nanostructure on the graphene sheet. Since oxygen atoms are uniformly distributed in the starting graphene oxide sheets, a uniform decoration of nanostructures on the graphene sheets is produced after the reaction. Some metal oxides such as Fe_3O_4 and SnO_2 can grow on the surface of rGO at room temperature [20, 39]. However, by increasing the reaction temperature, the high temperature calcination step may be able to be eliminated e.g. SnO_2/graphene [42]. The reduction of GO may not occur during the deposition of the nanostructures on the surface of the GO, for example in the synthesis of MnO_2/GO [61]. To ensure the GO is completely reduced, additional reducing agents can be used. For example, in the synthesis of nanocrystal Ag/graphene, hydrazine and ammonia solution in the presence of polyvinyl alcohol were used [27].

In situ chemical synthesis has also been described as a one-pot thermodecomposition route. When graphene sheets are directly used, a stabilizing agent such as sodium dodecylbenzenesulfonate (SDBS) is required along with a high reaction temperature to encourage the assembly of the nanostructure. The formation of Fe nanostructures on graphene is an example of this. [54]. Similarly, CuO/graphene was synthesized through the stabilization of graphene sheet with ethylene glycol at a high temperature [34]. For some materials, however, the high reaction temperature is not required to promote the assembly. Ag_2O/graphene nanocomposites are an example and these were prepared in the presence of N-Methyl-2-pyrrolidone (NMP) under ambient conditions [33]. The stabilizing agents also function as a size- and shape-controlling agent and a reducing agent. In FeS_2/graphene, the stabilization agent constraints the growth of FeS_2 to a gelatin micelles geometry, resulting in well-formed nanoparticles [56]. Variations of the in situ chemical synthesis include the electrode-less Ni-plating on graphene sheets [29] and the attachment of Bi_2WO_6 nanoparticles and nanoplates onto GO by refluxing [49].

5.2. Hydrothermal

Hydrothermal synthesis is an efficient inorganic synthesis approach for the formation of a variety of nanomaterials, catalysts, ion-conductors, and zeolites [62] under controlled temperature and pressure. This synthesis route overcomes the drawbacks of high processing temperatures and long reaction times compared to conventional aqueous chemical processing conditions [63]. It has been recognized as an environmentally friendly process because the reaction uses aqueous solutions as a reaction medium and it is carried out in an autoclave, which is a closed system. This method can also be used for the preparation of high-purity, highly crystalline, ultrafine and homogeneous powders of various single and multi-component powders [64, 65]. The autoclave used in the hydrothermal synthesis, as mentioned previously, is a closed system and so raising its temperature increases the pressure inside the vessel above the critical pressure for water which enhances the dissolution of thermodynamically unstable compounds. The high heat energy and pressure in the autoclave facilitates fracture of the macronucleus to form nano-sized particles [66]. This method has been used for the synthesis of CuO/graphene [21], NiO/graphene [37], ZnO/graphene [47], and FeS_2/graphene nano-composites [56].

The hydrothermal systhesis methology is not restricted to pure aqueous solutions. The addition of other solvents like ethanol can be used to enhance the dispersion of gel-like GO [43]. For example, TiO_2 nanoparticles could be chemically bonded to the surface of rGO [45, 46]. Likewise, a one pot synthesis of Fe_2O_3 nanoparticles, Zn salt and GO produced Fe_2O_3-ZnO/graphene nanocomposites, in which the Fe_2O_3 nanoparticles were chemically bonded to the graphene sheets. The 50 nm sized Fe_2O_3 nanoparticles were covered with ZnO nanoparticles that are less than 10 nm in size [55].

If the solutions are not aqueous based, the method is known as the solvothermal process, for instance absolute ethanol was used as a medium for the synthesis of SnSb/graphene [57]. For the synthesis of CdS/graphene, dimethyl sulfoxide (DMSO) not only acted as a reaction medium but also the source of sulphide and a reducing agent [67]. SnO_2/graphene was prepared via a gas–liquid interfacial reaction, in a process similar to hydrothermal method. During the reaction, a two-phase liquid was vaporized, which allowed the Sn^{4+} to react with the ammonia at gas/liquid interface to produce $Sn(OH)_4$ along with in situ deposited onto the graphene sheets. The $Sn(OH)_4$ subsequently decomposed to SnO_2 on the graphene sheets [41]. Fe_3O_4/graphene has also been prepared using the same process [35].

5.3. Microwave heating

Microwave heating is believed to be more depending on the molecular properties and the reaction conditions than conventional heating [68]. Microwave syntheses have been increasingly used in the preparation of high monodispersity nanoparticles of oxides such as SnO_2, CeO_2 and ZrO_2 [69]. Utilizing microwave energy for the thermal treatment generally leads to very fine particles in the nanocrystalline regime mainly caused by the shorter synthesis time and a highly focused localised heating. The particle size often falls in the range of 15 nm – 35 nm [70]. Microwave heating also has been widely used for the fabrication of inorganic nanostructure/graphene where the particles were less than 10 nm with a narrow size distribution. In this synthesis, the graphene sheets obviously played an important role in constraining the growth of nanostructures [32, 71]. Hydrothermal method has also been combined with microwave heating to ensure a complete reduction of GO [40].

5.4. Electrodeposition

Electrodeposition provides a facile procedure and offers precise control of the thickness of the resulting film [72]. In addition to the speed of polymerization, which can be controlled by the current density [73], this method also enables mild processing conditions at room temperature [74], without toxic or excess chemicals [75]. GO is coated on the surface of a glassy carbon electrode (GCE) before being immersed into a salt solution. By cycling the potential, the metal salt is oxidized to metal oxide and the GO is reduced to graphene. This synthesis methodology has been used in the preparation of PbO/graphene [38] and ZrO_2/graphene [48].

Although it is known to be hydrophilic, GO does not peel off from the GCE surface when it is placed in aqueous electrolytes because of the interaction of hydrophobic regions of GCE and the unoxidized portions of GO [38]. It does, however, swell in an aqueous solution as water molecules bond to the GO.

The formation of the actual graphene based GCE is problematic because it is difficult to obtain a uniform dispersion of graphene in a solvent. The graphene sheets when in solution tend to form irreversible agglomerates or even restack to graphite through strong π-π stacking and van der Waals interaction. If graphene is to be used for the modification of GCE, it must first be dispersed in a stabilizer to form a homogenous dispersion before being dropped cast on the GCE surface. The inorganic nanostructures on graphene are formed by cyclic voltammetry in the appropriate salt solution. An example of this method is the synthesis Ag/graphene [28].

Electrochemical co-deposition is another route to prepare inorganic nanostructure/graphene. Pd/graphene is an example, where a solution containing GO and a metal salt underwent voltammetry cycling simultaneously [30].

5.5. Other unusual strategic routes

There have been several novel processing routes that have been reported. A Ag precursor was physically grounded with dry GO before the product was heated at 1000°C for 20s to produce Ag/graphene [25]. This method produced high purity nanocomposite because the process did not require any reducing or stabilizing agents. The extreme condition spontaneously reduced the Ag ion to Ag metal, and GO to rGO. Inorganic nanostructure/graphene has also been prepared using irradiation (lasers [31], [26], and exposure to UV [50]). Layer-by-layer inorganic structure/graphene can be prepared by dipping-lifting in sol-gel solution [52]. Finally, ultrasonication of the starting solution was able to generate of Mn_3O_4/graphene nanocomposites [36].

6. Recent advances in the applications of inorganic nanostructure decorated graphene

6.1. Anode for lithium-ion batteries

Graphite-based lithium-ion batteries suffer from poor charge–discharge performance and consequently have poor power performance. In many applications like electric or hybrid cars, there will be periods of high power demand, for example accelerating to overtake a car, and this leads to increased heat dissipation in the cell and accelerated aging of the battery. The use of graphene electrodes significantly improves the power performance of the battery and has resulted in a significant number of papers. The versatility of graphene to accommodate lithium ions on both sides of its single atomic sheet provides high energy storage capacity above 600 $mAhg^{-1}$, which is higher than the theoretical capacity of graphite (372 $mAhg^{-1}$). Recently, the use of metal oxides and metal alloys graphene nanocomposite has improved the energy storage of graphene based electrodes even more (700–4000 $mAhg^{-1}$) [37]. The extra capacity has been attributed to the synergistic effect between the nanoparticles and graphene sheets in the nanocomposite, providing extra sites for the storage of lithium ions [35].

In any battery application, the degradation of the storage capacity with repeated charge-discharge cycles is extremely important. The cycling performance of graphene inorganic nanostructures far exceeds that of their individual counterparts [21, 35]. Moreover, graphene inorganic nanostructures also displayed excellent reversible specific capacities at a broad range of current densities [40-42]. The good lithium cycling performance is ascribed to the structural integrity of the composite electrodes. The nanocystals located on the surface of graphene prevented the agglomeration of graphene sheets. Likewise, the graphene sheets hindered the direct contact among the adjacent nanocrystals. Minimizing the aggregation of nanocrystals and graphene sheets during discharge/charge cycling gives rise to high surface area, excellent electronic conductivity of the electrodes by forming an efficient electrically conductive network, and high carrier mobility. This heterogeneous construction also provides buffering spaces against the volume changes of nanocrystals during the lithium insertion/extraction processes.

Regardless of the type of inorganic nanocrystal, the graphene nanocomposites have generally displayed a large irreversible capacity in the first cycle [51, 57]. It is widely reported that this phenomenon may be attributed to electrolyte decomposition and formation of solid electrolyte interface (SEI) film [76-78]. Irreversible capacity is the occurrence of naturally non-recoverable charge capacity, in which the kinetics will vary depending on the battery chemistry, electrode composition and design, electrolyte formulation and impurities, and on the storage temperature [79].

6.2. Supercapacitors

With the increase in affluence in developing countries, the energy needs are increasing and energy sustainability is of significant concern especially when the depletion of fossil fuels is also factored in [61]. Supercapacitors, also known as ultracapacitors or electrochemical supercapacitors, have several important characteristics, including prolonged life cycle [80, 81], higher power density than batteries [82] and higher energy density [83] than conventional capacitors which have driven their use in pulse power and power backup applications [84]. According to the mechanism of charge storage, supercapacitors can be classified as i) electric double layer capacitors (EDLCs) where charge is stored at the electrode/electrolyte interface, and ii) pseudocapacitors where the charge is stored mainly by Faradaic reactions on the surface of the electrode materials [85, 86].

Carbon-based materials are commonly used in EDLCs as electrodes because of its high electrical conductivity and outstanding long-term electrochemical stability as a result of the extraordinary chemical stability of carbon [87]. Carbon-based materials such as activated carbon [88], xerogels [89], carbon nanotubes [90], mesoporous carbon [91] and carbide-derived carbons [92] have all been investigated for use as electrodes in EDLC. However, the limited charge accumulation in electrical double layer restricts the specific capacitance of EDLCs to a relatively small range of values between 90 and 250 F/g [93]. Meanwhile, pseudocapacitors like metal oxides such as RuO_2 [94-96], NiO [96, 97], Co_3O_4 [98], MnO_2 [99], or conducting polymers including polypyrole [100] and polyaniline [101, 102] usually show poor capacitance behavior.

Recently, graphene has been used as a supercapacitor electrode material due to its high surface area, excellent stability and good conductivity [103-106]. To effectively overcome the shortage of low specific capacitance, the 2-D structured graphene has been hybridized with pseudocapacitors for the preparation of supercapacitors [32, 107-109]. The specific capacitance and electrochemical stability of graphene inorganic structures is enhanced tremendously in comparison to their individual counterparts [33, 36]. The improvement of the supercapacitive behaviour is attributable to the different double-layer and pseudocapacitive contributions.

6.3. Photocatalysts

Organic effluents from industries, agricultural activities and the rapid increase in human waste as a result of the rapid population increase represents some of the most serious environmental pollutants. It is estimated that around four billion people worldwide have no or little access to clean and sanitized water supply, resulting in death, by severe waterborne diseases, of millions of people annually [110]. Photodegradation of organic pollutants has attracted increasing attention during the past decade as it appears to be a viable decontamination process with widespread application, regardless of the state (gas or liquid) or chemical nature of the process target [45, 111]. Photocatalytic oxidation is an economical process owing to the fact that it involves only a photocatalyst and light source [112]. This process does not yield toxic intermediate product, making it suitable for cleaning water environment that contains low to medium contaminants concentration [113].

When a photocatalyst is illuminated with the light of sufficient energy, electrons in the valence band of the photocatalyst are excited into the conduction band, therefore, creating the negative electron-positive hole pairs ($e^{-}-h^{+}$). When the photocatalyst is in contact with water, the $e^{-}-h^{+}$ pairs will initiate a series of reactions and produce hydroxyl radicals, $HO^{\bullet}$ and superoxide radical anions, $O^{\bullet +}$, on the photocatalyst surface. Any organic contaminants at or near to the photocatalyst surface are oxidized by the generated radicals [114, 115]. The most commonly used photocatalysts are titanium dioxide and zinc oxide, which are semiconductors. The drawback of these photocatalysts is the quick electron-hole recombination on the surface of semiconductors, which hampers the hydrogen evolution efficiency.

To manipulate the rate of electron-hole recombination, the presence of non-metal species like carbon, nitrogen, boron and fluorine are targeted to minimize photogenerated electron-hole recombination rate, thus improving quantum efficiency and expanding their useful range of operation into visible light wavelengths [116, 117]. Graphene, a uniform and thin transparent conducting material, is a potential carbon source.

Nanocrystals are not stable and prone to aggregate, which results in a reduced surface area and so limits the likelihood of the photoinduced electron–hole pairs interacting with water molecules to produce the radicals thereby decreasing the application efficiency. This can be overcome by using graphene as a supporting matrix for the photocatalyst particles, as well as an electron acceptor, to improve the efficiency of the degradation of organic pollutants [67] and durability for consecutive photodegradation cycling [45, 49]. In addition, the giant π-conjugation of graphene and two-dimensional planar structure are the driving forces for the non-covalent adsorption between aromatic molecules and aromatic regions of the graphene [45].

Many photocatalysts have a wide band gap and so require UV light for operation. Since ultraviolet (UV) light accounts for only a small fraction (5%) of the Sun's energy as compared to visible light (45%); any shift in the optical response of a photocatalyst from the UV to the visible spectral range will have a profound positive effect on the photodegradation efficiency of a photocatalyst [118]. Incorporation of carbon is known to be able to reduce the band gap energy of a photocatalyst [116]. Photocatalysis using graphene inorganic structure could take place under the irradiation of visible light due to zero band semiconductor with symmetric K and K' [49].

On the other hand, the photoelectrocatalytic process takes advantage of the photocatalytic process by applying a biased voltage across a photo-electrode on which the photocatalysts are supported. An enhancement of the photocurrents using graphene nanocomposite electrodes is attributed to the enhanced migration efficiency of the photo-induced electrons and enhanced adsorption activity of the aromatic molecules [46]. Photoelectrocatalytic activity is dependent on the optimal value of graphene content, as superfluous graphene will reduce the absorption efficiency of light by a photocatalyst. Experiments have shown an increase in the degradation of the aromatic molecules by the graphene nanocomposite as the applied potential was increased relative to a reference electrode [119-124]. This was explained by the potential causing band bending close to the electrode surface reducing the potential barriers and improving the mobility of the carrier across the electrode, and hence minimizing the probability of recombination of electrons and holes and elevating the photoelectrocatalysis efficiency [124].

The overall excellent photocatalytic performance of graphene inorganic nanostructures is reportedly attributed to enhanced adsorptivity, extended light absorption range, efficient charge separation and transportation [45].

6.4. Sensing platform

Many analytical methods with high sensitivity and low detection limit have been established for determination of organic and inorganic matters [125]. However, they are time consuming, expensive, require complicated instruments and a skilled operator, which are unsuitable for on-line or in-field monitoring [126]. In contrast, electrochemical analysis, which has the advantages of quick response, cheap instrumentation, low power consumption, simplified operation, time-saving, high sensitivity and selectivity, is widely applied in applications such as gas sensing, chemical sensing and biosensing [27, 127-129].

Direct detection using bare electrode yields poor sensitivity to the target material [128]. Inorganic nanoparticles are versatile and sensitive tracers because of their high surface area, high mechanical strength but ultra-light weight, rich electronic properties, and excellent chemical and thermal stability [130] and when coupled with graphene, they provided an attractive nanocomposite for the fabrication of electrochemical sensors [33, 131]. In this regard, inorganic nanostructure decorated graphene has been proven to be an effective tool to detect low amounts of biomarker [27], pesticide [48], heavy metals [28, 38, 39], and glucose [34], with high sensitivity, selectivity, stability and reproducibility.

6.5. Other interesting applications

Besides the four major applications stated above, graphene inorganic nanostructures have also been investigated for other purposes. These include direct formic acid fuel cells, which are widely considered to be one of the most attractive power sources [30]; providing a favourable catalytic pathway for the formation of CO_2 [31] to form CO [132]; and producing electrocatalytic activity for methanol oxidation [32]. Other uses of the structures is to produce a photocurrent under UV light or visible light illumination to meet the demand of renewable and clean energy source [52], assisting the conversion of solar energy into hydrogen via the water splitting process [133], and removing chromium(IV) in water through adsorption [54].

7. What does the future hold for inorganic nanostructure decorated graphene?

The significant potential of graphene-based inorganic nanostructure in solving many of today's problems is evident from the amount of effort that has been devoted to exploring the synthesis of the materials and the investigation of the materials in real-life applications. The cutting-edge research on graphene-based nanoinorganic materials has yet to mature. Drawing comparisons to silicone research, the interest will continue to grow until commercial products using graphene are realized. The simple and scalable production of GO, a derivative of graphene, that is rich in oxygenous functional groups, is encouragement for researchers to modify the surface of the one-atom thick carbon layers with a variety of inorganic nanostructures to cater for the commercial demands and needs. The quality of inorganic nanostructure decorated graphene at atomic level is assured by systematic characterization using state-of-the-art instruments. The graphene research offers novel and exciting opportunities for the scientific community and industrialists who seek new partnerships and advances.

Acknowledgements

This work was supported by the Exploratory Research Grant Scheme (ER016-2011A), High Impact Research Grants from the University of Malaya (UM.C/625/1/HIR/030) and High Impact Research Grants from the Ministry of Higher Education of Malaysia (UM.C/625/1/HIR/MOHE/05).

Author details

Hong Ngee Lim[1,2], Nay Ming Huang[3], Chin Hua Chia[4] and Ian Harrison[5]

1 Department of Chemistry, Faculty of Science, Universiti Putra Malaysia, UPM Serdang, Selangor, Malaysia

2 Functional Device Laboratory, Institute of Advanced Technology, Universiti Putra Malaysia, UPM Serdang, Selangor, Malaysia

3 Department of Physics, Faculty of Science, University of Malaya, Kuala Lumpur, Malaysia

4 School of Applied Physics, Faculty of Science and Technology, Universiti Kebangsaan Malaysia, Bangi, Selangor, Malaysia

5 Faculty of Engineering, The University of Nottingham Malaysia Campus, Jalan Broga, Semenyih, Selangor, Malaysia

References

[1] Chen, F, & Tao, N. J. Electron transport in single molecules: From benzene to graphene. Accounts of Chemical Research (2009). , 42(3), 429-38.

[2] Novoselov, K. S, Geim, A. K, Morozov, S. V, Jiang, D, Zhang, Y, Dubonos, S. V, Grigorieva, I. V, & Firsov, A. A. Electric field effect in atomically thin carbon films. Science (2004). , 306(5696), 666-9.

[3] Nair, R. R, Blake, P, Grigorenko, A. N, Novoselov, K. S, Booth, T. J, Stauber, T. Peres NMR, & Geim AK. Fine structure constant defines visual transparency of graphene. Science (2008). , 320(5881): 1308.

[4] Yu, A, Ramesh, P, Itkis, M. E, Bekyarova, E, & Haddon, R. C. Graphite nanoplatelet-epoxy composite thermal interface materials. Journal of Physical Chemistry C (2007). , 111(21), 7565-9.

[5] Pumera, M. The electrochemistry of carbon nanotubes: Fundamentals and applications. Chemistry- A European Journal (2009). , 15(20), 4970-8.

[6] Stoller, M. D, Park, S, Zhu, Y, An, J, & Ruoff, R. S. Graphene-based ultracapacitors. Nano Letters (2008). , 8(10), 3498-502.

[7] Bunch, J. S, Van Der Zande, A. M, Verbridge, S. S, Frank, I. W, Tanenbaum, D. M, Parpia, J. M, Craighead, H. G, & Mceuen, P. L. Electromechanical resonators from graphene sheets. Science (2007). , 315(5811), 490-3.

[8] Pumera, M. Electrochemistry of graphene: New horizons for sensing and energy storage. The Chemical Record (2009)., 9(4), 211-23.

[9] Chen, X. M, Wu, G. H, & Jiang, Y. Q. Wang YRC, X. Graphene and graphene-based nanomaterials: The promising materials for bright future of electroanalytical chemistry. Analyst (2011)., 136, 4631-40.

[10] Liang, R, Deng, M, Cui, S, Chen, H, & Qiu, J. Direct electrochemistry and electrocatalysis of myoglobin immobilized on zirconia/multi-walled carbon nanotube nanocomposite. Materials Research Bulletin (2010)., 45(12), 1855-60.

[11] Chiu, W. S, Khiew, P. S, Isa, D, Cloke, M, Radiman, S, Abd-shukor, R, Abdullah, M. H, & Huang, N. M. Synthesis of two-dimensional ZnO nanopellets by pyrolysis of zinc oleate. Chemical Engineering Journal (2008)., 142, 337-43.

[12] Huang, N. M, Radiman, S, Lim, H. N, Yeong, S. K, Khiew, P. S, Chiu, W. S, & Kong, S. N. Saeed GHM. Synthesis and characterization of ultra small PbS nanorods in sucrose ester microemulsion. Materials Letters (2009)., 63, 500-3.

[13] Huang, N. M, Radiman, S, Lim, H. N, Yeong, S. K, Khiew, P. S, & Chiu, W. S. Saeed GHM, Nadarajah K. Gamma-ray assisted synthesis of Ni_3Se_2 nanoparticles stabilized by natural polymer. Chemical Engineering Journal (2009)., 147, 399-404.

[14] Chiu, W. S, Khiew, P. S, Cloke, M, Isa, D, Lim, H. N, Tan, T. K, Huang, N. M, Radiman, S, & Abd-shukor, R. Hamid MAA, Chia CH. Heterogeneous seeded growth: Synthesis and characterization of bifunctional Fe_3O_4/ZnO core/shell nanocrystals. The Journal of Physical Chemistry C (2010)., 114(18), 8212-8.

[15] Zhou, X, Huang, X, Qi, X, Wu, S, & Xue, C. Boey FYC, Yan Q, Chen P, Zhang H. In situ synthesis of metal nanoparticles on single-layer graphene oxide and reduced graphene oxide surfaces. The Journal of Physical Chemistry C (2009). , 113(25), 10842-6.

[16] Alivisatos, A. P. Semiconductor clusters, nanocrystals, and quantum dots. Science (1996)., 271(5251), 933-7.

[17] Geim, A. K. Graphene: Status and prospects. Science (2009)., 324(5934), 1530-4.

[18] Geim, A. K, & Novoselov, K. S. The rise of graphene. Nature Materials (2007)., 6(3), 183-91.

[19] Lim, H. N, Nurzulaikha, R, Harrison, I, Lim, S. S, Tan, W. T, Yeo, M. C, Yarmo, M. A, & Huang, N. M. Preparation and characterization of tin oxide, SnO_2 nanoparticles decorated graphene. Ceramics International (2012)., 38(5), 4209-16.

[20] Teo, P. S, Lim, H. N, Huang, N. M, Chia, C. H, & Harrison, I. Room temperature in situ chemical synthesis of Fe_3O_4/graphene. Ceramics International (2012)., 38, 6411-6.

[21] Mai, Y. J, Wang, X. L, Xiang, J. Y, Qiao, Y. Q, Zhang, D, Gu, C. D, & Tu, J. P. CuO/graphene composite as anode materials for lithium-ion batteries. Electrochimica Acta (2011). , 56(5), 2306-11.

[22] Singh, V, Joung, D, Zhai, L, Das, S, Khondaker, S. I, & Seal, S. Graphene based materials: Past, present and future. Progress in Materials Science (2011). , 56, 1178-271.

[23] Bai, S, & Shen, X. Graphene-inorganic nanocomposites. RSC Advances (2012). , 2, 64-98.

[24] Huang, N. M, Lim, H. N, Chia, C. H, Yarmo, M. A, & Muhamad, M. R. Simple room-temperature preparation of high-yield large-area graphene oxide. International Journal of Nanomedicine (2011). , 6, 3443-8.

[25] Zainy, M, & Huang, N. M. Vijay Kumar S, Lim HN. Simple and scalable preparation of reduced graphene oxide-silver nanocomposite via rapid thermal treatment. Materials Letters (2012). , 89, 180-3.

[26] Moussa, S, Atkinson, G, Shall, M. S, & Shehata, A. AbouZeid KM, Mohamed MB. Laser assisted photocatalytic reduction of metal ions by graphene oxide. Journal of Materials Chemistry (2011). , 21, 9608-19.

[27] Su, B, Tang, D, Li, Q, Tang, J, & Chen, G. Gold-silver-graphene hybrid nanosheets-based sensors for sensitive amperometric immunoassay of alpha-fetoprotein using nanogold-enclosed titania nanoparticles as labels. Analytica Chimica Acta (2011). , 692, 116-24.

[28] Gong, J, Zhou, T, Song, D, & Zhang, L. Monodispersed Au nanoparticles decorated graphene as an enhanced sensing platform for ultrasensitive stripping voltammetric detection of mercury(II). Sensors and Actuators B: Chemical (2010). , 150, 491-7.

[29] Hu, Q. H, Wang, X. T, Chen, H, & Wang, Z. Synthesis of Ni/graphene sheets by an electroless Ni-plating method. New Carbon Materials (2012). , 27(1), 35-41.

[30] Jiang, Y. Y, Lu, Y. Z, Li, F. H, Wu, T. S, Niu, L, & Chen, W. Facile electrochemical codeposition of "clean" graphene-Pd nanocomposite as an anode catalyst for formic acid electrooxidation. Electrochemistry Communications (2012). , 19, 21-4.

[31] Moussa, S, Abdelsayed, V, & Shall, M. S. Laser synthesis of Pt, Pd, CoO and Pd-CoO nanoparticle catalysts supported on graphene. Chemical Physics Letters (2011).

[32] Wang, S, Jiang, S. P, & Wang, X. Microwave-assisted one-pot synthesis of metal/metal oxide nanoparticles on graphene and their electrochemical applications. Electrochimica Acta (2011). , 56(9), 3338-44.

[33] Chen, S, Zhu, J, & Wang, X. An in situ oxidation route to fabricate graphene nanoplate-metal oxide composites. Journal of Solid State Chemistry (2011). , 184(6), 1393-9.

[34] Hsu, Y. W, Hsu, T. K, Sun, C. L, Nien, Y. T, Pu, N. W, & Ger, M. D. Synthesis of CuO/ graphene nanocomposites for nonenzymatic electrochemical glucose biosensor applications. Electrochimica Acta (2012). , 82, 152-7.

[35] Lian, P, Zhu, X, Xiang, H, Li, Z, Yang, W, & Wang, H. Enhanced cycling performance of Fe_3O_4-graphene nanocomposite as an anode material for lithium-ion batteries. Electrochimica Acta (2010). , 56(2), 834-40.

[36] Wang, B, Park, J, Wang, C, Ahn, H, Wang, G, & Mn, O. nanoparticles embedded into graphene nanosheets: Preparation, characterization, and electrochemical properties for supercapacitors. Electrochimica Acta (2011). , 55(22), 6812-7.

[37] Kottegoda IRM, Idris NH, Lu L, Wang JZ, Liu HK. Synthesis and characterization of graphene-nickel oxide nanostructures for fast charge-discharge application. Electrochimica Acta (2011). , 56(16), 5815-22.

[38] Ramesha, G. K, & Sampath, S. In-situ formation of graphene-lead oxide composite and its use in trace arsenic detection. Sensors and Actuators B: Chemical (2011). , 160, 306-11.

[39] Wei, Y, Gao, C, Meng, FL, Li, HH, Wang, L, Liu, JH, & Huang, XJ. SnO_2/reduced graphene oxide nanocomposite for the simultaneous electrochemical detection of cadmium(II), lead(II), copper(II), and mercury(II): An interesting favorable mutual interference. Journal of Physical Chemistry C 2011;116: 1034-41.

[40] Zhong, C, Wang, J, Chen, Z, & Liu, H. SnO_2-graphene composite synthesized via an ultrafast and environmentally friendly microwave autoclave method and its use as a superior anode for lithium-ion batteries. The Journal of Physical Chemistry C 2011;115: 25115-20.

[41] Lian, P, Zhu, X, Liang, S, Li, Z, Yang, W, & Wang, H. High reversible capacity of SnO_2/graphene nanocomposite as an anode material for lithium-ion batteries. Electrochimica Acta (2011). , 56(12), 4532-9.

[42] Yao, J, Shen, X, Wang, B, Liu, H, & Wang, G. In situ chemical synthesis of SnO_2-graphene nanocomposite as anode materials for lithium-ion batteries. Electrochemistry Communications (2009). , 11(10), 1849-52.

[43] Chang BYS, Huang NM, An'amt MN, Marlinda AR, Norazriena Y, Muhamad MR, Harrison I, Lim HN, Chia CH.Facile hydrothermal preparation of titanium dioxide decorated reduced graphene oxide nanocomposite. International Journal of Nanomedicine (2012). , 7, 3379-87.

[44] Guo, J, Zhu, S, Chen, Z, Li, Y, Yu, Z, Liu, Q, Li, J, Feng, C, & Zhang, D. Sonochemical synthesis of TiO_2 nanoparticles on graphene for use as photocatalyst. Ultrasonics Sonochemistry (2011). , 18(5), 1082-90.

[45] Zhang, H, Lv, X, Li, Y, Wang, Y, & Li, J. P25-graphene composite as a high performance photocatalyst. ACS Nano (2010). , 4(1), 380-6.

[46] Wang, P, Ao, Y, Wang, C, Hou, J, & Qian, J. Enhanced photoelectrocatalytic activity for dye degradation by graphene-titania composite film electrodes. Journal of Hazardous Materials (2012).

[47] Marlinda, A. R, Huang, N. M, & Muhamad, M. R. An'amt MN, Chang BYS, Yusoff N, Harrison I, Lim HN, Vijay Kumar S. Highly efficient preparation of ZnO nanorods decorated reduced graphene oxide nanocomposites. Materials Letters (2012)., 80, 9-12.

[48] Gong, J, Miao, X, Wan, H, & Song, D. Facile synthesis of zirconia nanoparticles-decorated graphene hybrid nanosheets for an enzymeless methyl parathion sensor. Sensors and Actuators B: Chemical (2012)., 162, 341-7.

[49] Min, Y. L, Zhang, K, Chen, Y. C, & Zhang, Y. G. Enhanced photocatalytic performance of Bi_2WO_6 by graphene supporter as charge transfer channel. Separation and Purification Technology (2012)., 86, 98-105.

[50] Wu, C. H, Zhang, Y. Z, Li, S, Zheng, H. J, Wang, H, Liu, J. B, Li, K. W, & Yan, H. Synthesis and photocatalytic properties of the graphene-$La_2Ti_2O_7$ nanocomposites. Chemical Engineering Journal (2011)., 178, 468-74.

[51] Fu, Y, Wan, Y, Xia, H, & Wang, X. Nickel ferrite-graphene heteroarchitectures: Toward high-performance anode materials for lithium-ion batteries. Journal of Power Sources (2012)., 213, 338-42.

[52] Li, G, Wang, T, Zhua, Y, Zhang, S, Mao, C, Wu, J, Jin, B, & Tian, Y. Preparation and photoelectrochemical performance of Ag/graphene/TiO_2 composite film. Aplied Surface Science (2011)., 257, 6568-72.

[53] Wang, X, Tian, X, H, Yang, H, Y, Wang, Y, H, Wang, H, S, Zhang, S, W, & Liu, W, Y. Y. Reduced graphene oxide/CdS for efficiently photocatalystic degradation of methylene blue. Journal of Alloys and Compounds (2012)., 524, 5-12.

[54] Zhu, J, Wei, S, Gu, H, Rapole, S. B, Wang, Q, Luo, Z, Haldolaarachchige, N, Young, D. P, & Guo, Z. One-pot synthesis of magnetic graphene nanocomposites decorated with core@double-shell nanoparticles for fast chromium removal. Environmental Science and Technology (2012)., 46, 977-85.

[55] Vijay Kumar, S, Huang, NM, Yusoff, N, Lim, HN. High performance magnetically separable graphene/zinc oxide nanocomposite. Materials Letters (2012)., 93, 411-4.

[56] Golsheikh, A. M, Huang, N. M, Lim, H. N, Chia, C. H, Harrison, I, & Muhamad, M. R. One-pot hydrothermal synthesis and characterization of FeS_2 (pyrite)/graphene nanocomposite. Chemical Engineering Journal (2012). DOI:http://dx.doi.org/10.1016/j.cej.2012.09.082.

[57] Xie, J, Song, W, Zheng, Y, Liu, S, Zhu, T, Cao, G, & Zhao, X. Preparation and Li-storage properties of SnSb/graphene hybrid nanostructure by a facile one-step solvothermal route. International Journal of Smart and Nano Materials (2011)., 2(4), 261-71.

[58] Xue, Y, Chen, H, Yu, D, Wang, S, Yardeni, M, Dai, Q, Guo, M, Liu, Y, Lu, F, Qu, J, & Dai, L. Oxidizing metal ions with graphene oxide: The in situ formation of magnetic nanoparticles on self-reduced graphene sheets for multifunctional applications. Chemical Communication (2011). , 47, 1689-91.

[59] Yusoff, N, Huang, NM, Muhamad, MR, Kumar, SV, Lim, HN, Harrison, I. Hydrothermal synthesis of copper oxide/functionalized graphene nanocomposites for dye degradation. Materials Letters (2012). , 93: 393-6.

[60] Huang, N. M, Marlinda, A. R, & Lim, H. N. inventors; Efficient fabrication of flower-like ZnO/reduced functionalized graphene oxide nanocomposites for gas sensing applications. Patent filed. (2012).

[61] Chen, S, Zhu, J, Wu, X, Han, Q, & Wang, X. Graphene oxide-MnO_2 nanocomposites for supercapacitors. ACS Nano (2010). , 4(5), 2822-930.

[62] Lim, H. N, Kassim, A, & Lim, S. P. RastamNizar NS, Huang NM. Microstructural changes of carbonaceous monoliths synthesized via hydrothermal. Journal of the Chilean Chemical Society (2011). , 56(1), 584-6.

[63] Zakarya, S. A, Kassim, A, Lim, H. N, Anwar, N. S, & Huang, N. M. Synthesis of titanium dioxide microstructures via sucrose ester microemulsion-mediated hydrothermal method. Sains Malaysiana (2010). , 39(6), 975-9.

[64] Anwar, N. S, Kassim, A, Lim, H. N, Zakarya, S. A, & Huang, N. M. Synthesis of titanium dioxide nanoparticles via sucrose ester micelle-mediated hydrothermal processing route. Sains Malaysiana (2010). , 39(2), 261-5.

[65] Lim, H. N, Kassim, A, & Huang, N. M. Preparation and characterization of calcium phosphate nanorods using reverse microemulsion and hydrothermal processing routes. Sains Malaysiana (2010). , 39(2), 267-73.

[66] Haw, C. Y, Mohamed, F, Radiman, S, Chia, C. H, Huang, N. M, & Lim, H. N. Hydrothermal synthesis of magnetite nanoparticles as MRI contrast agents. Ceramics International (2010). , 36(4), 1417-22.

[67] Wang, X, Tian, H, Yang, Y, Wang, H, Wang, S, Zhang, W, & Liu, Y. Reduced graphene oxide/CdS for efficiently photocatalystic degradation of methylene blue. Journal of Alloys and Compounds (2012). , 524, 5-12.

[68] Dressen, M. Microwave heating in fine chemical applications: Role of heterogeneity. Thesis (2009).

[69] Wang, Y, & Lee, J. Y. Microwave-assisted synthesis of SnO_2-graphite nanocomposites for Li-ion battery applications. Journal of Power Sources (2005). , 144, 220-5.

[70] Subramanian, V, Burke, W. W, Zhu, H, & Wei, B. Novel microwave synthesis of nanocrystalline SnO_2 and its electrochemical properties. Journal of Physical Chemistry C (2008). , 112, 4550-6.

[71] Zhang, M, Lei D, Du Z, Yin X, Chen L, Li Q, Wang Y, Wang T. Fast synthesis of SnO_2/graphene composites by reducing graphene oxide with stannous ions. Journal of Materials Chemistry (2010). , 21, 1673-6.

[72] Tully, E, Higson, S. P, & Kennedy, O. R. The development of a'labeless' immunosensor for the detection of Listeria monocytogenes cell surface protein, Internalin B. Biosensors and Bioelectronics (2008). , 23, 906-12.

[73] Zhang, J, Kong, L, Wang, B, Luo, Y, & Kang, L. In-situ electrochemical polymerization of multi-walled carbon nanotube/polyaniline composite films for electrochemical supercapacitors. Synthetic Metals (2009). , 159, 260-6.

[74] Torres, S, Neculqueo, G, & Martinez, F. Morphological and structural characterization of poly(3-decylthiophene) prepared by electropolymerization using 1-butyl-3-methyl-imidazoliumtetrafluorborate as solvent. Journal of the Chilean Chemical Society (2007). , 52(3), 1235-6.

[75] Tsai, M, Chen, P, & Do, J. Preparation and characterization of Ppy/Al_2O_3/Al used as a solid state capacitor. Journal of Power Sources (2004). , 133, 302-11.

[76] Li, Y, Lv, X, Lu, J, & Li, J. Preparation of SnO_2-nanocrystal/graphene-nanosheets composites and their lithium storage ability. Journal of Physical Chemistry C (2010). , 114, 21770-4.

[77] He, Y, Huang, L, Cai, J. S, Zheng, X. M, & Sun, S. G. Structure and electrochemical performance of nanostructured Fe_3O_4/carbon nanotube composites as anodes for lithium ion batteries. Electrochimica Acta (2010). , 55, 1140-4.

[78] Muraliganth, T, Murugan, A. V, & Manthiram, A. Facile synthesis of carbon-decorated single-crystalline Fe_3O_4 nanowires and their application as high performance anode in lithium ion batteries. Chemical Communication (2009). , 47, 7360-2.

[79] Yazami, R, & Reynier, Y. F. Mechanism of self-discharge in graphite-lithium anode. Electrochimica Acta (2002). , 47, 1217-23.

[80] Ran, L. Seung lC, Sang BL. Poly(3,4ethylenedioxythiophene) nanotubes as electrode materials for a high-powered supercapacitor. Nanotechnology (2008). pp).

[81] Ruiz, V, Blanco, C, Granda, M, & Santamaria, R. Enhanced life-cycle supercapacitors by thermal treatment of mesophase-derived activated carbons. Electrochimica Acta (2008). , 54, 305-10.

[82] Portet, C, Taberna, P. L, Simon, P, Flahaut, E, & Laberty-robert, C. High power density electrodes for carbon supercapacitor applications. Electrochimica Acta (2005). , 50, 4174-81.

[83] Hu, X, Deng, Z, Suo, J, & Pan, Z. A high rate, high capacity and long life ($LiMn_2O_4$ + AC)/$Li_4Ti_5O_{12}$ hybrid battery-supercapacitor. Journal of Power Sources (2009). , 187, 635-9.

[84] Jayalakshmi, M, & Balasubramaniam, K. Simple capacitors to supercapacitors- An overview. International Journal of Electrochemistry (2008)., 3, 1196-217.

[85] Simon, P, & Gogotsi, Y. Materials for electrochemical capacitors. Nature Materials (2008)., 7, 845-54.

[86] Lang, X, Hirata, A, Fujita, T, & Chen, M. Nanoporous metal/oxide hybrid electrodes for electrochemical supercapacitors. Nature Nanotechnology (2011)., 6, 232-6.

[87] Li, Q, Li, Z, Lin, L, Wang, X. Y, Wang, Y, Zhang, C, & Wang, H. Facile synthesis of activated carbon/carbon nanotubes compound for supercapacitor application. Chemical Engineering Journal (2010)., 156, 500-4.

[88] Frackowiak, E, & Beguin, F. Carbon materials for the electrochemical storage of energy in capacitors. Carbon (2001)., 39, 937-50.

[89] Mayer, S. T, Perkala, R. W, & Kaschmitter, J. L. The aerocapacitor: An electrochemical double-layer energy-storage device. Journal of the Electrochemical Society (1993)., 140(2), 446-51.

[90] Kaempgen, M, Chan, C. K, Ma, J, Cui, Y, & Gruner, G. Printable thin film supercapacitors using single-walled carbon nanotubes. Nano Letters (2009)., 9, 1872-6.

[91] Prabaharan SRS, Vimala R, Zainal Z.Nanostructured mesoporous carbon as electrodes for supercapacitors. Journal Power Sources (2006)., 161, 730-6.

[92] Chmiola, J, Yushin, G, Dash, R, & Gogotsi, Y. Effect of pore size and surface area of carbide derived carbons on specific capacitance. Journal Power Sources (2006)., 158, 765-72.

[93] Lu, X, Wang, G, Zhai, T, Yu, M, Gan, J, Tong, Y, & Li, Y. Hydrogenated TiO_2 nanotube arrays for supercapacitors. Nano Letters (2012)., 12, 1690-6.

[94] Wu, Z, Wang, D, Ren, W, Zhao, J, Zhou, G, Li, F, & Cheng, H. Anchoring hydrous RuO_2 on graphene sheets for high-performance electrochemical capacitors. Advanced Functional Materials (2010)., 20, 3595-602.

[95] Wang, H, Robinson, J. T, Li, X, & Dai, H. Solvothermal reduction of chemically exfoliated graphene sheets. Journal of the American Chemical Society (2009)., 131, 9910-1.

[96] Wang, H, Liang, Y, Mirfakhrai, T, Chen, Z, Casalongue, H, & Dai, H. Advanced asymmetrical supercapacitors based on graphene hybrid materials. Nano Research (2011)., 4, 729-36.

[97] Zhao, B, Song, J, Liu, P, Xu, W, Fang, T, Jiao, Z, Zhang, H, & Jiang, Y. Monolayer graphene/NiO nanosheets with two-dimension structure for supercapacitors. Journal of Materials Chemistry (2011)., 21, 18792-8.

[98] Zhou, W, Liu, J, Chen, T, Tan, K. S, Jia, X, Luo, Z, Cong, C, Yang, H, Li, C. M, & Yu, T. Fabrication of Co_3O_4-reduced graphene oxide scrolls for high-performance supercapacitor electrodes. Physical Chemistry Chemical Physics (2011). , 13, 14462-5.

[99] Cheng, Q, Tang, J, Ma, J, Zhang, H, Shinya, N, & Qin, L. Graphene and nanostructured MnO_2 composite electrodes for supercapacitors. Carbon (2011). , 49, 2917-25.

[100] Si, P, Ding, S, Lou, X, & Kim, D. An electrochemically formed three-dimensional structure of polypyrrole/graphene nanoplatelets for high-performance supercapacitors. RSC Advances (2011). , 1, 1271-127.

[101] Zhang, K, Zhang, L. L, Zhao, X. S, & Wu, J. Graphene/polyaniline nanofiber composites as supercapacitor electrodes. Chemistry of Materials (2010). , 22, 1392-401.

[102] Wang, H, Hao, Q, Yang, X, Lu, L, & Wang, X. A nanostructured graphene/polyaniline hybrid material for supercapacitors. Nanoscale (2010). , 2, 2164-70.

[103] Wang, Y, Shi, Z, Huang, Y, Ma, Y, Wang, C, Chen, M, & Chen, Y. Supercapacitor devices based on graphene materials. Journal of Physical Chemistry C (2009). , 113, 13103-7.

[104] Liu, C, Yu, Z, Neff, D, Zhamu, A, & Jang, B. Z. Graphene-based supercapacitor with an ultrahigh energy density. Nano Letters (2010). , 10, 4863-8.

[105] Kim, T. Y, Lee, H. W, Stoller, M, Dreyer, D. R, Bielawski, C. W, Ruoff, R. S, & Suh, K. S. High-performance supercapacitors based on poly(ionic liquid)-modified graphene electrodes. ACS Nano (2011). , 5, 436-42.

[106] Stoller, M. D, Park, S, Zhu, Y, An, J, & Ruoff, R. S. Graphene-based ultracapacitors. Nano Letters (2008). , 8, 3498-502.

[107] Dong, X, Xu, H, Wang, X, Huang, Y, Chan-park, M, Zhang, H, Wang, L, Huang, W, & Chen, P. 3D Graphene-cobalt oxide electrode for high-performance supercapacitor and enzymeless glucose detection. ACS Nano (2012). , 6, 3206-13.

[108] Rakhi, R. B, Chen, W, Cha, D, & Alshareef, H. N. High performance supercapacitors using metal oxide anchored graphene nanosheet electrodes. Journal of Materials Chemistry (2011). , 21, 16197-204.

[109] Su, H, Wang, T, Zhang, S, Song, J, Mao, C, Niu, H, Jin, B, Wu, J, & Tian, Y. Facile synthesis of polyaniline/TiO_2/graphene oxide composite for high performance supercapacitors. Solid State Sciences (2012). , 14, 677-81.

[110] Chong, M. N, & Jin, B. Chow CWK, Saint C. Recent developments in photocatalytic water treatment technology: A review. Water Research (2010). , 44(10), 2997-3027.

[111] He, Y, Wu, Y, Guo, H, Sheng, T, & Wu, X. Visible light photodegradation of organics over VYO composite catalyst. Journal of Hazardous Materials (2009). , 169, 855-60.

[112] Jiang, R, Zhu, H, Li, X, & Xiao, L. Visible light photocatalytic decolourization of C. I. Acid Red 66 by chitosan capped CdS composite nanoparticles. Chemical Engineering Journal (2009)., 152, 537-42.

[113] Fatin, S. O, Lim, H. N, Tan, W. T, & Huang, N. M. Comparison of photocatalytic activity and cyclic voltammetry of zinc oxide and titanium dioxide nanoparticles toward degradation of methylene blue. International Journal of Electrochemical Science (2012)., 7, 9074-84.

[114] Kropp, R, Tompkins, D, Barry, T, Zeltner, W, Pepping, G, Anderson, M, & Barry, T. A device that converts aqueous ammonia into nitrogen gas. Aquacultural Engineering (2009)., 41(1), 28-34.

[115] Soltaninezhad, M, & Aminifar, A. A. Study nanostructures of semiconductor zinc oxide (ZnO) as a photocatalyst for the degradation of organic pollutants. International Journal of Nanodimension (2011)., 2, 137-45.

[116] Im, J. S, Yun, S. M, & Lee, Y. S. Investigation of multielemental catalysts based on decreasing the band gap of titania for enhanced visible light photocatalysis. Journal of Colloid and Interface Science (2009)., 336, 183-8.

[117] Georgieva, J, Armyanov, S, Poulios, I, & Sotiropoulos, S. An all-solid photoelectrochemical cell for the photooxidation of organic vapours under ultraviolet and visible light illumination. Electrochemistry Communications (2009)., 11(8), 1643-6.

[118] Zhang, X, Jing, D, & Guo, L. Effects of anions on the photocatalytic H_2 production performance of hydrothermally synthesized Ni-doped $Cd_{0.1}Zn_{0.9}S$ photocatalysts. International Journal of Hydrogen Energy (2010)., 35(13), 7051-7.

[119] Tryk, D. A, Fujishima, A, & Honda, K. Recent topics in photoelectrochemistry: Achievements and future prospects. Electrochimica Acta (2000)., 45, 2363-76.

[120] Waldner, G, & Krysa, J. Photocurrents and degradation rates on particulate TiO_2 layers: Effect of layer thickness, concentration of oxidizable substance and illumination direction. Electrochimica Acta (2005)., 50, 4498-504.

[121] Carneiro, P. A, Osugi, M. E, Sene, J. J, & Anderson, M. A. Zanoni MVB. Evaluation of color removal and degradation of a reactive textile azo dye on nanoporous TiO2 thin film electrodes. Electrochimica Acta (2004)., 49(22-23), 3807-20.

[122] Solarska, R, Rutkowska, I, Morand, R, & Augustynski, J. Photoanodic reactions occurring at nanostructured titanium dioxide films. Electrochimica Acta (2006)., 51(11), 2230-6.

[123] Hepel, M, & Hazelton, S. Photoelectrocatalytic degradation of diazo dyes on nanostructured WO3 electrodes. Electrochimica Acta (2005).50(25-26): 5278-91.

[124] Hepel, M, & Hazelton, S. Photoelectrocatalytic degradation of diazo dyes on nanostructured WO_3 electrodes. Electrochimica Acta (2005)., 50(25-26): 5278-91.

[125] Ballesteros-gómez, A, Rubio, S, & Pérez-bendito, D. Analytical methods for the determination of bisphenol A in food. Journal of Chromatography A (2009). , 1216(3), 449-69.

[126] Yin, H, Cui, L, Chen, Q, Shi, W, Ai, S, Zhu, L, & Lu, L. Amperometric determination of bisphenol A in milk using PAMAM-Fe_3O_4 modified glassy carbon electrode. Food Chemistry (2011). , 125(3), 1097-103.

[127] Lim, H. N, Nurzulaikha, R, Harrison, I, Lim, S. S, Tan, W. T, & Yeo, M. C. Spherical tin oxide, SnO_2 particles fabricated via facile hydrothermal method for detection of mercury (II) ions. International Journal of Electrochemical Science (2011). , 6, 4329-40.

[128] Wang, F, Yang, J, & Wu, K. Mesoporous silica-based electrochemical sensor for sensitive determination of environmental hormone bisphenol A. Analytica Chimica Acta (2009). , 638(1), 23-8.

[129] Matin, B. M, Mortazavi, Y, Khodadadi, A. A, Abbasi, A, & Firooz, A. A. Alkaline- and template-free hydrothermal synthesis of stable SnO_2 nanoparticles and nanorods for CO and ethanol gas sensing. Sensors and Actuators B: Chemicals (2010). , 151, 140-5.

[130] Siangproh, W, Dungchai, W, Rattanarat, P, & Chailapakul, O. Nanoparticle-based electrochemical detection in conventional and miniaturized systems and their bioanalytical applications: A review. Analytica Chimica Acta (2011). , 690, 10-25.

[131] Yang, M, Javadi, A, & Gong, S. Sensitive electrochemical immunosensor for the detection of cancer biomarker using quantum dot functionalized graphene sheets as labels. Sensors and Actuators B: Chemical (2011). , 155, 357-60.

[132] Ligthart DAJM, van Santen RA & Hensen EJM. Supported rhodium oxide nanoparticles as highly active CO oxidation catalysts. Angewandte Chemie International Edition (2011). , 50, 5306-10.

[133] Biswal, N, Das, D. P, Martha, S, & Parida, K. M. Efficient hydrogen production by composite photocatalyst CdS-ZnS/zirconium-titanium phosphate (ZTP) under visible light illumination. International Journal of Hydrogen Energy (2011). , 36, 13452-60.

Chapter 13

Metal Chalcogenides Tetrahedral Molecular Clusters: Crystal Engineering and Properties

Chun-Chang Ou and Chung-Sung Yang

Additional information is available at the end of the chapter

1. Introduction

In recent years, the development of crystalline porous materials based on metal chalcogenides attracts scientific attention for their adjustable porous structures and potential applications in technology. In contrast to oxygen, for which only the di- and tri-nuclear homopolyatomic anions, i. e. O_2^{2-}, $O_2^{\cdot-}$and O_3^{2-}, are known in zeolite frameworks, the characteristic strong tendency of sulfur and the other elements of Group 16 is reflected in the wide range of polychalcogenide ions X_n^{2-} (X = S, Se, Te). The polychalcogenide$X_n^{2-}$are easily isolated as salts from polar solvents in the presence of suitable counter cations. [1]The choice of sulfides has many obvious advantages in the crystallization chemistry:[2] (a) In comparison with oxide and fluoride ions, the S^{2-}ion has a much largerionic radius, which favors the tetrahedral coordination withcationsand allows the discovery of sulfide homologues of zeolites. (b) The higher polar ability of the S^{2-}ion shows more flexibility for the structure of tetrahedra angles. For example, the tetrahedraT-S-T angle ranges from109°–161°. But the range of angle for tetrahedra T-O-T is 140°–145°. (T = tetrahedra metal atom, such as In). Obviously, the frameworks with higher flexibility will have better ability to accommodate various shapes of the templates, and the arrangement of tetrahedralunits in the dense matter can remain their original architectures.

Nowadays, chemists use inorganic clusters as molecular building blocks to create open framework with cavities and channels, including porous semiconductor, fast ion exchanger, shape- and size-selective catalysis, and optoelectronic applications. Among these clusters, only the metal chalcogenides tetrahedral molecular clusters can serve as artificial tetrahedral atoms, and assemble the tetrahedral clusters into porous open-framework through inorganic chalcogenides ligands.

The study of metal chalcogenides tetrahedral molecular clusters provides a valuable opportunity to explore the synthetic and structural chemistry at the interface of chalcogenide molecular chemistry and solid-state chemistry. [3,4]A general introduction and overview of metal chalcogenides tetrahedral molecular clusters is intended in this chapter. In the following sections, a description of four basic types of metal chalcogenides tetrahedral molecular clusters will be provided. Design, synthesis strategy, and crystal engineering of building open-framework chalcogenides materials will be discussed. In addition, the interrelated properties of metal chalcogenides tetrahedral molecular clusters will be emphasized as the highlight of this paper.

2. Classification of structure and mathematical extrapolation

2.1. Supertetrahedral clusters

The simplest tetrahedral clusters of metal chalcogenides tetrahedral molecular clusters is supertetrahedral clusters, i. e. Tn clusters, with tetrahedral shaped fragments similar to the cubic ZnS-type lattice. [3-5] The supertetrahedral clusters were first denoted as $2^{(n)}$by Dance et al. [6] Recently, these compounds are denoted as T*n* by Yaghi'sgroup. [4] In the formula, n is the number of metal layers. The mathematical of supertetrahedral clusters can be regarded as the analog of the ideal artificial tetrahedral atoms. The number of tetrahedra (T atoms) in a T*n*supertetrahedron is the nth tetrahedral number:$t_n = n(n+1)(n+2)/6$. The number of distinct vertexes (X atoms) in one supertetrahedron is t_{n+1}. The formulas for discrete T*n* clusters are given as follows: T1 (MX_4), T2 (M_4X_{10}), T3 ($M_{10}X_{20}$), T4 ($M_{20}X_{35}$) and T5 ($M_{35}X_{56}$),where M is a metal cation and X is a chalcogen anion. [5]The illustration figures for Tn clusters are shown in Figure 1.

In an open-continuous framework, each of the four outermost vertexes of a supertetrahedron is shared with another supertetrahedron. Therefore, the overall composition is T_xX_y with $x = t_n$and $y = t_{n+1}-2$. In Tn clusters, all the T atoms are 4-coordinated. Nevertheless, the X atoms possess 2-coordination sites (on the supertetrahedron edges and the outermost vertexes), 3-coordination sites (in the supertetrahedron faces), and 4-coordination sites (inside the cluster). In each corner linkedTn cluster (T_xX_y), the number of 2-coordinated X atoms is *6n-4*, and the number of 3-coordinated X atoms is *2(n - 1)(n - 2)*. The 4-coordinated sites will not appear, until the n value reaches 4or higher. When n = 4 or higher, the number of 4-coordinated X atoms is t_{n-3}. For example, a T2 cluster consists of only 2-coordinated anions (e. g., S^{2-}), and a T3 cluster has both 2- and 3-coordinated anions. Starting from the T4 cluster, tetrahedral coordination begins to adopt anions inside the cluster to create 4-coordinated anions, beside the existed 2- and 3-coordinated anions. [5]

In Table 1, the known T2 clusters, $[Ga_4S_{10}]^{8-}$,[7] $[In_4Q_{10}]^{8-}$,[8] $[Ge_4Q_{10}]^{4-}$ (Q=S, Se),[9] $[Sn_4Q_{10}]^{4-}$ (Q=S, Te),[10]and examples of T3 clusters, $[In_{10}S_{20}]^{10-}$,[11] $[Ga_{10}S_{16}(NC_7H_9)^4]^{2-}$,[12] and $In_{10}S_{18}(HPP)_6(H_2O)_{15}$. [4](HPP=1,3,4,6,7,8-hexahydro-2H-pyrimido[1,2-a]pyrimidine) are provided. On the other hand, the reported T4 clusters, $[M_4In_{16}S_{33}]^{10-}$ (M=Mn^{2+},Co^{2+},Zn^{2+},Cd^{2+})[13] and $[Cd_4In_{16}S_{35}]^{14-}$[5]are provided in Table 1, simultaneously. Up to date, the largest Tn clusters that has been prepared is T5 cluster. The examples are $Cu_5In_{30}S_{54}$,[14] $(In_{34}S_{54})^{6-}$ [15] and $(In_{28}Cd_6S_{56})^{12-}$. [16]

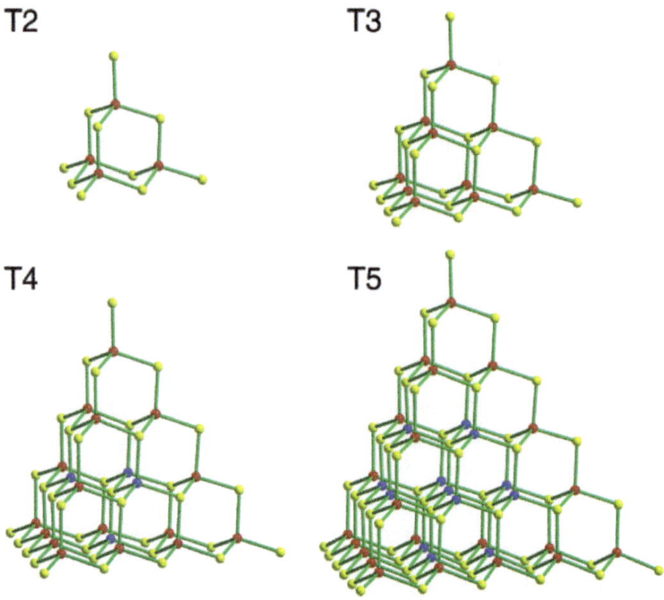

Figure 1. Illustration for*Tn* series clusters, from T2 to T5. [3]

Stoichiometry of Tn	examples	Ref.
T2 (MX_4)	$[Ga_4S_{10}]^{8-}$	[7]
	$[In_4Q_{10}]^{8-}$(Q=S, Se)	[8]
	$[Ge_4Q_{10}]^{4-}$ (Q=S, Se)	9
	$[Sn_4Q_{10}]^{4-}$ (Q=S, Te)	[10]
T3 (M_4X_{10})	$[In_{10}S_{20}]^{10-}$	[11]
	$[Ga_{10}S_{16}(NC_7H_9)^4]^{2-}$	[12]
	$In_{10}S_{18}(HPP)_6(H_2O)_{15}$	[4]
T4 ($M_{10}X_{20}$)	$[M_4In_{16}S_{33}]^{10-}$ (M=Mn^{2+},Co^{2+},Zn^{2+},Cd^{2+}),	[13]
	$[Cd_4In_{16}S_{35}]^{14-}$	[5]
T5 ($M_{35}X_{56}$)	$Cu_5In_{30}S_{54}$,	[14]
	$(In_{34}S_{54})^{6-}$	[15]
	$(In_{28}Cd_6S_{56})^{12-}$	[16]

Table 1. Supertetrahedral clusters base on metal chalcogenides tetrahedral molecular clusters

2.2. Pentasupertetrahedral clusters (Pn)

The second series of tetrahedral clusters is known as pentasupertetrahedral clusters, shown in Figure 2. This series cluster was denoted as 5 {n} by Dance et al. [6] and was named as P*n* by Feng'sgroup. [17] The P*n* cluster is composed of four same order T*n* clusters at the corner and one anti-T*n* cluster at the core. In comparison, a pentasupertetrahedral cluster is considerably larger than a supertetrahedral cluster of the same order. For example: the P1 cluster consists four T1 clusters (MX_4) and one anti-T1 (XM_4) cluster at the center, resulting in the composition ((MX_4)$_4$(XM_4) (i. e., M_8X_{17}). The supertetrahedral clusters as large as T5 are known, but the largest known cluster of the P*n* series is the P2 cluster. [17]

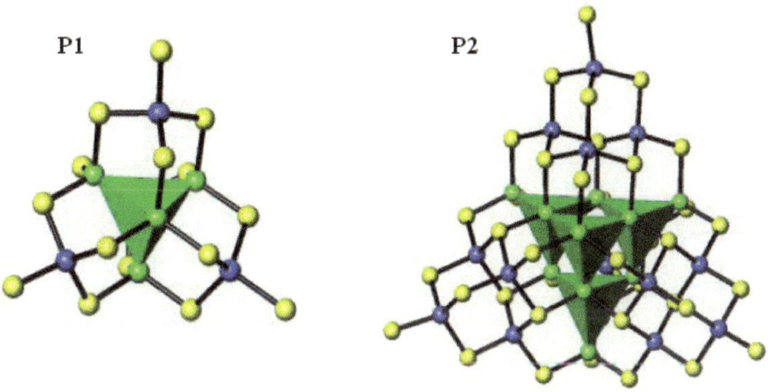

Figure 2. Illustration for P*n* series clusters, from P1 to P2. A polyhedral representation is used for the central antisupertetrahedral cluster. [30]

Some examples of P1 clusters, $[SCd_8(SBu)_{12}](CN)_{4/2}$,[18](SBu= n-butanethiolate) $[M_4Sn_4S_{17}]^{10-}$ (M=Mn^{2+}, Fe^{2+}, Co^{2+}, Zn^{2+}),[19] and $[M_4Sn_4Se_{17}]^{10-}$ (M=Mn^{2+}, Co^{2+}, Zn^{2+}),[20,21]and P2 clusters, $[Li_4In_{22}S_{44}]^{18-}$ in ICF-26[22](ICF= Inorganic ChalcogenideFramework) and $[Cu_{11}In_{15}Se_{16}(SePh)_{24}(PPh_3)_4]$. [23] (Ph= phenyl group, PPh3=triphenylphosphine group) are given in Table 2.

Stoichiometry of Pn	examples	Ref.
P1 (M_8X_{17})	$[SCd_8(SBu)_{12}](CN)_{4/2}$	[18]
	$[M_4Sn_4S_{17}]^{10-}$ (M=Mn^{2+}, Fe^{2+}, Co^{2+}, Zn^{2+})	[19]
	$[M_4Sn_4Se_{17}]^{10-}$ (M=Mn^{2+}, Co^{2+}, Zn^{2+})	[20, 21]
P2 ($M_{26}X_{44}$)	$[Li_4In_{22}S_{44}]^{18-}$	[22]
	$[Cu_{11}In_{15}Se_{16}(SePh)_{24}(PPh_3)_4]$.	[23]

Table 2. Pentasupertetrahedral clusters base on metal chalcogenides tetrahedral molecular clusters

2.3. Capped supertetrahedral clusters (Cn)

The third series of tetrahedral clusters is capped-supertetrahedral clusters, as shown in Figure 3. This series clusterwas denoted as $7^{(n)}$by Dance et al. [6] Recently, the series clusterwas named as C*n* by Feng's group. [3] The capped supertetrahedral clusters are defined as a regular supertetrahedral cluster (T*n*) at the core covered with a shell of atoms, which is also related to the Tn cluster. Accurately, each face of the T*n* core unit is covered with a single sheet of atoms called the T(*n* + 1) sheet and each corner of this cluster is covered with a MX group. The T(*n* + 1) sheet is defined as the bottom atomic sheet of a T(*n* + 1) cluster. [3]

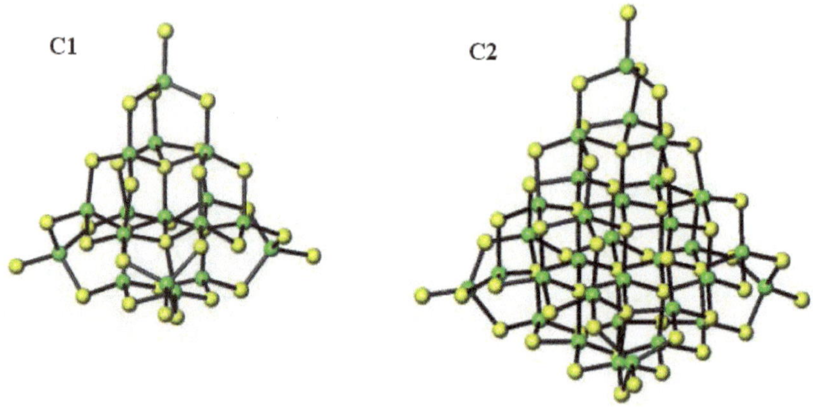

Figure 3. Illustration for C*n* series clusters, from C1 to C2. (adopted from Ref 30)[30]

As shown in Table 3, the first metal chalcogenides tetrahedral molecular clusters of C1 clusters is $[S_4Cd_{17}(SPh)_{28}]^{2-}$ [18], (SPh=benzenethiol ligand), by Dance group in 1988. In later days, two new C1 clusters were reported. They are $Cd_{17}S_4(SCH_2CH_2OH)_{26}$[24]and $[S_4Cd_{17}(SPh)_{24}(CH_3OCS_2)_{4/2}]_n$ nCH_3OH. [25]As for the C2 clusters, three clusters, $Cd_{32}S_{14}(SCH_2CH(OH)CH_3)_{36}$· $4H_2O$,[26] $Cd_{32}S_{14}(SC_6H_5)_{36}$· $(DMF)_4$,[27] and $Cd_{32}Se_{14}(SePh)_{36}(PPh_3)_4$[28] have been successfully synthesized, so far.

Stoichiometry of Cn	examples	Ref.
C1 ($M_{17}X_{32}$)	$[S_4Cd_{17}(SPh)_{28}]^{2-}$	[18]
	$Cd_{17}S_4(SCH_2CH_2OH)_{26}$	[24]
	$[S_4Cd_{17}(SPh)_{24}(CH_3OCS_2)_{4/2}]_n nCH_3OH$	[25]
C2 ($M_{32}X_{54}$)	$Cd_{32}S_{14}(SCH_2CH(OH)CH_3)_{36}$· $4H_2O$	[26]
	$Cd_{32}Se_{14}(SePh)_{36}(PPh_3)_4$	[27]
	$Cd_{32}S_{14}(SC_6H_5)_{36}$· $(DMF)_4$	[28]

Table 3. Capped-supertetrahedral clusters base on metal chalcogenides tetrahedral molecular clusters

2.4. Super-supertetrahedral clusters (Tp,q)

Besides theseries of tetrahedral molecular clusters, a special multi-series metal chalcogenides hollow cluster has been reported, simultaneously. These metal chalcogenides hollow clusters are known as super-supertetrahedral clusters, denoted as *Tp,q*. [29]This series of cluster is built in a T*q* supertetrahedron of T*p* supertetrahedra. The number of tetrahedra (T atoms) in a T*p* supertetrahedron is the *p*th tetrahedral number, $t_p = p(p+1)(p+2)/6$, and the number of X atoms is t_p+1. In a T*p*, *q* super-supertetrahedron the number of T atoms is *tqtp*,and the number of X atoms is $tq(t_{p+1}-2)+2$. The first metal chalcogenides tetrahedral molecular clusters of T*p,q* clusters is CdInS-420, i. e. T4,2 (given in Figure 4), prepared by Yaghi group. [29]

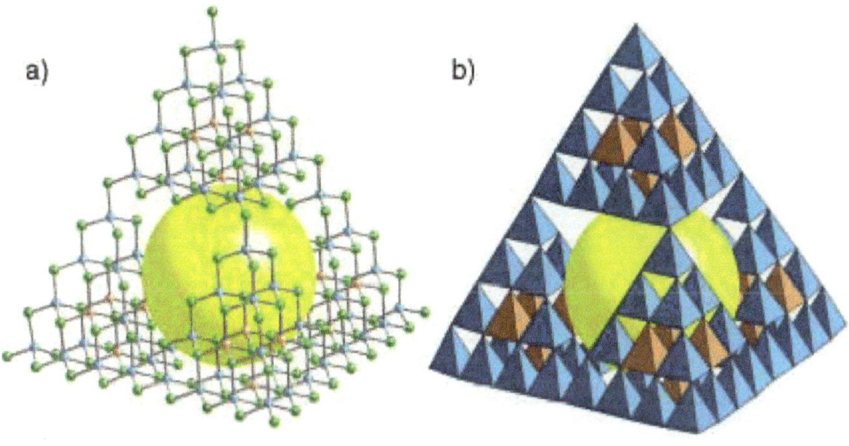

Figure 4. a) The CdInS-420 cluster as a ball and stick model. In (blue); Cd (orange); S (green). The large yellow sphere indicates the central cavity. (b) The same view as (a) shown as metal-centered tetrahedra. [29]

3. Design and synthesis: Crystal engineering

The explosive growth in the number of microporous and open framework materials is mainly contributed by the numerous variable synthetic and structural parameters. It is known that each Tn cluster behaves as an artificial tetrahedral atom (T atom). These T atoms link with others by four vertex chalcogenides (ex: sulfur) atoms into the nanoclusters to produce extended open frameworks. Among these clusters, the use of structure-directing agents with different charge, size, and shape is particularly effective to assist the formation of oxide frameworks. [3-5]Furthermore, the conventional synthetic methods to prepare the metal chalcogenides tetrahedral molecular clusters include the crystallization from solution[6, 30-31] and the hydrothermal synthetic route. [2-5, 13]

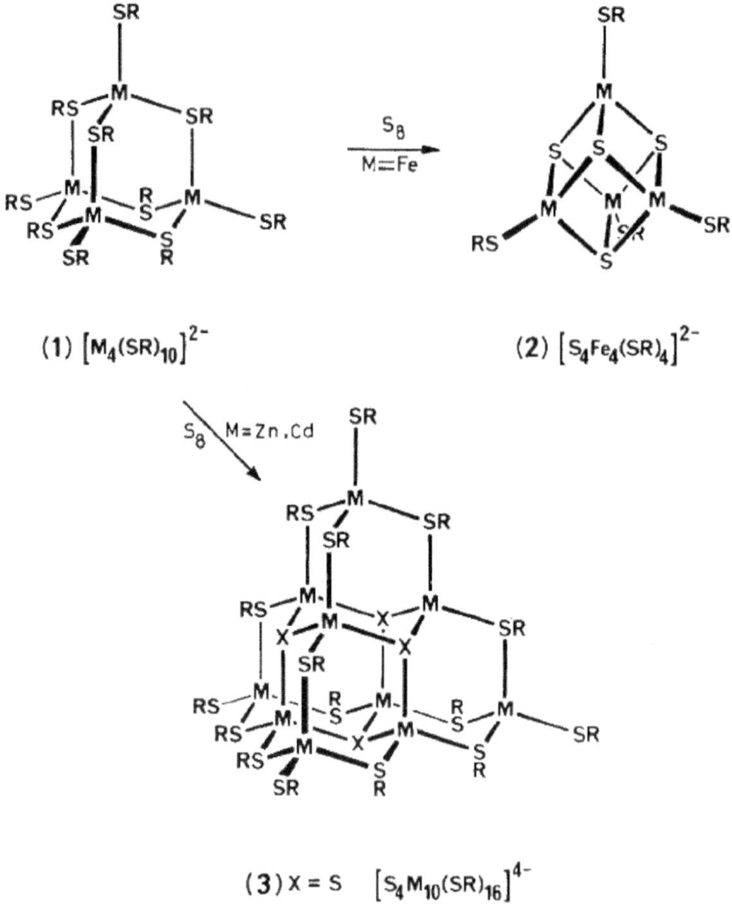

Figure 5. $[M_4(SPh)_{10}]^{2-}$, M=Fe, Co, Zn **(1)** and probably Cd react with sulphur and different metal cation (i. e., Fe, Zn, Cd) in acetonitrile to yield the $[Fe_4S_4(SPh)_4]^{2-}$ **(2)** and $[S_4M_{10}(SPh)_{16}]^{4-}$, M=Zn, Cd **(3)**. [31]

Prior to the development of open framework chalcogenides, tetrahedral clusters were not commonly encountered among open framework solids. In 1982, Dance group reported the extension reaction of T2 cluster, $[M_4(SPh)_{10}]^{2-}$ (M=Zn^{2+}, Cd^{2+}) with sulphur, including the metals zinc and cadmium in acetonitrile, to yield a different and unprecedented product $(Me_4N)_4[S_4M_{10}(SPh)_{16}]$ (M = Zn^{2+}, Cd^{2+}), shown in Figure 5. [31]A similar crystallization from solution reactions with elemental selenium yields analogous complexes (Me_4N_4) $[Se_4M_{10}(SPh)_{16}]$ (M = Zn^{2+}, Cd^{2+}). Therefore, a set of four homologous complexes, (Me_4N_4) $[X_4M_{10}(SPh)_{16}]$ (M = Zn^{2+}, Cd^{2+}; X = S^{2-}, Se^{2-}) have been prepared by this method. The four anions have the same molecular aggregation structure, i. e. a supertetrahedral 10-metal section of the cubic (sphalerite) metal chalcogenide structure. Nowadays, the unprecedented product $(Me_4N)_4[S_4M_{10}(SPh)_{16}]$ (M = Zn^{2+}, Cd^{2+}) is known as the T3 series clusters. The hydrothermal synthesis of open framework chalcogenidesis started with simple elemental forms (e. g., sulfur) and inorganic salts. The initial process usually involves redox chemistry in the formation of clusters. Clusters of various types and sizes could coexist in a solution. Equilibria between various clusters in solution would shift to the direction that favors the creation of one or more clusters, when crystallization involving these clusters occurs. [3]

3.1. Chalcogenides with different valence state of metal cations

Metal chalcogenides tetrahedral molecular clusters in the self-assembly process is critical for the synthesis of microporous and mesoporous oxides. [32]The known example is aluminophosphates open-frameworks with divalent metal cations (M^{2+}). The existence of M^{2+} can provide a rather flexible adjustment of the framework charge density and therefore makes it easier to achieve the charge matching of whole compound. In metal chalcogenides tetrahedral molecular clusters, the ratio between metal cations of different valences subjects to the limitation of the local charge balance within each cluster and may not be as flexible as that in the oxides analogue. [3, 29]

The classes of open framework materials are dominated by typical zeolites, such as ZSM-5, (named after Zeolite Socony Mobil by mobiloil company). The sodalite structure can be made in the neutral SiO_2 form. Neutral porous frameworks are also found in $AlPO_4$ and GeO_2 forms. Therefore, it is reasonable to expect the existence of open framework sulfides with framework composition of GeS_2 or SnS_2. The early development of metal chalcogenides tetrahedral molecular clusters is the preparation of open framework sulfides by using the mono- or divalent cations (e. g., Cu^+, Mn^{2+}) to join the metal chalcogenides tetrahedral molecular clusters together (i. e. $Ge_4S_{10}^{4-}$). [17] These low-valentcations can generate negative charges on the framework. Subsequently,the charge is balanced by structure-directing amine molecules. For example, the compounds prepared with the formula of $[(CH_3)_4N]_2[MGe_4S_{10}]$ (M= Mn^{2+}, Fe^{2+}, Cd^{2+}), by Yaghigroup. [7]In this case, the $MnGe_4S_{10}\cdot 2(CH_3)_4N$, has a non-interpenetrating diamond type lattice (the single diamond type) with alternating T2 and T1 clusters to occupy the tetrahedral nodes. In the Ge-S or Ge-Se system of metal chalcogenides tetrahedral molecular clusters, the largest metal chalcogenidessuper-tetrahedral molecular cluster is T2. As a result, a perspective charge-balance problem of metal chalcogenides tetrahedral molecular clusters system is proposed. [17]In the Ge-S or Ge-Se system of metal chalcogenides tetrahedral molecular clusters, Feng's group found that clusters larger than T2 cannot be prepared in this system. The reason is that the charge at cation sites is too high to satisfy the coordination environment of 3-coordinated anion sites in clusters larger than T2. For the same reason, it is not surprise to find that no regular T3 cluster can besuccessfully prepared in the pure Sn-S (or Sn-Se) system.

In In-S system of metal chalcogenides tetrahedral molecular clusters, Yaghi'sgroup used the In-S composition to build a unique porous sulfide-based frame work materials. [4-5] Its uniqueness comes from 50% or more framework cation sites in zeolite-like oxides with a valence ≧4. The linkage in In-O-In or Al-O-Al is not similar to that in zeolite-like oxides because of the Loewenstein rule. The Loewenstein rule states that the ratio of M^{4+}/M^{3+} should be larger or equal to one. [17] The most common in the In-S system is the occurrence of the T3 cluster, $[In_{10}S_{20}]^{10-}$. The lower charge on In^{3+}, compared with Ge^{4+} and Sn^{4+}, makes it possible to form the required 3-coordinated sulfur site for the formation of T3 clusters. Moreover, the In^{3+} composition is extended to Ga-S, Ga-Se, and In-Se compositions by Feng'sgroup. [11,17,33]The use of the nonaqueous synthesis method is responsible for the success in the syntheses of Ga-S composition. On the other hand, the synthesis of the $[Cd_4In_{16}S_{35}]_{14}$in T4 cluster shows that the access of regular clusters larger than T3 is possible by the help of divalent cations, in addition to the In-S composition. [5]Moreover, the combination of monovalent (eg., Cu^{+}) and trivalent cations (eg., In^{3+}) could provide the required local charge matching around the tetrahedral S^{2-} site. [14, 17]

In terms of chemical compositions, metal chalcogenides tetrahedral molecular clusters with tetravalent (M^{4+}) and trivalent (M^{3+}) metal cations closely resembles the structure of aluminosilicate zeolites. The M^{4+}/M^{3+} was not expected to be simple because eitherM^{4+} or M^{3+} could independently form amine-directed crystals with sulfur and thus the probability of phase separation was high. [7, 17]Nevertheless, the use of the nonaqueous synthesis method could lead to the integration of M^{4+} and M^{3+} ions into the same framework. [11, 17, 33]A series of open framework sulfides and selenides were made by the combination of tetravalent (i. e., Ge^{4+}, Sn^{4+}) and trivalent metal (i. e., Ga^{3+}, In^{3+}) ions. [34] The M^{4+}/M^{3+} ratio in these chalcogenides can be much smaller than that in zeolites. So far, the M^{4+}/M^{3+} ratio falls within the range from 1. 3 to 0. 21. Besides the low M^{4+}/M^{3+} ratio in this series, some sulfides possess adequate stability toward ion exchange and thermal treatment. [17,34]

As mentioned in the above paragraph, the type of metal cations existed in the metal chalcogenides tetrahedral molecular clusters growth system shows a limitation on the formation of individual clusters. It is worth nothing to mention that cluster larger than T3 can be formed, if only trivalent cation isused in solvothermal system. However, if the synthesis is employedwith trivalent and divalent cations, the system has the flexibility to create a variety of clusters, such as T3, T4, and T5. [17]

3.2. Structure-directing agent: Organic amines

In oxides compounds, the oxygen sites of the anionic framework can form strong hydrogen bonding with N-H groups of protonated amine molecules. [3] Such O…H-N bonding is an important factor in the directed assembly of oxide frameworks. The hydrogen bonding between chalcogenides frameworks and guest molecules (e. g., S…H-N) is very weak. [3]Based on the host–guest charge-density matching principle, proposed by Stucky et al., the content and distribution of heteroatoms in the framework can be adjusted by the guest species. [35] Thus, the co-assembly of metal chalcogenides tetrahedral molecular clusters with structure-directing agent (guest), such as amine molecules, depends to a large extent on the host-guest electrostatic interaction. This principle can explain that open framework metal chalcogenides tetrahedral molecular generally have a rather negative framework and few neutral or nearly neutral open framework metal chalcogenides tetrahedral molecular are known today.

Interestingly, Yang et al. employ 4,4-trimethylene dipiperidine (TMDP), and histidine as the structure-directing agents for the synthesis of compound with mixed supertetrahedralchalcogenide clusters T2, and pentasupertetrahedralchalcogenide clusters P1, denoted as NCYU-5, (NCYUis named after National Chia Yi University), shown in Figure 6. [36]The TMDP is widely used as a structure construction template in the synthesis of T*n* series chalcogenide clusters. [37] However, the use of amino acid in the hydrothermal synthesis for an inorganic tetrahedral cluster has not been reported. In this case, the porosity of the mixed chalcogenide clusters with a TMDP only template is about 55%, smaller than the one with TMDP and histidine as templates. On the basis of experimental data, the role of histidine in the formation of mixed chalcogenide clusters is to improve the pore size of the 2-D framework, and the porosity of the crystal. The potential cavity occupied 67. 1% of crystal cell volume can be calculated by the PLATON program. (a collection tool for single crystal structure analysis). The high percentage of cavity derived from NCYU-5 suggests that large amount of guest molecules, i. e. TMDP, and histidine, present in the structure of open frame work to make this material with high porosity. In each layer of the two dimensional open frame work of NCYU-5, alternating P1 and T2 clusters are linked together by bridging selenium atoms, Fig. 7(a). A triangle-shaped pore window is created by three T2 clusters, three P1 clusters, and six bridging Se atoms. The distance between the two corner Se atoms of the triangle-shaped window is 18. 467(3) Å^2, and the theoretical area of a triangle-shaped pore window is 147. 6 Å^2, Fig. 7(b). The T2 clusters in one layer are located above or below the center of the 15-ring-window of its adjacent layers. The orientation of these T2 clusters is consistent. Although, these two-dimensional layers are stacked along the c-axis, the P1 clusters are located above or below P1 clusters of the adjacent layers with skewed orientation, Fig. 7(c). [36]

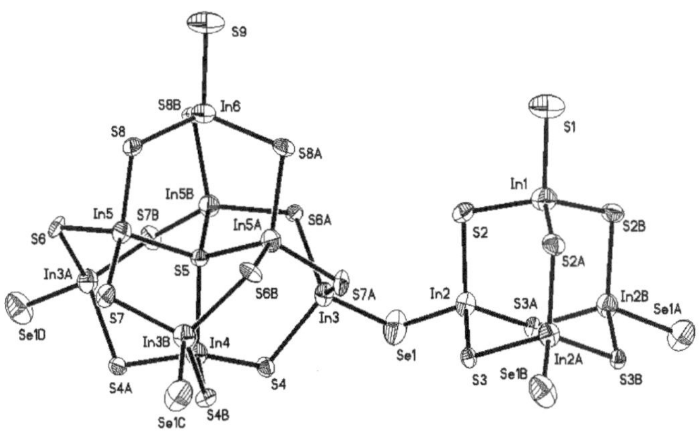

Figure 6. An illustrated unit cell structure for NCYU-5. The calculated occupancy possibility of the Se(1) site for Se to S is ~90%. The occupancy possibility of the S3 site in the T2 cluster for Se to S is ~10%. [36]

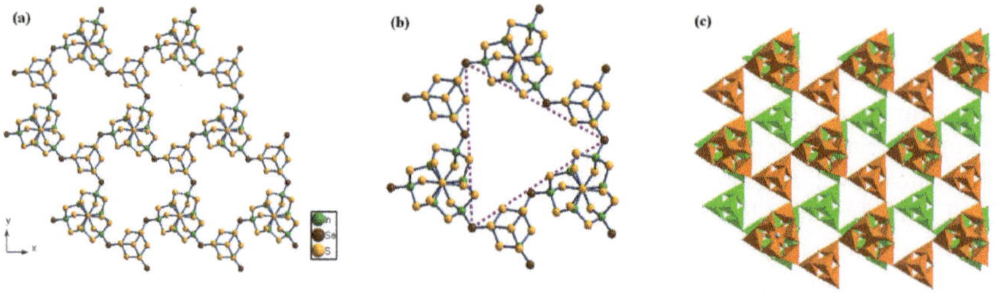

Figure 7. a) The 2-D open frame work of NCYU-5, alternating P1 and T2 clusters are linked together by bridging Se atoms. (b) The distance between the two corner Se atoms of the triangle-shaped window is 18. 467 Å^2. The theoretical area of a triangle-shaped pore window is 147. 6 Å^2. (c) Two adjacent layers stacked along c-axis. [36]

3.3. Selected properties of metal chalcogenides tetrahedral molecular clusters

3.3.1. Thermal characteristics

In the past decades, numerous metal chalcogenides tetrahedral molecular clusters have been successfully prepared. Although, these metal chalcogenides solids posses a wide range of compositions and topological features, the application potential is limited by the low thermal stability of these compounds. Up to present, not many metal chalcogenides tetrahedral molecular clusters can maintain the thermal stability in500℃ or higher, which is a required temperature to completely remove organic guest molecules through calcinations. [3]One example that can achieve the requirement is NCYU-5 that can maintain the crystal structure from room temperature (RT) to 505 °C.. The thermogravimetric analysis (TGA) data of NCYU-5 is given in Figure 8. [36]The total weight loss is 26. 72% from RT to 505 °C. The initial weight loss of 3. 09% from RT to 110 °C is because of the water desorption of surface. A sharp weight loss of 10. 46% starting from 305 °C to 405 °C is attributed by the decomposition of two histidine molecules (calcd. 11. 5%). The second sharp weight loss of 13. 17% observed from 405 °C to 505 °C is possibly contributed by the decomposition of two TMDP molecules and the removal of H_2S (calcd. 15. 6%). No further weight loss is observed after 505 °C. The total recorded weight loss of 26. 72% from 105 °C to 505 °C is in good agreement with the calculated weight loss (27. 1%) for the guest molecules, i. e. one TMDP and one histidine in each unit cell of NCYU-5. As for the direct calcination of as-synthesized samples by suitable temperature to remove a sizable fraction of extra-frame work organic components is shown in another example. In Feng et al. 's reported, ~77% of nitrogen and ~81% of hydrogen were removed from UCR-20GaGeS-TAEA, (TAEA = tris(2-aminoethyl)amine, UCR= initials of University of California Riverside), by direct calcination at 350 °C with nitrogen gas. [34] However, the coke formation made the removal of carbon difficult, only ~39% of carbon was removed from UCR-20GaGeS-TAEA in the same experiment. [34]

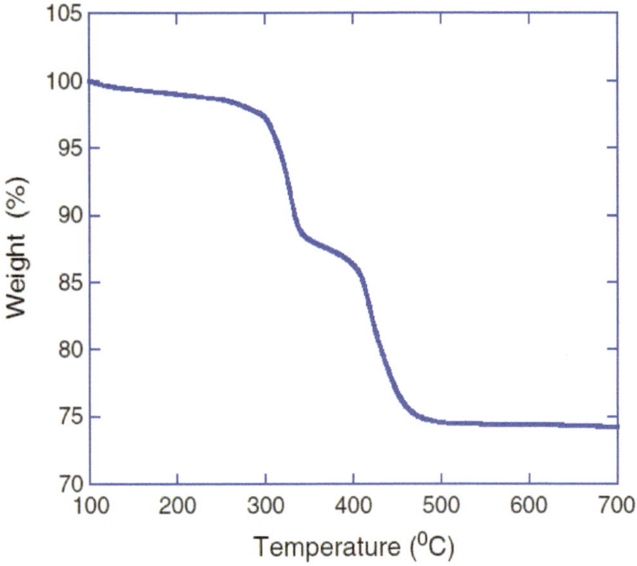

Figure 8. TGA data of NCYU-5. The total weight loss is 26. 72% from RT to 800 °C. [36]

3.3.2. Ion exchange

Ion exchange is the most common properties of open framework solids. [38] This property has been shown for a number of metal chalcogenides tetrahedral molecular clusters, in which the protonated guests can be exchanged in solution by inorganic monocations (i. e., Li^+, Na^+, K^+, Rb^+, Cs^+) and dications (i. e., Mg^{2+}, Ca^{2+}, Sr^{2+}, Ba^{2+}). After ion-exchange, microporosity of compound can be created by the removal of large organic cations. [34,39] For example, the ion-exchange with Cs^+ ions led to an almost complete removal of amine molecules from UCR-20GaGeS-TAEA. The Cs^+-exchanged UCR-20GaGeS-TAEA exhibits the type I isotherm characteristic of a microporous solid (Figure 9). This sample has a high BET surface area (807 m^2/g) and a micropore volume (0. 23 cm^3/g) despite the presence of much heavier elements (i. e., Cs, Ga, Ge, and S), compared with the analogues of aluminosilicate zeolites. [34]

3.3.3. Conductivity

An open-framework material has an inherent advantage for the applications in low-temperature fast-ion conductors. The existed open channels in these compounds provide the necessary paths for ions migration. Unfortunately, the sizeable open channels and cages contained in zeolites are not good fast-ion conductors because of the strong interaction between the oxygen framework and extra-framework charge carriers, such as Li^+ and Na^+. [38]

Open-framework metal chalcogenides tetrahedral molecular clusters are anticipated as better ion conductors than zeolites. The chalcogenides have higher anionic framework polarizability created by the large size of S^{2-} or Se^{2-}, as compared with O^{2-}. [1]The high polarizable anionic framework will facilitate the migration of mobile cations quick. Since the concentration of mobile cations is high in the open-framework metal chalcogenides tetrahedral molecular clusters. The chalcogenides clusters will have more negative frameworks for charge-balane than that of zeolites. The experimental data show that the framework M^{4+}/M^{3+} (where M is a tetrahedral atom) ratio in chalcogenides is smaller than one, whereas the ratio value is always larger or equal to one in zeolites or related oxides. [3]For example, the synthesis of hydrated sulphides and selenides with highly mobile alkali or alkaline earth metal cations as extra framework cations, such as ICF-26. [3,22] The ionic conductivity of ICF-26 (Figure 10) is comparable to or exceeds previously known crystalline sodium or lithium conductors at RT and under relative humidity of 30% or higher. The highest specific conductivity achieved among open framework chalcogenides is 0. 15 Ω^{-1} cm^{-1} at 27 °C and under 100% relative humidity. [3,40]

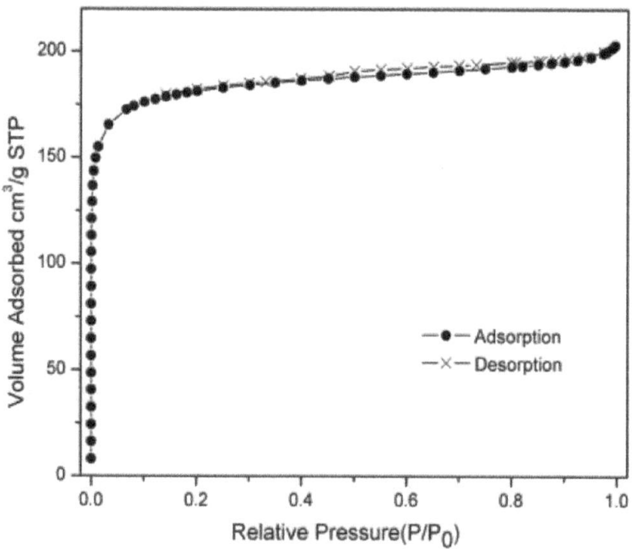

Figure 9. Nitrogen adsorption and desorption isotherms measured at 77 K for the Cs^+ exchanged UCR-20GaGeSTAEA. [34]

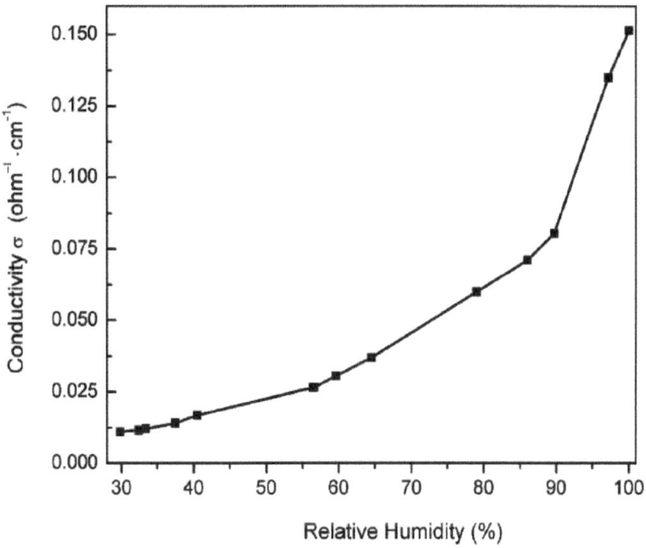

Figure 10. The ionic conductivity of ICF-26 under different relative humidity. Ionic conductivities were measured on a single crystal (cross section 0. 37 x 0. 43 mm^2, length 0. 63 mm). [3]

3.3.4. Optical properties

Most open framework chalcogenides without the incorporation with organic dyes or metal activators still can display photoluminescence with tunable emission wavelengths ranging almost continuously from 450 to 600nm. [41]The luminescence of open framework chalcogenidesis known to be related with the highly negatively charged inorganic framework and the presence of protonated guest amine molecules. [41, 42]On the top of this review point, Se exhibits a better induced optoelectronic property than S because the ionization energy barrier of Se is much lower than that of S. [36, 43, 44] Based on this understanding, Yang et al. prepared the Se doped metal chalcogenides tetrahedral molecular clusters, i. e. NCYU family, to study the optoelectronic luminescent phenomena induced by the quantum confinement of Se in these clusters. The clusters prepared by Yang's group include NCYU-1 (T4/Se), NCYU-3 (T3/Se), and NCYU-5 (mixed P1+T2/Se). [36, 43, 44]Two PL emission peaks, centered at about 457 nm and 538 nm, were revealed from the NCYU-3(T3/Se). But only the 457 nm peak is observed in the spectra of NCYU-InS-AEAE (T3)(AEAE= 2-(2-aminoethylamino)ethanol) and NCYU-4 (T3). The peak at 457 nm has been reported for the luminescence of open framework chalcogenides. Thus, the trace Se atoms confined in the NCYU-3 (T3/Se) supertetrahedral clusters is responsible for the newly discovered 538 nm emission peak (Figure 11). [44]

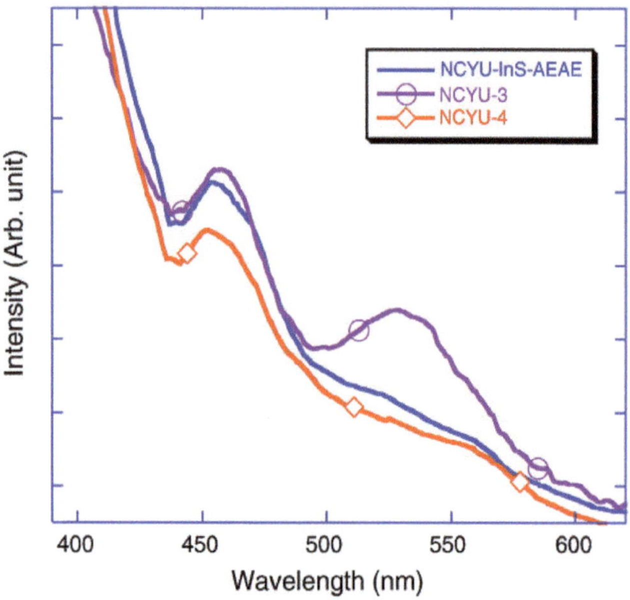

Figure 11. The RT PL spectra for NCYU-T3 family. Two emission peaks, centered at about 457 nm and 538 nm, were revealed from the NCYU-3. [44]

On the other hand, a family of discrete chalcogenide T4 clusters $[M_xGa_{18-x}Sn_2Q_{35}]^{12-}$[45] (x = 2 or 4; M = Mn, Cu, Zn; Q = S, Se), prepared by Feng's group, show an unusual phase transformation from a T4 covalent framework (3-D) into T4 molecular clusters (0-D), denoted as OCF-5s or OCF-40s (OCF stands for organically directed chalcogenide frameworks). In the case of OCF-40s, these iso-structural compounds show a remarkable effect of different d^{10} metal ions (Cu^+ and Zn^{2+}) and chalcogen anions(Se^{2-} and S^{2-}) on the sample colors and band gap of semiconductor materials. The UV–vis diffuse reflectance spectrum reveals that the dark red selenide sample of OCF-40-CuGaSnSe-PR (formula:$[Cu_2Ga_{16}Sn_2Se_{35}]$ 12(C_5NH_{12}), PR = piperidine) has a wider UV-vis adsorption peak thanthe yellow sample of OCF-40-ZnGaSnSe-PR. (formula:$[Zn_4Ga_{14}Sn_2Se_{35}]$ 12(C_5NH_{12})). A similar difference is also found in sulfides analogues between OCF-40-CuGaSnS-PR (formula:$[Cu_2Ga_{16}Sn_2S_{35}]$ 12(C_5NH_{12})) and OCF-40-ZnGaSnS-PR. (formula:$[Zn_4Ga_{14}Sn_2S_{35}]$ 12(C_5NH_{12})). The solid-state diffuse reflectance spectra, shown in Figure 12,show that OCF-40s are semiconductors with different band gaps(bg): OCF-40-CuGaSnSe-PR, bg = 1. 91 eV; OCF-40-ZnGaSnSe-PR, bg = 2. 71 eV; OCF-40-CuGaSnS-PR, bg = 2. 11 eV; and OCF-40-ZnGaSnS-PR, bg = 3. 59 eV. The experimental data show that metal chalcogenides tetrahedral molecular clusters with $Cu^+$and Se^{2-} in the framework will have much lower band gaps. [45]

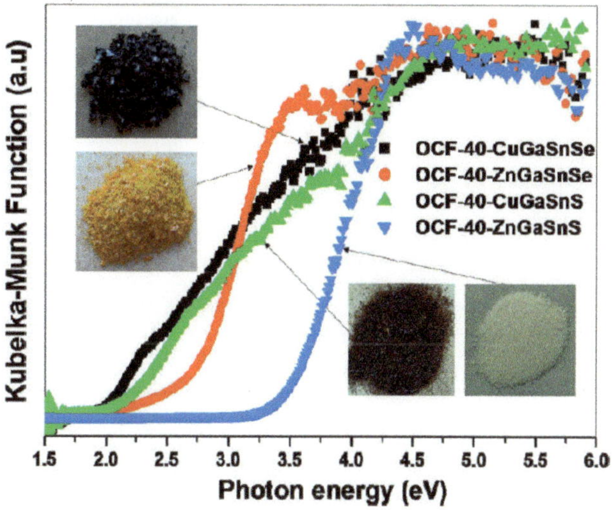

Figure 12. Effects of Cu^{+}vs Zn^{2+} and Se^{2-}vs S^{2-} on band structures in solid state: normalized solid-state UV-vis absorption spectra of OCF-40s. Insets are photos of the as-synthesized crystalline materials. [45]

3.3.5. Photocatalytic applications

Over the past few decades, a large family of crystalline porous materials based on metal chalcogenides were developed. [3-7,36,43,44]These materials integrate tunable band gaps with an open-framework architecture and are potential candidates for efficient photocatalysts due to their optical properties. By controlling framework architecture, it is possible to tune the band structure (both band positions and gap) of an open-framework solid within a given compositional domain. The open-framework construction can increase the number of active reaction sites by the high surface area. [46] In order to evaluate catalytic efficiency for hydrogen generation by metal chalcogenides tetrahedral molecular clusters, a series of porous crystalline open-framework sulfides, such as ICF-17MnInS-Na (formula: Na_{16}-$Mn_{13}In_{22}S_{54}$ xH_2O) or ICF-5CdInS-Na (formula: Na_{10}-$Cd_4In_{16}S_{33}$ xH_2O), are prepared by Feng's group. [46]Under the irradiation of visible light, ICF-5CdInS-Nais photocatalytically active without the use of a co-catalyst, such as Pt. As shown in Figure 13,about 18 mmol$h^{-1}g^{-1}$ of H_2 gas was produced over the ICF-5CuInS-Na catalyst under irradiation with the visible light. This activity was maintained for over 96 h and more than890 mmol of H_2 gas evolved during this period. The quantum efficiency for ICF-5CuInS-Na was about 3. 7% at 420 nm. Even though the number is lower than the quantum yield(~35%) of the well-known Pt/CdS photocatalyst, the efficiency is a considerable improvement on two anhydrous dense phases with similar compositions: $CuInS_2$ with the cubic ZnS structure, and $CuIn_5S_8$ with the spinel structure.

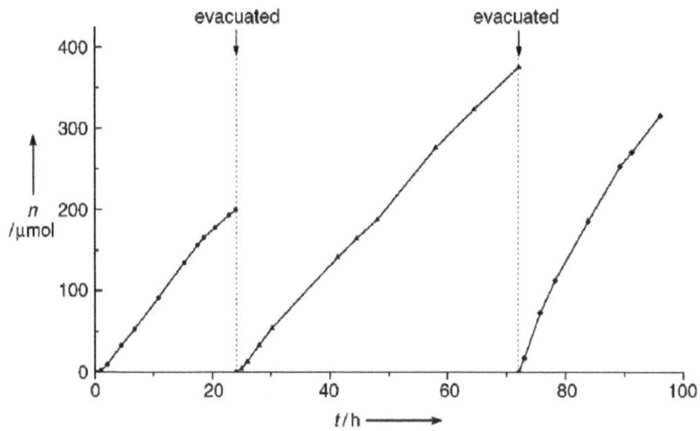

Figure 13. Photocatalytic H_2 evolution from an aqueous solution ofNa_2S (0. 5M) over ICF-5 CuInS-Na (0. 5 g); t: irradiation time, n:amount of H_2. [46]

Author details

Chun-Chang Ou and Chung-Sung Yang*

*Address all correspondence to: csyang@mail. ncyu. edu. tw

Department of Applied Chemistry, National Chia Yi University, Taiwan, ROC

References

[1] Sokolov, M. N. Metal Chalcogenides. In *Handbook of Chalcogen Chemistry: New Perspectives in Sulfur, Selenium and Tellurium*; Devillanova, F. A.; Royal Society of Chemistry: Cambridge U.K., 2007; p543.

[2] Férey, G. Supertetrahedra in Sulfides: Matter against mathematical series? *Angew. Chem., Int. Ed.* 2003, 42, 2576-257.

[3] Feng, P.; Bu, X.; Zheng, N. The Interface Chemistry between Chalcogenide Clusters and Open Framework Chalcogenides. *Acc. Chem. Res.* 2005, 38, 293-303.

[4] Li, H.; Laine, A.; O'Keeffe, M.; Yaghi, O. M. Supertetrahedral sulfide crystals with giant cavities and channels. *Science* 1999, 283, 1145-1147.

[5] Li, H.; Kim, J.; Groy, T. L.; O'Keeffe, M.; Yaghi, O. M. 20 Å $Cd_4In_{16}S_{35}^{14-}$ Supertetrahedral T4 Clusters as Building Units in Decorated Cristobalite Frameworks. *J. Am. Chem. Soc.* 2001, 123, 4867-6868.

[6] Dance, I. G.; Choy, A.; Scudder, M. L. Syntheses, Properties, and Molecular and Crystal Structures of $(Me_4N)_4[E_4M_{10}(SPh)_{16}]$ (E=S, Se; M=Zn, Cd) : Molecular Super-tetrahedral Fragments of the Cubic Metal Chalcogenide Lattice. *J. Am. Chem. Soc.* 1984, 106, 6285-6295.

[7] Yaghi, O. M.; Sun, Z.; Richardson, D. A.; Groy, T. L. Directed Transformation of Molecules to Solids: Synthesis of a Microporous Sulfide from Molecular Germanium Sulfide Cages. *J. Am. Chem. Soc.* 1994, 116, 807-808.

[8] Krebs, B.; Voelker, D.; Stiller, K. Novel Adamantane-lie Thio- and Selenoanions from Aqueous Solutions: $Ga_4S_{10}^{8-}$, $In_4S_{10}^{8-}$, $In_4Se_{10}^{8-}$. *Inorg. Chim. Acta* 1982, 65, L101-L102.

[9] Ahari, H.; Garcia, A.; Kirkby, S.; Ozin, G. A.; Young, D.; Lough, A. J. Self-assembling iron and manganese metal-germaniumselenide framework: $[NMe_4]_2MGe_4Se_{10}$, where M = Fe or Mn. *J. Chem. Soc., Dalton Trans.* 1998, 2023-2027.

[10] Pirani, A. M.; Mercier, P. A.; Dixon, D. A.; Borrmann, H.; Schrobilgen, G. J. Syntheses, Vibrational Spectra, and Theoretical Studies of the Adamantanoid $Sn_4Ch_{10}^{4-}$ (Ch = Se, Te) Anions: X-ray Crystal Structures of $[18\text{-Crown-6-K}]_4[Sn_4Se_{10}]\cdot 5en$ and $[18\text{-rown-6-K}]_4[Sn_4Te_{10}]\cdot 3en\cdot 2THF$. *Inorg. Chem.* 2001, 40, 4823- 4829.

[11] Zheng, N.; Bu, X.; Feng, P. Nonaqueous Synthesis and Selective Crystallization of Gallium Sulfide Clusters into Three Dimensional PhotoluminescentSuperlattices. *J. Am. Chem. Soc.* 2003, 125, 1138-1139.

[12] Vaqueiro, P.; Romero, M. L. $[Ga_{10}S_{16}(NC_7H_9)_4]^{2-}$: a hybrid supertetrahedralnanocluster. *Chem. Commun.*, 2007, 3282–3284.

[13] Wang, C.; Li, Y.; Bu, X.; Zheng, N.; Zivkovic, O.; Yang, C.; Feng, P. Three-Dimensional Superlattices Built from $(M_4In_{16}S_{33})^{10-}$ (M = Mn, Co, Zn, Cd) Supertetrahedral Clusters. *J. Am. Chem. Soc.* 2001, 123, 11506-11507.

[14] Bu, X.; Zheng, N.; Li, Y.; Feng, P. Pushing Up the Size Limit of ChalcogenideSupertetrahedral Clusters: Two- and Three-Dimensional Photoluminescent Open Frameworks from $(Cu_5In_{30}S_{54})^{13-}$ Clusters. *J. Am. Chem. Soc.* 2002, 124, 12646-12647.

[15] Wang, C.; Bu, X.; Zheng, N.; Feng, P. Nanocluster with One Missing Core Atom: A Three-Dimensional Hybrid Superlattice Built from Dual-Sized Supertetrahedral Clusters. *J. Am. Chem. Soc.* 2002, 124, 10268-10269.

[16] Su, W.; Huang, X.; Li, J.; Fu, H. Crystal of Semiconducting Quantum Dots Built Upon Covalently Bonded T5 $[In_{28}Cd_6S_{54}]^{-12}$: The Largest Supertetrahedral Cluster in Solid State. *J. Am. Chem. Soc.* 2002, 124, 12944-12945.

[17] Bu, X.; Zheng, N.; Feng, P. Tetrahedral Chalcogenide Clusters and Open Frameworks. Chem. Eur. J. 2004, 10, 3356-3362.

[18] Lee, G. S. H. ; Craig, D. C.; Ma, I.; Scudder, M. L.; Bailey, T. D.; Dance, I. G. $[S_4Cd_{17}(SPh)_{28}]^{2-}$, the First Member of a Third Series of Tetrahedral $[S_wM_x(SR)_y]^{z-}$ Clusters. *J. Am. Chem. Soc.* 1988, 110, 4864-4866.

[19] Palchik, O.; Iyer, R.G.; Liao, J. H.; Kanatzidis, M. G. $K_{10}M_4Sn_4S_{17}$ (M = Mn, Fe, Co, Zn): Soluble Quaternary Sulfides with the Discrete $[M_4Sn_4S_{17}]^{10-}$Supertetrahedral Clusters. *Inorg. Chem.* 2003, 42, 5052-5054.

[20] Dehnen, S.; Brandmayer, M. K. Reactivity of Chalcogenostannate Compounds: Syntheses, Crystal Structures, and Electronic properties of Novel Compounds Containing Discrete Ternary Anions $[M^{II}_4(_{\mu4}\text{-Se})(SnSe_4)_4]^{10-}$ (M^{II} = Zn, Mn). *J. Am. Chem. Soc.* 2003, 125, 6618-6619.

[21] Zimmermann, C.; Melullis, M.; Dehnen, S. Reactivity of Chalcogenostannate Salts: Unusual Synthesis and Structure of a compound Containing Ternary Cluster Anions $[Co_4(_{\mu4}\text{-Se})\text{-}(SnSe)_4]^{10-}$. *Angew. Chem., Int. Ed.* 2002, 41, 4269-4272.

[22] Zheng, N.; Bu, X.; Feng, P. Penta-Supertetrahedral Clusters as Building Blocks for Three-Dimensional Sulfide Superlattice. *Angew. Chem., Int. Ed.* 2004, 43, 4753-4755.

[23] Eichhofer, A.; Fenske, D. Syntheses and structures of new copper- (I)-indium(III)-selenide clusters. *J. Chem. Soc., Dalton Trans.* 2000, 941-944.

[24] Vossmeyer, T.; Reck, G.; Katsikas, L.; Haupt, E. T. K.; Schulz, B.; Weller, H. A Double Diamond Superlattice Built Up of $Cd_{17}S_4(SCH_2\text{-}CH_2OH)_{26}$. *Science* 1995, 267, 1476-1479.

[25] Jin, X.; Tang, K.; Jia, S.; Tang, Y. Synthesis and Crystal Structure of a polymeric Complex $[S_4Cd_{17}(SPh)_{24}(CH_3OCS_2)_{4/2}]n\cdot nCH_3OH$. *Polyhedron* 1996, 15, 2617-2218.

[26] Vossmeyer, T.; Reck, G.; Schulz, B.; Katsikas, L.; Weller, H. Double-Layer Superlattice Structure Built Up of $Cd_{32}S_{14}(SCH_2CH(OH)\text{-}CH_3)_{36}\cdot 4H_2O$ Clusters. *J. Am. Chem. Soc.* 1995, 117, 12881-12882.

[27] Herron, N.; Calabrese, J. C.; Farneth, W. E.; Wang, Y. Crystal structure and optical properties of $Cd_{32}S_{14}(SC_6H_5)_{36}\cdot DMF_4$, a cluster with a 15 angstrom cadmium sulfide core. *Science* 1993, 259, 1426-1428.

[28] Behrens, S.; Bettenhausen, M.; Deveson, A. C.; Eichhofer, A.; Fenske, D.; Lohde, A.; Woggon, U. Synthesis and Structure of the Nanoclusters $[Hg_{32}Se_{14}(SePh)_{36}]$, $[Cd_{32}Se_{14}(SePh)_{36}(PPh)_4]$, $[P(Et)_2\text{-}(Ph)C_4H_8OSiMe_3]_5[Cd_{18}I_{17}(PSiMe_3)_{12}]$, and $[N(Et)_3C_4H_8OSiMe_3]_5\text{-}[Cd_{18}I_{17}(PSiMe_3)_{12}]$. *Angew. Chem., Int. Ed.* 1996, 35, 2215-2218

[29] Li, H.; Kim, J.; O'Keeffe, M.; Yaghi, O. M. $[Cd_{16}In_{64}S_{134}]^{44-}$: 31-Å. tetrahedron with a large cavity. *Angew. Chem., Int. Ed.* 2003, 42, 1819-1821

[30] Vaqueiro, P. Hybrid materials through linkage of chalcogenide tetrahedral clusters. *Dalton Trans.* 2010, 39, 5965–5972.

[31] Choy, A.; Craig, D.; Dance, I. G.; Scudder, M. L. $[S_4M_{10}(SPh)_{16}]^{4-}$ (M=Zn, Cd), a Molecular Fragment of the Sphalerite MS Lattice: Structural Congruence of Metal Sulphides and Metal Thiolates. *J. Chem. Soc., Chem. Comm.* 1982, 21, 1246-1248.

[32] Feng, P.; Bu, X.; Stucky, G. D. Hydrothermal syntheses and structural characterization of zeolite analogue compounds based on cobalt phosphate. *Nature* 1997, 388, 735-741.

[33] Bu, X.; Zheng, N.; Wang, X.; Wang, B.; Feng, P. Three-Dimensional Frameworks of Gallium SelenideSupertetrahedral Clusters. *Angew. Chem. Int. Ed.* 2004, 43, 1502-1505.

[34] Zheng, N.; Bu, X.; Wang, B.; Feng, P. Microporous and PhotoluminescentChalcogenide Zeolite Analogues. *Science* 2002, 298, 2366-2369.

[35] Feng, P.; Bu, X.; Stucky, G. D. Hydrothermal syntheses and structural characterization of zeolite analogue compounds based on cobalt phosphate. *Nature* 1997, 388, 735-741.

[36] Chen,C.; Ou, C.; Huang, H.; Cheng, J; Yang, C. Mixed pentasupertetrahedral P1 and supertetrahedral T2 clusters as building units to create two-dimensional indium chalcogenides open framework. *Inorg. Chem. Commun.* 2011, 14, 1004–1009.

[37] Zhang, Q.; Bu, X.; Han, L.; Feng, P. Two-Dimensional Indium Sulfide Framework Constructed from Pentasupertetrahedral P1 and Supertetrahedral T2 Clusters. *Inorg. Chem.* 2006, 45, 6684-6687.

[38] Flanigen, E. M. Zeolites and Molecular Sieves. An Historical Perspective. In *Introduction to Zeolite Science and Practice*; van Bekkum, H., Flanigen, E. M., Jansen, J. C., Eds.; Elsevier: New York, 1991; 13-34.

[39] Li, H.; Eddaoudi, M.; Laine, A. O'Keeffe, M.; Yaghi, O. M. Noninterpenetrating Indium Sulfide SupertetrahedralCristobalite Framework. *J. Am. Chem. Soc.* 1999, 121, 6096-6097.

[40] Zheng, N.; Bu, X.; Feng, P. Synthetic design of crystalline inorganic chalcogenides exhibiting fast-ion conductivity. *Nature* 2003, 426, 428-432.

[41] Feng, P. Photoluminescence of open-framework phosphates and germinates. *Chem. Commun.* 2001, 1668–1669

[42] Zheng, N.; Bu, X.; Feng, P. Self-Assembly of Novel Dye Molecules and $[Cd_8(SPh)_{12}]^{4+}$ Cubic Clusters into Three-Dimensional PhotoluminescentSuperlattice. *J. Am. Chem. Soc.* 2002, 124, 9688-9689.

[43] Chen, C.; Yang, C.; Lin, X. Synthesis, characterization, and photoluminescence of quaternary $[Cd_4In_{16}S_{33-x}Se_x]^{10-}$supertetrahedral clusters: (0.33 < x < 0.45). *Inorg. Chem. Commun.* 2005, 8, 836–840.

[44] Yang, C.; Su, Y.; Wang, Y.; Cheng, J.; Lin, X. Synthesis of three dimensional photoluminescent $[In_{10}S_{20-x}Se_x]^{10-}$supertetrahedral clusters: (x<0.3). *Mater. Lett.* 2008, 62, 4015-4017.

[45] Wu, T., Wang, L., Bu, X., Chau, V., Feng, P. Largest Molecular Clusters in the Super-tetrahedralTn Series. *J. Am. Chem. Soc.*2010, 132, 10823–10831.

[46] Zheng, N., Bu, X., Vu, H., Feng, P. Open-Framework Chalcogenides as Visible-Light Photocatalysts for Hydrogen Generation from Water. *Angew. Chem.* 2005, 117, 5433 – 5437.

Permissions

The contributors of this book come from diverse backgrounds, making this book a truly international effort. This book will bring forth new frontiers with its revolutionizing research information and detailed analysis of the nascent developments around the world.

We would like to thank Dr. Sukarno Olavo Ferreira, for lending his expertise to make the book truly unique. He has played a crucial role in the development of this book. Without his invaluable contribution this book wouldn't have been possible. He has made vital efforts to compile up to date information on the varied aspects of this subject to make this book a valuable addition to the collection of many professionals and students.

This book was conceptualized with the vision of imparting up-to-date information and advanced data in this field. To ensure the same, a matchless editorial board was set up. Every individual on the board went through rigorous rounds of assessment to prove their worth. After which they invested a large part of their time researching and compiling the most relevant data for our readers. Conferences and sessions were held from time to time between the editorial board and the contributing authors to present the data in the most comprehensible form. The editorial team has worked tirelessly to provide valuable and valid information to help people across the globe.

Every chapter published in this book has been scrutinized by our experts. Their significance has been extensively debated. The topics covered herein carry significant findings which will fuel the growth of the discipline. They may even be implemented as practical applications or may be referred to as a beginning point for another development. Chapters in this book were first published by InTech; hereby published with permission under the Creative Commons Attribution License or equivalent.

The editorial board has been involved in producing this book since its inception. They have spent rigorous hours researching and exploring the diverse topics which have resulted in the successful publishing of this book. They have passed on their knowledge of decades through this book. To expedite this challenging task, the publisher supported the team at every step. A small team of assistant editors was also appointed to further simplify the editing procedure and attain best results for the readers.

Our editorial team has been hand-picked from every corner of the world. Their multi-ethnicity adds dynamic inputs to the discussions which result in innovative outcomes. These outcomes are then further discussed with the researchers and contributors who give their valuable feedback and opinion regarding the same. The feedback is then collaborated with the researches and they are edited in a comprehensive manner to aid the understanding of the subject.

Apart from the editorial board, the designing team has also invested a significant amount of their time in understanding the subject and creating the most relevant covers. They scrutinized every image to scout for the most suitable representation of the subject and create an appropriate cover for the book.

The publishing team has been involved in this book since its early stages. They were actively engaged in every process, be it collecting the data, connecting with the contributors or procuring relevant information. The team has been an ardent support to the editorial, designing and production team. Their endless efforts to recruit the best for this project, has resulted in the accomplishment of this book. They are a veteran in the field of academics and their pool of knowledge is as vast as their experience in printing. Their expertise and guidance has proved useful at every step. Their uncompromising quality standards have made this book an exceptional effort. Their encouragement from time to time has been an inspiration for everyone.

The publisher and the editorial board hope that this book will prove to be a valuable piece of knowledge for researchers, students, practitioners and scholars across the globe.

List of Contributors

Antonio Sánchez-Navas, Fernando Nieto and Antonio Ruiz-Bustos
Departamento de Mineralogía y Petrología, Facultad de Ciencias, Universidad de Granada, Granada, Spain

Agustín Martín-Algarra
Departamento de Estratigrafía y Paleontología, Facultad de Ciencias, Universidad de Granada, Granada, Spain

Mónica Sánchez-Román
Centro de Astrobiología, INTA, Torrejón de Ardoz, Madrid, Spain

Concepción Jiménez-López
Departamento de Microbiologia, Facultad de Ciencias, Universidad de Granada, Granada, Spain
Instituto Andaluz de Ciencias de la Tierra. C.S.I.C., Armilla, Granada, Spain

Arivanandhan and Y. Hayakawa
Research Institute of Electronics, Shizuoka University, Hamamatsu, Japan

Natarajan
Aditanar College of Arts and Science, Tirucherdur, Tamil Nadu, India

K. Sankaranarayanan
School of Physics, Alagappa University, Karaikudi, Tamil Nadu, India

Astrid Hoeppner
X-Ray Facility and Crystal Farm, Heinrich Heine University, Duesseldorf, Germany

Lutz Schmitt and Sander H.J. Smits
Institute of Biochemistry, Heinrich Heine University, Duesseldorf, Germany

Kui Chen, Li-Jun Ji and Yan-Yang Wu
Chemical Engineering Research Center, East China University of Science and Technology, Shanghai, China

Hidehisa Kawahara
Department of Life Science and Biotechnology, Kansai University, Yamate-cho, Suita-shi, Osaka, Japan

Ermanno Bonucci
Department of Experimental Medicine, La Sapienza University, Rome, Italy

J.S. Redinha, A.A.C.C. Pais and J. A. S. Almeida
University of Coimbra / Department of Chemistry, Portugal

A.J. Lopes Jesus
University of Coimbra / Faculty of Pharmacy, Portugal
University of Coimbra / Department of Chemistry, Portugal

Zhi Zhu He
Beijing Key Laboratory of Cryo-Biomedical Engineering, and Key Laboratory of Cryogenics, Technical Institute of Physics and Chemistry, Chinese Academy of Sciences, Beijing, China

Jing Liu
Department of Biomedical Engineering, School of Medicine, Tsinghua University, Beijing, China
Beijing Key Laboratory of Cryo-Biomedical Engineering, and Key Laboratory of Cryogenics, Technical Institute of Physics and Chemistry, Chinese Academy of Sciences, Beijing, China

Masato Sone, Tso-Fu Mark Chang and Hiroki Uchiyama
Precision & Intelligence Laboratory, Tokyo Institute of Technology, Yokohama, Japan

Peer Schmidt
Lausitz University of Applied Sciences, Faculty of Science, Senftenberg, Germany

Michael Binnewies
Leibniz-University Hannover, Inorganic Chemistry, Hannover, Germany

Robert Glaum
University Bonn, Inorganic Chemistry, Bonn, Germany

Marcus Schmidt
Max-Planck-Institute of Chemical Physics of Solids, Dresden, Germany

Huili Tang, Hongjun Li and Jun Xu
Shanghai Institute of Ceramics, Chinese Academy of Sciences, Shanghai, China

Chun-Chang Ou and Chung-Sung Yang
Department of Applied Chemistry, National Chia Yi University, Taiwan, ROC